Springer Proceedings in Physics **567**

Indexed by Scopus

The series Springer Proceedings in Physics, founded in 1984, is devoted to timely reports of state-of-the-art developments in physics and related sciences. Typically based on material presented at conferences, workshops and similar scientific meetings, volumes published in this series will constitute a comprehensive up to date source of reference on a field or subfield of relevance in contemporary physics. Proposals must include the following:

- Name, place and date of the scientific meeting
- A link to the committees (local organization, international advisors etc.)
- Scientific description of the meeting
- List of invited/plenary speakers
- An estimate of the planned proceedings book parameters (number of pages/articles, requested number of bulk copies, submission deadline).

Please contact:

For Americas and Europe: Dr. Zachary Evenson; zachary.evenson@springer.com
For Asia, Australia and New Zealand: Dr. Loyola DSilva; loyola.dsilva@springer.com

Paul Joseph D. · Jayalakshmi V. · K. Thangaraju
Editors

Proceedings of the International Conference on Emerging Multifunctional Materials and Devices for Sustainable Technologies

IEMDST-2024, 04–05 July, NIT Warangal, India

Springer

Editors
Paul Joseph D.
Department of Physics
National Institute of Technology Warangal
Warangal, Telangana, India

Jayalakshmi V.
Department of Physics
National Institute of Technology Warangal
Warangal, Telangana, India

K. Thangaraju
Department of Physics
National Institute of Technology Warangal
Warangal, Telangana, India

ISSN 0930-8989 ISSN 1867-4941 (electronic)
Springer Proceedings in Physics
ISBN 978-981-96-5862-6 ISBN 978-981-96-5863-3 (eBook)
https://doi.org/10.1007/978-981-96-5863-3

This Springer imprint is published by the registered company Springer Nature Singapore Pte Ltd.
The registered company address is: 152 Beach Road, #21-01/04 Gateway East, Singapore 189721, Singapore

Preface

The proceedings of the *"International Conference on Emerging Multifunctional Materials and Devices for Sustainable Technologies"* (IEMDST-2024) brought out under the title *"Emerging Multifunctional Materials and Devices for Sustainable Technologies"* is aimed to address and highlight the critical role of advanced materials and devices in the pursuit of sustainable technological solutions. The IEMDST-2024 is rooted in recognition of the pivotal role that advanced materials and innovative devices play in the development of sustainable materials and technologies. IEMDST-2024 was organized by the Dept. of Physics at NIT Warangal (4th and 5th July 2024) in association with the Dept. of Applied Sciences, NIT-Goa. This conference was intended to act as a catalyst to bridge the gap between current technological challenges and potential innovative solutions. Exchange of knowledge and ideas through effective scientific interactions among more than two hundred and fifty participants including researchers, scientists, engineers, industrialists, and other international delegates from several areas of research on multifunctional materials and devices has contributed significantly towards sustainable technologies.

This proceedings constitutes peer review of selected manuscripts to disseminate knowledge and viewpoints from diverse domains of Science and Engineering to ensure to track progress over time and to adapt to emerging trends, and maintain momentum in addressing the ever-evolving challenges of sustainability. Contents of this proceedings captures the advancements in the development, characterization, and testing of multifunctional materials by elucidating the structure-property-function relationships. To mention a few, materials like ZnSe, ZnO, SnS, and WS_2 are explored in their nano-dimensional forms for several optoelectronic properties. CuO and Fe_2O_3 in thin form and NiO/CuO composites emphasize towards sustainability especially for water purification from widespread pollutants based on photocatalysis and antimicrobial activity using appropriate experimental procedure. Currently trending artificial intelligence (AI) and machine learning (ML) applied to various domains of Science and Engineering in terms of simulation and first principles study addressing the scope to predict newer trends for utilization in relevant material systems and devices.

Manuscripts in this proceedings submitted by researchers, scientists, and engineers, presenting their findings will be highly beneficial for the readers. Manuscripts encompass wide range of topics and areas of interest and thus offer the readers a perspective on how different fields can contribute for multifunctional materials and devices for the benefit of sustainable technologies and the society.

The Editors

Sponsors

The original version of the book has been revised. The Series Name "Springer Proceedings in Physics" and the volume number "567" have been updated. A correction to this book can be found at https://doi.org/10.1007/978-981-96-5863-3_48

Acknowledgement

In the recent decades, research on Materials Science has become highly interdisciplinary nature by the involvement of materials scientists, chemists, physicists, engineers and data scientists. The advancements in synthesis of various multifunctional materials and fabrication of highly efficient devices based on newer and modified materials have high relevance for sustainable technologies. Addressing the research gaps and overcoming the challenges are crucial for the advanced future technologies. To emphasize these aspects, the *"International Conference on Emerging Multifunctional Materials and Devices for Sustainable Technologies"* (IEMDST-2024) was conducted with support from scientific community and various scientific institutions, funding agencies and industry partners.

We thank the co-organizing secretaries Dr. P. Saidi Reddy and Dr. Velavan Kathirvelu from the associated institute, Dept. of Applied Sciences, NIT-Goa, for their support towards the successful conduction of IEMDST-2024. We are grateful to the plenary and invited speakers who have enlightened and facilitated the exchange of knowledge through IEMDST-2024. We appreciate and thank the participants and delegates for partaking enthusiastically by submitting more than 250 number of contributed abstracts. We also thank our institution NIT Warangal, faculty members, organizing committee members and research scholars of our department for their support in various aspects. We also immensely thank the institute post-doctoral fellow Dr. Benjamin Hudson Baby (NIT Warangal) for his efforts in tuning, formatting and proof-reading the proceedings carefully and to bring it out in good format. We also thank the publisher "Springer-Nature" and their team for publishing the proceedings of IEMDST-2024 and taking it to larger scientific audience.

Finally, we gratefully acknowledge the generous financial support from SERB (ANRF), Govt. of India, DRDO, Govt. of India, Indian National Science Academy (INSA) and industry sponsors. The financial support rendered has actually materialized the international conference IEMDST-2024 and to bring out this proceedings for widespread dissemination of knowledge.

The Editors

Contents

About the Editors

Dr. Paul Joseph D. is currently working as an Associate Professor in the Department of Physics at the National Institute of Technology Warangal (NIT Warangal). He obtained his Ph.D. in Physics-Materials Science (Interdisciplinary) from the Department of Nuclear Physics, University of Madras, in 2009. He was a post-doctoral fellow at the Centre for Condensed Matter Sciences (CCMS) at National Taiwan University, Taipei, Taiwan. His active research areas are alternative transparent conducting electrodes (both metals and metal oxides in thin film form), proximity effect between superconductors and half-metallic ferromagnets, alternative components for dye-sensitized solar cells, nanostructured materials for batteries, swift heavy-ion-beam irradiation effects on materials, and photocatalytic activity for degradation of pollutants.

Dr. Jayalakshmi V. is currently an Associate Professor in the Department of Physics at the National Institute of Technology, Warangal. Prior to joining NIT Warangal, she was a postdoc at Queen's University, Canada, and Georgia Institute of Technology, USA. She obtained her Ph.D. in Physics from the Centre for Nano and Soft Matter Sciences, Bangalore. Her research interests include synthesis of nematic and chiral nematic emulsions, the development of technologies for detecting Alzheimer's biomarkers using liquid crystals and ZnO tetrapods for novel functional applications.

Dr. K. Thangaraju is currently working as a Professor in Department of Physics at National Institute of Technology, Warangal, India. He obtained his Ph.D. from Anna University, Chennai, India. His research area includes organic semiconductors, organic optoelectronic devices such as OLEDs, OPVs, OTFTs, photodetectors, and organic/inorganic hybrid nanostructured thin films and their optoelectronic devices.

Time Saving Synthesis Approach of WS_2 Nanoparticles by Ethylene Glycol Assisted Solvothermal Method

Yuvashree Jayavelu, Benjamin Hudson Baby, Gouranga Maharana, and D. Paul Joseph(✉)

Department of Physics, National Institute of Technology, Warangal, Telangana 506004, India
paul@nitw.ac.in

Abstract. Tungsten disulphide (WS_2) is an emerging 2D transition metal dichalcogenide (TMDCs), explored widely due to its interesting optoelectronic and, thermal properties. This study reports synthesis attempt of WS_2 using ethylene glycol medium with a reduced time period of 6 h at 170 °C where predominantly it takes around 12–48 h in general. The formation of WS_2 is confirmed using Raman spectroscopy and optical bandgap is calculated to be 2.1 eV from UV-Vis DRS spectroscopy. Morphological studies predominantly showed disrupted hexagonal morphology. This study can be further fine-tuned towards time saving synthesis approach of single phase WS_2.

1 Introduction

Transition metal dichalcogenides (TMDCs) have garnered extensive attention recently as promising materials for wide range of applications due to their excellent physico-chemical properties [1, 2]. These are semiconductors with transition metal atom (Mo, W, Pt, Pd, etc.,) binded to a chalcogen atom (S, Se, Te, O etc.,) where their alignment is one layer of metal atom sandwiched in between two layers of chalcogens [3]. Various TMDCs including graphene, MoS_2, WS_2, $MoSe_2$, WSe_2, etc., monolayers have distinct physico-chemical properties than their bulk counterparts making them versatile and efficient [2, 4]. Most TMDC monolayers have direct bandgap, crystal structure with no inversion center, high spin orbit coupling, making them exclusive for emitters, detectors, spintronics, photocatalysis, fuel cells, solar cells etc. [2]. Among various TMDCs, WS_2, is recently extensively explored for photovoltaic applications due to their optimum bandgap (1–2 eV), high absorption coefficient (105 cm^{-1}), abundance and cost effectiveness [5, 6]. WS_2 is intriguing due to its monolayered feature and tunable opto-electronic aspects. WS_2 possess two distinct symmetries based on the positioning of S atoms where enhanced electrical conductivity and electrocatalytic activity depends on them. Tungsten having larger mass shows strong spin orbit coupling and valence band splitting resulting in valley Hall effect (electrons in 2D possess valley degree of freedom exhibiting anomalous hall effect) than most TMDCs. Several physical and chemical methods like thermal exfoliation, sol-gel, colloidal synthesis, hydrothermal, solvothermal, and microwave assisted process are reported for WS_2 synthesis [7]. Among which,

P. Joseph D. et al. (Eds.): IEMDST 2024, SPPHY 567, pp. 1–6, 2025.
https://doi.org/10.1007/978-981-96-5863-3_1

solvothermal and hydrothermal are cost-effective techniques involving facile synthesis steps. 2D nanomaterials are mostly synthesized by hydrothermal or solvothermal methods due to their high yield producing bulk quantities with desirable morphologies [4, 6, 7]. The solvothermal technique enables synthesis of high-quality crystalline samples with narrow size distribution having versatile morphologies with suitable organic solvents [8]. This technique is highly scalable due to their straight forward and easy synthesis and reproducibility.

In general, solvothermal method enables synthesis of WS_2 by considering surface-active species for treatment at higher temperature like 190 °C for prolonged duration of about 48 h [9]. This study reports synthesis of WS_2 by solvothermal method using ethylene glycol solvent in a minimal time period of 6 h at 170 °C. The synthesized WS_2 was further characterized for structural formation by Raman analysis and morphological confirmation using FESEM. The morphologies of WS_2 can be further controlled by varying their reaction temperature and time. This study can be further explored for emerging field of WS_2 photothermal therapy-based sensors, photocatalysts, detectors and other applications.

2 Materials

All the procured precursor chemicals were of analytical grade and were utilized used as it is for solvothermal synthesis of WS_2 nanoparticles. Tungsten hexachloride (WCl_6) from Alfa Aesar (MW. 396.57, CAS: 13283-01-7), Sodium sulphide flakes (Na_2S) from FINAR, (MW. 78.04, CAS: 1313-84-4) were dissolved in ethylene glycol was procured from FINAR (MW. 62.07, CAS: 107-21-1) for the one step synthesis process.

3 Synthesis Procedure

WS_2 was synthesised by one-step solvothermal synthesis using Na_2S with ethylene glycol as solvent. A 3.3 M of Na_2S and 0.1 M of WCl_6 were mixed separately in 50 ml of ethylene glycol by stirring at room temperature. Then, Na_2S solution was added drop wise into the WCl_6 solution with constant stirring. The formed mixture underwent a reaction in a stainless-steel autoclave with teflon lining at 170 °C for 6 h. Upon cooling, black precipitates formed were then collected and washed with methanol and deionized (DI) water (Fig. 1) to obtain the WS_2 powder.

4 Characterization Measurements

Raman spectroscopy is performed (RENISHAW InVia Raman Microscope, UK) by exciting with 532 nm wavelength of using a semiconductor laser and scanned in the range 50–350 cm^{-1} for an acquisition period and power as 150 s and 0.0025 mW respectively. The optical bandgap is calculated by UV-Vis spectroscopy, SPECORD-210 Plus, Analytik Jena. JEOL JSM IT-800 Field Emission-Scanning Electron Microscopy (FE-SEM) measurements were carried out to explore the morphology and concentration of the elements present in WS2 using Energy Dispersive X-Ray spectroscopy (EDS).

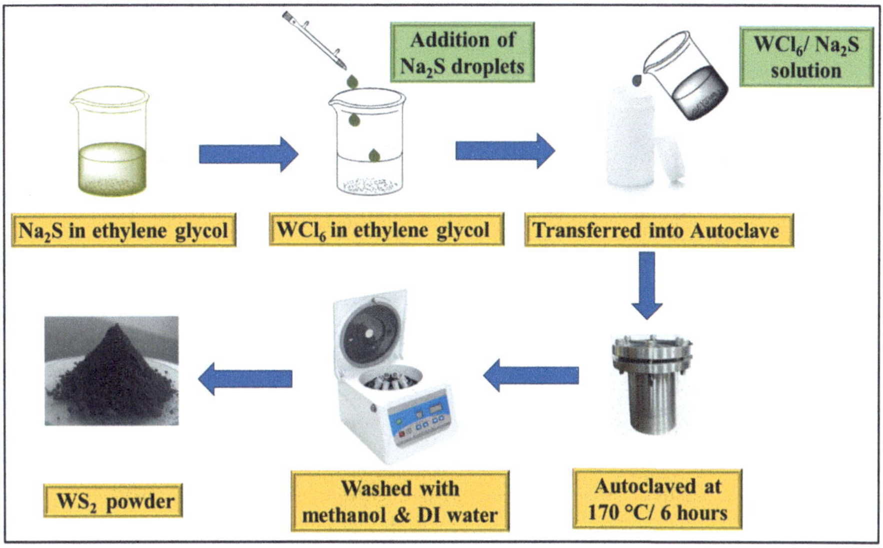

Fig. 1. Schematics of solvothermal synthesis of WS_2 nanoparticles in ethylene glycol medium.

5 Results and Discussion

Firstly, Raman analysis of the samples performed with 532 nm laser excitation reveals the aggregation and structural formation of WS_2. Two characteristic peaks 367 cm^{-1} and 570 cm^{-1} are assigned to the reported standard WS_2 peaks [9]. Mostly, different morphologies of WS_2 shows different intensities due to the scattering and increase in density. The peaks corresponding to the second-order modes, 2LA (M) at about 367 cm^{-1} and combination of first order modes at A1g(M) + LA(M) at about 570 cm^{-1} (Fig. 2) [7], confirms the formation of WS_2. The slight shift in the peak value is attributed to the synthesized WS_2 in bulk form. The absence of any WO_3 secondary phase is confirmed from the reported Raman peaks (68, 129, 175, 269, 320, 430 and 444 cm^{-1}) validating the formation of phase pure WS_2 nanoparticles [6, 10].

The FESEM images of the synthesized nanomaterials showed predominantly distorted hexagonal shaped particles of WS_2 (Fig. 3). The WS_2 nanostructured particles are mildly agglomerated together showing distorted morphology indicating the necessity to alter reaction temperature, time or solvent used. This observed morphology of WS_2 might be due to the controlled nucleation induced by the use of ethylene glycol solvent [11, 12]. The particle size calculated from Fig. 3a is ~ 1.1 μm indicating their potential use in catalysis, optical and electronic sensors [13]. The particles show hexagonal morphology due to transition favoured by directional growth, crystal nucleation or recrystallization during the synthesis process (Fig. 3a) [11]. This distorted morphology could be influenced by a complex interplay of thermodynamic and kinetic factors during the synthesis process [11, 12, 14]. The EDS data confirms the presence of near stochiometric ratio of W and S in the synthesised WS_2 nanostructures (Table inset of Fig. 3a). This is further confirmed by the existence of W and S in the elemental mapping images (Fig. 3b).

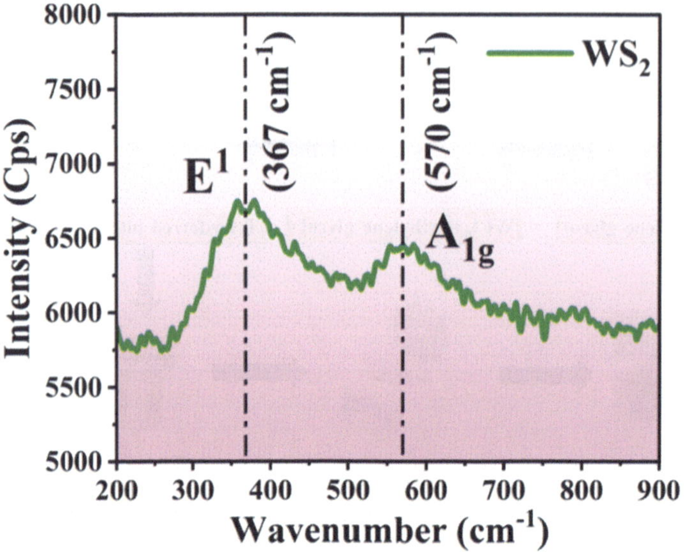

Fig. 2. Raman spectra of the WS_2 nanostructures prepared by solvothermal method.

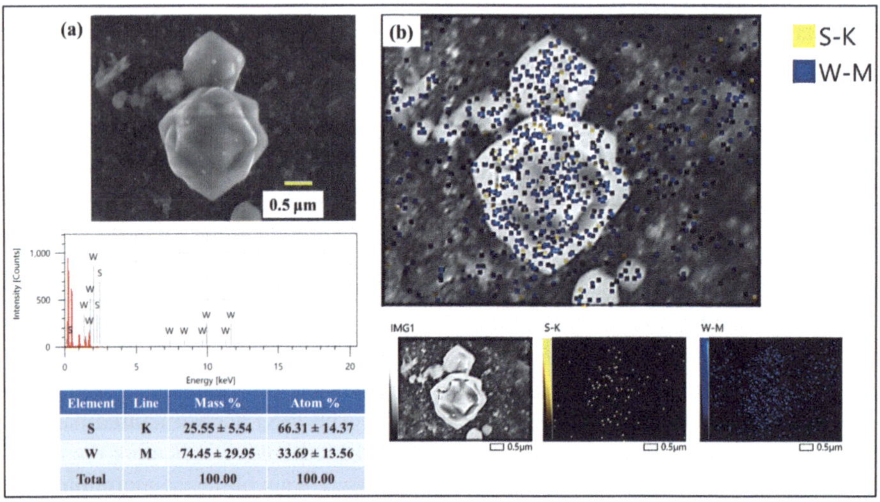

Element	Line	Mass %	Atom %
S	K	25.55 ± 5.54	66.31 ± 14.37
W	M	74.45 ± 29.95	33.69 ± 13.56
Total		100.00	100.00

Fig. 3. FESEM images of WS_2 nanoparticles having (a) distorted hexagonal shaped morphology at 0.5 µm with their EDS data and (b) elemental mapping confirming the near stochiometric ratio of the elements W and S.

The absorbance spectrum of WS_2 shows strong absorption from 300–400 nm and gradual decrease in the IR region (750–1000 nm) [9]. Optical direct energy band gap of WS_2 is estimated using UV-VIS spectrum measured in the range between 200–1000 nm (Fig. 4a). The bandgap of 2.17 eV (Fig. 4b) is calculated by Tauc plot by extrapolating

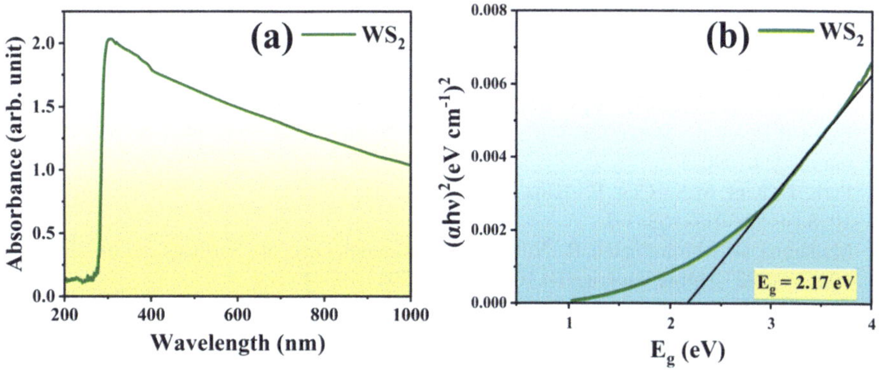

Fig. 4. (a) Absorbance spectra and (b) optical bandgap obtained from UV-Vis spectra for the prepared WS_2 nanoparticles

the linear fit of $(\alpha h\nu)^2 = 0$ from the Tauc's equation given below [10, 15],

$$(\alpha h\nu)^2 = A\left(h\nu - E_g\right)^n \tag{1}$$

where, A = constant, hν = photon energy, E_g = bandgap, n = 1/2 for direct transition. The obtained value is matched well with the literature reports (2.0–2.2 eV) of WS_2 nanostructures [2, 7].

6 Conclusion

Time saving WS_2 synthesis process was attempted using ethylene glycol as the reaction medium with a reduced time period of 6 h at 170 °C. The WS_2 peaks of Raman spectroscopy corroborate with literature confirming no oxide formation. The optical bandgap of WS_2 is calculated to be 2.1 eV using Tauc plot. Mixed distorted hexagonal morphology was observed from FE-SEM data along with their EDS and mapping data confirming their near stoichiometric ratio with an average particle size of 140 nm. This study can be further fine-tuned for time efficient single phase WS_2 formation for versatile catalytic and optoelectronic applications.

References

1. Hussain, S., Akbar, K., Vikraman, D., Rabani, I., et al.: J. Market. Res. **12**, 385–398 (2021). https://doi.org/10.1016/j.jmrt.2021.02.097
2. Huang, H.H., Fan, X., Singh, D.J., Zheng, W.T.: Nanoscale **12**, 1247–1268 (2020). https://doi.org/10.1039/c9nr08313h
3. Wang, Y., Chhowalla, M.: Nat. Rev. Phys. **4**, 101–112 (2022). https://doi.org/10.1038/s42254-021-00389-0
4. Sebastian, A., Pendurthi, R., Choudhury, T.H., Redwing, J.M., Das, S.: Nat. Commun. **12**, 693 (2021). https://doi.org/10.1038/s41467-020-20732-w
5. Bin Rafiq, Md.K.S., et al.: Sci. Rep. **10**, 771 (2020). https://doi.org/10.1038/s41598-020-57596-5

6. Durai, L., Kong, C.Y., Badhulika, S.: Mater. Sci. Eng., C **107**, 110217 (2020). https://doi.org/10.1016/j.msec.2019.110217
7. Berkdemir, A., Gutiérrez, H.R., Botello-Méndez, A.R., Perea-López, N., Elías, A.L., et al.: Sci. Rep. **3**, 1755 (2013). https://doi.org/10.1038/srep01755
8. Baby, B.H., et al.: Mater Today Commun **39**, 109041 (2024). https://doi.org/10.1016/j.mtcomm.2024.109041
9. Park, J., Kim, M.S., Cha, E., Kim, J., et al.: Sci. Rep. **7**, 16121 (2017). https://doi.org/10.1038/s41598-017-16251-2
10. Maharana, G., Muniramaiah, R., Yuvashree, J., Mandal, D., et al.: Surfaces and Interfaces **42**, 103413 (2023). https://doi.org/10.1016/j.surfin.2023.103413
11. Shahabuddin, S., Mehmood, S., Ahmad, I., Sridewi, N.: Nanomaterials **12** (2022). https://doi.org/10.3390/nano12122090
12. Mohan, V.V., Manuraj, M., Anjana, P.M., Rakhi, R.B.: Energy Technology **10** (2022). https://doi.org/10.1002/ente.202100976
13. Jayavelu, Y., Maharana, G., Joseph, D.P., Divyadharshini, S.: Electrochim Acta, 144777 (2024). https://doi.org/10.1016/j.electacta.2024.144777
14. Wang, Z., et al.: Nat. Commun. **12**, 332 (2021). https://doi.org/10.1038/s41467-020-20515-3
15. Jubu, P.R., Yam, F.K., Igba, V.M., Beh, K.P.: J. Solid State Chem. 290 (2020). https://doi.org/10.1016/j.jssc.2020.121576

A Facile Synthesis of TGA Stabilized ZnSe Quantum Dots for Energy Harvesting Applications

Udoshi Basavaraj[1,2(✉)], B. Jai Kumar[1], and H. M. Mahesh[1]

[1] Thin Film and Solar Cell Laboratory (TFSCL), Department of Electronic Science, Bangalore University, Bengaluru 560056, India
basavaraj.udoshi@gmail.com
[2] PES University, EC Campus, Bengaluru 560100, India

Abstract. Thioglycolic Acid (TGA) stabilized aqueous Zinc Selenide (ZnSe) Quantum Dots (QDs) were synthesized using a facile, Single Injection Hydrothermal (SIH) method in complete ambient environment. The synthesis technique eliminates the use of inert atmosphere for reducing elemental Selenium power to Selenium precursor avoiding the oxidation. The synthesized ZnSe QDs were characterized for its structural and optical properties. The short chained TGA capping on the surface of ZnSe QDs was confirmed using FTIR spectrum. The XRD confirmed the formation of ZnSe QDs. The ZnSe QDs, absorption spectra shifted to higher energy (blue shifted) with respect to the bulk ZnSe absorption peak, due to strong quantum confinement effect. The absorption peak wavelength red shifted with increasing reflux time from 0th minute to 500th minutes, this indicates the QDs have grown in size with increasing reaction time. However, the largest sized QDs were still below Bohr's excitons radius of the material; the bandgap calculated using Tauc's plot showed blue shift with respect to bulk ZnSe bandgap energy, hence confirming strong quantum confinement effect. These synthesized QDs can be used for wide range of energy harvesting applications.

1 Introduction

Quantum dots (QDs) are semiconductor nanoparticles distinguished by their size-dependent optical and electrical features resulting from quantum confinement effects. Quantum dots, which typically range in diameter from 2 to 10 nm, can be precisely tailored to emit specific wavelengths of light by altering their size, shape, and chemical composition. This tunability arises from the quantum mechanical principles that influence the energy levels of electrons within the material, making QDs indispensable in an array of uses including as biomedical imaging, optoelectronics, and environmental monitoring. [1, 2]. Quantum dots have exceptional luminous features, including high quantum yield, narrow emission spectra, and broad absorption bandwidths, enabling nanoparticles to be utilized in a multitude of applications, notably optoelectronics and nanomedicine. Quantum dots are used in quantum dot light-emitting diodes (QLEDs) in displays, which improve color purity and energy efficiency in televisions and monitors [3]. Likewise, QDs offer significant potential in solar devices, where they can help

P. Joseph D. et al. (Eds.): IEMDST 2024, SPPHY 567, pp. 7–15, 2025.
https://doi.org/10.1007/978-981-96-5863-3_2

boost light absorption and the effectiveness of energy conversion [4]. In the field of biomedicine, quantum dots serve as efficient fluorescent tags for imaging and diagnosis. Their multiplexing capability, which allows them to mark several targets at the same time, combined with their variable fluorescence, enables improved imaging techniques [5]. Research has revealed that by conjugating QDs with specific biomolecules, such as antibodies or peptides, it is feasible to accomplish targeted distribution and imaging of cells, making QDs invaluable for both diagnostic and therapeutic purposes [6]. Despite its impressive qualities, the widespread use of quantum dots is currently limited by a number of issues, including toxicity and stability concerns. Traditional QD materials, such as cadmium selenide (CdSe), include toxic heavy metals, which pose environmental and health hazards [7]. As a consequence, there is a significant focus on creating non-toxic alternatives, such as copper-indium-selenide (CIS), carbon quantum dots (CQDs) and zinc selenide (ZnSe) QDs, which potentially offer safer and more sustainable solutions for applications in biomedicine and environmental sensor developments [8].

Zinc selenide (ZnSe) quantum dots (QDs) have emerged as crucial elements in the fabrication of advanced biosensors due to their distinctive optical attributes, biocompatibility, and flexibility for functionalization. ZnSe QDs, which range in size from 2 to 10 nm, exhibit quantum confinement characteristics that allow for size-tunable photoluminescence, producing crisp emission peaks that are ideal for sensitive detection applications [9]. ZnSe's direct bandgap enhances its fluorescence efficiency, making these quantum dots perfect choices for real-time biomolecular detection in diagnostic biosensors. Researchers may fine-tune their emission wavelengths throughout a broad spectrum by altering their size and composition, allowing for the fabrication of multicolor imaging techniques and powerful sensors [10]. ZnSe QDs' capacity to modify optical characteristics makes them vital in developing highly sensitive detection tools for an assortment of applications, including biosensing and diagnostic procedures.

Thioglycolic acid (TGA)-capped zinc selenide (ZnSe) quantum dots (QDs) have received a lot of consideration because of their superior optical characteristics and stability, which are important for a variety of applications ranging from biomedical imaging to optoelectronics. TGA-capped ZnSe QDs are commonly synthesized using a variety of colloidal, hydrothermal, and solvent-free processes. Each approach has significant advantages for controlling quantum dot size, optical characteristics, and surface chemistry. Colloidal synthesis is a popular method for producing TGA-capped ZnSe QDs. This approach comprises the reaction of zinc and selenium precursors in the presence of TGA as a stabilizer. Zhang et al. (2023) reported a modified colloidal synthesis approach that improves the emission efficiency and stability of TGA-capped ZnSe QDs by adjusting the precursor concentrations and reaction temperature [11].

According to the systematic literature references, the best suited technique is simple hydrothermal ZnSe QD synthesis with TGA as a capping agent. As a result, this study uses a single injection hydrothermal approach to easily synthesize TGA-stabilized ZnSe quantum dots by varying the concentration of the capping agent. This paper aims to investigate the synthesis methods for ZnSe quantum dots capped with thioglycolic acid, evaluate their optical and structural properties, and assess their performance in biosensing applications. By given that a complete analysis of these aspects, this study seeks to advance the understanding and application of TGA-capped ZnSe quantum dots in various scientific and technological fields.

2 Materials and Methods

2.1 Materials

Zinc Acetate hydrate ($Zn(CH_3COO)_2H_2O$, 98%, Aldrich, India) was used as zinc precursor, Selenium powder (Se, −200 mesh, 99.8%, Aldrich) as selenium precursor, Sodium Borohydride ($NaBH_4$, 98%, Aldrich) as reducing agent, Thioglycolic acid (TGA, $C_2H_4O_2S$, 99%, Sd fine chem. Limited) 99%, Spectrochem) as capping agent, Sodium Hydroxide (NaOH pellets, 97%, Central Drug house, India) was used for balancing pH of the solution. Distilled water was used as a solvent and all chemicals were of AR grade, and used without further purification.

2.2 Experimental Details

A simple single injection hydrothermal (SIH) synthesis technique [12] that was devised at room temperature, it was used to create the TGA-capped ZnSe QDs, yielding ultra-small QDs with great purity. Three steps can be used to explain the synthesis process. First, the Zn^{2+} precursor was prepared and stabilized using TGA. Next, the Se^{2-} precursor, NaHSe, was prepared and Se was kept from oxidizing. Lastly, the TGA-capped ZnSe QDs grew (Fig. 1). By monitoring the reaction temperature and the amount of time that passed, the size of the QDs was controlled. To regulate the pH level, zinc acetate, selenium powder, sodium hydroxide, and thioglycolic acid (TGA, capping agent) were combined in a 6:4:1 ratio to create ZnSe QDs. The initial step was starting the synthesis of TGA-capped Zn precursor with 0.06 M MPA in a conical flask, to which 0.04 M zinc acetate solution was added and stirred for 10 min, which forms the transparent solution. Subsequently, the reaction liquid is gradually supplemented with 1 M NaOH until the pH level reaches 9. At a pH of 5, zinc hydroxide, which is produced by combining zinc acetate with NaOH, precipitates white. Addition of additional 1 M NaOH drop by drop causes zinc hydroxide to dissolve, yielding a transparent zinc ion precursor that is pH-corrected to 9 and capped with TGA. The precursor solution of TGA- Zn^{2+} was agitated until the NaHSe solution was ready.

In the second step, the Se source was created by adding an equivalent weight quantity of selenium powder directly to the $NaBH_4$ dissolved in distilled water. The flask was placed in a water bath that was heated to 90 °C and sealed firmly. Few punch holes were done to flask cap so that the reaction gas can be released. In a minute, the dark gray solution goes transparent as H_2 gas escape; as the reaction proceeds, the little bubbles of gas grow larger. The precursor is ready when the translucent solution progressively takes on a dark pinkish-violet hue. This Se precursor was injected into the Zn precursor during the third stage. The solution first turns golden, indicating that ZnSe QD monomers are formed, and then it becomes clear again. This solution is mixed at room temperature for 10 min. It's then put on a hot plate that's been preheated to 90 °C. In the end, aliquots of ZnSe QDs with TGA capped were gathered and examined for various reflux times.

The synthesis was repeated for 0.08 M TGA concentration, keeping all other parameters constant. The synthesized QDs are further characterized and results are discussed in further sections. To eliminate surplus and unreacted bi-products, acetone was utilized to precipitate the produced QDs. Next, the liquid was centrifuged at 3000 rpm to create the dots. After going through the procedure described above twice, the solid residues of QDs were re-dispersed in water. ZnSe QDs powder was created using the vacuum drying technique.

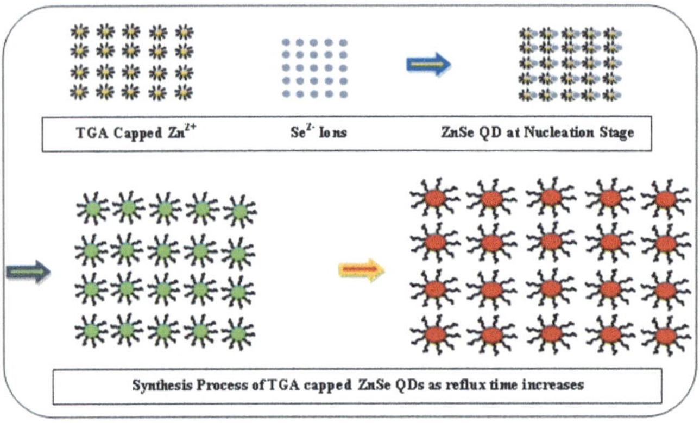

Fig. 1. Schematics showing the growth of TGA capped ZnSe QDs.

2.3 Instrumentation

The synthesized TGA capped ZnSe QDs with different concentrations of capping agent characterized for various characterization techniques. FTIR spectrum was obtained using Bruker-Alpha Fourier Transform Infrared Spectroscopy (FTIR) for solid samples in transmittance mode, in the range 4000 cm^{-1} to 400 cm^{-1}. Optical absorbance and transmittance spectra were obtained using UV-Visible-NIR spectrometer (USB4000-XR, Ocean optics), in wavelengths ranging from 200 nm to 1000 nm, with step size of 0.25 nm. All spectra were compared and calibrated with the reference sample (cuvette filled with solvent with path length of 1 cm, deducting ambient light in room temperature conditions). Photoluminescence measurements conducted using Perkin Elmer (LS-45) fluorescence spectrometer.

3 Results and Discussion

The traditional nucleation and growth model may describe the growth of ZnSe QDs; nucleation is the process of forming ZnSe monomers by the reaction between the cation and anion of Zn and Se precursors, respectively. The zinc cation is created by decomposing zinc acetate with the strong base NaOH, and the selenium anion is obtained by reducing elemental selenium powders. The degraded Zn is additionally stabilized by

TGA. When these monomers are combined it results into the rapid nucleation step, also known as the ZnSe monomer preparation stage, occurs as a result of ionic interaction between Zn^{2+} and Se^{2-}, resulting in the formation of ZnSe monomers. If the ZnSe monomer is refluxed at room temperature, it produces the smallest nanocrystals known as QDs; the size of these QDs is mostly determined by the cation and anion concentrations and also when the concentration of monomers decreases, the growth of QDs stops. Furthermore, to increase the size of QDs, the reagent must be exposed to continuous energy in the form of heat, irradiation, or illumination, which triggers the Ostwald ripening (OR) stage. The OR stage begins as Zn^{2+} and Se^{2-} concentrations fall, causing small QDs to dissolve and subsequently redeposit on the surface of large-sized QDs, enabling the growth of larger-sized QDs. The increased surface energy of small-sized QDs, which implies a high total Gibbs energy, drives this phase of reaction, resulting in more solubility than larger-sized QDs. As the process continues, more stable crystals emerge with lower total Gibbs energy. Thus, larger QDs are more stable in solution than smaller ones. As a result, the size of QDs grows rapidly in the early part of the reaction and slows as the reaction develops, implying that the reaction time required to obtain large-sized QDs increases. The synthesized QDs were thoroughly characterized for their structural and optical properties, and the results were analyzed in detail to gain comprehensive insights into their characteristics.

3.1 FTIR Study

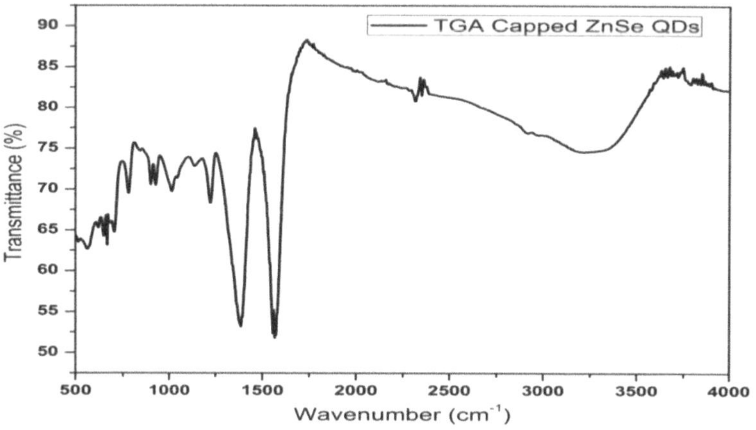

Fig. 2. FTIR spectra of TGA capped ZnSe QDs.

One versatile analytical technique for figuring out which functional groups are in a material is Fourier transform infrared (FTIR) spectroscopy. ZnSe QDs with thioglycolic acid (TGA) confirmed by FTIR spectra (Fig. 2). The absence of a broad and strong absorption peak in the 3200–3300 cm^{-1} region, which indicates the presence of the thiol (S–H) functional group covalently attaching to the Zn cation surface, confirms the high stability of ZnSe QDs. TGA capping also has the advantage of allowing the QDs to dissolve in an aqueous solution.

3.2 Optical Properties of Synthesized ZnSe QDs

A material's optical properties play a major role in identifying its behavior and creating strategies for using it for certain applications. The transmittance and absorption spectra of the TGA-stabilized ZnSe QDs aliquots obtained at zero, 20, 60, 110, 200, 320, and 500 min of reflux time are shown in Fig. 3a and 3b. The formation of QDs is demonstrated by the clear blue shift in the sample's absorption peak wavelength when compared to the bulk ZnSe material's 460 nm absorption peak wavelength, which is caused by the quantum confinement effect. The absorption peak color changed from 360 nm to 415 nm as the reflux time while growing from zero to 500 min, revealing that the QDs' size increased. The transmittance edge red shift from 365 nm to 425 nm indicates that the generated QDs have strong transmittance above 430 nm. It can be modified the size of the QDs to absorb a specific wavelength within the range of 360 nm to 420 nm for a given purpose.

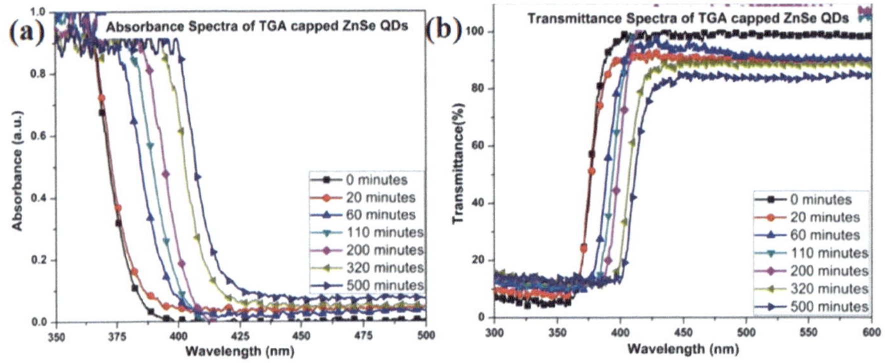

Fig. 3. a) Absorption spectra of TGA capped ZnSe QDs and b) Transmittance spectra of TGA capped ZnSe QDs.

It is important to keep in mind that this absorption occurs at spectral wavelengths ranging from ultraviolet to almost blue. The transmittance edge and absorption red shift amply demonstrated that the QDs' size rapidly increased in the initial stages of the reaction and then slowed down as it went on, primarily due to the rapid growth of reaction.

3.3 Photoluminescence Spectra

The photoluminescence spectrum indicates that the TGA-capped ZnSe QDs exhibit robust photoluminescence, despite the different sizes of the produced QDs. The PL spectra show the higher-energy electronic transitions of QDs. The photoluminescence spectra of ZnSe QDs capped with TGA are shown in Fig. 4 for aliquots taken between 0 and 500 min, the PL peak shifts from 463 to 467 nm. The red shift in PL peaks is clearly caused by QD growth; furthermore, the PL peak shift with respect to reflux time shows that QD size increases rapidly during the nucleation stage and gradually climbs slowly during the OR phase.

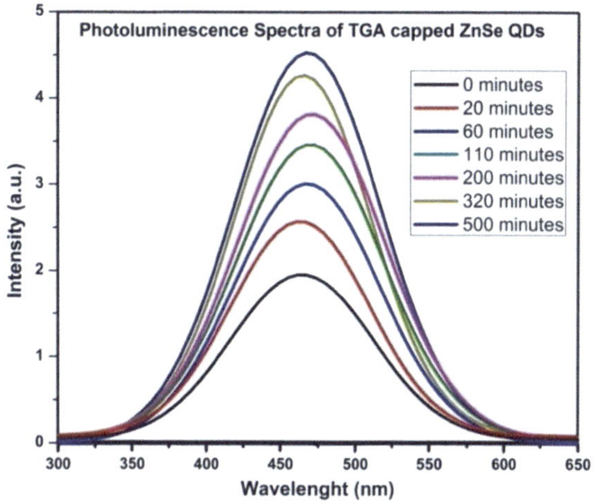

Fig. 4. Photolumiscence spectra of TGA capped ZnSe QDs

3.4 Determination of Optical Band Gap of ZnSe QDs

One of the key elements that control the material's electrical conductivity and, consequently, its use in optoelectronics is the ZnSe QDs' optical band gap. The Tauc plot is used to calculate the band gap [13].

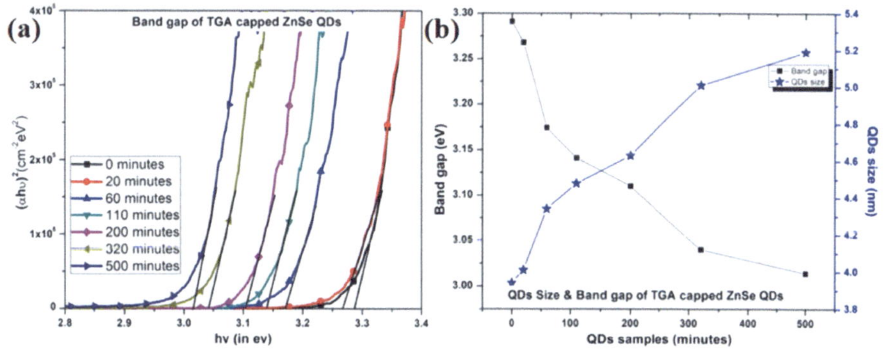

Fig. 5. a) TGA capped ZnSe QDs Tauc plot. b) TGA capped ZnSe QDs Band gap and QDs size Vs reflux time.

The Tuac's plot of as-synthesized ZnSe QDs is displayed in Fig. 5a and 5b. As reflux time extended from 0 to 500 min, the band gap red shifted from 3.291 eV to 3.014 eV; still, the band gap of each sample is less than the bulk band gap of ZnSe (2.7 eV). Quantum confinement energy and QD development are indicated by the band gap energy drops that occur with an increase in reflux time clearing.

3.5 Determination of QDs Size

The Brus equation can be used to compute QD size using the ZnSe optical band gap [14]. A straightforward technique for figuring out the size of spherical QDs is the Brus equation. The produced QDs ranged in size from 3.94 nm to 5.19 nm, all of which were significantly smaller than the ZnSe's Bohr exciton size. This demonstrates that there is high quantum confinement energy on the produced ZnSe QDs. Early in the reaction, QD size increases quickly. This is followed by a sluggish OR stage and a fast nucleation, which causes the size of QDs to gradually decrease. These materials with specifically tuned band gap are best suitable resources for optoelectronic and biological labels etc.

High-quality water-soluble TGA-stabilized ZnSe quantum dots have been developed successfully using a single injection hydrothermal approach in a completely ambient environment. The process has eliminated the oxidation of elemental Se powder, which acts as a selenium precursor. This mechanism is thoroughly analyzed together with its effects on other properties. Rapid nucleation is the first stage of growth, which is followed by a slow OR phase. It is evident that the current QD synthesis procedure, which could potentially be scaled up for industrial production, has numerous optoelectronic, photovoltaic, and biological applications.

4 Conclusion

ZnSe QDs of the different size were successfully synthesized by single injection hydrothermal method. The currently employed synthesis method shows prominent advantage over organometallic technique due to synthesis carrying at ambient temperature and inexpensive need. Particle size of the ZnSe QDs prepared by SIH method can be further controlled with variation in the concentration of capping agent – TGA. The concentrations of capping agent – TGA and Zinc precursor employed in this research paper as mentioned in the experimental part have resulted the average size of the QDs as 4.5 nm; the size of the QDs can be tuned to a specific dimensions based on its applications. The studies of optical characterizations have indicated the high luminescence QDs with bandgap between 3.291 eV to 3.014 eV and the size of QDs between 3.94 nm to 5.19 nm, showing the synthesized ZnSe QDs are in strong quantum confinement regime. Thus the synthesized QDs can be exploited in the sensor development of biological applications. The developed method of synthesis is easy to implement, cost effective and advantageous to perform at room temperature without the need of high facility laboratory.

References

1. Brus, L.E.: Nat. Rev. Mater. **4**(5), 307–320 (2019)
2. Kershaw, S.V., et al.: Journal of Materials Chemistry C **9**(24), 8054–8075 (2021)
3. Kim, J.S., et al.: Nat. Nanotechnol. **15**(1), 1–11 (2020)
4. Todosiev, M., et al.: Sol. Energy Mater. Sol. Cells **231**, 111308 (2022)
5. Patel, S.D., et al.: Biomaterials Science **10**(4), 897–920 (2022)
6. Liu, Y., et al.: Mol. Pharm. **18**(5), 1501–1521 (2021)
7. Wagner, C., et al.: Environ. Sci. Technol. **55**(3), 1200–1214 (2021)
8. Liu, Q., et al.: Advanced Energy Materials **12**(28), 2201060 (2022)
9. Patel, S.D., et al.: Materials Today: Proceedings **44**, 3604–3608 (2021)
10. Lee, J.H., Park, J.S., Lee, Y.: Journal of Materials Chemistry C **11**(5), 1598–1608 (2023)
11. Yu, W.W., et al.: Nano Lett. **3**(3), 385–388 (2003)
12. Jai Kumar, B., Sumanth Kumar, D., Mahesh, H.M.: J. Lumin. **178**, 362–367 (2016)
13. Deerinck, T.J.: Toxicol. Pathol. **36**(1), 112–116 (2008)
14. Brichkin, S.B., Razumov, V.F.: Russ. Chem. Rev. **85**, 1297 (2016)

Investigation of Structural and Magnetic Property of Fe_3O_4 and $CoFe_2O_4$ Nanoclusters for Biomedical Application

Akash Marsalin and Rajaboopathi Mani(✉)

Key Laboratory of Emergent Materials, Department of Physics and Nanotechnology, SRM Institute of Science and Technology, Kattankulathur, Chengalpattu, Tamil Nadu 603203, India
rajaboom@srmist.edu.in

Abstract. Ferrite and ferrite-based materials are widely used for biomedical and energy storage applications, due to their biocompatibility and magnetic property. In this study, Fe_3O_4 and $CoFe_2O_4$ nanoclusters were synthesised using a modified solvothermal method. The structure and morphology were analysed using X-ray diffraction (XRD), scanning electron microscopy (SEM), and the magnetic property of the samples were determined using vibrating sample magnetometry (VSM). The phases observed through the experiments were well matched with the standard JCPDS data observed for both Fe_3O_4 and $CoFe_2O_4$. SEM images revealed a uniform spherical morphology of the nanoclusters, while the VSM measurement indicates the ferromagnetic nature with the coercivity (H_c) values are 190 and 1132 Oe and the saturation magnetization (M_s) values of 87.17 and 59.12 emu/g corresponds to Fe_3O_4 and $CoFe_2O_4$ nanoclusters. The higher coercivity of $CoFe_2O_4$ is attributed to its enhanced magneto-crystalline anisotropy. These unique hierarchical structure and better magnetic response of these nanocluster make them a promising candidate for advanced biomedical applications especially in magnetic hyperthermia.

1 Introduction

Nanostructured magnetic particles in the form of nanoclusters and mesocrystals have received significant attention in recent decades. Due to their remarkable properties, which offer potential applications across diverse fields ranging from energy storage to biomedical technologies [1]. Among these, ferrite and ferrite-based materials such as Fe_3O_4, $NiFe_2O_4$, $CoFe_2O_4$ and $ZnFe_2O_4$ have widely been utilized for various applications which includes electromagnetic shielding, magnetic resonance imaging, magnetic memory, magnetic hyperthermia and targeted drug delivery [2–4]. This is attributed to their strong magnetic response to external magnetic field, biocompatibility, chemical stability, and low toxicity [5].

To obtain reliable and high-performance nanoparticle for these applications, magnetic particle with homogeneous shape and narrow size distribution, high saturation magnetization, good dispersibility in liquid medium are important [1, 6]. Therefore, the synthesis of nanostructured materials should possess monodispersity, size tuneable,

P. Joseph D. et al. (Eds.): IEMDST 2024, SPPHY 567, pp. 16–23, 2025.
https://doi.org/10.1007/978-981-96-5863-3_3

uniform shape, excellent magnetic properties and water dispersibility [7]. These characteristics can be achieved by modifying synthesis methods such as adding surfactant and adjusting reaction conditions. However, the correlation between the synthesis condition, structural property and functional performance of these materials is not completely understood, presenting an ongoing challenge for researchers.

In the recent years, focus of the research has been shifted from nanomaterials to secondary structure of ferrite-based nanoparticles named nanoclusters, mesocrystals, superlattices, core-shell structures, and so on. While significant efforts have been devoted to developing magnetic Fe_3O_4 dendrite like patterns, arrays of nanotube, and nanodiscs [8, 9], relatively few articles have focused on nanoclusters of Fe_3O_4 and $CoFe_2O_4$. For instance, Xia and his co-workers proposed a facile approach to synthesize aggregated magnetic nanoparticle using various surfactants in water [10]. Furthermore, Fe_3O_4 colloidal nanocrystalclusters (CNCs) were synthesised via a high-temperature hydrolysis method pioneered by Yin and his co-workers [11]. These CNCs secondary structures, often exhibit emergent properties that differ from those individual nanocrystals and expected to have unique properties in comparison with nanocrystals. Nonetheless Fe_3O_4 is an established material that has a large area of application however nanocluster has been less focused due to challenges synthesis process.

This study involves the synthesis, and characterization of Fe_3O_4 and $CoFe_2O_4$ nanoclusters and compared their structural and magnetic property without changing the experimental conditions. A solvothermal method was chosen to prepare the well-dispersed nano clusters, using carefully controlled ratio of ethylene glycol to di-ethylene glycol as solvents, which plays a vital role in tuning the size of both nanoclusters and constituent nanoparticles [12]. The results provide valuable insight into how cobalt substitution influences the physiochemical properties of this material, which can pave the way for their potential application in the biomedical field.

2 Materials and Methods

Analytical grade, Iron (III) chloride hexahydride ($FeCl_3{\cdot}6H_2O$, 97%), sodium acetate (NaOAc, 99%), cobalt chloride tetrahydride ($CoCl_2{\cdot}4H_2O$, 99%), ethylene glycol (EG, 99%), and diethylene glycol (DEG, 99%) were purchased from SRL Chemical Pvt. Ltd., and used without any additional purification. Milli-Q water exhibiting a resistivity of 18 $M\Omega\ cm^{-1}$ and analytical-grade ethanol were used throughout the experiments.

The samples were synthesized using a modified solvothermal method. For the preparation of Fe_3O_4, 0.1 mol of $FeCl_3{\cdot}6H_2O$ and 0.9 mol of NaOAc were added in a 1:1 ratio of ethylene glycol and diethylene glycol and vigorously stirred for 30 min. The final solution mixture was then transferred to Teflon lined autoclave (100 ml), sealed and in the hot air oven it was heated at 200 °C for 6 h. After reaction period it was cooled to ambient temperature and the black precipitates were extracted by centrifugation, and then washed multiple times with milli Q water and followed by ethanol. The final precipitate was then dried at 60 °C for 10 h.

The same experimental procedure was followed in the synthesis of $CoFe_2O_4$, except that 0.063 mol of $FeCl_3{\cdot}6H_2O$ and 0.034 mol of $CoCl_2{\cdot}4H_2O$ were mixed in the EG/DEG solution mixture in a stoichiometric ratio of 2:1.

3 Experiment and Characterization

For XRD analysis, used Benchtop X-Ray diffractometer, Malvern Panalytical Aeris, Netherlands was employed, utilizing Copper Kα radiation as the source (λ = 1.5406 Å) over 2θ angular range scanning from 10 to 70°,

The samples crystallite size (D_{hkl}) corresponds to a plane were estimated using Scherrer's approximation [13],

$$D_{hkl} = \frac{K\lambda}{\beta cos\theta} \tag{1}$$

where, λ represents the wavelength corresponds to Copper Kα radiation, K denotes the broadening constant (K = 0.9), β indicates the full width at half maximum (FWHM in radian) and θ refers to the diffraction angle (in degrees).

Micro-strain (ε_{hkl}) caused due to imperfection in crystal, distortion in the plane was calculated using Eq. (2) and the X-ray density of the material (density of a material based on its atomic structure) can be obtained by the relation [14]:

$$\varepsilon_{hkl} = \frac{\beta_{hkl}}{4tan\theta} \tag{2}$$

$$\rho_X = \frac{8M}{N_A a^3} \tag{3}$$

Additionally, we have used Williamson-Hall plot (W-H plot) to find the crystallite size and micro-strain (ε) for the particle [14, 15].

$$\beta\cos\theta = \frac{K\lambda}{D} + 4\varepsilon\sin \tag{4}$$

The scanning electron microscope (SEM) image was taken from Apreo 2 SEM, Thermo Fisher scientific in nanometre range. The sample preparation was as follows: 1 mg of the sample were dissolved in 2 ml of ethanol in a microcentrifuge vial and ultrasonicated for 10 min. Then 200 μl of the solution was cast onto aluminium stud ($\approx$ 1 × 1 cm) using micropipette and dried at 60 °C for 1 h before SEM analysis.

The magnetic property was determined at ambient temperature using vibrating sample magnetometer, Lake Shore 7400 series, USA.

The magnetic moment (m) were found using the formula [16];

$$m = \frac{M_S \times M_W}{5585} \tag{5}$$

where, M_s represents the saturation magnetization (in emu/g) M_w denotes the molecular weight of samples (in g/mol), and 5585 correspond to the magnetic factor.

In high field region of M/H plot, the Law of approach to saturation (LAS) were used to estimate the magneto-crystalline anisotropy constant (κ_1) of the sample [17]:

$$\kappa_1 = \mu_0 M_S\sqrt{\frac{105 \times b}{8}} \tag{6}$$

where κ_1, μ_0, and b are the corresponding magneto crystalline anisotropy constant, permeability of free space (1.257 × 10^{-6} $mKgS^{-2}A^{-2}$) and saturation magnetization in A/m respectively. 8/105, a constant value in the formula represents the anisotropy of polycrystalline materials [17].

4 Results and Discussion

4.1 XRD Analysis

The XRD pattern shown in Fig. 1, where obtained pattern closed match with the standard JCPDS data for Fe_3O_4 (19–0629) and $CoFe_2O_4$ (22–1086), both of which have cubic structure [18, 19]. The characteristic peaks corresponding to (220), (311), (222), (400), (422), (511), and (440) planes were noted, with no additional peaks, indicating the formation of spinel ferrite exhibiting Fd-3m space group. For $CoFe_2O_4$, we observed a significant peak broadening and a minor shift, which might be due to reduction in particle size and crystallinity [17]. The light shift toward lower angles suggests an increase in the lattice constant of $CoFe_2O_4$, is account for the difference in radii of ions between Fe^{2+} and Co^{2+} cations [16]. Furthermore, this shift confirms the successful incorporation of cobalt into the ferrite structure.

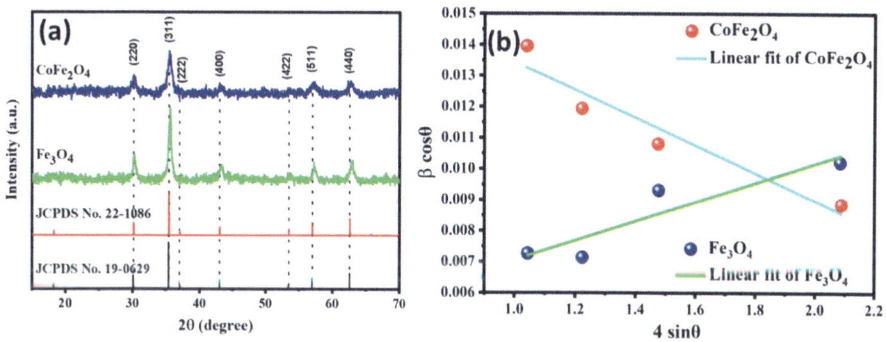

Fig. 1. Illustrates (a) XRD pattern of the synthesised Fe_3O_4 and $CoFe_2O_4$ samples along with its corresponding JCPDS data (black for Fe_3O_4 and red for $CoFe_2O_4$) (b) W-H plot for both Fe_3O_4 and $CoFe_2O_4$ and its corresponding linear fitting.

Justin et al. have reported the corelation between the lattice parameter and the ionic radii in the dopant ions, noting a reduction in the lattice parameter when the dopant with small ionic radii, followed by an increase beyond a certain doping level. This change in lattice behaviour may be impacted by synthesis conditions, such as surface distortion, cation distribution and lattice strain due to surface-to-volume ratio or surface oxidation [20]. In our study, the lattice parameter increases along with a decrease in crystallite size was observed, as reported earlier.

The samples crystallite size was determined by using both Debye-Scherrer method as well as Williamson-Hall analysis. The Debye method gives estimate of the crystallite size based on the peak broadening. In contrast, the Williamson-Hall approach accounts for both the effect of crystallite size and strain induced broadening (the fitted R^2 value are ≈ 0.9 for both samples). These methods provide comprehensive analysis by correlating the peak broadening with both factors. The lattice constant ($a = b = c$), X-ray density of the samples, crystallite size, micro-strain (ε_{ms}) of 311 plane (ε_{hkl}) were calculated using standard formulae and the obtained values are presented in Table 1.

Table 1. Shows the values obtained from XRD analysis of the Fe_3O_4 and $CoFe_2O_4$.

Sample	Lattice Constant (Å)	Scherrer's method (D_{hkl}) (nm)	Micro-strain ($\varepsilon_{hkl} * 10^{-3}$) (%)	W-H method (D) (nm)	Micro-Strain ($\varepsilon_{ms} * 10^{-3}$) (%)	X-ray density (g/cm^3)
Fe_3O_4	8.365	19.47	5.8	34.93	3.1	5.347
$CoFe_2O_4$	8.371	11.17	10.2	7.74	4.47	5.242

4.2 SEM Analysis

Size distribution and the morphology of Fe_3O_4 and $CoFe_2O_4$ samples were characterized using SEM. As shown in Fig. 2, both samples exhibit well-dispersed, spherical-shaped nanoclusters along with its particle distribution plot. The average particle size of Fe_3O_4 and $CoFe_2O_4$ was 235.6 ± 8.54 nm and 160.43 ± 2.48 nm, respectively. The samples uniformity in size and shape may be influenced by the combination of EG and DEG. The aggregation of nanoparticles to form nanoclusters suggests non-classical crystallization in these samples, leading to hierarchical structures and enhanced properties [21]. The physical and chemical properties of nanoclusters, including magnetic properties and surface area are directly influenced by the size and shape distribution. The presence of hierarchical morphology in both Fe_3O_4 and $CoFe_2O_4$ nanoclusters which could potentially impart unique functional properties to the materials.

4.3 VSM Analysis

Figure 3 (a) showed the room temperature magnetization plot for the Fe_3O_4 and $CoFe_2O_4$, measured in the applied magnetic field ranges from +2 T to −2 T. Characteristic hysteresis loops is produced by the two samples with well-defined saturation magnetization (M_s), and coercivity (H_c), which confirms their ferromagnetic nature. The saturation magnetization (M_s) values are 83.68 emu/g for Fe_3O_4 and 50.11 emu/g for $CoFe_2O_4$ nanoclusters, while the corresponding coercivity (Hc) values were 190 Oe and 1132 Oe, respectively. The contribution of high crystalline anisotropy and inhomogeneity in particle size resulted in higher coercivity for $CoFe_2O_4$ nanoclusters despite reduced particle size [5, 22].

The magnetic moment (m) was calculated using Eq. (5). For the inverse spinel structure such as Fe_3O_4 and $CoFe_2O_4$, the magnetic moment depends on the distribution of cations between the octahedral (B-site) and the tetrahedral (A-site) sites. The overall theoretical magnetic moment of Fe_3O_4 and $CoFe_2O_4$ are 4 and 3.87 magnetic moment per formula unit (μ_B/f.u.), respectively [5, 23]. However, the observed moment (m) for Fe_3O_4 and $CoFe_2O_4$ nanoclusters is 3.469 and 2.105 μ_B per formula unit, respectively, which are lower than the standard values. The mismatch between the theoretical and experimental values may arise from various factor including cation distribution, grain disorder, and grain size in the synthesized samples [24].

Figure 3 b, illustrates the LAS equation fitting for both samples. The magneto crystalline anisotropic constant (κ_1) was found to be 0.721×10^4 J/m^3 for Fe_3O_4, and 1.068×10^4 J/m^3 for $CoFe_2O_4$ indicating a higher anisotropy in $CoFe_2O_4$ compared

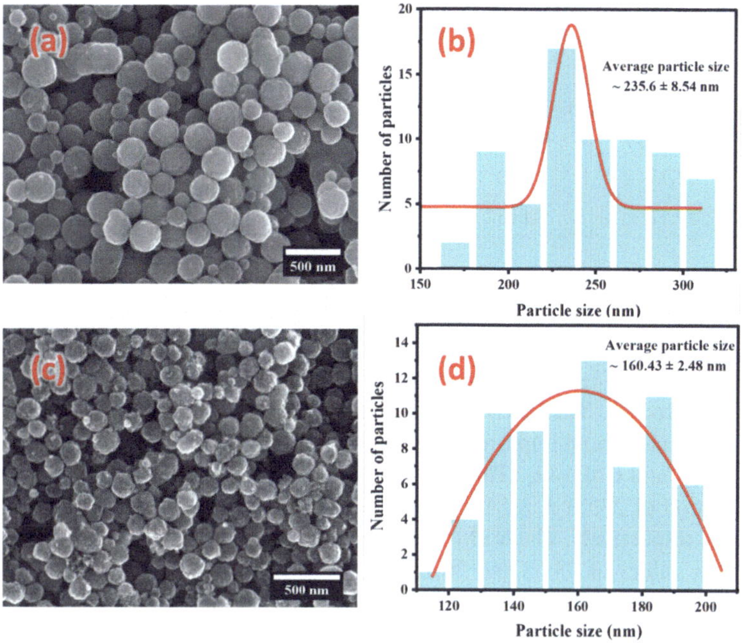

Fig. 2. (a, c) represents the SEM image of Fe_3O_4 and $CoFe_2O_4$ nanocluster and (b, d) illustrates its corresponding particle size distribution.

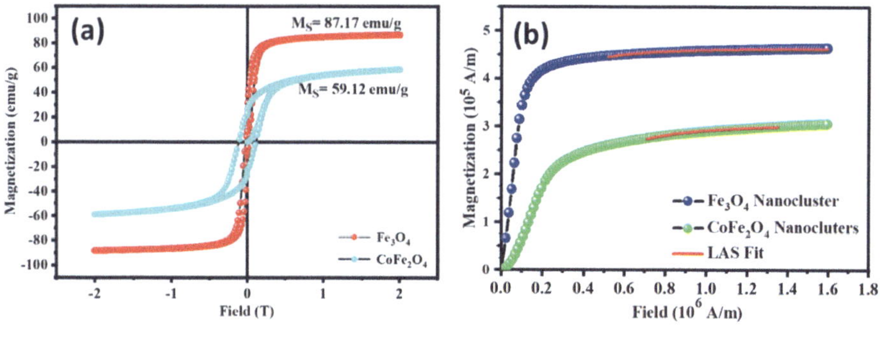

Fig. 3. Illustrates (a) M-H plot at room temperature and (b) LAS fit of the synthesized nanoclusters, (magnetization Vs magnetic field of the first quadrant is plotted in terms of A/m).

with Fe_3O_4. This can be ascribed the existence of Co^{2+} ions occupying the B-site of $CoFe_2O_4$, coupled with strong spin-orbit coupling of the Co^{2+} ions [25]. The calculated and observed value VSM analysis were listed in the Table 2. Thus, obtained results highlight the distinct magnetic properties of $CoFe_2O_4$, emphasizing the impact of cobalt substitution on enhancing magnetic anisotropy and, consequently, the overall magnetic performance of the nanoclusters.

Table 2. Shows the calculated magnetic parameters obtained from VSM measurement.

Sample	Saturation magnetization (emu/g)	Coercivity (Oe)	Magnetic moment (μ_B/f.u.)	Magneto crystalline anisotropy (κ_1) (J/m^3)
Fe_3O_4	87.17	190	3.469	0.721×10^4
$CoFe_2O_4$	59.12	1132	2.105	1.068×10^4

5 Conclusion

The Fe_3O_4 and $CoFe_2O_4$ nanoclusters were successfully synthesized using facile solvothermal process. The XRD analysis of the sample confirmed the formation of pure nanoclusters without impurity evidenced by their agreement with standard JCPDS data for Fe_3O_4 and $CoFe_2O_4$. The SEM images show uniform morphologies with the size distribution of 235.6 ± 8.54 nm and 160.43 ± 2.48 nm for Fe_3O_4 and $CoFe_2O_4$ nanoclusters respectively. The VSM measurements demonstrate ferromagnetic properties, with M_S values of 83.68 and 50.11 emu/g for Fe_3O_4 and $CoFe_2O_4$ nanoclusters, respectively. The incorporation of cobalt into the ferrite resulted in change in lattice parameters which in turn enhanced coercivity and magneto crystalline anisotropy of the $CoFe_2O_4$ nanocluster. The uniform morphology and controlled size distribution, along with the ferromagnetic properties, make these nanoclusters promising for biomedical applications including magnetic hyperthermia and drug delivery. Future work will focus on optimizing synthesis conditions to further improve these properties, enabling their application in the biomedical field.

Acknowledgments. The authors express gratitude to the SRM Central Instrumentation Facility and Nanotechnology Research Centre at SRMIST for their provision of research facilities. The author AM wishes to convey his sincerest gratitude to SRM IST for granting institutional fellowship.

References

1. Wang, W., Tang, B., Ju, B., Zhang, S.: RSC Adv. **5**(92), 75292–75299 (2015). https://doi.org/10.1039/C5RA14354C
2. Liu, Y., Wu, N., Wang, Z., Cao, H., Liu, J.: New J. Chem. **41**(14), 6241–6250 (2017). https://doi.org/10.1039/C7NJ00230K
3. Wu, N., et al.: Carbon N Y **145**, 433–444 (2019). https://doi.org/10.1016/J.CARBON.2019.01.028
4. Sharifi, I., Shokrollahi, H., Amiri, S.: J. Magn. Magn. Mater. **324**(6), 903–915 (2012). https://doi.org/10.1016/J.JMMM.2011.10.017
5. Karthickraja, D., et al.: New J. Chem. **43**, 13584–13593 (2019). https://doi.org/10.1039/C9NJ02510C
6. Li, Q., Kartikowati, C.W., Horie, S., Ogi, T., Iwaki, T., Okuyama, K.: Sci. Rep. **7**(1), 1–7 (2017). https://doi.org/10.1038/s41598-017-09897-5

7. Liu, J., et al.: Angew. Chem. Int. Ed. **48**(32), 5875–5879 (2009). https://doi.org/10.1002/ANIE.200901566
8. Hwang, S.O., et al.: J. Phys. Chem. C **112**(36), 13911–13916 (2008). https://doi.org/10.1021/JP802943Z
9. Chu, Y., Hu, J., Yang, W., Wang, C., Zhang, J.Z.: J. Phys. Chem. B **110**(7), 3135–3139 (2006). https://doi.org/10.1021/JP056506R
10. Xia, H.B., Yi, J., Foo, P.S., Liu, B.: Chem. Mat. **19**(16), 4087–4091 (2007). https://doi.org/10.1021/CM070918Q
11. Ge, J., Hu, Y., Biasini, M., Beyermann, W.P., Yin, Y.: Angew. Chem. Int. Ed. **46**(23), 4342–4345 (2007). https://doi.org/10.1002/ANIE.200700197
12. Choi, K.H., et al.: IEEE Trans. Magn. **47**(10), 3369–3372 (2011). https://doi.org/10.1109/TMAG.2011.2152375
13. Munusamy, G., Mani, R., Varadharajan, K., Narasimhan, S., Munusamy, C., Chandrasekaran, B.: Res. Chem. Intermed. **46**, 715–736 (2020). https://doi.org/10.1007/s11164-019-03986-y
14. Kim, Y.Y., et al.: Nat. Commun. **5**(1), 1–14 (2014). https://doi.org/10.1038/ncomms5341
15. Mote, V., Purushotham, Y., Dole, B.: J. Theor. Appl. Phys. **6**(1), 1–8 (2012). https://doi.org/10.1186/2251-7235-6-6
16. Singhal, S., Chandra, K.: J. Solid State Chem. **180**(1), 296–300 (2007). https://doi.org/10.1016/J.JSSC.2006.10.010
17. Anantharamaiah, P.N., Joy, P.A.: Phys. Chem. Chem. Phys. **18**(15), 10516–10527 (2016). https://doi.org/10.1039/C6CP00369A
18. Wei, Y., Han, B., Hu, X., Lin, Y., Wang, X., Deng, X.: Procedia Eng **27**, 632–637 (2012). https://doi.org/10.1016/J.PROENG.2011.12.498
19. Mitra, S., Veluri, P.S., Chakraborthy, A., Petla, R.K.: Chem. Electro. Chem. **1**(6), 1068–1074 (2014). https://doi.org/10.1002/CELC.201400026
20. Anandhi, J.S., Arun, T., Joseyphus, R.J.: Physica B Condens. Matter **598**, 412429 (2020). https://doi.org/10.1016/J.PHYSB.2020.412429
21. Mani, R., et al.: Cryst. Growth Des. **23**, 236–245 (2023). https://doi.org/10.1021/acs.cgd.2c00971
22. Anikin, A., et al.: J. Magn. Magn. Mater. **595**, 171507 (2024). https://doi.org/10.1016/J.JMMM.2023.171507
23. Dmitriev, A.I., Alekseev, S.I., Kostyuchenko, S.A.: Inorg. Mater. **55**(6), 576–581 (2019). https://doi.org/10.1134/S0020168519060037
24. Kamble, R.B., Varade, V., Ramesh, K.P., Prasad, V.: AIP Adv. **5**, 1 (2015). https://doi.org/10.1063/1.4906101
25. Regmi, S., et al.: Appl. Phys. Lett. **121**(10), 102401 (2022). https://doi.org/10.1063/5.0107242/2834227

Study of Structural, Optical and Magnetic Properties of $Ni_{0.8}Zn_{0.2}Fe_2O_4$ Ferrites Calcined at Different Temperatures

Sk. Sameena[1], P. Naveen[1,2], G. Chanti[3], C. Shiva Shankar[4], Y. Vasudeva Reddy[4], Ch. Swapna[5], and N. Pavan Kumar[1(✉)]

[1] Matrusri Engineering College, Saidabad, Hyderabad, Telangana 500059, India
pallakash@gmail.com
[2] Department of Physics, ISBM University, Nawapara, Chhattisgarh 493996, India
[3] School of Physics, University of Hyderabad, Gachibowli, Hyderabad 500046, India
[4] Department of Physics, Government City College, Nayapul, Hyderabad 500002, India
[5] Department of Sciences and Humanities, Dr. BR Ambedkar Government Model Residential Polytechnic for Women, Karimnagar 505001, India

Abstract. In this work, we examined the effect of different calcination temperatures on the structural, optical, and magnetic properties of $Ni_{0.8}Zn_{0.2}Fe_2O_4$ (NZFO) prepared through the sol-gel method. As X-ray diffraction (XRD) and Raman spectroscopy studies confirmed, all compounds demonstrated the spinel phase. It has been observed that the lattice parameter values 8.3932 Å, 8.4015 Å, and 8.4040 Å are increased as the heating temperature is raised from 450 °C to 750 °C. The material's optical bandgap was estimated through UV-visible spectroscopy, which revealed its potential for optoelectronic applications. M-H hysteresis loops depict the softness of the compounds, and calculated magnetic parameters. Overall, the coercivity values 150 Oe, 170 Oe, and 110 Oe are decreased due to an increase in grain size. Moreover, NZFO calcined at 600 °C showed remarkable magnetic properties compared to the other compounds.

1 Introduction

Ferrites are a large family of metal oxides with various characteristics, including high electrical resistivity, low electrical losses, considerable saturation magnetization, and excellent chemical stability—in-depth research following World War II [1]. Since ferrites employ the transition element Fe, as in Fe_2O_3, they are naturally ferromagnetic and display magnetism below the Curie temperature, T_C. Depending on the properties of the ferrites, they are used in transformers, inductors, and devices of gyromagnetic electronics like resonance isolators, circulators, etc. [2]. Based on their structure and magnetic properties, ferrites are classified into four types: garnets, spinel, hexagonal, and ortho ferrites. In those, spinel ferrites have the common chemical formula MFe_2O_4 here M refers to the transition metal. The unit cell in the spinel structure is made up of eight smaller unit cells with two different sites: tetrahedral interstitial site (A) and octahedral interstitial site (B). In one unit cell, a total of 32 octahedral and 64 tetrahedral interstitial sites are

P. Joseph D. et al. (Eds.): IEMDST 2024, SPPHY 567, pp. 24–34, 2025.
https://doi.org/10.1007/978-981-96-5863-3_4

present. The arrangement of oxygen atoms in the spinel structure is closely packed, with 32 oxygen atoms making up a unit cell [3, 4]. In Nickel ferrite ($NiFe_2O_4$), Ni^{+2} occupies the octahedral site (B), whereas Fe^{+3} is equally distributed at both the tetrahedral site (A) and octahedral site (B) and possesses the inverse spinel structure [5]. Because of its great electrochemical stability, high electrical resistivity [6], high Curie temperature [7], low magnetic coercivity, low magnetic anisotropy [8], low magnetostriction [9], high Neel temperature [10], high permeability in RF region and reduced eddy current losses, nickel ferrite is used in many applications [11]. V.A. Bharati et al. reported the saturation magnetization (M_S), remanence magnetization (M_r), coercivity (H_C), remanence ratio (M_r/M_s) and magneton number (μ_B) for $NiFe_2O_4$ nanoparticles as 53.16 emu/g, 12.9 emu/g, 1.97 Oe, 0.037, 2.234 μ_B respectively [12]. In zinc ferrite ($ZnFe_2O_4$), Zn^{+2} occupies the tetrahedral site, and Fe^{+3} occupies the octahedral site, exhibiting the standard spinel structure. They are suitable for use as a soft magnet and low-loss material at high frequencies because of their low coercivity, mild magnetic saturation, electromagnetic solid performance, excellent chemical stability, moderate thermal coefficient, energy transfer efficiency, and mechanical hardness [13]. Ni-Zn ferrites are soft ferrimagnetic materials with low magnetic coercivity, high resistivity values, and minimal eddy current loss in high-frequency operations (10–500 MHz). Magnetic saturation, Curie temperature, permeability, and dielectric loss are significant properties of these ferrites in many applications, such as power transformers and telecommunications.

Ferrites can be prepared through different synthesis methods such as sol-gel, ball milling, hydrothermal method, co-precipitation method, solid-state reaction method, microwave-assisted method, and solvothermal method [14]. Among these methods, sol-gel techniques have been widely used in recent decades to prepare various mixed-metal oxides, nanomaterials, nano-scale architectures, nano-porous oxides, and organic-inorganic hybrids. The sol-gel process has been proven to offer significant advantages, such as improved mixing of the starting materials and superior chemical homogeneity in the final product [15]. Rimi Sharma et al., synthesized the Zn doped nickel ferrite via the sol-gel method, and results reported the enhancement of saturation magnetization, coercivity, remanence magnetization, remanence ratio for $x = 0.2$ compared to the pure nickel zinc ferrite [16]. M. Sertkol et al., prepared the $Zn_xNi_{1-x}Fe_2O_4$(where $x = 0.0$, 0.2, 0.7) ferrites and reported the highest saturation magnetization value 49.2 emu/g, and coercivity value 317 Oe for $x = 0.2$ [17]. K. Anu et al., synthesized the $Zn_xNi_{(1-x)}Fe_2O_4$ through the co-precipitation method and obtained the maximum magnetization value of 30 emu/g for $x = 0.2$ [18]. From the above references, it was observed that the magnetic properties of the nickel ferrites increase when the Zinc is doped. Costa et al., synthesized the $Ni_{0.5}Zn_{0.5}Fe_2O_4$ via the combustion reaction method sintered at 1100 °C, 1200 °C, 1300 °C, and 1400 °C in results reported the enhancement of magnetic properties [19]. N Pavan Kumar et al., prepared the $NiFe_2O_4$ and sintered sample through the different sintering methodologies such as (i) regular press and sintering (NPS), (ii) magnetic field press and sintering (MFPS), and (iii) spark plasma sintering (SPS), and got highest saturation magnetization for the MFPS and lowest coercivity for NPS samples. From these, it can be concluded that sintering methodologies and temperatures also affect the magnetic, electrical, and optical properties of spinel ferrites [20]. Therefore, based on the above studies, it has been planned to prepare Ni-Zn ferrite through the sol-gel method, and the effect of different calcination temperatures on structural, magnetic and optical properties has been studied. In this present study, $Ni_{0.8}Zn_{0.2}Fe_2O_4$ (NZFO) ferrite was prepared and sintered at different temperatures i.e. 450 °C, 600 °C, and 750 °C. The samples were given codes as NZFO450, NZFO600 and NZFO750.

2 Experimental Procedure

The sol-gel process was used to produce nanocrystalline $Ni_{0.8}Zn_{0.2}Fe_2O_4$ (NZFO) ferrite. The chemicals $Fe(NO_3)_3.9H_2O$, $Ni(NO_3)_2.6H_2O$, and $Zn(NO_3)_2.6H_2O$ are considered as the primary precursors, with 99.9% purity taken from the Sigma-Aldrich company. An aqueous solution was formed by dissolving nickel nitrate, zinc nitrate, and iron nitrate in distilled water at a stoichiometric ratio. Citric acid was added as a chelating agent, and the pH was corrected to 7 with an ammonia solution. The mixture was then heated to 80 °C while constantly swirling to facilitate gel-to-powder development. The resulting powder was pounded into a fine powder with a mortar and pestle. The powder was then calcined at three different temperatures (450 °C, 600 °C, and 750 °C) for two hours in a muffle furnace with a heating rate of 5 °C/min. Following synthesis, the samples were characterized using a variety of methods. The prepared samples were analyzed using a Rigaku Mini Flex 600 powder X-ray diffractometer, utilizing Cu-K_α radiation with a wavelength of 1.54 Å. Optical experiments were performed in the 200–900 nm wavelength range using a T90 UV-Vis spectrophotometer. The Raman spectra of prepared compounds were captured at room temperature using a Horiba micro-Raman system, with a wavenumber range of 100 to 800 cm^{-1} and a 532 nm He-Ne laser in operation. Magnetic hysteresis (M-H) measurements were taken with a Vibrating sample magnetometer (VSM) to explore the magnetic characteristics of the produced ferrite nanoparticles.

3 Results and Discussion

3.1 X-ray Diffraction

X-ray diffraction (XRD) confirmed the prepared material's crystal structure. XRD patterns of NZFO at different calcination temperatures are shown in Fig. 1(a). The materials show the diffracted peaks at $2\theta = 18, 30, 35, 37, 43, 53, 57, 62$, and 74 degrees, which are corresponding to (111), (220), (311), (222), (400), (422), (511), (440), and (533) planes confirm the cubic spinel structure with the space group *Fd-3m* [21]. A small number of secondary phases was observed and are indicated with the symbol '*' in the Fig. 1(a), and these secondary phases are due to the presence of α-Fe_2O_3 [22]. It is also observed that the intensity of the peaks increases with the calcination temperature, suggesting an improvement in crystallinity [23].

Rietveld refinement is carried out using the Full-prof suite to obtain the structural parameters. The resulting XRD fit and residual plots are shown in Fig. 1 (b-d). Table 1 displays the structural parameters and the fit R-values: R_p represents the "unweighted profile factor," R_{wp} stands for the "weighted profile factor," and R_{exp} indicates the "experimental factor." The goodness of fit (χ^2) values demonstrate that the theoretical and

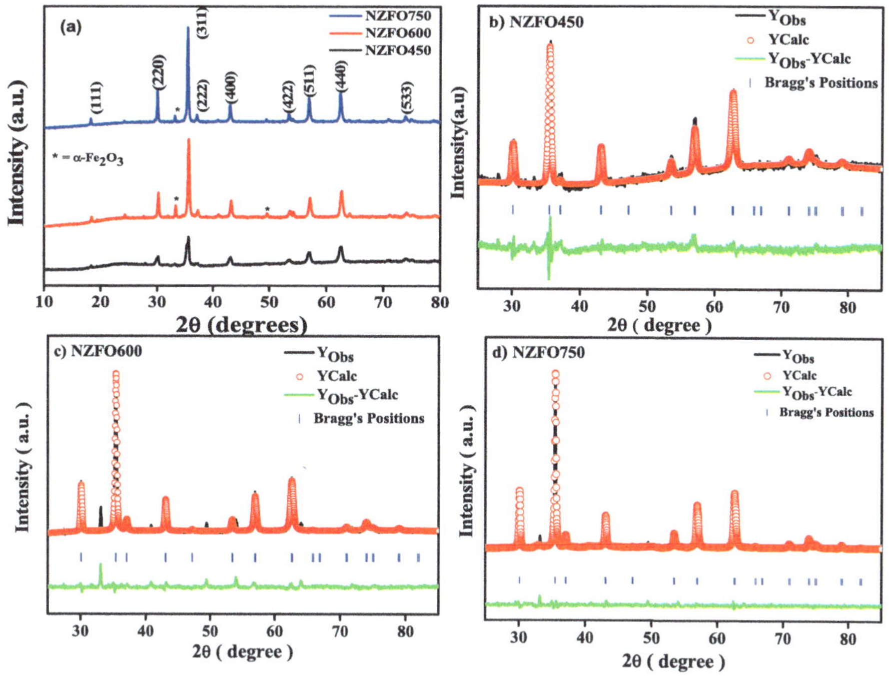

Fig. 1. (a) XRD patterns of NZFO at various calcination temperatures, (b-d) XRD patterns of NZFO at various calcination temperatures after refinement.

Table 1. Crystallography data from XRD and Rietveld refined values of NZFO at different calcination temperatures.

S. No.	Parameter	NZFO450	NZFO600	NZFO750
1	R_p	44.1	33.5	15.1
2	R_{wp}	24.8	20.5	17.7
3	R_{exp}	15.1	10.9	10.8
4	χ^2	2.69	3.53	2.69
5	Lattice parameters (a = b = c)Å	8.3932	8.4015	8.4040
6	Crystallite size D (nm) from Scherrer's formula	13	22	31
7	Density (g/cc)	5.297	5.282	5.277
8	Crystallite size from W-H plots D (nm)	16	61	69
9	Strain (10^{-4})	7.09	20.7	20.7
10	FWHM (degree)	0.6163	0.4215	0.3245

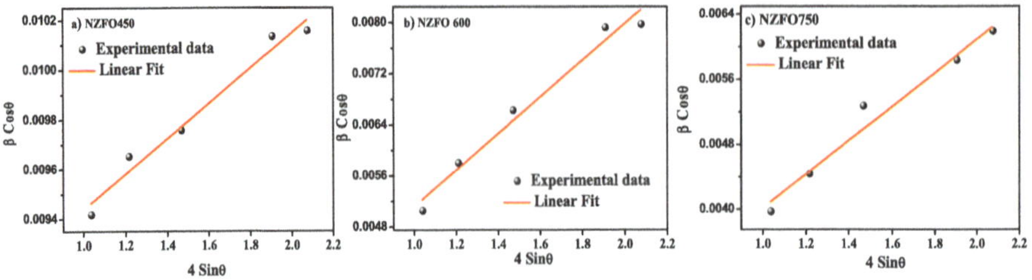

Fig. 2. W-H plots of NZFO at different calcination temperatures.

experimental data agree. We got slightly higher goodness of fit values, which may be due to the secondary phases in XRD.

The NZFO's crystallite size (D) was measured using Scherrer's formula $D = \frac{k\lambda}{\beta cos\theta}$, where k is 0.93, λ is the wavelength, and β is the full-width half maximum. Calculated crystallite size, lattice parameter, and density values are mentioned in Table 1. The crystallite size increases as the calcined temperature rises to 750 °C. This growth in crystallite size with temperature is attributed to the coalescence process [24]. Minor differences in lattice parameters may be related to oxidation states and cation arrangement inside the spinel structure's lattice sites. From NZFO450 to NZFO750, the slight increase in the lattice parameters is primarily due to the reduction of iron from Fe^{3+} to Fe^{2+} during the calcination process, as the Fe^{2+} ion has a larger ionic radius (0.76 Å) compared to Fe^{3+} (0.64 Å) [25]. From Table 1, density shows decreasing trend with increasing calcined temperature which is attributed to an increase in intragranular porosity caused by discontinuous grain growth [26].

The Williamson-Hall (W-H) equation has been used to calculate the strain value in our samples: $\beta^*(\cos\theta)/\lambda = k/D_{WH} + 4\epsilon^*\sin\theta$, where ϵ represents the effective strain, D_{WH} is the grain size, and the remaining symbols are the same as in the Debye-Scherrer formula [27]. Figure 2 shows the W-H plots of NZFO at different calcined temperatures. Crystallite size increases from NZFO450 to NZFO750 which is similar to the crystallite size variation from Debye Scherrer formula. Strain values also increase with increased calcined temperature.

3.2 UV-Vis. DRS Spectroscopy

It is recognized that the basic mechanisms of UV-visible absorption, reflection, or transmission of light in semiconducting materials are closely linked to their morphology and crystallite size [28]. Therefore, UV diffuse reflectance spectroscopy studies are crucial for estimating band gaps. Figure 3(a) shows the UV absorbance spectra of the compounds of the present investigation. Band gap energy was calculated by using Tauc's method.

$$(\alpha h\nu)^{\frac{1}{n}} = A\left(h\nu - E_g\right) \quad (1)$$

The obtained Tauc plots are shown in Fig. 3(b-d) and it has been found that the energy gap values are increasing with increasing the calcination temperature. The increase in the energy gap is due to the increase in crystallinity and grain size with the temperature. Larger grains usually result in fewer grain boundaries and consequently less scattering of electrons, which can contribute to a higher energy gap.

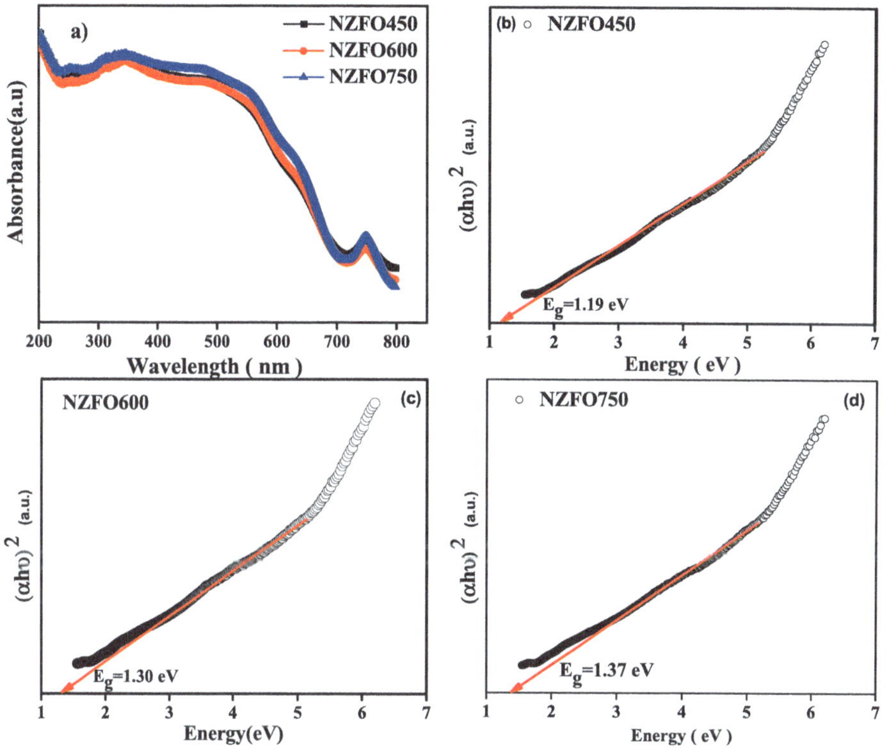

Fig. 3. a) UV-Vis. DRS spectra of NZFO samples. b-d) energy band gaps of NZFO450, NZFO600, and NZFO750.

3.3 Raman Spectroscopy

The Raman spectra of the synthesized spinel ferrites are illustrated in Fig. 4 to gain a deeper insight into their structure. Nickel spinel ferrites generally belong to the class of inverse spinel with the space group *Fd-3m* [31]. The structure of spinel ferrites is expected to exhibit five active Raman modes. It was found that only four active Raman modes, denoted as F_{2g}, E_g, and A_{1g}. As illustrated in Fig. 4, the characteristic Raman peaks are well-defined and located at their anticipated positions, confirming the structural purity of the synthesized ferrites. The vibrational modes observed in the Raman spectra of ferrite materials include the E_g, F_{2g}, and A_{1g} modes. The E_g mode arises from the symmetric bending of oxygen atoms for iron (Fe) ions. The F_{2g} mode is linked to the translation of metal ions at tetrahedral sites coordinated with four oxygen atoms. Meanwhile, the A_{1g} peak corresponds to the unit cell involving all Fe ions. These modes are essential for understanding ferrite materials' structural and vibrational properties [32].

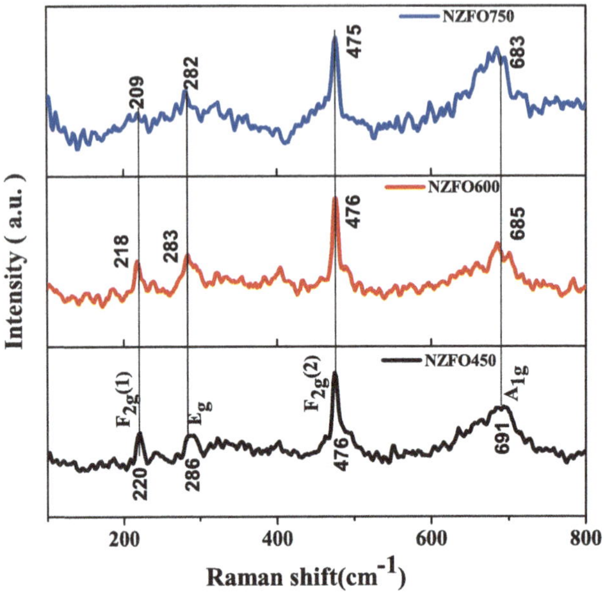

Fig. 4. Raman spectra of NZFO450, NZFO600, and NZFO750.

3.4 M-H Studies

M-H hysteresis loops of the compounds are depicted in Fig. 5. The slim loop demonstrates the soft magnetic nature of the compounds, and their magnetic parameters are listed in Table 2. Table 2 shows that when the calcination temperature increases, the coercivity value increases for NZFO600 and decreases for NZFO750. Previous researchers detected this type of behaviour in similar systems [34]. The drop in H_c for calcined samples can be attributed to the change from a magnetic single domain to a magnetic multidomain within a particle and the decrease in the anisotropic field. NZFO600 compound showed higher coercivity than other compounds.

The value of saturation magnetization increases as the calcination temperature rises i.e. for NZFO600 and NZFO750, showing that the magnetic moment of the B site is growing and that Fe^{+3} ions are migrating from the A to the B site. Change in magnetization values with sintering temperature, is ascribed to a change in cation distribution between tetrahedral and octahedral sites [35]. This is also supported by Raman spectroscopy studies. Because of their high saturation magnetization at higher temperatures, these compounds are ideal for high density information storage and ferrofluid technology devices. From magnetic parameters it was clearly observed that NZFO600 showed better values than others.

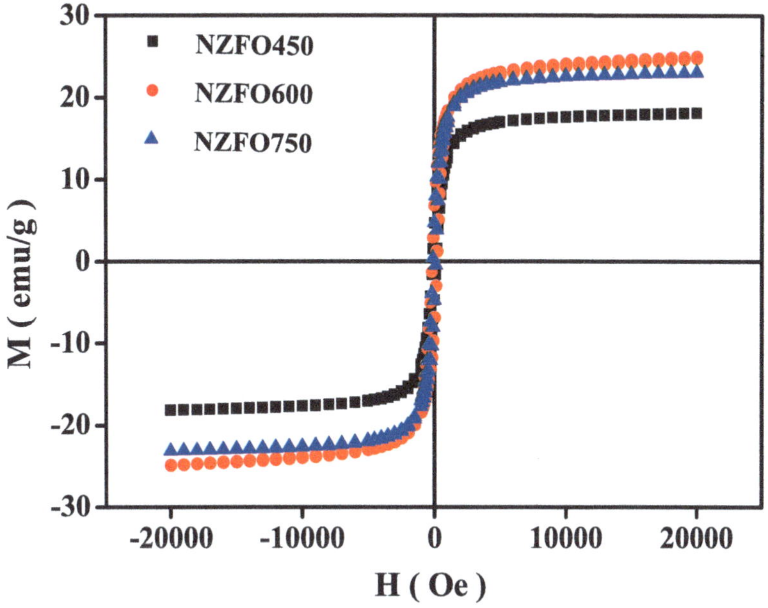

Fig. 5. M-H hysteresis loops of NZFO at different calcination temperatures.

Table 2. Magnetic parameters from M-H loops.

S. No.	Parameters	NZFO450	NZFO600	NZFO750
1	H_c (O_e)	149.6	169.7	110.5
2	M_r (emu/g)	4.638	6.818	4.798
3	M_s (emu/g)	18.1	24.8	23.3
4	Anisotropy constant K (10^2) (erg/cm^3)	21.93	57.16	23.61

Figure 6 shows the M vs $\frac{1}{H^2}$ of NZFO at different calcined temperatures. The magnetic anisotropy constants are coefficients in the energy expansion of spherical harmonics that vary with the moment direction. Due to symmetry, lower-order anisotropy constants vanish in cubic crystals, resulting in a low anisotropy energy. Tetragonal is the most important non-cubic symmetry with a high magnetic anisotropy. The magnetocrystalline anisotropy was also calculated using the 'Law of Approach (LA)' technique. The connection below illustrates how this method connects the magnetization (M) to the applied magnetic field (H)

$$M = M_s\left[1 - \left(\left(\frac{8}{105}\right) \times \left(\frac{K^2}{\mu_0^2 M_s^2 H^2}\right)\right)\right] \tag{2}$$

$$M = M_S\left[1 - \frac{b}{H^2}\right] \tag{3}$$

where b represents a constant correction factor for magnetization. M_S denotes the saturation magnetization, which is the maximum magnetization a material can achieve in an external magnetic field, μ_0 is the permeability of free space, a fundamental physical constant that describes how a magnetic field interacts with a vacuum. Lastly, K stands for the magnetic crystalline anisotropy constant, which quantifies the dependence of a material's magnetic properties on the direction of magnetization within the crystal lattice. Furthermore, the following equation has evaluated the values of K

$$K^2 = -\left[\frac{M_s \times Slope}{\beta}\right] \tag{4}$$

The β value of 8/105 indicates the cubic anisotropy of random polycrystalline samples. Table 2 presents the estimated magnetocrystalline anisotropy (K) values.

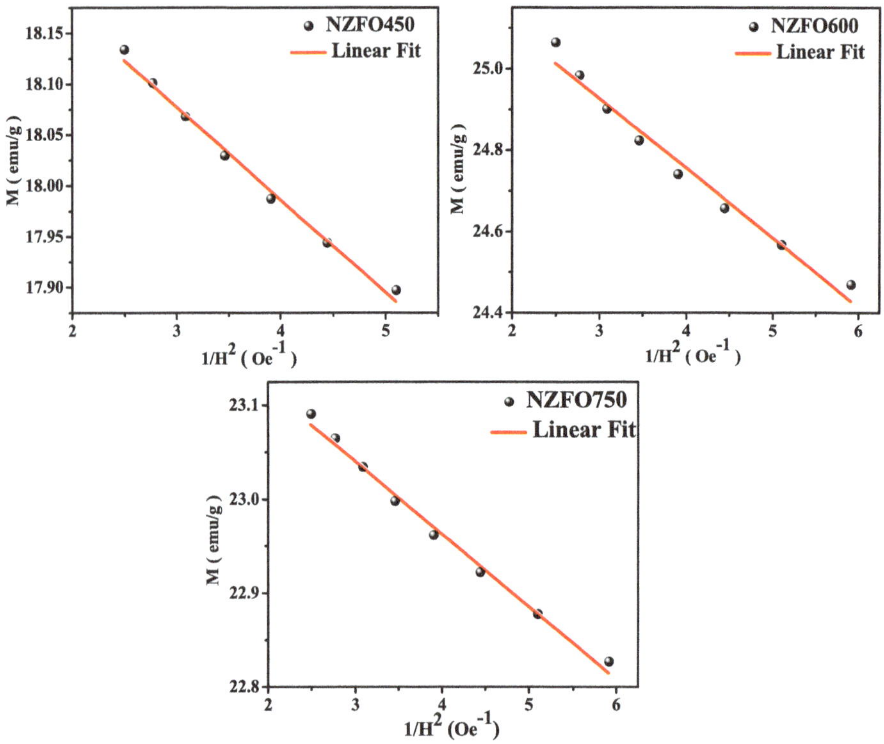

Fig. 6. M vs $1/H^2$ of NZFO at different calcination temperatures.

Generally, the anisotropy constant K increases with different calcined temperatures; the values are 21.93, 57.16, and 23.61 (x 10^2 erg/cm^3), indicating significant changes in magnetic anisotropy at different temperatures. Here also NZFO600 showed higher anisotropy value than others. This could be due to variations in cation distribution, increased crystallinity, or changes in particle shape. Overall, these trends highlight the importance of calcination temperature in tailoring the magnetic properties of

NZFO nanoparticles, providing a means to fine-tune their characteristics for a variety of technological applications such as spintronics, magnetic storage, and electromagnetic interference shielding [36].

4 Conclusion

The structural, magnetic, and optical properties of compound $Ni_{0.8}Zn_{0.2}Fe_2O_4$ are affected by different calcination temperatures 450 °C, 600 °C, and 750 °C. NZFO compounds were prepared by sol-gel method and calcined at temperatures 450 °C, 600 °C, and 750 °C. All the prepared compounds exhibited the spinel phase, and confirmed by XRD and Raman spectroscopy analyses. The crystallite size rose at different calcined temperatures. The band gap values increase when the material is calcined at higher temperatures, indicating that more energy is needed for charge transfer transitions between oxygen ions and metal ions. This shows that processing conditions have a significant impact on the optical properties of NZFO, which is important for customizing the material for specific applications in optoelectronics or photocatalysis. The magnetic properties of NZFO also change with calcined temperatures. Overall, with calcination temperature, the coercivity, remanence, saturation magnetization show varying trends. These changes are due to factors such as grain growth, reduction in defects, changes in magnetic anisotropy, phase composition, and stress relief. Additionally, the magnetic anisotropy constant values increased at higher calcined temperatures. It was observed that NZFO600 showed higher magnetic parameters and magnetic anisotropy values than other compounds NZFO450 and NZFO750. These variations highlight the importance of controlled synthesis conditions in tailoring the magnetic properties of NZFO for applications such as spintronics, magnetic storage, or electromagnetic interference shielding.

Acknowledgements. The authors acknowledged the Principal and management of Matrusri Engineering College, Hyderabad for providing research lab facilities.

References

1. Thakur, P., Taneja, S., Chahar, D., Ravelo, B., Thakur, A.: J. Magn. Magn. Mater. **530**, 167925 (2021)
2. Tolani, Smita C., Golhar, A.R., Rewatkar, K.G.: AIP conference proceedings, vol. 2104, no. 1. AIP Publishing (2019)
3. Thakur, A., Thakur, P., Hsu, J.-H.: Scripta Mater. **64**(2), 205–208 (2011)
4. Moradmard, H., Farjami Shayesteh, S., Tohidi, P., Abbas, Z., Khaleghi, M.: Journal of Alloys and Compounds **650**, 116–122 (2015)
5. Rana, G., et al.: Chem. Eng. Res. Des. **175**, 182–208 (2021)
6. Gunjakar, J.L.: Surf. Sci. **254**, 5848 (2008)
7. Vladikova, D., Yonchev, H., Ilkov, L., Karbanov, S.: J. Magn. Magn. Mater. **78**(3), 420–426 (1989)
8. Von Aulock, W.H.: Handbook of Microwave Ferrite Materials (1965)

9. Data, N.: Numerical data and functional relationships in science and technology, vol. 3. Springer, Berlin (2009)
10. Gorter, E.W.: Philips Res. Rep. **9**, 295–320 (1954)
11. Bozorth, R.M., Davis, D.D., Williams, J.: J. Williams. Physical Review **119**(5), 1570 (1960)
12. Bharati, V.A., Supriya, R.P., Bajaj, S., Parlikar, R., Keche, A.P., Sondur, V.V.: Journal of Physics: Conference Series vol. 1644, 1, 012005. IOP Publishing (2020)
13. Qin, M., Shuai, Q., Guanglei, W., Zheng, B., Wang, Z., Hongjing, W.: Mater. Sci. Eng., B **224**, 125–138 (2017)
14. Hong, Y.Z., Cheng, Z., Huang, H., et al.: J. Mater. Sci.: Mater. Electron. **31**, 16975–16982 (2020)
15. Sedlář, M., Matějec, V., Grygar, T., Kadlecová, J.: Ceram. Int. **26**(5), 507–512 (2000)
16. Sharma, R., Singhal, S.: Physica B **414**, 83–90 (2013)
17. Sertkol, M., Köseoğlu, Y., Baykal, A., Kavas, H.Ü.S.E.Y.İ.N., et al.: Journal of Alloys and Compounds **486**(1–2), 325–329 (2009)
18. Anu, K., Hemalatha, J.J.C.I.: Ceram. Int. **48**(3), 3417–3425 (2022)
19. Costa, A.C.F.M., Tortella, E., Morelli, M.R., Kiminami, R.H.G.A.: J. Magn. Magn. Mater. **256**(1–3), 174–182 (2003)
20. Kumar, N.P., et al.: Materials Research Express **6**(8), 086112 (2019)
21. Cheng, Y., Zheng, Y., Wang, Y., Bao, F., Qin, Y.: J. Solid State Chem. **178**(7), 2394–2397 (2005)
22. Hankare, P.P., Jadhav, S.D., Sankpal, U.B., Chavan, S.S., Waghmare, K.J., Chougule, B.K.: J. Alloy. Compd. **475**(1–2), 926–929 (2009)
23. Thakur, P., Sharma, R., Sharma, V., Sharma, P.: Mater. Chem. Phys. **193**, 285–289 (2017)
24. Zabotto, F.L., et al.: Materials Research **15**, 428–433 (2012)
25. Murthy, S.: Ramana. Bull. Mater. Sci. **24**, 379–383 (2001)
26. Nandan, B., Bhatnagar, M.C.: AIP Conference Proceedings **1728**, 1 (2016)
27. Judith Vijaya, J., Bououdina, M.: Journal of Nanoscience and Nanotechnology **16**(1), 689–697 (2016)
28. Wani, T.A., Suresh, G., Masrour, R., Batoo, K.M., Rasool, A.: Materials Chemistry and Physics **307**, 128169 (2023)
29. Jadhav, S.A., et al.: Journal of Materials Science: Materials in Electronics **31**, 11352–11365 (2020)
30. Pubby, K., Meena, S.S., Yusuf, S.M., Narang, S.B.: Journal of Magnetism and Magnetic Materials **466**, 430–445 (2018)
31. Haque, M.M., Huq, M., Hakim, M.A.: J. Magn. Magn. Mater. **320**, 2792 (2008)
32. Maaz, K., Karim, S., Mumtaz, A., Hasanain, S.K., Liu, J., Duan, J.L.: J. Mag. Mag. Mater. **321**, 1838 (2009)
33. B.D. Cullity-Introduction to Magnetic Materials. Addision-wesley, Newyork (1972)
34. Hunpratub, S., Phokha, S., Kidkhunthod, P., Chanlek, N., Chindaprasirt, P.: Results in Physics **24**, 104112 (2021)
35. Naveen, P., et al.: J. Supercond Nov. Magn. 0123456789 (2024). https://doi.org/10.1007/s10948-024-06809-3

Influence of Calcination Temperature on Structural, Optical and Magnetic Properties of Nanocrystalline $Ni_{0.5}Zn_{0.5}Fe_2O_4$ Ferrite

Y. Vasudeva Reddy[1], M. Sharath Kumar[2], C. Mahen[1], S. E. Naina Vinodini[1], P. Sowjanya[3], K. N. Prasanna Kumari[4], and N. Pavan Kumar[2](✉)

[1] Department of Physics, Government City College, Nayapul, Hyderabad 500002, India
[2] Department of Sciences and Humanities, Matrusri Engineering College, Saidabad, Hyderabad 500059, India
pallakash@gmail.com
[3] Government Degree College, Medak District, Narsapur, Telangana 502313, India
[4] Government Polytechnic, Maheshwaram, Telangana 506331, India

Abstract. This study investigates the structural, optical, and magnetic properties of $Ni_{0.5}Zn_{0.5}Fe_2O_4$ nanoparticles calcined at different temperatures, i.e., 450 °C, 600 °C, and 750 °C. X-ray diffraction (XRD) analysis confirmed the formation of a single-phase cubic spinel structure. The lattice parameter obtained from Rietveld refinement is decreased from 8.373 Å to 8.357 Å as the calcination temperature increased from 450 °C to 750 °C. UV-visible spectroscopy was employed to estimate the optical band gap of the material, revealing its potential for optoelectronic applications. Magnetic characterization through M-H loop measurements elucidated the ferromagnetic behaviour of the nanoparticles. The magnetic properties exhibited interesting trends with calcination temperature: coercivity (H_c) and remanence (M_r) decreased from 83.5 Oe to 46.4 Oe and 5.46 emu/g to 2.78 emu/g, respectively, indicating softer magnetic behaviour at higher calcination temperatures. The combined analysis of structural, optical, and magnetic properties offers a comprehensive understanding of $Ni_{0.5}Zn_{0.5}Fe_2O_4$, highlighting its potential for various technological applications in fields such as spintronics, magnetic storage, and catalysis.

1 Introduction

Ferrite materials have been the subject of extensive research due to their unique magnetic, electrical, and chemical properties, which make them indispensable in a wide range of electronic and electromagnetic applications. Ferrites are a class of ceramic materials composed of iron oxide (Fe_2O_3) combined with other metal oxides, typically divalent metal ions such as nickel, zinc, manganese, or magnesium. The general formula for ferrites can be expressed as MFe_2O_4, where M represents the divalent metal ion [1]. Among the various ferrite compositions, nickel-zinc ferrites (Ni-Zn ferrites) have garnered significant attention, as they exhibit superior high-frequency performance, low power losses, and good thermal stability [2]. The Ni-Zn ferrite system, with the general

P. Joseph D. et al. (Eds.): IEMDST 2024, SPPHY 567, pp. 35–44, 2025.
https://doi.org/10.1007/978-981-96-5863-3_5

formula $Ni_xZn_{1-x}Fe_2O_4$, offers the flexibility to tailor the material properties by adjusting the nickel and zinc content, making it a versatile choice for various technological applications. In recent years, the development of nanocrystalline ferrite materials has gained momentum, as the reduction in particle size can lead to remarkable changes in the structural, magnetic, and electrical characteristics of the material. Nanocrystalline ferrites often exhibit enhanced magnetization, coercivity, and permeability compared to their bulk counterparts, making them attractive for applications such as high-frequency transformers, microwave devices, and high-density data storage media [3].

This study is mostly about looking into the structure, spectral and magnetic properties of nanocrystalline $Ni_{0.5}Zn_{0.5}Fe_2O_4$ ferrite, a mix that has shown promise in many different situations. Of particular importance is how calcination temperature affects this nanocrystalline ferrite's magnetic activity [4]. This study also aims to provide valuable insights into the underlying mechanisms governing the material's performance. The findings of this investigation can contribute to the development of advanced ferrite-based technologies and enhance our understanding of the complex interplay between nanostructure, composition, and temperature in magnetic materials. Therefore, in this study, the $Ni_{0.5}Zn_{0.5}Fe_2O_4$ has been prepared through sol-gel method and calcined at 450 °C, 600 °C and 750 °C temperatures. Structural, optical and magnetic properties were studied on these compounds and the results are presented here.

2 Experimental Procedure

The nanocrystalline $Ni_{0.5}Zn_{0.5}Fe_2O_4$ ferrite was synthesized using the sol-gel method [4]. The chemical precursors used in this synthesis were nickel nitrate ($Ni(NO_3)_2{\cdot}6H_2O$), zinc nitrate ($Zn(NO_3)_2{\cdot}6H_2O$), and ferric nitrate ($Fe(NO_3)_3{\cdot}9H_2O$), all sourced from Sigma-Aldrich with a purity of $> 99\%$ was dissolved in distilled water to create an aqueous solution, citric acid was added as a chelating agent, and the p^H was adjusted to 7 using an ammonia solution. The mixture was then heated at 80 °C under constant stirring to promote gel-to-powder formation. Using a mortar and pestle, the resulting powder was ground into a fine powder. The powder was then calcinated at three different temperatures 450 °C, 600 °C, and 750 °C for 2 h in a muffle furnace with a heating rate of 5 °C/min. Here on words, the compounds calcined at 450 °C, 600 °C, and 750 °C temperatures are labelled as NZ450, NZ600, and NZ750, respectively.

After synthesis, the samples were characterized using various techniques. The prepared samples were characterized by a Rigaku Mini Flex 600 powder X-ray diffractometer with Cu K_α radiation of wavelength 1.54 Å. By using a T90 UV-Vis Spectrophotometer, optical properties were studied in the wavelength range of 200–800 nm. Room temperature magnetic characteristics were examined using EV9 ADE, USA model Vibrating Sample Magnetometer.

3 Results and Discussion

3.1 X-ray Diffraction Analysis

The X-ray diffractograms of $Ni_{0.5}Zn_{0.5}Fe_2O_4$ compounds calcined at different temperatures are shown in Fig. 1. The Miller indices (220), (311), (222), (400), (422), (511), and (440) at respective diffraction angles are indicative of the formation of nickel-zinc ferrite with cubic spinel structure and also which is consistent with the JCPDS data card #380647 [5]. The small peak between the (220) and (311) planes is attributed to the existence of α-Fe_2O_3 [6]. Typically, as shown in Fig. 1, the most intense (311) peak is often used as a reference for spinel ferrites.

Rietveld refinement is performed using the FullProf suite to obtain the structural parameters. The resulting XRD fits are displayed in Fig. 2. The Rietveld refinement parameters, such as R_p, R_{wp}, and χ^2 (given in Table 1), provide insights into the quality of the fit between the observed and calculated XRD patterns. The values of R_p and R_{wp} initially decrease from 450 °C to 600 °C, indicating an improved fit due to enhanced crystallinity and reduced defects. However, from 600 °C to 750 °C, there is a slight increase in these values, possibly due to minor secondary phases or slight deviations in stoichiometry at higher temperatures. Despite this, the overall goodness of fit (χ^2) continues to improve, reflecting the robustness of the structural model.

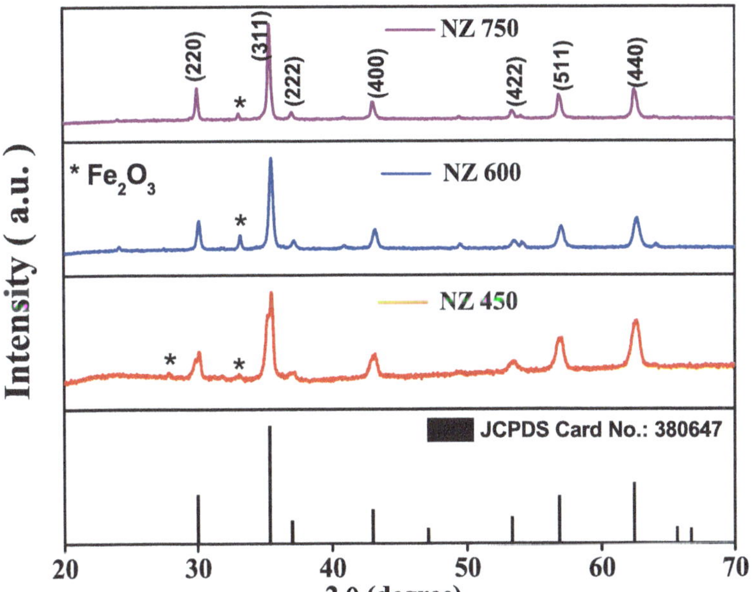

Fig. 1. XRD patterns of $Ni_{0.5}Zn_{0.5}Fe_2O_4$ calcined at different temperatures.

In summary, higher synthesis temperatures lead to improved crystallinity, enhanced grain growth, reduced porosity, and increased phase purity, resulting in a more closely packed and well-ordered crystal structure [7]. The lattice parameter (a) represents the unit cell dimension of the crystal structure and slightly decreases with increasing temperature, from 8.373 Å at 450 °C to 8.357 Å at 750 °C. This decrease can be attributed to the improved crystallinity and reduced lattice strain at higher temperatures, leading to a more compact and stable structure. At higher temperatures, the nanoparticles become denser, reaching 6.511 g/cm^3 at 750 °C compared to 5.574 g/cm^3 at 450 °C and 3.062 at 600 °C. This increase in density is due to improved crystallinity, enhanced grain growth,

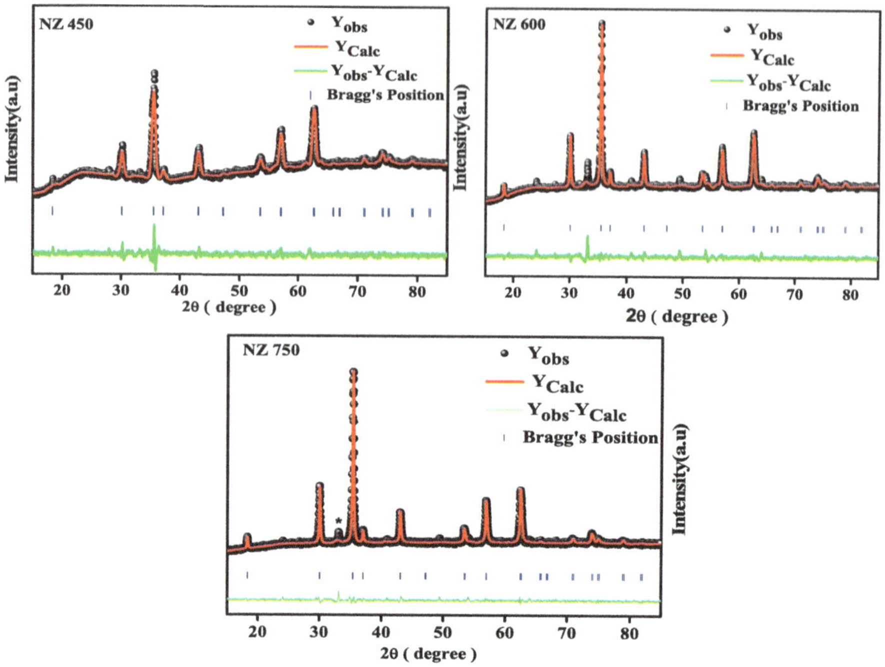

Fig. 2. Rietveld refined patterns of NZ450, NZ600, and NZ750 compounds.

reduced porosity, and potentially increased phase purity. These factors contribute to a more closely packed and well-ordered crystal structure. Additionally, the R_p and R_{wp} values decreases initially from 450 °C to 600 °C, indicating an improved fit of the structural model. This is likely a result of enhanced crystallinity, reduced strain and defects, and improved cation ordering within the spinel structure [8]. However, from 600 °C to 750 °C, there is a slight increase in these values despite an overall improvement in the goodness of fit (χ^2), and the values are shown in Table 1.

The XRD analysis revealed variations in peak intensities, widths, and slight shifts in peak positions among NZ450, NZ600, and NZ750 samples. A decrease in the full width at half maximum (FWHM) of the peaks indicates an increase in crystallite size and improved crystallinity with higher synthesis temperatures. This trend suggests that heat treatment enhances the crystallite growth and reduces lattice strain, leading to sharper peaks. Using the Scherrer equation, the crystallite size of $Ni_{0.5}Zn_{0.5}Fe_2O_4$ was determined for all three synthesis temperatures. The equation is given by,

$$D = \frac{K\lambda}{\beta Cos\theta} \tag{1}$$

where the Scherrer constant (0.89) is denoted by K, the wavelength of the X-ray beam is represented as λ, the FWHM of the diffraction peak is symbolized by β, and the Bragg angle is indicated as θ. A notable increase in the crystallite size from 14 to 31 nm has been observed with temperature. This growth can be attributed to several interrelated

Table 1. Crystallography data from XRD and Rietveld refined values of NZ450, NZ600, and NZ750 compounds.

S. No.	Parameter	NZ450	NZ600	NZ750
1	R_p	61.7	31.5	39.1
2	R_{wp}	25.2	17.8	22.7
3	R_{exp}	17.2	10.5	11.3
4	χ^2	2.83	2.14	1.63
5	a (Å)	8.373	8.364	8.357
6	Crystallite size D (nm)	14	25	31
7	Density (g/cm^3)	5.574	3.062	6.511
8	Crystallite size from W-H plots D (nm)	67	71	91
9	Strain (10^{-4}) from W-H plots	14.8	28.7	23.5

factors. As temperature rises, atomic mobility increases, allowing atoms to migrate more freely, which facilitates the growth of existing crystallites.

Additionally, higher temperatures promote sintering, a process where particles bond together, reducing overall surface area and enabling the formation of larger crystallites. Elevated temperatures also reduce defects and grain boundaries, supporting the growth of larger, more stable crystallites. While new crystallites may still form at higher temperatures, the growth rate of existing ones tends to outpace nucleation, resulting in larger average sizes. Collectively, these factors explain the observed increase in crystallite size as the calcination temperature rises [9].

The Williamson-Hall (W-H) plot is a valuable method for analyzing the crystallite size and strain in materials. The W-H plots of NZ450, NZ600, and NZ750 are shown in Fig. 3. In case of $Ni_{0.5}Zn_{0.5}Fe_2O_4$ the W-H plots at different temperatures (NZ450 to NZ750) show varying slopes and corresponding crystallite sizes and strains. The increase in slope with temperature can be attributed to the enhanced thermal energy, which facilitates the movement of dislocations and defects within the crystal lattice. This increased movement leads to a higher strain in the material, as observed in the strain values of 14.8×10^{-4}, 28.7×10^{-4} and 23.5×10^{-4} for NZ450, NZ600, and NZ750 compounds respectively. As the temperature increases, the crystallite size also increases from 67 nm to 91 nm. This growth in crystallite size is due to the coalescence of smaller crystallites into larger ones, driven by the reduction in surface energy. The larger crystallites have fewer grain boundaries, reducing the material's overall strain. Consequently, the strain initially increases with temperature due to enhanced defect mobility but eventually decreases as the crystallites grow larger and more stable [10].

3.2 UV-Vis Spectroscopy

The optical properties of the synthesized NZ450, NZ600, and NZ750 ferrite powders were investigated using UV-Vis spectroscopy. The samples were dispersed in a suitable solvent, and the UV-Vis absorption spectra were recorded in the wavelength range of

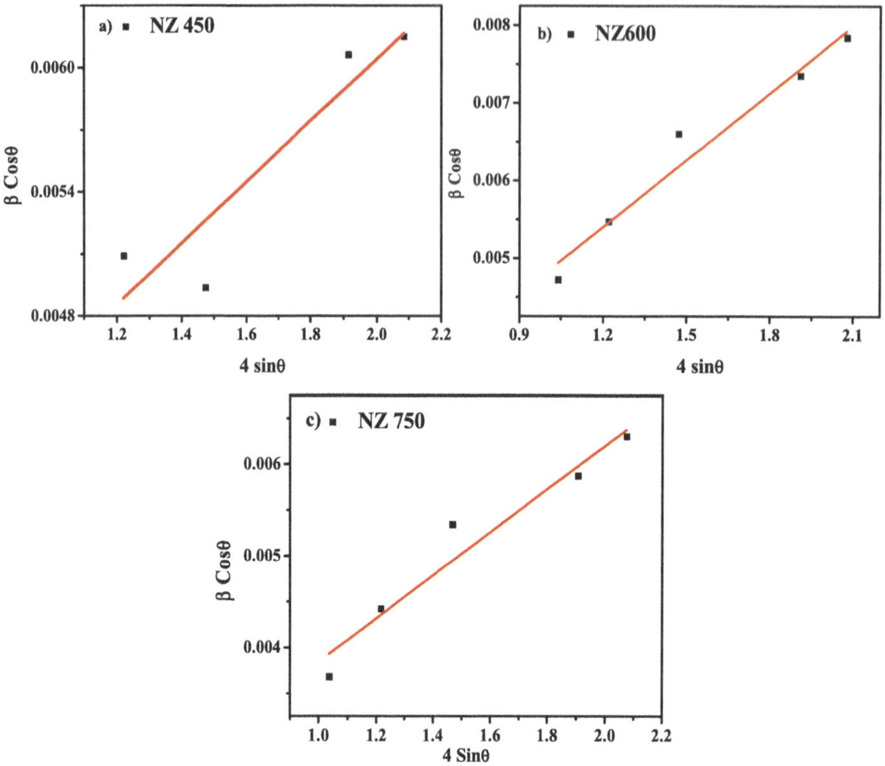

Fig. 3. W-H plots of the compounds of NZ450, NZ600, and NZ750.

200 to 800 nm, as shown in Fig. 4. The optical bandgap of the ferrite materials was determined from the UV-Vis data using appropriate methods. NZ450 shows a band gap of 1.30 eV, which increases to 1.34 eV for NZ600 and 1.49 eV for NZ750. This progressive increase in band gap energy with annealing temperature can be attributed to several factors. Firstly, higher temperatures likely promote better crystallization and ordering of the material's structure, reducing defects and impurities that could contribute to mid-gap states. Secondly, the elevated temperatures may cause subtle changes in the material's composition, potentially altering the distribution of cations (Ni^{2+}, Zn^{2+}, Fe^{3+}) within the crystal lattice [11]. These structural and compositional refinements can widen the energy gap between the valence and conduction bands. Additionally, the increased thermal energy during annealing might facilitate the oxidation of any residual lower-valence iron, further modifying the electronic structure. The observed band gap values are associated with charge transfer transitions between oxygen (O^2) ions and metal ions, and their increase suggests that higher energy is required for these electronic transitions as the material is annealed at higher temperatures. This phenomenon highlights the significant influence of processing conditions on the optical and electronic properties of $Ni_{0.5}Zn_{0.5}Fe_2O_4$, which could be crucial for tailoring the material for specific applications in optoelectronics or photocatalysis [12].

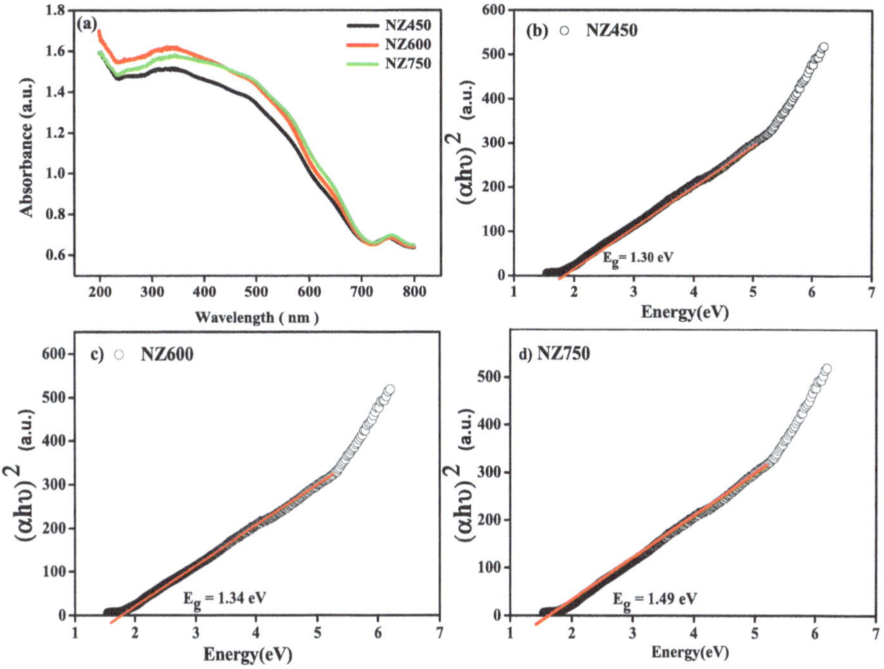

Fig. 4. (a) UV-Visible Spectra and (b-d) Bandgap Analysis of NZ450, NZ600, and NZ750.

3.3 Magnetic Measurements

The magnetic hysteresis (M-H) behavior of NZ450, NZ600, and NZ750 ferrites shows significant changes with varying annealing temperatures as shown in Fig. 5 (a). For the sample annealed at 450 °C, the coercivity (H_c) is 83.5 Oe, the remanence (M_r) is 5.4 emu/g, and the saturation magnetization (M_S) is 28.7 emu/g [13]. As the annealing temperature increases to 600 °C, the coercivity decreases to 78.5 Oe, the remanence drops to 3.23 emu/g, and the saturation magnetization decreases to 23.4 emu/g. At 750 °C, the coercivity further reduces to 46.4 Oe, with a remanence of 2.78 emu/g and an increased saturation magnetization of 29.7 emu/g and the values are shown in Table 2. The saturation magnetization (M_S) of materials like Ni-Zn ferrites can exhibit random variations with temperature due to several underlying factors. As temperature increases, the thermal energy disrupts the alignment of magnetic moments, generally leading to a decrease in M_S [14]. However, this is not always a straightforward relationship, as changes in the material's structure or phase at certain temperatures can cause fluctuations. Additionally, temperature variations can affect the magnetic anisotropy of the material, influencing M_S differently along various crystallographic directions. Some materials may also undergo phase transitions at specific temperatures, resulting in abrupt changes in their magnetic properties. Furthermore, temperature-induced spin reorientation transitions can alter the direction of magnetic moments, thereby affecting M_S. These combined factors contribute to the observed random changes in M_S with temperature. The slight increase in saturation magnetization at 750 °C might be due to improved crystallinity or changes in the cation distribution within the ferrite structure.

The decrease in coercivity and remanence with increasing annealing temperature can be attributed to factors such as grain growth, reduction in defects, changes in magnetic anisotropy, phase composition, and stress relief. Higher annealing temperatures lead to larger grains, fewer defects, and reduced internal stresses, all of which facilitate easier movement of magnetic domain walls, resulting in lower coercivity [15]. These variations highlight the impact of annealing temperature on the microstructure and magnetic properties of $Ni_{0.5}Zn_{0.5}Fe_2O_4$ samples.

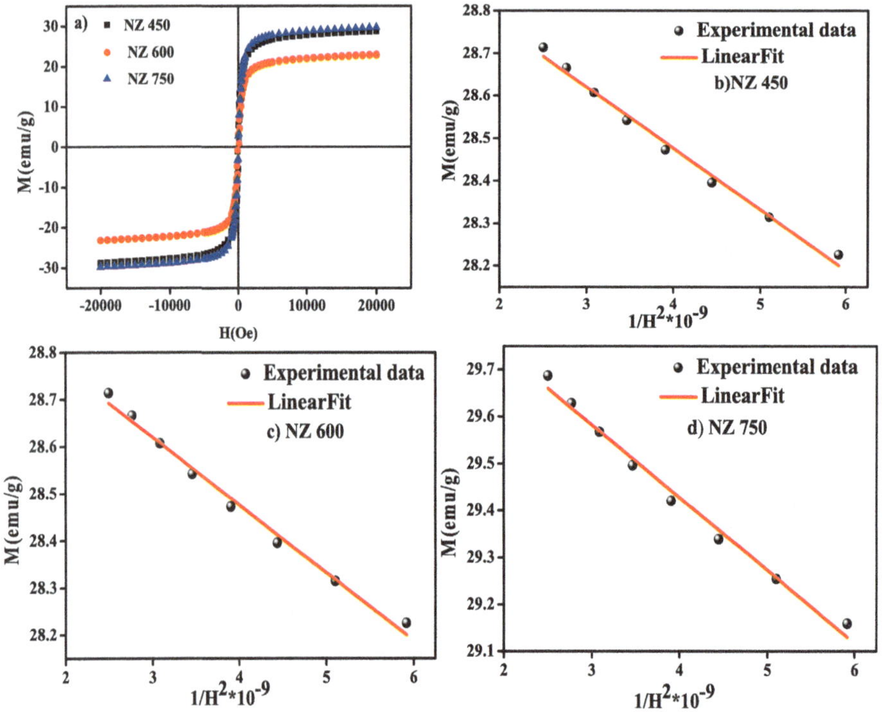

Fig. 5. (a) M-H hysteresis loops and (b-d) M vs $1/H^2$ Linear fitting for NZ450, NZ600 and NZ750

The magnetic anisotropy constants are the coefficients in the expansion of the energy in terms of spherical harmonics which depend upon the moment direction. In cubic crystals, the lower-order anisotropy constants vanish by symmetry; consequently, the anisotropy energy is small. In the present work, the 'Law of Approach (LA)' approach has been used to calculate the magneto crystalline anisotropy. The following relationship describes how this method relates the magnetization (M) to the applied magnetic field (H).

$$M = M_s\left[1 - \left(\left(\frac{8}{105}\right) \times \left(\frac{K^2}{\mu_0^2 M_s^2 H^2}\right)\right)\right] \tag{2}$$

Table 2. Magnetic parameters from M-H data

S. No.	Parameters	NZ450	NZ600	NZ750
1	H_c (Oe)	83.5	78.5	46.4
2	M_r (emu/g)	5.46	3.23	2.78
3	M_S (emu/g)	28.73	23.4	29.7
4	Anisotropy constant K (10^3) (erg/cm^3)	54.90	54.90	60.9

$$M = M_S\left[1 - \frac{b}{H^2}\right] \tag{3}$$

In this equation, b represents the magnetization correction factor, M_s represents saturation magnetization, μ_0 represents free space permeability, and K represents the magnetic crystalline anisotropy constant. The intercept and slope of linear fitted M vs. $1/H^2$ plots are used to derive the values of M_S and b, respectively [10]. Furthermore, the following equation has evaluated the values of K

$$K^2 = -\left[\frac{M_s \times Slope}{\beta}\right] \tag{4}$$

where the β value is 8/105 representing the cubic anisotropy of random polycrystalline samples. The estimated magneto crystalline anisotropy (K) values of the materials are listed in the below Table 2.

The anisotropy constant K remains stable at 54.90×10^3 erg/cm^3 for the lower temperatures before increasing to 60.9×10^3 erg/cm^3 at 750 °C, implying significant changes in magnetic anisotropy at higher preparation temperatures [14]. These could be attributed to alterations in cation distribution, enhanced crystallinity, or changes in particle morphology. Collectively, these trends underscore the critical role of preparation temperature in tailoring the magnetic properties of $Ni_{0.5}Zn_{0.5}Fe_2O_4$ nanoparticles, offering a means to fine-tune their characteristics for various technological applications such as spintronics, magnetic storage, or electromagnetic interference shielding. The complex interplay between structural and magnetic properties highlights the importance of carefully controlled synthesis conditions in optimizing these materials for specific uses.

4 Conclusion

In conclusion, the XRD analysis of $Ni_{0.5}Zn_{0.5}Fe_2O_4$ samples synthesized by the sol-gel method at different temperatures NZ450, NZ600, and NZ750 confirmed the formation of a single-phase cubic spinel structure. The crystallite size varies with temperature, ranging from 14 to 31 nm. The optical band gap was found to fluctuate with calcination temperature, with a tendency of rising band gap energy from 1.30 eV at NZ450 to 1.34 eV at NZ600 and then decreasing to 1.25 eV at NZ750. As the heating temperature increases from NZ450 to 750, both coercivity (H_c) and remanence (M_r) decrease. The saturation magnetization (M_s) shows a decrease from 28.7 emu/g at 450 °C to 23.4 emu/g at 600 °C, followed by an increase to 29.7 emu/g at 750 °C. The anisotropy constant K remains stable at 54.90 × 10^3 erg/cm^3 for the lower temperatures before increasing to 60.9 × 10^3 erg/cm^3 at 750 °C, implying significant changes in magnetic anisotropy at higher preparation temperatures. As the temperature increased from 450 °C to 750 °C, crystallite size and density increased, while the optical band gap showed a non-linear trend. Magnetic properties such as coercivity, remanence, saturation magnetization, and anisotropy constant also exhibited temperature-dependent variations. These findings highlight the crucial role of synthesis conditions in tailoring the structural, optical, and magnetic properties of $Ni_{0.5}Zn_{0.5}Fe_2O_4$ nanoparticles for various applications.

Acknowledgements. The authors acknowledged the Principal and management of Matrusri Engineering College, Hyderabad for providing research lab facilities.

References

1. Deraz, N.M., Abd-Elkader, O.H.: J. Anal. Appl. Pyrolysis **106**, 171–176 (2014). https://doi.org/10.1016/j.jaap.2014.01.017
2. Gao, P., et al.: Journal of Applied Physics 107 (2010). https://doi.org/10.1063/1.3309767
3. Ganesh, G., Sandeep, A., Chanti, G., et al.: Phys Status Solidi Appl Mater Sci. **220**, 1 (2023). https://doi.org/10.1002/pssa.202200864
4. Sowjanya, P., Thirupathi, G., Kumar, M.S., et al.: J Alloys Compd. **2024**(971), 172623 (2023). https://doi.org/10.1016/j.jallcom.2023.172623
5. Sertkol, M., Köseoğlu, Y., Baykal, A., Kavas, H., Bozkurt, A., Toprak, M.S.: J. Alloys Compd. **486**, 325–329 (2009). https://doi.org/10.1016/j.jallcom.2009.06.128
6. Hunpratub, S., Phokha, S., Kidkhunthod, P., Chanlek, N., Chindaprasirt, P.: Results Phys. 24 (2021). https://doi.org/10.1016/j.rinp.2021.104112
7. Ganesh, K., Kumar, N.P., Joseph, A., Raju, K.C.J., Lalitha, G.: Indian. J. Phys. **97**(13), 3941–3948 (2023). https://doi.org/10.1007/s12648-023-02735-7
8. Naveen, P., Shekharam, T., Reddy, Y.V., Pallavi, A., Reddy, V.R.: AIP Conf. Proc. **3149**, 020028 (2024). https://doi.org/10.1063/5.0224762
9. Sharada, G., Pavan Kumar, N., Sowjanya, P., Sreenivasu, D.: Mater. Today Proc. **92**, 440–444 (2023). https://doi.org/10.1016/j.matpr.2023.03.589
10. DamodarReddy, K., Pavan Kumar, N., Xavier, D., Sowjanya, P., Subramanian, V., Siva, C.: Ceram. Int. **50**(7), 10402–10415 (2024). https://doi.org/10.1016/j.ceramint.2023.12.352
11. Sowjanya, P., Kumar, N.P., Chelvane, A., Reddy, M.V.R.: Mater. Sci. Eng. B. **2022**(279), 115674 (2021). https://doi.org/10.1016/j.mseb.2022.115674
12. Yousaf, M., Akhtar, M.N., Wang, B., Noor, A.: Ceram. Int. **46**(4), 4280–4288 (2020). https://doi.org/10.1016/j.ceramint.2019.10.149
13. Sandeep, A., et al.: Ceram. Int. **48**(19), 29493–29501 (2022). https://doi.org/10.1016/j.ceramint.2022.07.071
14. Naveen, P., et al.: J. Supercond Nov. Magn. 2024. (0123456789). https://doi.org/10.1007/s10948-024-06809-3
15. Siddeshwar, M., Naveen, P., Srinivas, K., Boda, N., Chittury, K.L.: ECS Journal of Solid State Science and Technology **13**(8), 083001 (2024)

Chemical Precipitation Synthesis of MoS_2 Nanoflowers and Their Structural and Optical Properties

G. Sreedevi[1], K. Venkatarao[2](✉), K. Bhargavi[3], P. Vijaya sirisha[4], and V. Jayalakshmi[5]

[1] Department of Physics, P.V.P. Siddhartha Institute of Technology College, Vijayawada, India
[2] Department of Physics, Government Institute of Textile Technology, Guntur, A.P., India
g.sridevimsc@gmail.com
[3] Department of Physics, Government College (A), Rajamundry, A.P., India
[4] Department of Physics, Lakireddy Bali Reddy College of Engineering, Mylavaram, India
[5] Department of Physics, National Institute of Technology, Warangal 506004, India

Abstract. An economical and environmentally acceptable chemical precipitation approach was used to create MoS_2 nanoparticles. The structural, morphological, optical absorption, and infrared spectral characteristics of the prepared nanopowder at room temperature were investigated using instruments such as Powder XRD, SEM, DRS and FTIR. According to analysis of XRD. MoS_2 exhibits a hexagonal phase structure. The prepared nanopowder's average crystallite size is around 17 nm. Additionally, Density of dislocations of the sample and due to the synthesis process amount of strain occurred in the lattice were computed. The generated nanopowder has the morphology of spherical hexagonal flowers with agglomerations, according to the SEM investigation. The prepared sample's predicted optical energy band gap energy value is 1.34 eV. The distinctive frequencies associated with precursors and MoS_2 were revealed from the FTIR spectra.

1 Introduction

In the twenty-first century, nanotechnology is a cutting-edge technology that has only lately gained popularity. The properties of the material are altered and new principles are created when the structure of the material is manipulated to create nanoscale structures. Due to quantum mechanical phenomena and unique size-dependent material features, UV-Vis-NIR spectrum responses of nanoparticles with semiconducting nature have been studied theoretically and experimentally for the past thirty years. In three dimensional aspect semiconducting materials exhibit the phenomena of quantum confinement of charges that result in substantial changes to their electrical, structural, and optical properties when a particle's size approaches the Bohr radius of an exciton (electron and hole separation distance). The preferred materials for upcoming technology advancements in various display panel manufacturing, Fabrication of sensor materials, electronic devices, solar cells, lightning sources, and other applications are widely regarded as widely tunable energy band gap semiconductors.

P. Joseph D. et al. (Eds.): IEMDST 2024, SPPHY 567, pp. 45–51, 2025.
https://doi.org/10.1007/978-981-96-5863-3_6

A 2D direct band gap transition-metal dichalcogenide semiconductor, molybdenum disulfide (MoS_2) stands out among other semiconductors due to its distinctive layered structure, which allows for high electron mobility, stable chemical characteristics, a large surface area, and excellent electrical conductivity [1, 2]. There are several optoelectronic uses for MoS_2 due to its high absorption rates in the visible and infrared light spectrum. This absorption capability is explained by MoS_2's band gap, which is approximately 1.3 eV to 1.8 eV [3]. In addition, the electrical and optoelectronic properties of MoS_2 can be customized to suit specific application requirements because to its layer-dependent nature. This study made sulfur-based MoS_2 nanopowder using the chemical precipitation method [4]. Research on MoS_2's optical and structural properties was carried out.

2 Synthesis of MoS_2 Nanopowder

Under chemical precipitation process, MoS_2 nanopowder was created. The following amounts of CH_4N_2S and $Na_2MoO_4.2H_2O$ were measured and added to 100 ml of double deionized water in a 250 ml conical flask: 2.25 mmol and 1 mmol, respectively. The mixture was then agitated for two hours. The chemical mixture was mixed and then stirred magnetically for 48 h. during stirring time a black precipitate tooks. We collected this precipitate, spun it at 10,000 r.p.m., and then rinsed it four times, alternating between ethyl alcohol and double deionized water, to remove any remaining impurities. After letting the final black precipitate dehydrate for 6 h at 120 °C, it was ground using an agate mortar. At long last, powdered MoS_2 was obtained. The following chemical equation depicts the steps used in the production of MoS_2 nanopowder.

$$4Na_2MoO_4.2H_2O + 9CH_4N_2S + 4H_2O \rightarrow 4MoS_2 + Na_2SO_4 + 6NaOH + 9CO_2 + 18NH_3$$

3 Characterization Techniques

The XRD model of Miniflex 600 of Rigaku company radiated with copper-Kα of wavelength of 1.5406 Å was scanned from 3° to 90° to record XRD spectrum of the prepared MoS_2 nano powder sample's at room temperature. The reflection spectrum (DRS) of the produced sample was recorded using a SPECORD-210 plus, an Analytik Jena UV/VIS/NIR spectrophotometer, in the 200–1000 nm range. To know the morphological changes of generated nanopowder were conducted with the aid of a SEM equipped with energy dispersive X-ray analysis (EDAX), specifically a Jeol JSM IT-500. Using KBr pellet, the FTIR spectra were acquired using Brukar ALPHA II model FTIR equipment, spanning from 4000 cm^{-1} to 400 cm^{-1}, of the produced sample.

4 Results and Discussion

4.1 XRD Studies

The XRD pattern provides insights into the phase purity of the synthesized MoS_2 nanopowder. Figure 1 illustrates the diffraction patterns obtained within the range of 3° to 90°. The XRD pattern of MoS_2 nanopowder indicates the hexagonal phase of MoS_2, with relevant diffraction pattern peaks observed at 14.02°, 28.52°, 32.59°, 36.42°, 43.44°, 44.81°, and 58.96°. These peaks correspond to the lattice planes (002), (004), (100), (102), (006), (104), and (110) diffractions of the hexagonal phase of MoS_2, showing a strong correlation with the standard XRD-ICDD-JCPDS card (Number: 01-075-1539) [5–7]. The XRD pattern shows intense peaks with a broad nature, clearly indicating that the crystallite sizes of the synthesized powder are in the nanometre range. The mean crystallite size (D) can be determined using the FWHM (full width at half maximum) values obtained from the intensity of the diffraction peaks. The Debye-Scherrer formula serves as a method for estimating the average crystallite size (D) of the prepared sample.

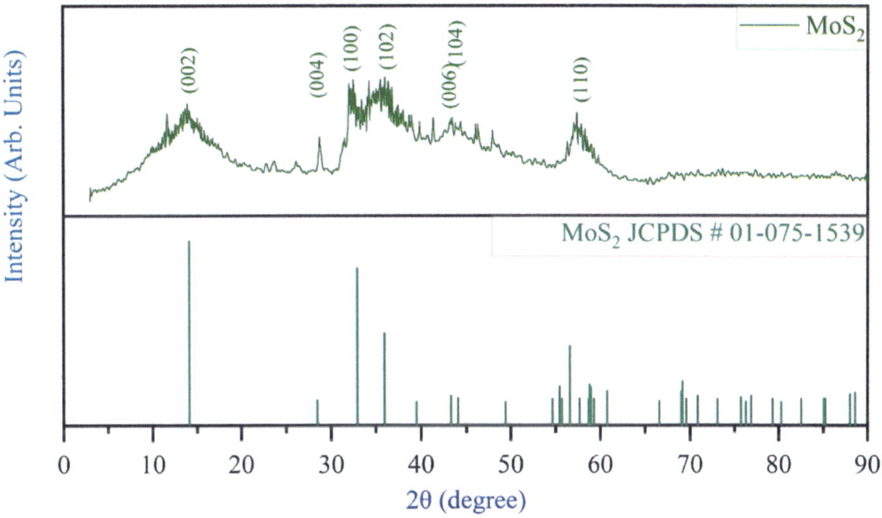

Fig. 1. XRD pattern for MoS_2 nanopowder.

The Debye-Scherrer equation, commonly utilized in analysis [8], is expressed as

$$D = k\lambda/(\cos\theta)\beta \tag{1}$$

Additionally, the Stokes-Wilson relation $\varepsilon = \beta/4\tan\theta$ and the Williamson-Smallman relation $\delta = 1/D^2$ are used to determine the δ-dislocation density and ε micro-strain, respectively [9]. The computed values for dislocation density (δ), micro-strain (ε), and crystallite size (D) are 3.29×10^{15} (lines/m^2), and 163.23×10^{-4}, respectively.

4.2 SEM and EDAX Studies

Under normal ambient temperature and pressure, surface morphologies and nanoparticle distribution of produced nanoparticle samples were investigated by means of scanning electron microscopy research. SEM micrographs and EDAX pattern of MoS_2 nanoparticle are shown in Fig. 2. SEM pictures of MoS_2 exhibit spherical flower like form. The EDAX approach allowed for the confirmation of the composition of the MoS_2 nanopowder by revealing the presence of the elements S and Mo that were sought in the samples.

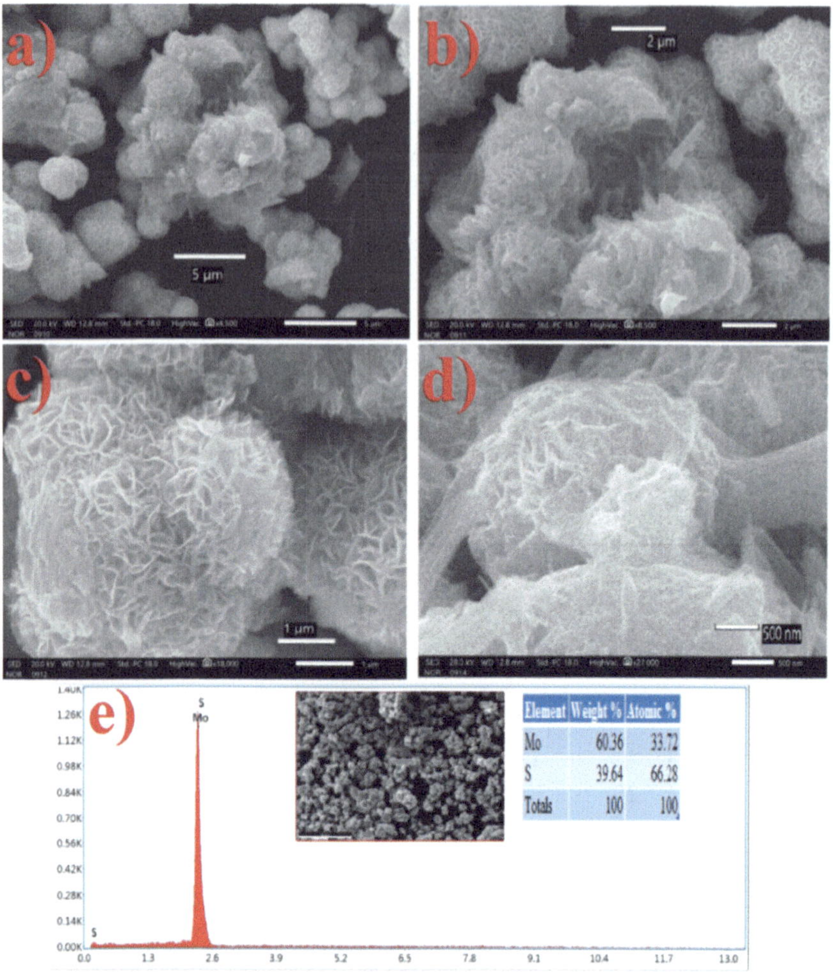

Element	Weight %	Atomic %
Mo	60.36	33.72
S	39.64	66.28
Totals	100	100

Fig. 2. Morphological SEM images and EDAX pattern of MoS_2 (a) 5 µm (b) 2 µm (c) 1 µm and (d) 500 nm resolutions; (e) EDAX spectrum of MoS_2.

4.3 Optical Properties

DRS serves as an efficient method for powdered samples since it eliminates the need for a suitable medium during measurement. The calculations for the bandgap were conducted with assistance. Using the gathered data of reflection spectra within the range of 200 to 1000 nm, calculations for the bandgap were conducted. The established Kubelka-Munk relation is employed to determine the band gap values of the synthesized MoS_2 nanopowder sample:

$$F(R) = \frac{(1-R)^2}{2R} \tag{2}$$

The reflectivity of the sample is represented as R. By applying a transformation [10, 11],

A graph is presented in Fig. 3, illustrating the relationship between $(F(R)h\nu)^2$ and hν. The graph allows for the extrapolation of bandgap values for the MoS_2 nanopowder. This method effectively characterizes semiconducting nanoparticles that exhibit quantum size phenomena resulting from electrons and holes pairs generated by light [12]. The reflection spectrum in the UV-Visible region is influenced by the size of semiconductor nanoparticles, with the maxima of reflection decreasing as the size of the nanoparticles diminishes. The bandgap energy for MoS_2 is estimated to be 1.34 eV.

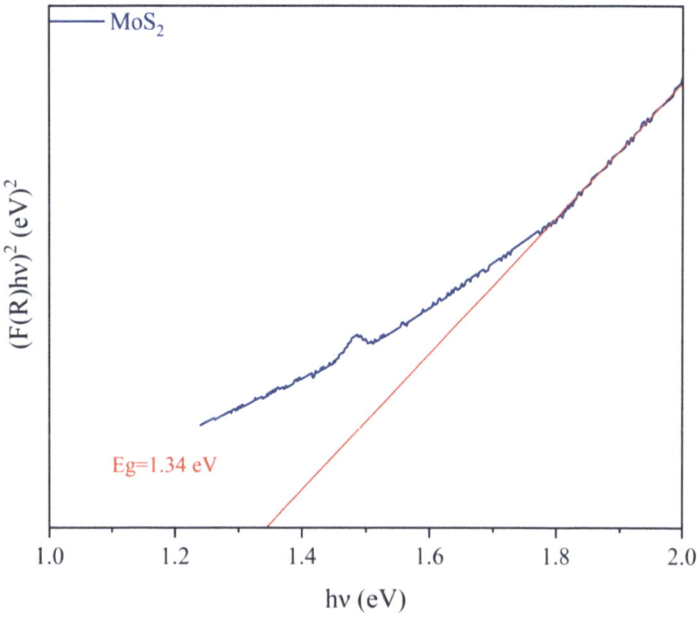

Fig. 3. Munk plot for MoS_2 nanopowder.

4.4 FTIR Study

FTIR measurements were conducted in the range of 400–4000 cm^{-1} in transmittance mode to confirm the existence of precursor residuals and to examine the presence of various characteristic functional groups in the MoS_2 nanopowder sample. The resulting FTIR spectrum is shown in Fig. 4. The observed peaks are meticulously examined, and the peak positions along with their distinctive characteristics are compiled in Table 1.

A broad dip is observed at 3408 cm^{-1}, attributed to O-H stretching, resulting from the strong interaction between H_2O and the MoS_2 lattice surface. This phenomenon can be linked to the residuals of precursors present during the synthesis or characterization processes [13]. Around 1637 cm^{-1} attributed to the H-O-H bending mode [13]. The bending vibration observed at approximately 1637 cm^{-1} indicates the presence of H_2O.

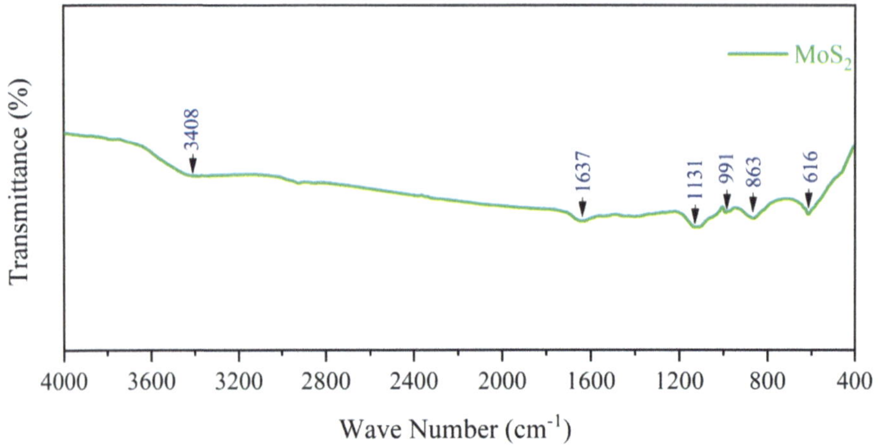

Fig. 4. FTIR spectrum of MoS_2 nanopowder.

The bands observed between 1131 cm^{-1} and 991 cm^{-1} provide confirmation of the complex sulfur formation in MoS_2, indicating the presence of active sites [14]. The band observed around 863 and 616 cm^{-1} was attributed to S-Mo-S vibration [15].

Table 1. FTIR spectrum of MoS_2 nanopowder along with IR band assignment.

Vibrational frequency (cm^{-1})	Assignment of various Bands
616	Vibrations of S-Mo-S bonds (Stretching mode)
863	S-Mo-S bonds vibrations (Stretching mode)
991	MoS_2 active sites
1131	MoS_2 active sites
1637	H_2O molecules on the particle surface
3408	O-H stretching mode vibarations

5 Conclusion

Molybdenum disulfide nanopowder particles are synthesized by chemical precipitation of method. The XRD results indicated the identification of the hexagonal phase of MoS_2. The synthesized nanopowder, exhibited an average crystallite size of approximately 17 nm. The SEM images displays hexagonal shaped flower like network. The determined energy band gap value of the synthesized nanopowder is 1.34 eV. FTIR spectra indicate the potential stretching and bending vibration modes of the prepared sample.

Acknowledgments. The author, G, Sreedevi o, expresses his profound gratitude to the "Nanotechnology Research Lab", Department of Freshman Engineering, Prasad V Potluri Siddhartha Institute of Technology, Vijayawada, A.P. The author also expresses special gratitude to the Siddhartha Academy of General and Technical Education in Vijayawada, Andhra Pradesh, India, for encouraging the continuation of the research as well as providing invaluable financial support.

References

1. Li, C., et al.: Chem. Front. **4**, 309–314 (2017)
2. Qin, S., Lei, W., Liu, D., et al.: Sci. Rep. **4**, 7582 (2014)
3. Wang, L., et al.: Adv. Funct. Mater. **25**(19), 2910–2919 (2015)
4. Sreedevi, G., Srinivas, K., Subbarao, M., Cole, S.: J. Mol. Struct. **1222**, 128903 (2020)
5. Saber, M.R., Khabiri, G., Maarouf, A.A., Ulbricht, M., Khalil, A.S.G.: RSC Advances **8**(46), 26364–26370 (2018)
6. Zhan, W., Yuan, Y., Yang, B., Jia, F., Song, S.: Chem. Eng. J. **394**, 124866 (2020)
7. Zheng, J., Zhang, R., Wang, X., Yu, P.: J. Mater. Sci **54**(21), 13509–13522 (2019)
8. Munimasthani, S., Sarathkumar, S., Udaychandran Thampy, U.S., Ravikumar, R.V.S.S.N.: Appl. Phys. A **126**(1), 2 (2019)
9. Joyce Stella, R., et al.: J. Alloys Compd **628**, 39–45 (2015)
10. Muswareen, S.K.K., Rao, M.S., Sridevi, G., Cole, S.: Mater. Sci. Semicond. Proces **102**, 104588 (2019)
11. Ali, H., Mansor, E.S.: Colloid and Interface Sci. Commun. **39**, 100330 (2020)
12. Stroyuk, A.L., Kryukov, A.I., Kuchmii, S.Y., Pokhodenko, V.D.: Theo. Exp. Chem. **41**(4), 207–228 (2005)
13. Ayodhya, D., Veerabhadram, G.: JS: AMD **4**(3), 381–391 (2019)
14. Zhao, D., Wu, T., Zhou, Y.: A European Journal **25**(41), 9710–9720 (2019)
15. Yi, M., Zhang, C.: RSC Advances **8**(17), 9564–9573 (2018)

Comparative Analysis of Different Nanofillers in Polyvinylidene Fluoride-Hexa-Fluoro Propylene Polymer for Electromagnetic Shielding Applications

Gayathri Thanasekaran and M. Vijayajayanthi(✉)

Department of Physics, B.S. Abdur Rahman Crescent Institute of Science and Technology, Chennai, Tamil Nadu 600048, India
vijaya@crescent.education

Abstract. Polyvinylidene fluoride-hexa-fluoropropylene (PVDF-HFP) polymer nanocomposites (PNCs) with various nanofillers, namely iron oxide (Fe_2O_3), graphene nanoplatelets (GNPs), and nanodiamonds (NDs) @ 10% wt were synthesized using the solution blending technique for electromagnetic interference shielding applications in satellite communication.We employed the X-ray diffraction (XRD) and Fourier-transform infrared spectroscopy (FTIR) techniques to evaluate the structural properties and chemical functionalities of the synthesized nanocomposites containing different nanofillers at equal weight percentage loadings. The thermal behaviour of the prepared polymer nanocomposites (NCs) was examined by the thermo gravimetric analyser (TGA). It reveals that the thermal stability of polymer NCs has been enhanced by adding different nanofillers in PVDF-HFP matrix. Using the vector network analyser technique, the electromagnetic interference shielding effectiveness (EMI SE) of GNPs@PVDF-HFP, Fe_2O_3@PVDF-HFP, and NDs@PVDF-HFP are found to be 27.33 dB, 9.77 dB, and 5.15 dB, respectively. It is observed that among the three distinct synthesized polymer nanocomposites, PVDF-HFP with GNPs exhibits the greatest EMI SE and is found to be the most promising material for potential application in satellite communications.

1 Introduction

As technology progresses, electronic equipment generates more heat and electromagnetic waves interference (EMI). Improper handling of these elements can cause heatbuild-up and increased EMI, compromising electronic device operation and human health [1]. This issue is extremely relevant to cutting-edge technologies such as 5G communication, artificial intelligence, batteries, and semiconductors [2]. Materials with outstanding thermal conductivity and EMI shielding effectiveness (EMISE) may tackle these difficulties. In connections with this, numerous investigations have shown that nanocomposites provide good EM wave shielding performance [2, 3]. Also, several studies have investigated nanocomposites with capabilities beyond EM wave shielding effectiveness [4, 5]. In general, EMI shielding materials can have significant electrical

P. Joseph D. et al. (Eds.): IEMDST 2024, SPPHY 567, pp. 52–60, 2025.
https://doi.org/10.1007/978-981-96-5863-3_7

conductivity, making them difficult to use in applications that require insulation. To address the difficulty, materials must have high thermal conductivity and EMI SE while preserving electrical isolation [4]. Polymer-based nanocomposites have been widely studied for this purpose. To improve the low thermal conductivity of polymers, fillers with high conductivity, such as iron oxide, zinc oxide, silicon dioxide, zirconium oxide, titanium dioxide, and carbon-based materials are largely used [5].

Polyvinylidene fluoride-hexa-fluoro propylene is a lightweight, flexible, and chemically stable polymer. In addition to this, PVDF-HFP as a matrix material in graphene nanocomposite films results in lightweight EMI shielding solutions [6]. Likewise, Graphene-based composites are durable and corrosion-resistant due to their intrinsic qualities, which include graphene and PVDF-HFP components. For long term EMI shielding in a variety of backgrounds, the graphene/PVDF-HFP composite thin film is appropriate [7]. Zhang et al. investigated the improved absorption of electromagnetic waves and dielectric properties of a PVDF/reduced graphene oxide composite film with a thickness of 4 mm [8]. They also found that the absorption of composite films was more than 25.6 dB. In particular, graphene/PVDF composites were shown to provide effective EMI shielding at greater graphene loadings [8]. Rengaswamy et al. determined that the solution-cast PVDF/rGO/MWCNT (1:1) composite film of thickness 0.5 mm exhibited a highest EMI shielding of 20 dB at the frequency of 12 GHz [9]. Shilpee Chauhan et al. found that the PVDF-HFP/rGO composite film can shield up to 98.5% of incident EM waves with an EMI SE of 18.3 dB [6]. Nanodiamonds (NDs) are a fairly unknown family of carbon nanoparticles with high surface-to-volume ratios. NDs unique features enable multifunctional integration including imaging, bio sensing, drug delivery, and cell tracking. NDs have been demonstrated efficacy in drug delivery, gene therapy, bio-sensing, tissue engineering, and bio-imaging [10]. Kaftelen-Odabaşi et al. studied the EMI wave shielding performance of GO/Fe_3O_4/PPy/TPU nanocomposites in the 12 GHz–18 GHz frequency range. They found that the increase in the GO: Fe_3O_4 ratio resulted in a maximum EMI efficiency of 4.81 dB and a 66.72% blockage of the incident electromagnetic wave [11].

Here we present the comparative study on PVDF-HFP with different nanofillers such as Fe_2O_3, GNPs, and NDs. The processing conditionssuch as nanofiller distribution method, magnetic bar stirring, stirring ratio, temperature and ultrasonication have an impact on the composite network of polymer chains, also the percolation network will beformed from distributed nanoparticles. These elements eventually decide the promising qualities of the prepared polymer NCs film [12]. The PVDF-HFP matrix has excellent mechanical qualities, making it ideal for effective EMI shielding. This study introduces a novel technique for dynamically switching EM responses with different nanofillers, opening up new possibilities for constructing advanced smart EM devices.

2 Materials and Methods

A pellet of PVDF-HFP with a molecular weight of 400,000 g/mol was acquired from Sigma Aldrich. Fe_2O_3, GNPs, and NDs with a particle size of less than 50 nm were purchased from Aldrich. N-Methyl-2-pyrrolidone (NMP), an organic solvent was procured from Sisco Research Laboratories. In this present work, we prepare three different polymer NC films namely GNPs@PVDF-HFP, Fe_2O_3@PVDF-HFP, and NDs@PVDF-HFP by using solution casting approach. Figure 1 shows the synthesis procedure for polymer nanocomposite films with different nanofillers. In a hot air oven, PVDF-HFP pellets were subjected to heat at 80 °C for three hours, and then the mixture was stirred for an hour to get a clear texture. Nanofillers were ultrasonicated with NMP in individual beakers for an hour, and then stirred individually at ambient temperature. Then the nanofillers were mixed with the PVDF-HFP polymer solution at appropriate intervals and stirred all-night to produce a homogeneous solution. Finally, the solution was transferred into a glass petri dish and placed in a hot air oven at 80 °C for 12 h of casting. The prepared NC films were detached from the glass petri dish and used for future investigations. Table 1 illustrates the feed compositions of polymer NCs for various nanofillers.

The Fourier-transform infrared spectroscopy (FTIR) spectra of PVDF-HFP, GNPs@PVDF-HFP, Fe_2O_3@PVDF-HFP, and NDs@PVDF-HFP-based polymer NCs in the spectral range of 400 cm^{-1} to 3500 cm^{-1} were studied. X-ray diffraction (XRD) of PVDF-HFP, GNPs@PVDF-HFP, Fe_2O_3@PVDF-HFP, and NDs@PVDF-HFP polymer NCs films were investigated using a D8 Bruker X-ray diffractometer with Cu Kα radiation of wavelength ($\lambda = 1.5406$ Å) in the 2θ range from 10° to 70° with 0.01° step size and 1°/min scanning rate. The thermal characteristics of PVDF-HFP, GNPs@PVDF-HFP, Fe_2O_3@PVDF-HFP and NDs@PVDF-HFP films were studied by heating the samples upto 800 °C in the TA Q 500 Thermogravimetric Analyser (TGA) with N_2 gas at a rate of 20 °C/min. EMI SE measurements of GNPs@PVDF-HFP, Fe_2O_3@PVDF-HFP, and NDs@PVDF-HFP were performed at the Ku-band range with an Agilent Technologies 8510C vector network analyser.

Table 1. The feed composition for the polymer nanocomposite films for various nanofillers.

Samples	PVDF-HFP (wt %)	Fe_2O_3 (wt %)	GNPs (wt %)	NDs (wt %)
1	100	0	0	0
2	90	0	0	10
3	90	0	10	0
4	90	10	0	0

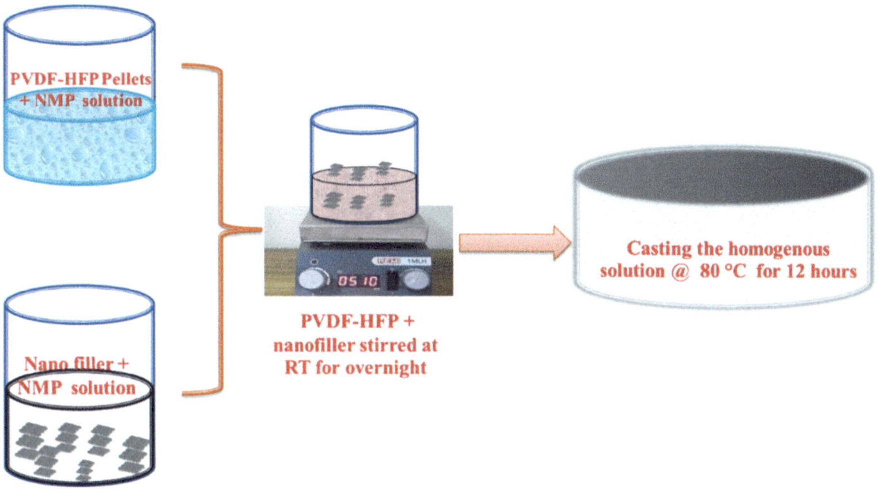

Fig. 1. The schematic diagram for the solution casting method of polymer nanocomposite films.

3 Theory and Calculation

EMI shielding effectiveness (EMI SE) is evaluated in far-field radiation by using vector network analyser technique. The network analyser measures the scattering parameters (S-parameters) such as the reflection coefficients S_{11} and S_{21}, absorption coefficients S_{12} and S_{22} and also used to determine reflectance (R), absorbance (A), transmittance (T) coefficients, and EMI SE. The total EMI shielding effectiveness (EMI SE) is the addition of reflection (SE_R), absorption (SE_A), and multiple internal reflections (SE_{MR}) [13]. SE_{MR} is considered as SE_A when the total electromagnetic shielding effectiveness surpasses 15 dB. The total EMI SE is determined using the below Eqs. (1–7) and its pictorial depiction of EMI shielding process is shown in Fig. 2.

$$R = |S_{11}|^2 \tag{1}$$

$$T = |S_{21}|^2 \tag{2}$$

$$A + R + T = 1 \tag{3}$$

$$SE_T(dB) = 10\log(1 - R) \tag{4}$$

$$SE_A(dB) = 10\log\left(\frac{1 - (S_{11})^2}{(S_{12})^2}\right) \tag{5}$$

$$SE_R(dB) = 10\log\left(\frac{1}{1 - (S_{11})^2}\right) \tag{6}$$

$$\text{EMI SE (dB)} = SE_R + SE_A + SE_{MR} \tag{7}$$

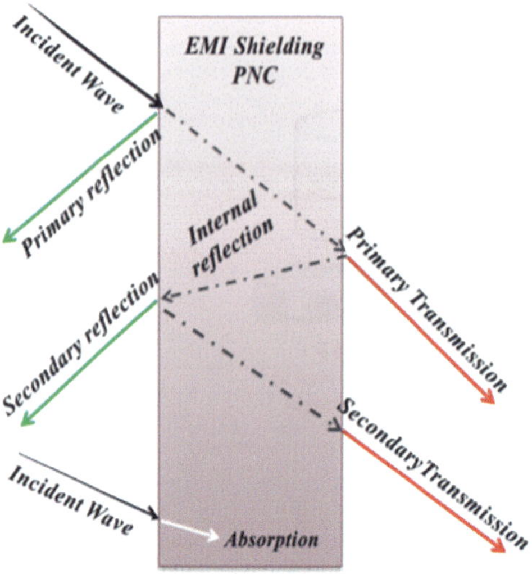

Fig. 2. Pictorial depiction of the electromagnetic interference shielding mechanism.

4 Results and Discussion

FTIR graphs of PVDF-HFP, GNPs@PVDF-HFP, Fe_2O_3@PVDF-HFP, and NDs@PVDF-HFP are presented in Fig. 3(a). PVDF-HFP FTIR characteristic peaks suggest that the polymer has both an amorphous phase and a crystalline phase. At the vibrational bands of 610 cm^{-1}, 762 cm^{-1}, and 1070 cm^{-1}, the crystalline phase can be found. At 835 cm^{-1} and 875 cm^{-1}, the amorphous phase is detected. Furthermore, the peaks of the CF_2 groups (wagging vibrations) at 483 cm^{-1}, deformation vibrations at 1401 cm^{-1}, and vibrations of the CH_2 group (asymmetric stretching) and C-F group at 3025 cm^{-1} and 2981 cm^{-1}, respectively are identified. The crystallinity is changed with the addition of Fe_2O_3, GNPs, and NDs to the PVDF-HFP matrix. The spectra of GNPs@PVDF-HFP peaks shifted to lower wavenumber. These changes may occur due to the GNPs vibration along with the PVDF-HFP matrix [14, 15]. The FTIR characteristicpeaks of Fe_2O_3@PVDF-HFP at 483 cm^{-1}, 610 cm^{-1}, 762 cm^{-1}, 2981 cm^{-1}, and 3025 cm^{-1} are shifted to the lower wavenumber, and also the band at 875 cm^{-1} is shifted towards the higher wavenumber. This plentiful modification of the wavenumber might be the cause for the dehydration or dehydroxylation of Fe_2O_3 [16]. As observed from the NDs@PVDF-HFP, the intensity of peaks at 483 cm^{-1}, 610 cm^{-1}, 3025 cm^{-1} is slightly reduced while the intensity peak at 875 cm^{-1} is increased. These shifts in wavenumber suggest the successful preparation of PVDF-HFP polymer nanocomposite with different nanofillers.

The X-ray diffraction (XRD) study was performed to observe the crystalline behaviour of the polymer matrix.The XRD results were observed for the pure PVDF-HFP polymer, and PVDF-HFP matrix filled with various fillers @ 10% wt loadings. Figure 3(b) reveals the XRD pattern for PVDF-HFP, GNPs@PVDF-HFP, Fe_2O_3@PVDF-HFP, and NDs@PVDF-HFP. The XRD typical peak of PVDF-HFP at angle 2θ = 20.08°, denote the (200) plane of PVDF and also corresponds to the crystalline α phase of VDF [17]. The average crystallite size of PVDF-HFP films was computed as 16.78 nm using Scherrer's equation and the Bragg peaks are found to bematching with the JCPDS card number (00-041-1517). The XRD peaks were observed for GNPs@PVDF-HFP, and the characteristic peak angles 2θ are found to be 26.57°, and 54.58° which correspond to the crystalline lamellae of graphite (002) and carbon peak crystal plane (004), respectively [18]. The above study indicates that there exist more overlapped graphene particles in the form of stacked and continuous layer as graphene nanoplatelets within the PVDF-HFP matrix. The XRD distinctive peaks of Fe_2O_3@PVDF-HFP nanocomposite are found at angles of 2θ = 30.09°, 35.47° [19], 57.15° and 62.86° which correspond to the crystal plane of Fe_2O_3 (220), (110), (018) and (214), respectively [20]. It is observed that the PVDF-HFP crystalline peak tends to decline with the addition of Fe_2O_3 due to the interaction of the intermolecular chain, which leads to an upsurge of the amorphous phase in the polymer matrix [18].

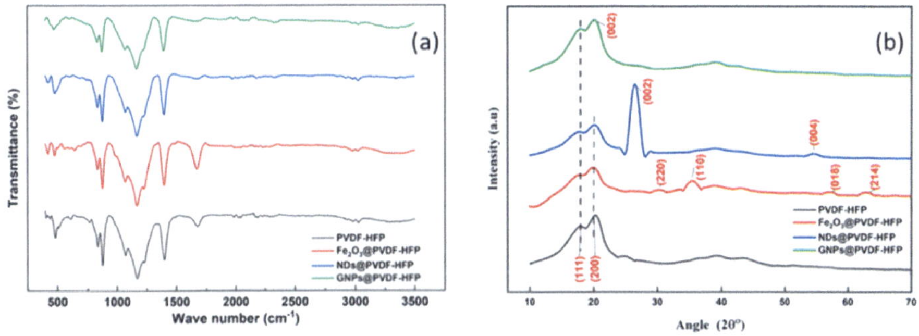

Fig. 3. (a) FTIR graph for PVDF-HFP, GNPs@PVDF-HFP, Fe_2O_3@PVDF-HFP, and NDs@PVDF-HFP polymer nanocomposites. (b) XRD graph for PVDF-HFP, GNPs@PVDF-HFP, Fe_2O_3@PVDF-HFP, and NDs@PVDF-HFP polymer nanocomposites.

Table 2. The thermo gravimetric analysis of polymer nanocomposites.

Samples	T_{in} (°C)	T_{max} (°C)	Char residue (%)		
			500 (°C)	600 (°C)	700 (°C)
Pure PVDF-HFP	403	464	23.5	18.73	15.56
GNPs@PVDF-HFP	437	489	32.16	30.65	28.12
Fe_2O_3@PVDF-HFP	424	553	26.09	17.56	17.31
NDs@PVDF-HFP	455	504	27.96	26.54	22.94

By comparing the patterns determined by XRD of the pure PVDF-HFP and the NDs@PVDF-HFP nanocomposite, it is noted that the addition of NDs did not considerably alter the packing of polymers in the solid state. It should be emphasized that the concentrations of NDs were insufficient to see their distinctive diffraction peaks [13]. Hence, The XRD results suggest PVDF-HFP nanocomposites with different nanofillers are synthesized under good conditions.

Figure 4(a) shows the thermo gravimetric analysis (TGA) curves of pure PVDF-HFP, GNPs@PVDF-HFP, Fe_2O_3@PVDF-HFP, and NDs@PVDF-HFP. Pure PVDF-HFP nanocomposite decomposes at 468 °C and exhibits asingle-step degradation. The TGA curve of GNPs@PVDF-HFP follows a similar polymer decomposition tendency and starts its decomposition at 493 °C. The Fe_2O_3@PVDF-HFP curve also shows the single-step degradation and decomposes at a temperature range of 496 °C. As seen from the TGA curve of NDs@PVDF-HFP, it is observed that the film also undergoes single-step degradation and decomposes with a 27% weight loss at 504 °C. The thermogravimetric plots of GNPs@PVDF-HFP, Fe_2O_3@PVDF-HFP, and NDs@PVDF-HFP differed significantly from that of PVDF-HFP, further indicating that all the three different polymer NCs are thermally stable when compared to pure PVDF-HFP polymer. On comparing PVDF-HFP nanocomposite with different nanofillers, the thermal stability of GNPs@PVDF-HFP is the highest, which also indicates that the interaction of GNPs with PVDF-HFP begins to degrade at higher temperatures which enhance the thermal stability in the polymer matrix. Table 2 shows the char residue weight percentage of different nanofillers along with PVDF-HFP.

Further, the EMI shielding effectiveness of GNPs@PVDF-HFP, Fe_2O_3@PVDF-HFP, and NDs@PVDF-HFP nanocomposites are evaluated in detail. Figure 4(b) shows the EMI SE of different nanocomposite films at 12 GHz–18 GHz (Ku band), where the highest EMI SE is found to be 27.33 dB which corresponds to GNPs@PVDF-HFP. The EMI SEvalue of Fe_2O_3@PVDF-HFP is 9.77 dB, which can block 79% of incident electromagnetic waves. The EMI SE value for the NDs@PVDF-HFP polymer nanocomposite is found to be 5.15 dB which can shield 99% of incoming EM waves. From the above studies, it is observed that GNPs@PVDF-HFP polymer nanocomposite film is the best material for effective EMI shielding applications among the other nanofillers such as Fe_2O_3 and NDs.

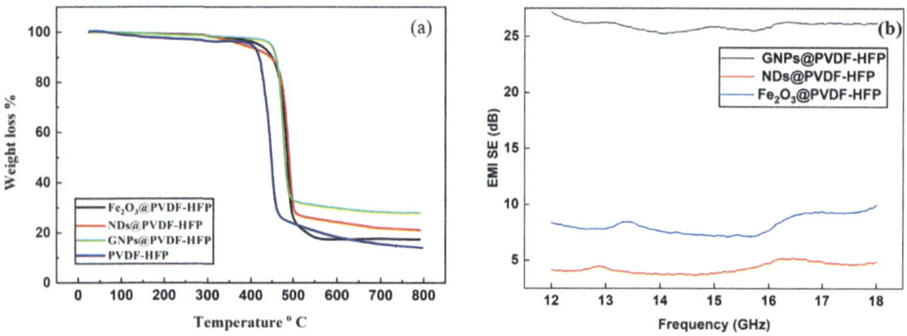

Fig. 4. (a) The TGA graph for PVDF-HFP, GNPs@PVDF-HFP, Fe_2O_3@PVDF-HFP, and NDs@PVDF-HFP polymer nanocomposites. (b) Comparative EMI shielding results for GNPs, Fe_2O_3, and NDs nanofillers inPVDF-HFP matrix.

5 Conclusion

In this study, the efficacy of three distinct nanofillers, such as GNPs, Fe_2O_3, and NDs in a PVDF-HFP polymer matrix was compared and examined in the Ku band region for the applications of satellite communication and radar systems. For this purpose, GNPs@PVDF-HFP, Fe_2O_3@PVDF-HFP, and NDs@PVDF-HFP polymer nanocomposite films were synthesizedby using the solution casting technique. The FTIR and XRD results indicate the uniform dispersion of distinct nanofillersin the PVDF-HFP polymer matrix. Furthermore, the thermal stability of the PVDF-HFP matrix has been improved by adding the nanofillers. Finally, for thesynthesized polymer nanocomposite films such as GNPs@PVDF-HFP, Fe_2O_3@PVDF-HFP, and NDs@PVDF-HFP the values of EMI SE at 12 GHz - 18 GHz are determined as 27.33 dB, 9.77 dB, and 5.15 dB, respectively, by using a vector network analyser. Interestingly, GNPs@PVDF-HFP polymer nanocomposites stand out because of their exceptional intrinsic qualities, which include high specific surface area, low density, superior conductivity, and dielectric characteristics. Hence the prepared GNPs@PVDF-HFP polymer nanocomposite film can be considered as the potential material for effective EMI shielding application devices.

Acknowledgments. G. T. would like to thank the management of B.S. Abdur Rahman Crescent Institute of Science and Technology for providing JRF to carry out this research work. M. V and G. T sincerely thank the organizers of IEMDST- 2024, NIT Warangal/NIT Goa.

References

1. Cheng, J., et al.: Nano Micro Lett. **14**, 4705–4718 (2022). https://doi.org/10.1007/s40820-022-00823-7
2. Jia, X., Yang, L., Shen, B., Zheng, W.: Composites Part B. **233**, 109652 (2022). https://doi.org/10.1016/j.compositesb.2022.109652
3. Lee, T.-W., Lee, S.-E., Jeong, Y.G., Appl, A.C.S.: Mater. Interfaces. **8**, 13123–13132 (2016). https://doi.org/10.1021/acsami.6b02218
4. Tian, K., Hu, D., Wei, Q., Fu, Q., Deng, H.: J. Mater. Sci. Technol. **134**, 106–131 (2023). https://doi.org/10.1016/j.jmst.2022.06.031
5. Wang, Z., Bockstaller, M.R., Krzysztof, M.: ACS Mater. Lett. **3**, 599–621 (2021). https://doi.org/10.1021/acsmaterialslett.1c00145
6. Chauhan, S., et al.: Surf. Interfaces. **46**, 103961–103961 (2024). https://doi.org/10.1016/j.surfin.2024.103961
7. Kumar, P., Kumar, A., Cho, K.Y., Das, T.K., Sudarsan, V.: AIP Adv. **7**, 015103 (2017). https://doi.org/10.1063/1.4973535
8. Zhang, X.-J., Wang, G.-S., Cao, W.-Q., Wei, Y.-Z., Cao, M.-S., Guo, L.: RSC Adv. **4**, 19594–19601 (2014). https://doi.org/10.1039/c4ra02040e
9. Rengaswamy, K., et al.: Polym. Adv. Technol. 1–11 (2021). https://doi.org/10.1002/pat.5395
10. Lazovic, J., Goering, W., Schützendübe, S., Löffler, W., Sitti, M.: Adv. Mater. 36 (2023). https://doi.org/10.1002/adma.202310109
11. Kaftelen-Odabaşi, H., Ruiz-Perez, F., Odabaşı, A., Helhel, S., López-Estrada, S.M., Caballero-Briones, F.: Eng. Sci. Technol. Int J. **55**, 101753 (2024). https://doi.org/10.1016/j.jestch.2024.101753

12. Charan, C.P., Sengwa, R.J., Mukul Saraswat, S.: Chem. Phys. Impact. **8**, 100410–100410 (2024). https://doi.org/10.1016/j.chphi.2023.100410
13. Chen, M., et al.: Nano Lett. **24**, 2643–2651 (2024). https://doi.org/10.1021/acs.nanolett.3c05021
14. Sethy, D., Balasubramaniam, K.: J. Nanopart. Res. **24**, 118 (2022). https://doi.org/10.1007/s11051-022-05504-5
15. Shalu, S.K., Chaurasia, R.K., Singh, S.C.: J. Phys. Chem. B **117**, 897–906 (2013). https://doi.org/10.1021/jp307694q
16. Mimouni, I., et al.: Environ. Sci. Pollut. Res. **29**, 7984–7996 (2021). https://doi.org/10.1007/s11356-021-16146-w
17. Vickraman, P., Senthilkumar, V.: Ionics **16**, 763–768 (2010). https://doi.org/10.1007/s11581-010-0467-5
18. Borkar, T., et al.: J. Alloys Compd. **646**, 135–144 (2015). https://doi.org/10.1016/j.jallcom.2015.06.013
19. Badawi, A., Ahmed, E.M., Mostafa, N.Y., Abdel-Wahab, F., Alomairy, S.E.: J. Mater. Sci. Mater. Electron. **28**, 10877–10884 (2017). https://doi.org/10.1007/s10854-017-6866-x
20. Zhang, N., Huang, Y., Wang, M.: J. Colloid Interface Sci. **530**, 212–222 (2018). https://doi.org/10.1016/j.jcis.2018.06.088

Influence of Annealing Temperature on Structural, Morphological and Optical Properties of Nickel Ferrite ($NiFe_2O_4$) Nanoparticles

P. Sundararajaperumal, P. Velusamy, V. Aravindan, and M. Mahendran(✉)

Smart Materials Laboratory, Department of Physics, Thiagarajar College of Engineering, Madurai, India
manickam-mahendran@tce.edu

Abstract. In this study, we examined into the synthesis of nickel ferrite ($NiFe_2O_4$) nanoparticles utilizing the sol-gel auto-combustion method, followed by heat treatment at different temperatures. The structural properties of $NiFe_2O_4$ were found to be a spinel structure with the space group Fd-3m and a face-centered cubic arrangement. The crystallites size and lattice parameter increased, while the density decreased. SEM was utilised to analyse the surface structure morphology of the synthesised $NiFe_2O_4$ nanoparticles. The functional groups and their accompanying strong absorption bands at 534 cm^{-1} and 375 cm^{-1} were identified using Fourier transform infrared analysis (FTIR). The sample's optical absorbance was measured between 200 to 800 nm, and the size of the bands detected in the test sample was subsequently investigated. This study examines the ultraviolet (UV) area, specifically the energy band gap (E_g) values of 2.1 and 2.21 eV. The aforementioned technology has a wide range of applications in the areas of electromagnetic wave absorbers, electrodes, sensors, and magnetic recording media devices.

1 Introduction

Ferrites are magnetic nanoparticles classified according to their magnetic characteristics and crystal structures. Spinel-structured ferrites, with the formula MFe_2O_4 (M = Ni, Co, Zn, Cu), have received the most interest due to their distinctive characteristics [1]. These ferrites, which can occupy tetrahedral or octahedral sites, are widely employed in electronic devices due to their high frequency permeability, electrical resistivity, chemical inertness, and mechanical stability [2, 3]. The number of electronic devices in the globe is increasing. So is the need for more energy, and we must save and utilise it. As a result, scientists are inventing novel storage technologies. Electrode substitutes for capacitors are manufactured by combining several components [4, 5]. The elements that are produced should be composed in a straightforward and effective manner. $NiFe_2O_4$ is a more suitable alternative material for energy storage applications, such as Li-ion batteries.

Transition metal oxides (MOs) with a spinel structure [AB_2O_4] are discovered to be feasible [6]. Among these MOs, $NiFe_2O_4$ (NFO) emerges as an intriguing contender for

P. Joseph D. et al. (Eds.): IEMDST 2024, SPPHY 567, pp. 61–68, 2025.
https://doi.org/10.1007/978-981-96-5863-3_8

electrode applications. $NiFe_2O_4$ is an inverse spinel ferrite in which Fe^{3+} and Ni^{2+} ions occupy the octahedral and tetrahedral positions, respectively. Nickel ferrite ($NiFe_2O_4$) is a well-known inverse spinel synthesised utilising several methods [7]. Several chemical synthesis methods have been devised, including sol-gel techniques, freeze drying, co-precipitation, spray drying, precursor, and hydrothermal methods for the creation of stoichiometric and chemically pure spinel ferrites [8]. The sol-gel technique, in particular, is a very useful and appealing method for the preparation of nano-sized particles. It provides benefits such as superior stoichiometric control and the ability to produce micro particles with a restricted size distribution in a relatively short processing time. Taurig Vvzzaman et al. prepared $NiFe_2O_4$/AC using the sol-gel method at 700 °C for 6 h [1]. Neetu Dhanda et al. produced Ag-doped Ni-Co Nano ferrites using an eco-friendly approach method at 800 °C for 3 h. The lattice constants were determined between 8.314 and 8.324 Å in this example [2]. Amulya et al. prepared $NiMnFe_2O_4$ by sonication at 450 °C for 6 h [9]. Bera et al. prepared $ZnFe_2O_4$ via a 3 h solid-state process at 1200 °C [10].

Our present research focuses on synthesizing the large surface area and dense $NiFe_2O_4$ nanoparticles by using the sol-gel process with highly crystalline. Due to their high thermal and chemical stability, MOs pay greater attention to electrode materials for energy storage applications. As a result, we look into the temperature at which the formation rises. The literature has no information on acid-based (pH - 3) $NiFe_2O_4$ production or its magnetic and structural properties [11, 12]. Sensors, catalysis, ferrofluids, and magnetic resonance imaging are further applications for $NiFe_2O_4$. Calcined is one of the most critical steps in the process. The structure of $NiFe_2O_4$ from the COD data chip file is clearly displayed in Fig. 1 [22].

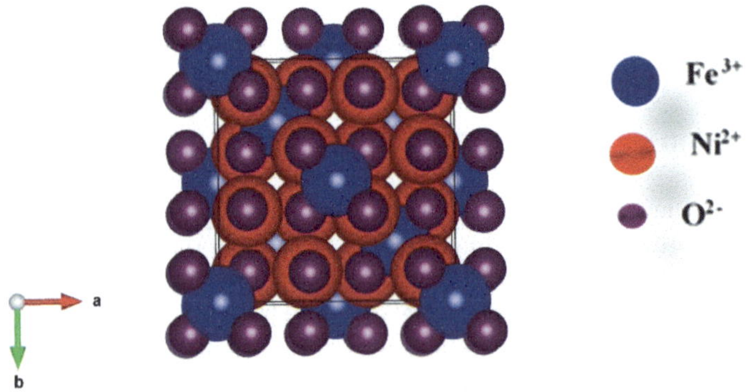

Fig. 1. Schematic diagram of nickel ferrite ($NiFe_2O_4$) crystal structure.

2 Experimental

The sol-gel method was utilised to create nickel ferrites nanoparticles. The stoichiometric amounts of analytical grades of iron nitrate [$Fe(NO_3)_3{\cdot}9H_2O$], nickel nitrate [$Ni(NO_3)_2{\cdot}6H_2O$], and citric acid [$C_6H_8O_7$] were dissolved in 25 mL of deionized water in three separate beakers and swirled for 10 min each. The pH of citric acid, which is commonly used as a fuel agent, is 3.0. Following that, the components were stirred for 30 m or until a homogeneous solution was achieved, and the mixture was kept on a hot plate at 70 °C. After the first four hours, a thick gel was created and heated to 90 °C for ten hours. The prepared sample was calcined in the muffle furnace for three hours at 600 and 800 °C. A mortar and a ball mill were used to grind the sample into a fine powdered consistency. The structural properties were determined using XRD (XRD model-BRUKER ECO D8 ADVANCE), the morphological analysis was done using SEM, and the absorption band and functional group were found using FTIR. An ultraviolet (UV) investigation was carried out to assess the energy band gap, transitions, and absorption.

3 Results and Discussion

3.1 Structural Properties

The XRD investigation confirmed the structural properties of the nickel ferrite $NiFe_2O_4$, as shown in Fig. 2. A Cuk1 radiation source [1.54060 Å] was tested with a scan range of 10.00 to 80.008° and a step size of 0.02° per second. The current and voltage ratings were 25 kV and 40 mA. Figure 2a shows the crystal structure and XRD pattern produced at room temperature (25 °C) from an annealed nickel ferrite ($NiFe_2O_4$) sample. The generated nickel ferrite has a cubic spinel phase, as evidenced by the relative intensities and positions of the identified peaks (220), (311), (222), (400), (511), and (440), which correspond to the (JCPDS card no. 10–325). A description of the space group Fd-3m can be found in the prepared sample. The (311) plane's prominent peak appearance serves as proof of its existence. That have a specific peak X-ray angle range of (33.87° to 36.41°) for a calcined ferrite at 600 °C and (35.14° to 36.16°) for a calcined ferrite at 800 °C. As a result, calcined ferrites at 800 °C are narrower than those at 600 °C. Throughout the calcining process, temperatures ranged from 800 to over 600 °C. Use the Scherrer formula to calculate the size of the crystallite [13].

$$D = k\lambda/\beta\cos\theta \quad (1)$$

where the wave length of the x-ray $\lambda = 1.5406$, k is the crystalline shape factor (0.89) [9], is the Bragg's diffraction angle, and is the FWHM. To calculate inter-planner spacing (d) using Bragg's law

$$d = n\lambda/2\sin\theta \quad (2)$$

where n = 1 order of reflection, to calculate the lattice parameter using the following equation,

$$a = d\sqrt{h^2 + k^2 + l^2} \quad (3)$$

hkl denotes the Miller indices of the diffraction plane.

Cell volume $V = a^3$, X-ray density was measured by means of the equation

$$\rho x = ZM/VNa \quad (4)$$

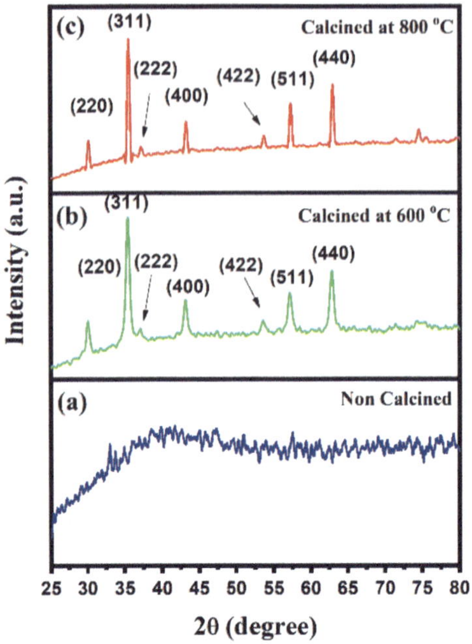

Fig. 2. XRD patterns of the $NiFe_2O_4$ nanoparticle: (a) NFO 00 non-calcined, (b) NFO 600 calcined at 600 °C and (c) NFO 800 calcined at 800 °C.

Table 1. Structural parameter such as crystallite size (D), cell volume (Å^3), lattice constant (Å), crystalline density (ρ), dislocation density, micro-strain values are calculated from XRD data

Sample ID	Crystallite size (D) (nm)	Lattice constant (a) (Å)	Volume of unit cell (V) (Å)^3	Density (ρ_{xrd}) (g/cm^3)	Dislocation density × 10^{14} (line/m^2)	Micro-strain (Ɛ) × 10^{-3} (line^{-2}/m^{-4})
NFO-600 °C	23.48	8.28	569.18	5.47	20.4	4.47
NFO-800 °C	34.69	8.34	580.71	5.36	8.5	2.82

The molecular weight is represented by (M), the number of Avogadro's units by (Na), the volume of the unit cell by (V), and the number of formula units contained in the unit cell by (Z). The maximum intensity peak (311) in calcined ferrite at 600 °C is 35.48°, with a d-spacing of 2.52. Ferrite crystallites have an average size of 23.48 nm, a lattice parameter of 8.28, a volume of 569.18 cubic units, and a density of 5.47 g/cm^3. The greatest intensity peak (311) in calcined ferrite heated to 800 °C is 35.64°, with a d-spacing of 2.51. The average crystallite size is 34.69 nm, with a lattice parameter of 8.34, unit volume of 580.71, and density of 5.36 g/cm1. As a result, raising the calcined temperature increased the crystallite size, the lattice parameter, and decreased the density. Table 1 shows the structural parameters from the XRD data, including the crystallite size (D), cell volume (Å^3), lattice constant, crystalline density, dislocation density, and micro-strain values [14].

3.2 Surface Morphology

Fig. 3. SEM microstructure of the $NiFe_2O_4$ nanoparticle. (a) non-calcined, (b) calcined at 600 °C and (c) calcined at 800 °C.

Figure 3a shows that the surface micromorphology appears to be agglomerated, which could be related to the weak crystalline quality of the non-calcined sample. When the sample is calcined at a temperature of 600 °C, there is a minor alteration in the morphology, but the crystalline structure or shape is not repaired in any manner. Figure 3b. Calcinating the sample at high temperatures 800 °C considerably improves its surface shape and crystallinity. Figure 3c. As a result of our investigation, we discovered that the temperature of the annealing process has a significant impact on the crystallinity of ferrites. The sample's powder appears to be made up of coarse grains of various sizes and shapes, as shown in the images [2, 13].

3.3 FTIR Analysis

Figure 4 (a) shows the FTIR spectrum of ferrite $NiFe_2O_4$ heat treated at different temperatures (room temperature, 600 °C, and 800 °C). The non-calcined ferrite contains more absorption bands. The absorption at 3370 cm^{-1}, 1570 cm^{-1}, and 1400 cm^{-1} is attributed to O-H, C = O C-N, and C-H. Fe-O and Ni-O had the highest absorption at 534 cm^{-1} and 375 cm^{-1}, respectively. As a result, single phase $NiFe_2O_4$ was effectively synthesised; nevertheless, the sample was amorphous due to the presence of water and acidity. XRD tests proved that it was true, substantial absorptions at 534 cm^{-1} and 375 cm^{-1} in the calcined ferrites reveal the presence of a Ni-O and Fe-O bonds [16].

3.4 UV Analysis

We measured the optical absorption spectrum of the prepared $NiFe_2O_4$ Nano ferrites using the UV-DRS spectrometer in the series of 200 nm to 800 nm. The nickel ferrite showed an absorption peak in the UV region Fig. 4 (b) and it has major absorption

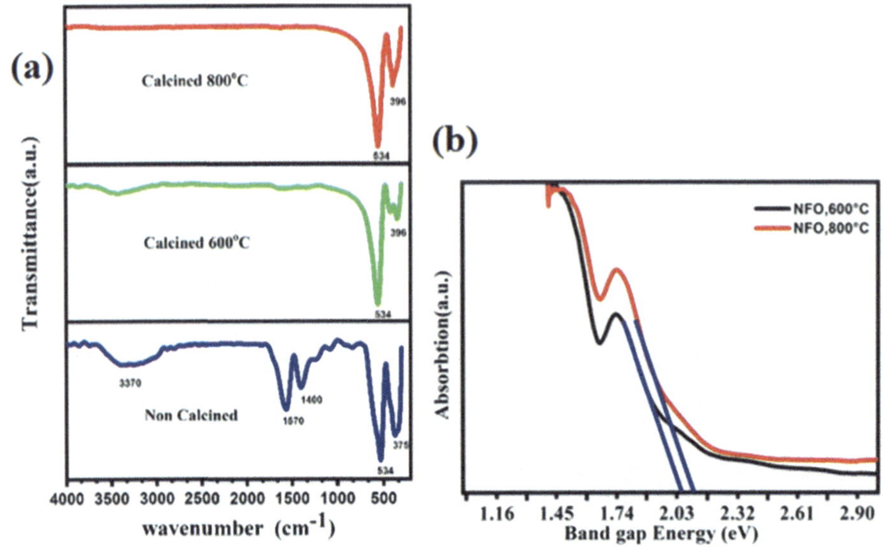

Fig. 4. (a) FTIR spectra of $NiFe_2O_4$ at non-calcined, calcined at 600 °C and calcined at 800 °C. (b) UV-visible spectra of the annealed at 600 °C and 800 °C nickel ferrite ($NiFe_2O_4$).

wave lengths of 229.5 nm and 226.5 nm for samples heat treated at 600 °C and 800 °C, respectively [15, 16].

$$Eg = hc/\lambda \tag{5}$$

where the Planck's constant h and the velocity of light c, respectively. λ correlate to the minimum dip in the absorption wavelength [20, 21]. This optical character of absorption of the synthesized ferrite found at the edge of the UV spectrum correlates with band gaps of 2.1 eV and 2.21 eV. we have observed that the energy (E_g) value increases with an increase in the calcined temperature, it may due to high crystallinity [11, 17].

4 Conclusion

A low-cost sol-gel auto-combustion method was employed to successfully synthesise pure phase nickel ferrite. Samples are investigated using XRD, SEM, UV, and FTIR to assess their structure, surface morphology, optical absorption, and molecular vibration. The estimated crystallite size values varied with annealing temperature (23.48 nm at 600 °C and 34.69 nm at 800 °C), which was mirrored in the XRD analysis of the samples' crystalline quality. Crystallinity is strongly associated with annealing temperature but only marginally influenced by annealing time. According to statistics, the amount of crystalline perfection rises as the annealing time increases. Surface analysis (SEM) shows that annealing the sample at high temperatures (800 °C) significantly enhances surface form and crystallinity. The major molecular vibrations in $NiFe_2O_4$ are Fe-O and Ni-O, which occur at 534 cm^{-1} and 375 cm^{-1}, respectively. The band gap values derived from UV analysis of optical absorption in the manufactured samples have a limited range of 2.1 to 2.21 eV. The findings of this study imply that annealed $NiFe_2O_4$ could be used in a variety of optoelectronic and sensor applications.

Acknowledgments. One of the authors, P. Sundararajaperumal, would like to thank Thiagarajar College of Engineering, Madurai for providing the Thiagarajar Research fellowship (TRF).

References

1. Uzzaman, T., et al.: Ceram. Int. **47**, 10733–10741 (2021). https://doi.org/10.1016/j.ceramint.2020.12.189
2. Dhanda, N., Thakur, P., Aidan Sun, A.-C., Thakur, A.: J. Magneti. Magn. Mat. **572**, 170598 (2023). https://doi.org/10.1016/j.jmmm.2023.170598
3. Niu, J.Y., Hong, B., Xu, J.C., Han, Y.B., Jin, H.X., et al.: Physica E **143**, 115383 (2022)
4. Veisi, P., Seyed Dorraji, M.S., Rasoulifard, M.H., et al.: J. Photochem. Photobiol. Chem. **414**, 113268 (2021). https://doi.org/10.1016/j.jphotochem.2021.113268
5. Abdel, N.S., Ihsan, H., Noori, F.T., Pallon, J.: Nucl. Instrum. Methods Phys. Res., Sect. B **511**, 57–63 (2022)
6. Momin, A.A., Parvin, R., Islam, M., Hossain, A.K.M.A.: J. Magn. Magn. Mater. **526**, 167708 (2021). https://doi.org/10.1016/j.jmmm.2020.167708
7. Tian, C., Yao, Q., Tong, Z., Rao, G., Deng, J., Wang, Z., et al.: J. Alloy. Compd. **859**, 157757 (2021). https://doi.org/10.1016/j.jallcom.2020.157757
8. Huang, T., Liu, B., Yang, P., Qiu, Z., Hu, Z.: Int. J. f Electrochem. Sci. **13**, 6144–6154 (2018). https://doi.org/10.20964/2018.06.60
9. Amulya, M.A.S., Nagaswarupa, H.P., Kumar, M.R.A., Ravikumar, C.R., Kusuma, K.B.: Appl. Surf. Sci. Adv. **2**, 100038 (2020). https://doi.org/10.1016/j.apsadv.2020.100038
10. Bera, S., Prince, A.A.M., Velmurugan, S., Raghavan, P.S., Gopalan, R.: G. J. Mater. Sci. **36**, 5379–5384 (2001). https://doi.org/10.1023/A:1012488422484
11. Ahilandeswari, E., Sakthipandi, K., Rajesh Kanna, R., Hubálovská, M., Vigneswaran, D.: Physica B: Condensed Matter. **635**, 413849 (2022). https://doi.org/10.1016/j.physb.2022.413849
12. Ahilandeswari, E., Rajesh Kanna, R., Sakthipandi, K.: Physica B: Condensed Matter **599**, 412425 (2020). https://doi.org/10.1016/j.physb.2020.412425
13. Rajesh Kanna, R., Sakthipandi, K., Maraikkayar, S.M.S.M.A., Lenin, N., Sivabharathy, M.: J. of Rare Earths **36**, 1299–1309 (2018). https://doi.org/10.1016/j.jre.2018.03.033
14. Ahmed, S.M., Szymanski, P., El-Nadi, L.M., El-Sayed, M.A., Appl, A.C.S.: Mater. Interfaces. **6**, 1765–1772 (2014). https://doi.org/10.1021/am404662k
15. Aslan, E., Zarbali, M.: Opt. Mater. **125**, 112030 (2022). https://doi.org/10.1016/j.optmat.2022.112030
16. Castro-Hurtado, I., Mandayo, G.G., Castaño, E.: Thin Solid Films **548**, 665–676 (2013)
17. Velusamy, P., Babu, R.R., Ramamurthi, K., Viegas, J., Elangovan, E.: J. Mater. Sci. Mater. Electron. **26**, 4152–4164 (2015). https://doi.org/10.1007/s10854-015-2960-0
18. Velusamy, P., et al.: Sens. Actuators, B Chem. **297**, 126718 (2019). https://doi.org/10.1016/j.snb.2019.126718
19. Bai, J., et al.: Sens. Actuators, B Chem. **324**, 128755 (2020). https://doi.org/10.1016/j.snb.2020.128755
20. Naz, K., Khan, J.K., Khalid, M., Akhtar, M.S., Gilani, Z.A., et al.: Mater. Chem. Phys. **285**, 126091 (2022). https://doi.org/10.1016/j.matchemphys.2022.126091

21. Assili, K., Gonzalez, O., Alouani, K., Vilanova, X.: Arabian J. Chem. **13**, 1229–1246 (2020). https://doi.org/10.1016/j.arabjc.2017.10.004
22. El-Hafiz, D.R.A., Ebiad, M.A., Sakr, A.A.E.: Ultrasonic-assisted nano-nickel ferrite spinel synthesis for natural gas reforming. J. Inorg. Organomet. Polym Mater. **31**(1), 292–302 (2021). https://doi.org/10.1007/s10904-020-01718-z

Structural, Optical and Magnetic Properties of Ni-Doped ZnO Nanoparticles Prepared by Sol-Gel Method

Sunena Subhash(✉) and Sudheendran Kooriyattil

Department of Physics, Sree Kerala Varma College, University of Calicut, Thrissur, Kerala 680011, India
sunenamidhunam827@gmail.com

Abstract. The objective of this work is to produce pure and Ni-doped $Zn_{1-x}Ni_xO$ (where x = 0, 0.05, 01, or 0.15) nanoparticles using the sol-gel method. The samples were described using X-ray diffraction, UV-Vis spectroscopy, and vibrating sample magnetometer investigation. X-ray diffraction (XRD) was used to determine the hexagonal wurtzite structure of both pure and Ni-doped samples, showing no secondary phases or impurities. Pure and doped ZnO nanoparticles revealed a blue shift in the absorption edge in UV-Visible absorption edge. Moreover, magnetic characterization confirms room temperature ferromagnetism in $Zn_{0.95}Ni_{0.05}O$ and $Zn_{0.9}Ni_{0.1}O$ samples.

1 Introduction

In the emerging field of optoelectronics, spin polarized Light emitting diodes (LEDs), photo-voltaic, and sensors, diluted magnetic semiconductors (DMSs) have potential applications [1, 2]. In future spintronic devices, these materials play an important role which manipulate both charge and spin [3]. ZnO is a flexible multifunctional candidate among II–VI semiconductors that can be used in electronics and optoelectronics because of its large excitonic binding energy (60 meV), direct wide band gap (3.37 eV at 300 K), excellent chemical stability, and exceptional electro-optic and piezoelectric properties [4, 5]. ZnO nanostructures have potential uses like varistors, lasers, optoelectronic devices and even luminous screens. More precisely, a variety of substrates with suitable chemical and thermal stabilities can be used to produce high-grade ZnO at room temperature (RT). Because of all these characteristics, ZnO is an intriguing and trustworthy semiconductor that may be used as a versatile material [6–11]. RT ferroelectricity and ferromagnetism in doped ZnO reported concurrently in studies. The optical, electrical, and magnetic properties of wide gap semiconductor, such as ZnO changes dramatically by addition of impurities into it [12–14]. For enhancing and controlling its performance, it is crucial to select suitable doping element into ZnO [15–18]. In the magnetic materials, Ni is a significant dopant. Moreover, Ni^{2+} (0.69 Å) has a radius that is similar to Zn^{2+} (0.74 Å) and the same valence as Zn^{2+}, suggesting that Ni^{2+} might replace Zn^{2+} in the ZnO lattice. By doping Ni into ZnO, a composite material with magnetic and optical properties can be produced. Dar et.al observed red shift in the optical band gap [19] in addition to changes in

P. Joseph D. et al. (Eds.): IEMDST 2024, SPPHY 567, pp. 69–76, 2025.
https://doi.org/10.1007/978-981-96-5863-3_9

magnetic behavior [20] by Ni-doping into ZnO. Due to the combined effects of its optical and magnetic features, it has optical integrated circuit applications with a larger threshold [21]. Chemical stability of Ni^{2+} in Zn^{2+} sites improves optical and electrical qualities of ZnO [22]. Consequently, optical, electrical, and magnetic characteristics of ZnO can be adjusted using Ni ions [23]. The material's unique combination of semiconducting and ferromagnetic properties makes it a promising candidate for non-volatile memory technologies, including FeRAM, FeFETs, and RRAM. Its potential for integration into flexible, radiation-hardened, and neuromorphic devices further expands its importance in the future of electronics and computing.

2 Materials and methods

Nanoparticles of $Zn_{1-x}Ni_xO$ (x = 0, 0.05, 0.1, and 0.15) were prepared by the sol-gel technique. Two acetic acid solutions were combined: one containing zinc acetate dihydrate (50 mL, 0.095 M) and the other containing nickel acetate tetrahydrate (50 mL, 0.005 M). To attain the desired ratios, nickel acetate tetrahydrate (50 mL, 0.005 M) was added to this mixture as the doping source. The resultant mixture was then added to 5 mL of ethylene glycol and agitated continuously for 35 min. After that, a clear gel was generated by gradually heating the mixture to 80 °C while stirring it. After being kept at room temperature for twenty-four hours, the resulting gel was heated to 250 °C for two hours. The final transformation of the solid product was into white ZnO nanoparticles and moss green Ni-doped ZnO nanoparticles following a rigorous grinding procedure and a two-hour annealing to 750 °C.

Utilizing CuKα(λ = 1.5406 Å) radiation in the 2θ range of 20°–80°, the produced nanoparticles were subjected to an X-ray diffraction technique (PANALYTICAL, Aeris Research) for structural investigation. By using JASCO V-670 double beam spectrophotometer, the UV absorption spectra of the samples were recorded. Using a vibrating sample magnetometer (Lake Shore), the magnetic hysteresis loop at room temperature was measured.

3 Results and discussion

3.1 Structural Analysis: X-ray Analysis

The X-ray diffraction patterns of both pure and Ni-doped ZnO nanoparticles are displayed in Fig. 1. All XRD samples display the pure hexagonal wurtzite phase, with no secondary phases formed. The XRD pattern shows that the diffractions of the hexagonally-structured (100), (002), (101), (102), (110), (103), (200), (112), and (201) planes are clearly indexed to all of the typical diffraction peaks. Figure 1 shows the X-ray diffraction patterns of pure and Ni-doped ZnO nanoparticles without the formation of any secondary phases, all XRD samples show the pure hexagonal wurtzite phase. All of the characteristic peaks are clearly indexed to the hexagonally-structured (100), (002), (101), (102), (110), (103), (200), (112), and (201) planes, according to the XRD pattern. Variations in the FWHM are consistent with the crystallite size because of the ionic radius similarity between Zn^{2+} and Ni^{2+}. It is determined using Debye Scherrer's formula [23].

$$D = \frac{k\lambda}{\beta cos\theta} \tag{1}$$

where FWHM is the full width at half maximum of the diffraction peak, λ is the incoming X-ray wavelength, and θ is the diffraction angle. Pure ZnO NPs are discovered to have a crystallite size of roughly 47.543 nm. Table 1 shows that when the dopant concentrations increase to 0.05, 0.10, and 0.15, the size of the crystallites in Ni-doped ZnO NPs decreases. The distortion caused by foreign contaminants in the ZnO lattice is the main reason for decrease in crystallite size. Reduction in the nucleation, growth rate, and slight shift in crystal lattice orientation that results an increase in Ni-doping percentage. This is because of the ionic radius mismatch between the Zn and Ni ions.

Micro-strain was calculated using the formula [23],

$$\varepsilon = \frac{\beta \cos\theta}{4} \tag{2}$$

The micro-strain values are shown in Table 1. The strain value of Ni^{2+} doped ZnO NPs is higher than that of pure ZnO NPs. The change in size and shape of the particle also caused by strain on individual unit cells.

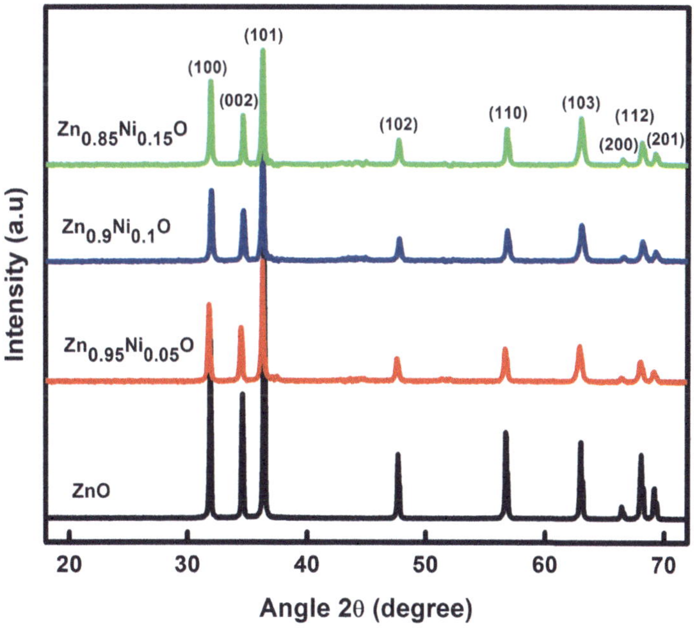

Fig. 1. XRD pattern of pure and Ni-doped ZnO nanoparticles.

The length of dislocation lines per unit volume of the crystal or amount of defects in the sample is known as dislocation density (δ), and is calculated using the Eq. (3), where

D is the crystallite size. a and c (Lattice parameters) of the samples were calculated by using Eq. (4), where the Miller indices are h, k, and l, and d is the interplanar distance. For (1 0 0) plane the lattice parameter a is calculated using following Eq. (5). The lattice parameter c for the plane (0 0 2) is calculated using following Eq. (6). When dopant concentration increased, lattice characteristics somewhat redused. Replacement of different ionic radii for the Ni ion is the cause of this small variation in the lattice parameters. The dopant ions appear to be thoroughly integrated into the ZnO crystal lattice based on the small changes in (c/a) aspect ratio.

$$\delta = \frac{1}{D^2} \tag{3}$$

$$\frac{1}{d^2} = \frac{4}{3}\frac{\left(h^2 + hk + k\right)^2}{a^2} + \frac{l^2}{c^2} \tag{4}$$

$$a = \frac{\lambda}{\sqrt{3}\sin\theta} \tag{5}$$

$$c = \frac{\lambda}{\sin\theta} \tag{6}$$

Table 1. Structural and geometrical parameters of pure and Ni-doped ZnO nanoparticles

Samples	a = b (Å)	c (Å)	c/a	Crystallite size D (nm)	Strain ε (× 10^{-3})	Dislocation density δ (× 10^{-4}) (nm^{-2})
ZnO	3.237	5.182	1.6006	47.543	0.00078	5.193
ZnNiO 5	3.249	5.203	1.6012	22.077	0.0017	18.447
ZnNiO 10	3.228	5.172	1.6020	21.181	0.0019	20.225
ZnNiO 15	3.233	5.174	1.6003	21.125	0.0018	34.532

3.2 Optical Characterization: UV-Vis Analysis

The absorbance spectra of undoped and Ni-doped ZnO plotted in Fig. 2 (a). Peak in the absorption can be observed between 200 and 400 nm. Moreover, it is clear that as Ni^{2+} rises, the absorbance value also found to be increasing. ZnO nanostructures can thus absorb UV light more efficiently. According to certain theories, the absorbance of metal oxides is influenced by a number of variables, such as the bandgap, surface roughness, oxygen deficiency, and impurity centres [24]. The diffuse reflectance spectroscopy is very useful in estimating the optical band gap. Kubelka - Munk function F(R) expressed in terms of reflectance (R) in Eq. (7). The absorbance, reflectance or transmittance of light influences the electronic structure of the material.

$$F(R) = \frac{(1 - R)^2}{2R} \tag{7}$$

By using Tauc relation as in Eq. (8), the optical band gap can be calculated

$$F(R)h\upsilon = A\left(h\upsilon - E_g\right)^n \quad (8)$$

For indirect and direct transitions, Where n = ½ and 2, respectively. $(F(R)\ h\upsilon)^2$ on Y-axis and hυ on X-axis for undoped and Ni-doped ZnO NPs plotted in Fig. 2 (b). The straight part of these graphs extrapolated to $(F(R)\ h\upsilon)^2 = 0$ in order to determine the direct bandgap. Extrapolating the linear vertical curve to intercept the energy axis, which shows the band gap energy value, is a routine process [25]. Strain in the crystal lattice may be the cause of the band gap's initial drop from 3.23 eV of pristine ZnO 5% Ni^{2+} with 2.75 eV [24]. This decrease in the band gap suggests that Zn^{2+} is replaced by Ni^{2+} in the ZnO lattice [26].

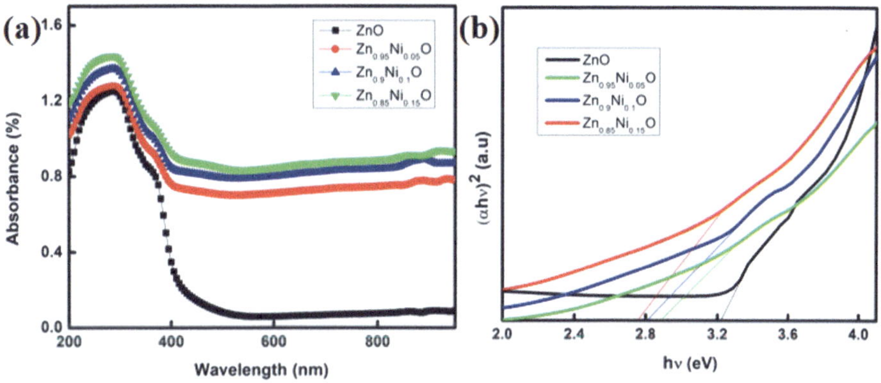

Fig. 2. (a) Absorbance graph and (b) Band gap graph of pure and Ni-doped ZnO nanoparticles.

Due to sp-d exchange interactions between the band electrons and localized spin of transition metal ions, the band gap increases from 2.75, 2.81, 2.91 eV, for 5, 10 to 15% as Ni^{2+} increases. [27]. Samples show Burstein-Moss effect, which leads to increase in band gap. Rise in carrier density due to shift of Fermi level close to the conduction band occurs [28]. Therefore, band gap of ZnO nanostructures can be tuned by Ni^{2+} doping.

3.3 Magnetic Properties: VSM Analysis

The magnetic hysteresis (M–H) curve was utilized to assess the magnetization of synthesized samples. For every measurement, ambient temperature was used. The diamagnetic characteristics of pure ZnO have also been documented in the literature [29]. The ferromagnetic behaviour of all the samples is evident from the M–H curves for the $Zn_{0.95}Ni_{0.05}O$ and $Zn_{0.9}Ni_{0.1}O$ samples, which are shown in Fig. 3 (a) and (b), respectively. However, Table 2 clearly shows that each sample's measured coercivity is incredibly low, showing that the samples are soft ferromagnetic. The fundamental magnetic properties, such as coercive field (Hc), saturation magnetization (Ms), remnant magnetization (Mr) can be tuned by changes in particle size, shape, and composition. The

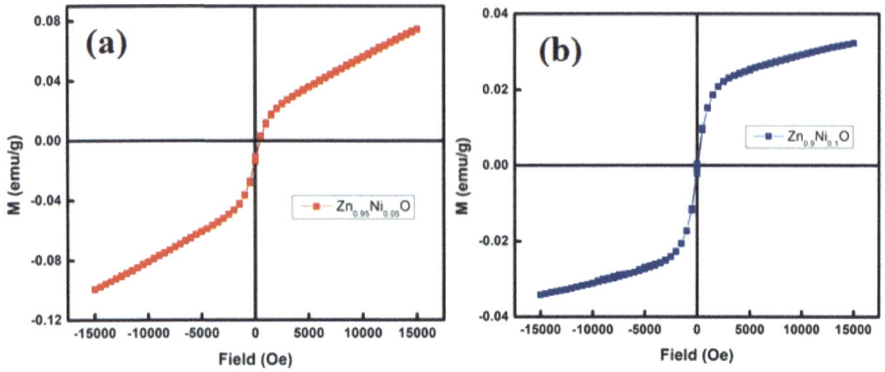

Fig. 3. (a) M-H hysteresis curve of $Zn_{0.95}Ni_{0.05}O$ and (b) $Zn_{0.9}Ni_{0.1}O$ nanoparticles.

magnetic characteristics of $Zn_{0.95}Ni_{0.05}O$ and $Zn_{0.9}Ni_{0.1}O$ systems are listed in Table 2. 10% Ni^{2+} showed the maximum remanence Mr (0.0016 emu/g). The results suggested that, lesser amount of Ni^{2+} into the ZnO induces and enhances the magnetic moment of the entire sample.

Table 2. Magnetic properties of pure and Ni-doped ZnO nanoparticles.

Sample	H_C (Oe)	M_r (emu/g)	M_s (emu/g)
$Zn_{0.95}Ni_{0.05}O$	54.96	0.0012	0.033
$Zn_{0.9}Ni_{0.1}O$	23.31	0.0016	0.087

Anisotropies in crystal structure or combination of crystal and form could be the cause of this decrease in saturation magnetization. The evolution of magnetic characteristics of material can also be attributed to surface structure, crystallinity, and size factor, according to some previously published study [30, 31]. Microstructural and structural parameters, such as the formation of lattice strain, the reduction of crystallite (particle) size with Ni-doping level, the partial or full occupancy of Ni^{2+} ions in Zn^{2+} sites, and the creation of specific point defects (vacancies), may be related to The values of coercivity and remanence may be related to microstructural and structural parameters such as the formation of lattice strain, the decrease in crystallite (particle) size, the partial or full occupancy of doped ions. The origin of ferromagnetism has been explained by a number of theories thus far, however there is still debate on the matter. One of the hypotheses states that a variety of intrinsic and extrinsic processes can cause a sample to exhibit room-temperature ferromagnetism. Exchange interactions give rise to intrinsic phenomena, while secondary phases or clusters of dopant semiconductor atoms arise to cause extrinsic phenomena [32]. A few studies also address the connection between magnetic and inherent defects such oxygen or zinc vacancies [33, 34]. Without the formation of metallic clusters or secondary phases, Ni^{2+} ions are effectively incorporated into ZnO wurtzite structure, which is evident from XRD. Thus, Ferromagnetism due to metallic clusters and secondary phases is completely ruled out.

4 Conclusion

$Zn_{1-x}Ni_xO$ nanoparticles (x = 0, 0.05, 0.1, 0.15, and 0.2) have been successfully synthesized using the sol-gel process and then annealed for two hours at 750 °C. From XRD examination, the hexagonal wurtzite structure and the trend of crystallite size decreases as the concentration of Ni^{2+} increases is revealed. Ni-doped ZnO nanoparticles have somewhat smaller lattice parameters ("a" and "c") than pure ZnO nanoparticles. With an increase in Ni^{2+} concentration, the bandgap is blue shifted. The current study additionally validates that ZnO nanoparticles' band gap and optical absorbance can be regulated using Ni-doping. From room-temperature ferromagnetic studies, we observe that it is the inherent characteristics of the sample and is not the product of any metallic clusters or additional phases developing, it is reasonable to assume from the XRD and VSM study of our samples.

Acknowledgements. I'm grateful to the Kerala State Council for Science, Technology, and Environment (KSCSTE) for funding a fellowship that made my work easier. We really appreciate the facilities made available by DST - FIST, UGC, SERB, and KSCSTE SARD.

References

1. Klingshirn, C., Fallert, J.: Phys. Status Solidi B **247**, 1424–1447 (2010). https://doi.org/10.1002/pssb.200983195
2. Mc Cluskey, M.D., Jokela, S.J.: J. Appl. Phys. **106**, 071101–071113 (2009). https://doi.org/10.1063/1.3216464
3. Wu, D., Xu, Q.Y., Zhang, F.M., Liu, X.S.: AAPPS Bull. **18**, 52 (2008). https://doi.org/10.1063/1.3216464
4. Dietl, T.: J. Phys. Condens. Matter **19**, 165204 (2007). https://doi.org/10.1088/0953-8984/19/16/165204
5. Pearton, S.J., Norton, D.P., Ip, K., Heo, Y.W.: Prog. Mater. Sci. **50**, 293 (2005). https://doi.org/10.4236/njgc.2014.44008
6. Grigorjeva, L., Millers, D., Smits, K.: Solid State Phenom. **128**, 135–140 (2007)
7. Yan, L., Ong, C.K.: J. Appl. Phys. **96**, 508–511 (2004). https://doi.org/10.1063/1.1757652
8. Kim, D.H., Woo, S.I.: Solid State Commun. **136**, 554–558 (2005). https://doi.org/10.1016/j.ssc.2005.09.012
9. Raj Mohan, R., Sambath, K., Rajendran, K.: Experimental investigation on structural and optical properties of ZnO: AZO nano particles by hydrothermal synthesis. J. Mater. Sci.: Mater. Electron. **26**(3), 1748–1755 (2014). https://doi.org/10.1007/s10854-014-2603-x
10. Ueda, K., Tabata, H., Kawai, T.: Appl. Phys. Lett. **79**, 988–990 (2001). https://doi.org/10.1063/1.1384478
11. Nunes, P., Fortunadeo, E., Martins, R.: Thin Solid Films **383**, 277–280 (2001). https://doi.org/10.1063/1.1384478
12. Chen, K.X., Wang, H.: J. Colloid Interface Sci. **330**, 380 (2009). https://doi.org/10.1016/j.jcis.2008.10.067

13. Lia, B.B., Xiua, X.Q., Zhanga, R., Taoa, Z.K.: Mater. Sci. Semicond. Proc. **9**, 141 (2006)
14. Wu, D., Yang, M., Huang, Z., Yin, G.: J. Colloid Interface Sci. **330**, 380 (2009). https://doi.org/10.1021/acsnano.9b03766
15. Cong, C.J., Hong, J.H., Liu, Q.Y., Liao, L.: Solid State Commun. **138**, 511 (2006). https://doi.org/10.1088/0022-3727/47/4/045308
16. Wu, D.W., Yang, M., Huang, Z.B., Yin, G.F.: J. Colloid Interface Sci. **330**, 380 (2009). https://doi.org/10.3390/mi10090622
17. El-Nahass, M.M., Farag, A.A.M., Atta, A.A.: Synth. Met. **159**, 589 (2009). https://doi.org/10.3390/mi10090622
18. Banerjee, P.P.: Proc. IEEE **73**, 1859–1860 (2005). https://doi.org/10.1109/PROC.1985.13378
19. Dar Tanveer, A., Agrawal, A.: Thin Solid Films **589**, 817e821 (2015). https://doi.org/10.1016/j.tsf.2015.07.028
20. Dar, T.A., Agrawal, A., Misra, P., Kukreja, L.M.: Curr. Appl. Phys. **14**, 171e175 (2014). https://doi.org/10.1016/j.cap.2013.10.017
21. Wakano, T., Fujimura, N., Morinaga, Y., Abe, N., Ashida, A.: Phys. E **10**, 260 (2001). https://doi.org/10.1016/S1386-9477(01)00095-9
22. Raja, K., Ramesh, P.S., Geetha, D.: Spectrochim. Acta A **120**, 19e24 (2014). https://doi.org/10.1016/j.saa.2013.09.103
23. Vijayaprasath, G., Murugan, R., Ravi, G.: Appl. Surf. Sci. **313**, 870–876 (2014). https://doi.org/10.1016/j.apsusc.2014.06.093
24. Karthika, K., Ravichandran, K.: J. Mater. Sci. Technol. **31**, 1111e1117 (2015). https://doi.org/10.1016/j.jmst.2015.09.001
25. Shinde, K.P., Pawar, R.C., Sinha, B.B., Kim, H.S.: Optical Int. **40**, 16799e16804 (2014). https://doi.org/10.1016/j.jallcom.2016.10.022
26. Yanmei, L., Tao, W., Xia, S.: Appl. Surf. Sci. **257**, 6540e6545 (2011). https://doi.org/10.1016/j.apsusc.2011.02.074
27. Yan, X., Hu, D., Li, H., Li, L., Xiaoyu: Phys. B **406**, 3956e3962 (2011)
28. Ashokkumar, M., Muthukumaran, S.: Opt. Mater. **37**, 671e678 (2014). https://doi.org/10.1016/j.optmat.2014.08.012
29. Zhou, S., Potzger, K., Reuther, H., Kuepper, K.: J. Appl. Phys. **101**, 09H109 (2007). https://doi.org/10.1063/1.2710802
30. Manikandan, A., Judith Vijaya, J.: J. Ind. Eng. Chem. **20**, 2077e2085 (2014). https://doi.org/10.1016/j.jiec.2013.09.035
31. Fabbiyola, S., John Kennedy, L., Ratnaji, T.: Ceram. Int. **42**, 1588e1596 (2016). https://doi.org/10.1166/jnn.2016.12094
32. Sharma, V.K., Varma, G.D.: Adv. Mat. Lett. **3**, 44–49 (2012). https://doi.org/10.5185/amlett.2011.5264
33. Pal, B., Giri, P.KJ.: Nanosci. Nanotechnol. **11**, 1–8 (2011). https://doi.org/10.1166/jnn.2011.4293
34. Santara, B., Giri, P.K., Dhara, S., Imakita, K.: J. Appl. Phys. Phys. D **47**, 235304 (2014). https://doi.org/10.1088/0022-3727/47/23/235304
35. Das, J., Mishra, D.K., Sahu, D.R.: Roul, Mater. Lett. **65**, 598–601 (2011). https://doi.org/10.1016/j.matlet.2010.11.044

Impact of ZnO Nanosphere on Ionic Conductivity and Dielectric Relaxation of Nematic Liquid Crystal

Samarth Mishra, Shweta Mishra, V. Manjuladevi(✉), and R. K. Gupta

Department of Physics, Birla Institute of Technology and Science, Pilani, Pilani Campus, India
manjula@pilani.bits-pilani.ac.in

Abstract. Liquid crystal (LC) devices, including displays, switchable optical components, and sensors, are widely used today. The fundamental physical properties of LCs can be controlled through external physical fields, allowing for dynamic configuration, and can also be modified by incorporating nanoparticles. Ions present in a minuscule amount can affect the reorientation of liquid crystals through the screening effect at the interfaces. Dielectric relaxation, which impacts switching time through the dipole dispersion phenomenon, gets altered by the incorporation of nanoparticles in the bulk regime. Therefore, understanding the effects of nanoparticle dispersion in liquid crystals is crucial for both existing and emerging technologies that rely on LC materials. Here, we present the dielectric spectroscopy results of ZnO nanosphere-incorporated nematic liquid crystal (E7), using an electric field as an external stimulus. In this study, we have synthesized the ZnO nanospheres and dispersed these ZnO nanospheres in E7 at various concentrations. We found that 0.02 wt.% of ZnO nanosphere dispersed in the E7 mixture exhibits a reduction of ionic conductivity as well as relaxation frequency.

1 Introduction

Liquid crystals (LCs) are an essential class of materials that are widely used in display technology. They exhibit fluidity like liquids but retain the anisotropic properties of crystals viz. They flow while still showing directional dependence. This duality makes them an intermediate state between solid crystals and liquid matter [1]. Nanotechnology, in the last two decades, has seen exponential growth both as a research field and as an applicational avenue. Still, in its nascent stage, nanotechnology has already brought revolutionary developments utilizing the magnificent nanoscale properties of materials and the fabrication of devices. To obtain optimized physical parameters of LCs, many research groups have made attempts to alter dielectric, electro-optical properties, and structural rearrangements of LCs by incorporating suitable nanomaterial dopants into LC host material [2–8]. Conventionally, till now, various types of nanoparticles (NPs) like metal oxide nanoparticles (MO-NPs), carbon-based NPs, and other forms of inorganic NPs have all been studied and explored as effective LC dopants. However, among all these dopants, MO-NPs have garnered a great deal of interest because of their faster electro-optical response and other diverse properties [9–12].

P. Joseph D. et al. (Eds.): IEMDST 2024, SPPHY 567, pp. 77–84, 2025.
https://doi.org/10.1007/978-981-96-5863-3_10

Since the beginning of research in the field of LCs, smaller values of ionic conductivity have been reported in various articles [13]. The significance of ions in liquid crystals (LCs) decreased with the widespread use of twisted nematic mode LCDs, as their impact on display performance was considered negligible. However, this changed with the advent of active-matrix (AM) LCDs, where LC materials are now required to exhibit much higher electrical resistivity compared to those used in passive matrix displays to ensure optimal performance [14–17]. The behavior of ions in the nematic matrix is crucial for display applications, which has attracted researchers to understand the ionic conduction in nematic LCs, especially nanoparticles, to greater extents.

In this work, we have studied the impact of the ZnO nanosphere (ZNS) in the E7 mixture, which shows a wide nematic temperature range (−10 °C–59 °C) and is widely used in display technologies. ZnO is a distinctive semiconductor (Type II-VI) that has a wide and direct band gap of 3.37 eV [18]. Apart from this, ZnO-based NPs are also known to exhibit excellent piezoelectric properties at room temperature [19]. These distinctive properties have garnered great interest from researchers in ZnO nanomaterials in the recent past.

We have studied the dielectric response and the behavior of mobile ions in the nematic LC matrix for the E7 mixture incorporated with ZNS. We fitted the experimental data obtained from dielectric spectroscopy using Jonscher's equation to analyze the degree of interaction among the mobile ions and their surroundings, which include the nanospheres and liquid crystal molecules.

2 Experimental Technique

The nematic LC (NLC) mixture, E7 with wide nematic temperature range was prepared by mixing 51 wt.% of 4′-pentyl-4-biphenylcarbonitrile (5CB), 25 wt.% of 4′-heptyl-4-biphenylcarbonitrile (7CB), 16 wt.% of 4-cyano-4′-n-octyloxybiphenyl (8OCB) and 8wt.% of 4-Cyano-4″-pentyl-p-terphenyl (5CT). These LCs were procured from TCI (India), Pvt. Ltd., and used without further purification. ZnO Nanosphere (ZNS) with a diameter of 50 nm was synthesized in the lab and characterized [20]. To prepare the ZnO nanospheres-E7 composite, ZNS was dispersed into the chloroform solvent (0.05 mg/ml) and then ultrasonicated for 2 h to obtain uniform dispersion. After that, the calculated quantity of dispersed solution is added to the E7 to get the composites of 0.02 and 0.05 wt.%.

For the dielectric spectroscopy measurement, we have prepared the planar-aligned cells. Planar cells were prepared by spin-coating polyimide (PI2555) layer onto an indium tin oxide-coated (ITO) glass substrate, followed by heating at 250 °C for an hour. After this, the polyimide-coated layer was rubbed unidirectionally and stuck together in an antiparallel direction of rubbing. The thickness of the cell was approximately 14 μm, measured by using a fiber optic spectrometer using the interferometry technique, and later, E7 and its composite samples were filled by using the capillary tube through capillarity.

The optical textures of pure E7 and its composites were examined using a polarizing optical microscope (Olympus BX53M), with the sample placed between crossed polarizers at a 45° angle relative to the rubbing direction of both the polarizer and analyzer. We performed dielectric spectroscopy measurements by using an LCR meter (Hioki IM3536-01). The LC cells, filled with E7 and its composites, were positioned within a hot stage coupled with a precise temperature controller (MicrOptik MDTC-600) to maintain a temperature of 30 °C consistently. We measured the parameters like capacitance (C), impedance (Z), resistance (R), and dielectric loss (tan δ) as a function of frequency (20 Hz to 10 MHz) and from the data available, we calculated parameters like real part of dielectric permittivity (ϵ'), the imaginary part of dielectric permittivity (ϵ"), conductivity (σ) and relaxation frequency (R_f) by using the established equation.

3 Results and Discussion

Figure 1 displays the textures of pure E7 and its ZNS nanocomposites observed between the crossed polarizers of the polarizing optical microscope (POM). The POM image shows that ZnO nanospheres are uniformly and homogeneously dispersed into the E7 nematic matrix.

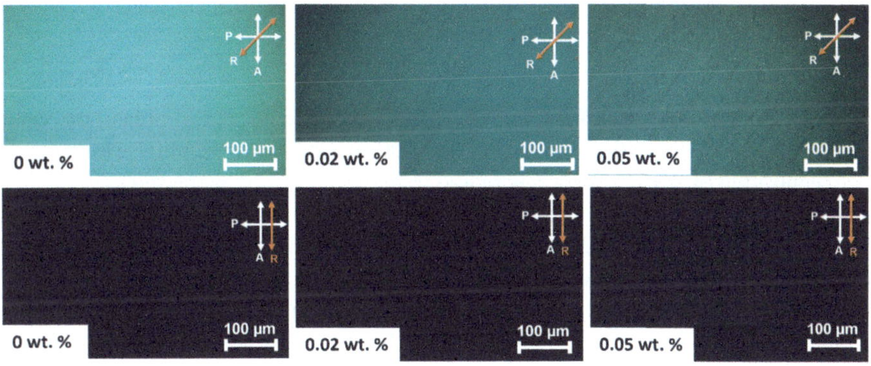

Fig. 1. POM images of the pure E7 and its ZNS nanocomposites of 0, 0.02 and 0.05 wt.% of ZnO nanosphere E7 mixture. (P: polarizer, A: analyzer, R: rubbing direction).

The real (ε') and imaginary (ε'') parts of the dielectric permittivity of E7 and its ZNS composites are shown in Fig. 2. Dielectric permittivity is a characteristic that emerges from the response of dipoles within systems when subjected to an electric field. The results presented in Fig. 2 show the behavior of dipoles under the influence of the applied field. The findings show that the dielectric response of the NLC is significantly affected by the quantity of ZNS present in the LC matrix.

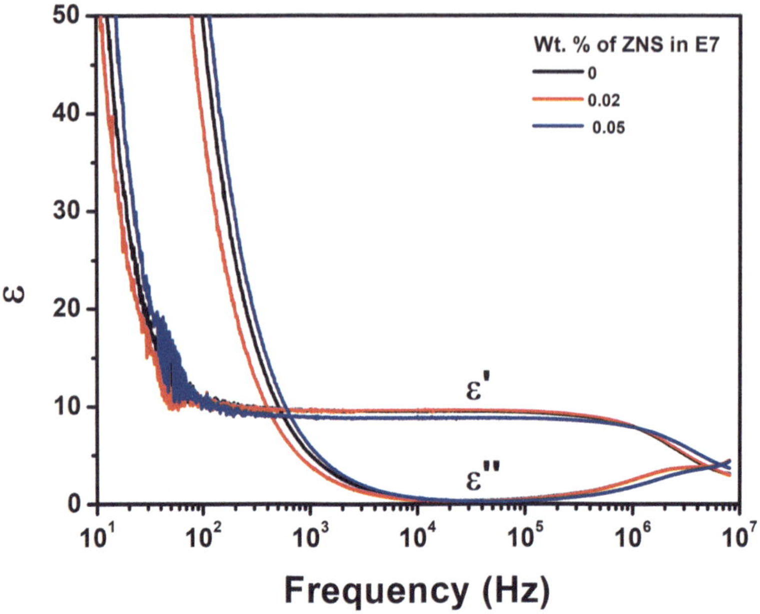

Fig. 2. Frequency-dependent variation of real (ε') and imaginary (ε'') components of dielectric permittivity for E7 and its ZNS composites at 30 °C.

In the low-frequency regime (f < 300 Hz), where ionic effects dominate, the value of ε' initially decreases with the incorporation of 0.02 wt. % of ZNS in E7. However, as the concentration of ZNS is further increased to 0.05 wt.%, the ε' increased as compared to pure E7. In a high-frequency regime (f > 300 Hz), the value of ε' is slightly changed with the incorporation of 0.02 wt. % of ZNS in E7 but when concentration of ZNS is further increased to 0.05 wt.%, the ε' decreases as compared to pure E7. This may be because, in the plateau region, upon incorporation of ZNS into E7, the dominance of the LC-NPs interaction increases, leading to the anchoring of NPs at the interface. This, in turn, reduces the orientational ordering of the neighboring LC molecules, which causes a decrease in the order parameter and, consequently, the permittivity [21].

From Fig. 2, the relaxation frequency (R_f), which is defined as the point of intersection between the ε' and ε'' curve can be found for E7 and its ZNS nanocomposites. The value of R_f shows a minimal increase for 0.05 wt.% as compared to pure E7, and 0.02 wt.% shows a significant decrease in relaxation frequency as compared to pure E7. The increase in the relaxation frequency with a higher concentration of ZNS might be because of the increased mutual interaction between ZNS as this increased interaction reduces the constraints on pure E7 molecules, causing the E7 molecules to relax freely [22].

In Fig. 3, we have shown the changes in conductivity with frequency for E7 and its ZNS composites. We analyzed the experimentally measured data using Jonscher's universal power law [23], as this is commonly used to study how the real part of complex conductivity changes with frequency and migration of ions under the application of the electric field. Jonscher's equation [23] can be expressed as,

$$\sigma = \sigma_{DC} + Af^{n} \tag{1}$$

where considering $Af^{\mathrm{n}} = \sigma_{DC}(\frac{f}{fch})^{n}$, it can be further expressed as

$$\sigma = \sigma_{DC}(1 + (\frac{f}{fch})^{n}). \tag{2}$$

Here, σ_{DC} represents the DC conductivity, f_{ch} denotes the characteristic frequency, and n representing a parameter that indicates the degree of interaction between mobile ions and their surroundings. The curve in Fig. 3 can be characterized by a plateau region at low frequencies corresponding to σ_{DC} and a growth region at higher frequency where conductivity is governed by Af^{n}. According to Jonscher, the frequency dependence of conductivity arises from the relaxation phenomena associated with the ionic atmosphere, which is caused by the movement of mobile charge carriers [24]. The fitting results are given in Table 1.

Table 1. Fitting parameters (on Experimental data) from Fig. 3 were derived using Jonscher's law.

Sample	$\sigma_{DC} \times 10^{-7}$ (S/m)	f_{ch} (Hz)	n
0 Wt.% of ZNS in E7	2.913	28800	1.58
0.02Wt.% of ZNS in E7	2.09	22800	1.57
0.05 Wt.% of ZNS in E7	3.43	39093.98	1.640

As from Table 1, σ_{DC} decreases for 0.02 wt.% ZNS-E7 composites whereas for 0.05 wt.% ZNS-E7 composite, it increases compared to pure E7. The change in conductivity is essentially because of the alteration in the number of mobile ions when the ZNS is incorporated into E7 mixture. The reason for the decrease in conductivity might be because of the adsorption of ions onto the spherical ZNS. However, with the higher concentration of 0.05 wt% ZNS in E7, ZNS acts as additive ions due to the exceeding number of ZNS and hence mobile ions in the nematic matrix, increasing the bulk conductivity of the composite as compared to pure E7.

Here, n values are greater than 1 ($n > 1$) for pure E7 and its composites. The parameter n depends on the mechanism involved in the AC conductivity and temperature. In the present case, although the temperature was maintained at the same value for all experiments, the viscosity of the E7 and its ZNS composites varied with the addition of ZNS [25]. The value $n > 1$, this can be attributed to the conduction of mobile charge carriers through the asymmetric double well potential, as proposed by Kılıç et al. [26]. Many other research groups reported values with $n > 1$ [27–29].

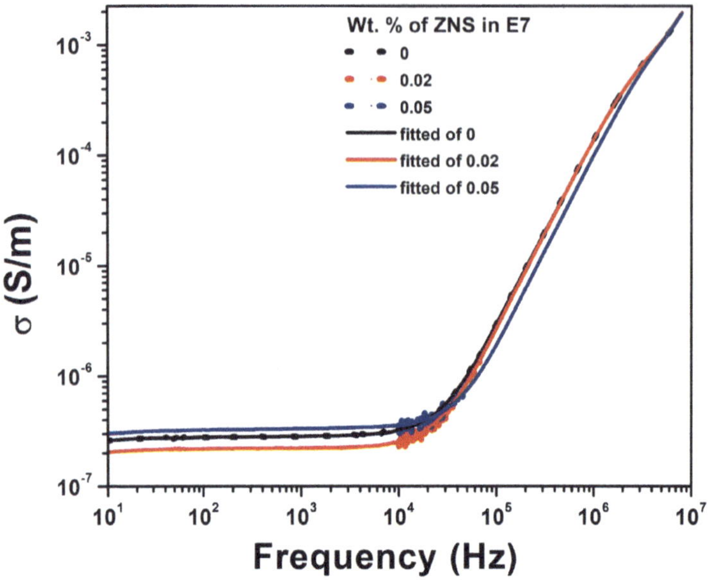

Fig. 3. Frequency-dependent variation of conductivity (σ) of E7 and its ZNS nanocomposites at 30 °C. The experimental data are denoted by a dashed line, and the solid line shows the curves computed by applying Jonscher's model.

Complex impedance (Z*) and phase angle were obtained as a function of frequency (f) of the applied AC electric field to the sample-filled planar cell using dielectric spectroscopy. The Nyquist plots (Z'' versus Z') for the E7 and its ZNS composites demonstrated semi-circular arcs, as shown in Fig. 4. The diameter of the semicircle in the plots is related with the bulk resistance (R_b) of the samples [30, 31]. The value R_b for 0.02 wt.% of ZNS in E7 is lower as compared to pure E7, while for 0.05 wt. %, it is higher as compared to pure E7.

The values of the ionic conductivity (σ) obtained from Fig. 3 of E7 and its ZNS nanocomposites can be related through bulk resistance (R_b), as R_b is inversely proportional to σ. Conductivity in the bulk regime is associated with the low-frequency region, where the dielectric behavior results from space-charge polarization. Hence, the lower σ resulting from 0.02 wt.% of ZNS present in the E7 matrix might be because ZNS is effective in capturing mobile ions and reducing space charge polarization and vice versa for the 0.05 wt. %.

The variation in the ionic conductivity of the nematic liquid crystal E7 can be attributed to the ion-releasing or trapping effects caused by the presence of ZNS, a phenomenon previously reported in nematic liquid crystals incorporated with relatively low concentrations of nanomaterials [31]. Thus, the behavior of both conductivity and permittivity in the studied frequency range can be related to the amount of ZNS present in the nematic matrix. ZNS influences the orientation of LC molecules within the LC matrix and also affects the number of mobile ions and their mobility within the LC cell.

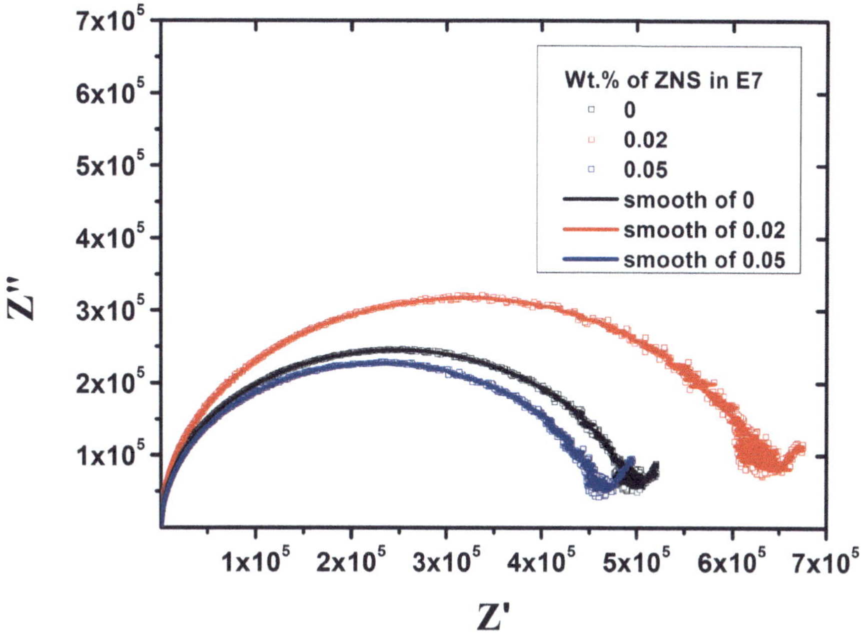

Fig. 4. Nyquist plot (Z″ versus Z′) of E7 and its ZNS nanocomposites at 30 °C.

4 Conclusion

The dielectric permittivity and the conductivity of the E7 and its ZnO nanosphere composites were investigated at 30 °C by using the dielectric spectroscopy technique. The addition of the ZnO nanosphere alters the dielectric properties and conductivity significantly. We observed that the quantity of the the ZnO nanosphere in the E7 matrix decides whether the interaction between liquid crystal and nanoparticle will be dominant or interaction between nanosphere and nanosphere is dominant. We also observe that ZnO nanosphere have ion trapping effect for the lower concentration and ion releasing effect for the higher concentration. This result can be substantial information if further studies are done to understand the physical properties of the system from the application point of view.

Acknowledgments. We are thankful to SERB, India (SPG/2021/002260) for funding & BITS-PILANI for the lab facility.

References

1. Gray, W.: HWSVV Handbook of Liquid Crystals
2. Rajh, D., et al.: Phys. Status Solidi A **210**, 2328–2334 (2013). https://doi.org/10.1002/pssa.201329363
3. (1978) Biogr. Mems Fell. R. Soc. 24:443–471. https://doi.org/10.1098/rsbm.1978.0014

4. Schadt, M.: Ann. Rev. Mater. Sci. **27**, 305–379 (1997). https://doi.org/10.1146/annurev.matsci.27.1.305
5. Katiyar, R., et al.: J. Theor. Appl. Phys. **14**, 237–243 (2020). https://doi.org/10.1007/s40094-020-00374-5
6. Shivaraja, J.S., Gupta, R.K., Kumar, S., Manjuladevi, V.: Liq. Cryst. **47**, 1678–1690 (2020). https://doi.org/10.1080/02678292.2020.1755901
7. Vardanyan, K.K., Daykin, A., Kilmer, B.: Liq. Cryst. **44**, 1240–1252 (2017). https://doi.org/10.1080/02678292.2016.1276223
8. Kumar, J.: Effect of Nanomaterials on Physical Properties of Liquid Crystals, pp. 27–44. PhD Thesis Submitted to BITS PILANI (2016)
9. Prakash, J., Khan, S., Chauhan, S., Biradar, A.M.: J. Mol. Liq. **297**, 112052 (2020). https://doi.org/10.1016/j.molliq.2019.112052
10. Goel, P., Arora, M., Biradar, A.M.: Appl. Phys. (2014). https://doi.org/10.1063/1.4869740
11. Manohar, R., et al.: Jpn. J. Appl. Phys. **48**, 101501 (2009). https://doi.org/10.1143/JJAP.48.101501
12. Eskalen, H., Özğan, Ş, Alver, Ü., Kerlı, S.: Acta Phys. Pol. A **127**, 756–760 (2015). https://doi.org/10.12693/APhysPolA.127.756
13. Rondelez, F.: Solid State Commun. **11**, 1675–1678 (1972). https://doi.org/10.1016/0038-1098(72)90769-7
14. Murakami, S., Iga, H., Naito, H.: Appl. Phys. **80**, 6396–6400 (1996). https://doi.org/10.1063/1.363658
15. Shuichi Murakami, S.M., Hiroyoshi Naito, H.N.: Jpn. J. Appl. Phys. **36**, 2222 (1997). https://doi.org/10.1143/JJAP.36.2222
16. Mada, H., Masahiro Ryuzaki, M.R.: Jpn. J. Appl. Phys. **34**, L1134 (1995). https://doi.org/10.1143/JJAP.34.L1134
17. Hitoshi Mada, H.M., Akira Nishikawa, A.N.: Jpn. J. Appl. Phys. **32**, L1009 (1993). https://doi.org/10.1143/JJAP.32.L1009
18. Samuel, S., et al.: Optical Properties of ZnO Nanoparticles
19. Elastic, K.I.B.: Solid State Commun. **35**, 305–310 (1980). https://doi.org/10.1016/0038-1098(80)90502-5
20. Mishra, S., Manjuladevi, V., Gupta, R.K.: J. Mol. Liq. **386**, 122482 (2023)
21. Johri, M., Saxena, A., Johri, S., Saxena, S., Singh, D.P.: Pramana **76**, 621–628 (2011). https://doi.org/10.1007/S12043-011-0060-0
22. Jonscher, A.K.: Nature **267**, 673–679 (1977). https://doi.org/10.1038/267673a0
23. Barsoukov, E., Macdonald, J.R.: Impedance Spectroscopy. Wiley (2005)
24. Prasad, S.K., Kumar, M.V., Shilpa, T., Yelamaggad, C.V.: RSC Adv. **4**, 4453–4462 (2014). https://doi.org/10.1039/C3RA45761C
25. Gilroy, K.S., Phillips, W.A.: Philos. Mag. B **43**, 735–746 (1981). https://doi.org/10.1080/01418638108222343
26. Patari, S.: Nath A Mater Today Proc **5**, 2105–2111 (2018). https://doi.org/10.1016/j.matpr.2017.09.207
27. Kılıç, M., et al.: Ferroelectrics **505**, 102–113 (2016). https://doi.org/10.1080/00150193.2016.1254586
28. Kamaliya, B., Vijay Kumar, M., Yelamaggad, C.V., Krishna Prasad, S.: Appl. Phys. Lett. (2015). https://doi.org/10.1063/1.4913608
29. Grant, F.A.: J. Appl. Phys. **29**, 76–80 (1958). https://doi.org/10.1063/1.1722949
30. Macdonald, J.R., et al.: Impedance Spectroscopy: Theory, Experiment, and Applications, pp. 424–458. Wiley (2018)
31. Garbovskiy, Y., Glushchenko.: Crystals (Basel) **5**, 501–533 (2015). https://doi.org/10.3390/cryst5040501

A New Perspective into the Optical Properties of Tellurite Glasses Doped with Nd^{3+} Nanoparticles

M. N. Raghavendra[1], J. S. Ashwajeet[1(✉)], K. M. Sandeep[1], and S. O. Manjunatha[2]

[1] Department of Studies in Physics, Davangere University, Shivagangotri, Davanagere, Karnataka 577007, India
ashphysics358@gmail.com

[2] Department of Physics, B.M.S. College of Engineering, Bengluru 560019, India

Abstract. The synthesis of Neodymium nanoparticle (Nd^{3+} NPs) doped borotellurite glasses $40B_2O_3$-(30-x)TeO_2-15Y_2O_3-15Na_2O-xNd_2O_3 (where x = 2, 6, and 10 mol%) was carried out using the conventional melt-quench technique. These glasses were then labelled as TN1, TN2, and TN3, respectively. The sample's glassy structure was verified using X-ray diffraction analysis. The glasses' densities ranged from 3.12 to 3.34 g/cm^3, value was established via the use of Archimedes' principle with toluene as the liquid used for immersion. Furthermore, a negative association was seen between the molar volume and the density. Analysed were the structural characteristics of borotellurite glasses including Nd^{3+} nanoparticle using FTIR. Wavelength dependent UV–visible optical absorption spectra were obtained between the 100 to 900 nm range. To calculate the Urbach energy and optical band gap, Tauc's model was used in conjunction with the Absorption Spectrum Fitting (ASF) and Derivative Absorption Spectrum Fitting (DASF) methodologies. In the optical bandgap values derived from the ASF and DASF approaches, correlation and comparable patterns were seen.

1 Introduction

Glasses are being prominently used in the design of photonic and optoelectronic devices due to their inherent advantages over crystalline materials [1]. The optical characteristics of doped glass systems with rare earth oxide have been extensively explored for applications in lasers, telecommunications, sensors, display devices, fibre amplifiers and waveguide amplifiers for optical transmission network, among other optical devices [2, 3]. The combination of TeO_2 and B_2O_3 is better than silicate and phosphate glasses in a number of ways. Borotellurite glasses are a beneficial compromise between the advantages and disadvantages of B_2O_3 and TeO_2 [4]. Optical devices can utilize borate glasses due to their ability to dissolve rare-earth ions, ease of fabrication, and UV-resistant transparency. Moreover, adding borate oxide to tellurite oxide enhances glass stability and quality, improves transparency and reduces hygroscopic nature [5].

Activated glasses containing trivalent neodymium (Nd^{3+}) are very exciting rare-earth ions for solid-state lighting devices. The optimization of spectrum properties of

P. Joseph D. et al. (Eds.): IEMDST 2024, SPPHY 567, pp. 85–95, 2025.
https://doi.org/10.1007/978-981-96-5863-3_11

rare-earth ions may be accomplished by development of novel host glass materials [6]. The distinctive optical and structural characteristics of neodymium nanoparticles in glass compositions make them tremendously useful in a wide range of sophisticated applications. In addition to its use in colour displays, lighting systems, and fibre-optic communication, neodymium-doped glasses provide adaptable solutions in other industry sectors [7].

The major aim of this study is to produce and examine borotellurite glasses that have been doped with Nd^{3+} nanoparticles. Our study included a thorough analysis of the optical and physical characteristics of newel TN glasses. To attain this goal.

1. Have synthesized three TN glass samples.
2. The density, oxygen packing density (OPD), and molar volume (V_m) have all been studies using the Archimedes method.
3. Computed the physical characteristics of the specified glasses.
4. XRD study verified the amorphous behaviour of the proposed glasses, whereas FTIR investigations established the molecular composition and arrangement of the samples.
5. UV–Visible data was gathered for every pair of glasses between 200 and 800 nm in wavelength.
6. The optical energy bandgaps, Urbach energy and refractive indices have been determined using Tauc's, ASF and DASF techniques.

2 Preparation and Characterization Details

Neodymium (III) oxide nanoparticles (NPs), Tellurium (IV) oxide, Sodium Carbonate, (Na_2CO_3), Yttrium (III) oxide, and Boric acid (H_3BO_3) were used synthesize samples using high-purity chemical powders without additional purification. Conventional melt quenching produced the sample. 15 g of material were mixed in agate mortar, placed in a silica crucible, and heated to 1200 °C for 60 min in an electric furnace for 40 min. a molten sample is poured onto the brass plate and then quenched using a different brass plate during the casting process. Thermal and mechanical stresses of sample were alleviated by subjecting the glasses to annealing at 600 °C for duration of 1 h. After treatment, samples were labelled TN1, TN2, and TN3.

TN glasses in powder state were investigated for their amorphous nature using XRD. A chemical component and structural investigation for current TN glasses was performed using FTIR. Spectral absorbance of the synthesized glasses was measured in the region of 400 to 4000 cm^{-1}. To determine the density Archimedes' principle was used employing Toluene as the immersing liquid. The produced TN glasses were subjected to a UV–Vis. Absorption measurement between 200 and 800 nm in wavelength.

3 Results and Discussion

Figure 1 displays the XRD spectra of TN glass doped with Nd^{3+} nanoparticles. X-ray diffraction spectra of all glass samples exhibited significant diffusion at lower scattering angles. The observed pattern provide evidence of the amorphous nature of the glass samples [8].

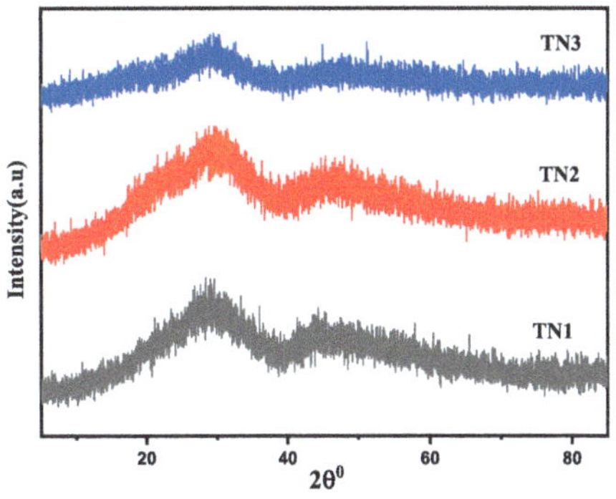

Fig. 1. Nd^{3+} NPs TN glasses-XRD spectra.

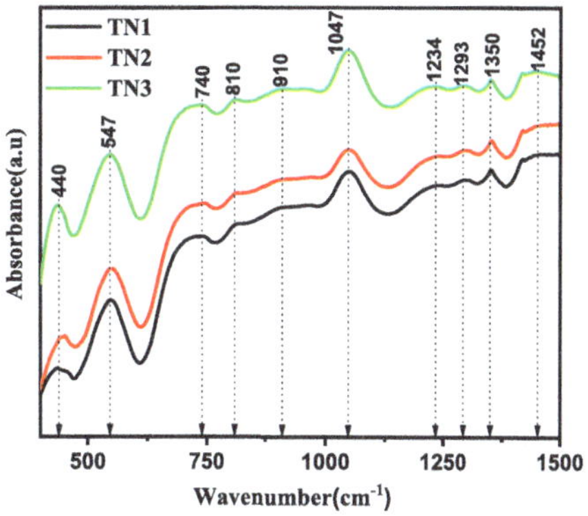

Fig. 2. Nd^{3+} NPs TN glasses-FTIR Spectra

Figure 2 displays the FTIR spectra of glass samples all over the range of 400 to 1500 cm^{-1}. The spectra include many peaks indicating local structure. Typical tellurite oxide structural units are about ~ 540 cm^{-1}. Tellurite oxide forms trigonal pyramidal (TeO_3) and trigonal bipyramidal (TeO_4) structures during glass production. The glass samples display a band representing the trigonal pyramidal TeO_3 structural unit at approximately 720 cm^{-1}. Trigonal pyramidal TeO_3 in glass exhibit non-bridging oxygen atom distribution. Around 800 cm^{-1}, three spectral areas display a band group indicating the existence of bending vibrations caused by different combinations of borate. The band group that was detected at around 1200 cm^{-1} is associated with the stretching vibrations of the tetrahedral BO_4 units. In trigonal BO_3 units, the elongation of B-O bonds leads to a distinct emphasis of the third band group at about 1400 cm^{-1} [8–10]. Using the Archimedes method, density of TN glasses was determined and subsequently calculated.

$$\rho = \frac{w_a}{w_a - w_t} \times \rho_t \tag{1}$$

where W_a represents the weight of the glass in air, W_t represents its weight in toluene, and ρ_t represents the density of the toluene (0.867 g/m^3) [11].

The following equation calculates molar volume (V_m) from density.

$$V_m = \frac{M_{glass}}{\rho_{glass}} \tag{2}$$

where M_{glass} represents the average molecular weight of glass composition, while ρ_{glass} represents sample density [12].

The cation coordination number and oxide packing density (OPD). Its formula is:

$$OPD = 1000 \times \frac{N_o}{M_{glass}} \times \rho \tag{3}$$

M_{glass} represents average molecular weight of the glass sample, whereas N_o denotes the number of oxygen atoms per oxide per unit formula [13].

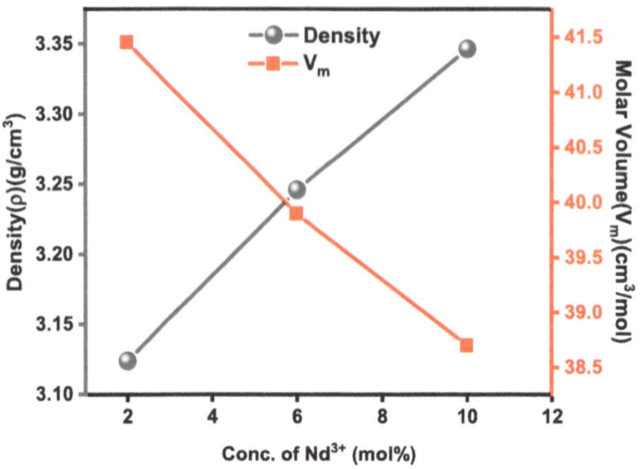

Fig. 3. Density versus concentration of Nd^{3+} with molar volume: TN glasses.

Table 1. Physical parameters: density (ρ), molar volume (V_m) and oxygen packing density (OPD) measurements of produced TN glasses

Glasses	Nd^{3+}(%)	Physical parameters		
		Density (g/cm³)	V_m(cm³/mol)	OPD (g.atm/L)
TN1	2	3.124	41.459	58.372
TN2	4	3.246	39.900	61.654
TN3	6	3.346	38.702	64.596

The association between each glass sample's density and molar volume and the Nd_2O_3 concentration (x wt%) is shown in Fig. 3. The results of the experiment demonstrate that when the concentration of Nd_2O_3 in the samples increases, their density rises and molar volume decreases as presents in Table 1. The observed behaviour may be ascribed to the disparity in the molecular weights of TeO_2 (159.59 g/mol) and Nd_2O_3 (336.47 g/mol) reported in reference [14].

Furthermore, Table 2 lists the physical parameters derived from the density data and refractive index. We analyse structural modifications in glass samples caused by variations in the concentration of Nd^{3+} NPs. The following physical characteristics of TN glasses were determined by analysing densities and refractive indices: The electrical dielectric constant, ionic concentration, polaron radius, inter-nuclear distance, molar polarizability, field strength, reflection loss, optical transmittance, metallization criterion, energy gap, optical basicity, and dielectric constant of TN glasses were calculated using formulas found in the literature [15, 16]. The findings presented in Table 2 show that although Nd^{3+} NPs cause a reduction in molar polarizability, polaron radius, inter-nuclear distance, and molar refraction, they also enhance ionic concentration, field strength, and optical basicity. Electronic polarizability, dielectric constant, reflection loss, electronic susceptibility, and optical dielectric constant indicate a decrease in TN2 and an increase in TN3 with an increase in Nd^{3+} NPs, while optical transmittance, metallization criterion, and energy gap indicate an increase in TN2 and an increase in TN3 with an increase in Nd^{3+} NPs [17].

Table 2. Various physical parameters of TN glasses

Sl. No.	Physical property Parameters	Glasses			Ref.
		TN1	TN2	TN3	
		X = 2	X = 6	X = 10	
1	Ionic concentration N_i ($\times$ 10^{21} ions/cc)	0.290	0.905	1.556	[18]
2	Polaron Radius, $r_p(\times 10^{-8})$	0.320	0.222	0.187	
3	Inter nuclear distance, $r_i(\times 10^{-7})$	0.794	0.550	0.464	[19]
4	Field strength, $F \times 10^{15}$	29.320	60.988	85.732	[20]
5	Molar refractivity, R_m (cm^3/mol)	27.276	26.038	25.535	[21]
6	Electronic polarizability , $\alpha_e\left(\times 10^{-25} cm^3\right)$	2.226	2.203	2.235	–
7	Molar polarizability , $\alpha_m\left(\times 10^{-23} cm^3\right)$	1.081	1.032	1.012	–
8	Reflection losses, R_L (%)	0.141	0.138	0.142	–
9	Optical Transmission, (T_o)	0.753	0.758	0.751	–
10	Dielectric constant, ε	4.847	4.575	4.879	–
11	Metallization criterion (M)	0.342	0.347	0.340	–
12	Optical Basicity (^)	0.778	0.794	0.810	[20]
13	Energy Gap (E_g)	2.340	2.414	2.315	–
14	Electronic susceptibility (χ)	0.306	0.299	0.309	–
15	Optical Dielectric Constant	3.847	3.757	3.879	–

3.1 UV–Vis. Studies

Figure 4 illustrates the absorbance of present TN glasses. The relative concentration of Nd_2O_3(x wt%) plays a significant role in determining the cutoff wavelength for each absorbance spectra. On the other hand, a shift in the absorption edges' positions causes the optical system's energy band gaps to vary. The origin of these phenomena is usually ascribed to the observed fluctuations in the concentration of Nd_2O_3.

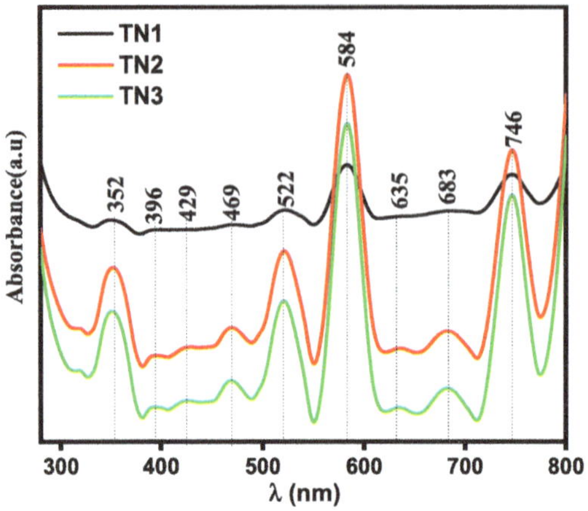

Fig. 4. Absorption spectra of Nd^{3+} NPs TN glasses.

3.2 Optical Energy Band Gap (E_{gap})

Tauc's model and the α(ν) optical absorption coefficient have been used to calculate the optical energy band gap for TN glasses [22]. Mott and Davis revised the optical absorption coefficient α(ν) in Equations [22]. The power (r) is a precise measure of electronic transition characterised by a direct allowed transition at r = 0.5 and an indirect permissible transition at r = 2.0.

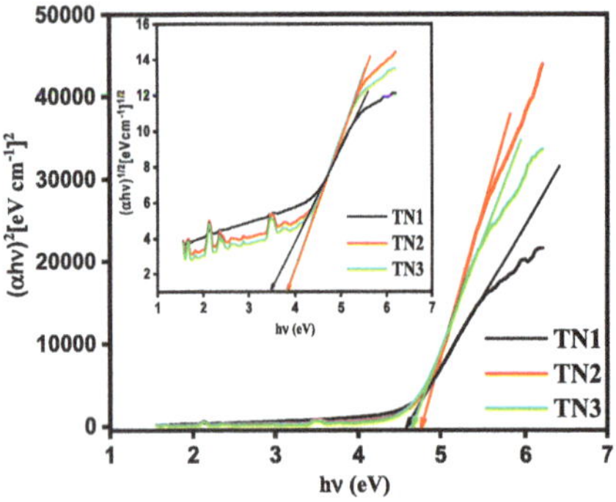

Fig. 5. Variation of $(\alpha h\nu)^2$ (inset $(\alpha h\nu)^{1/2}$) versus (hν) TN glasses.

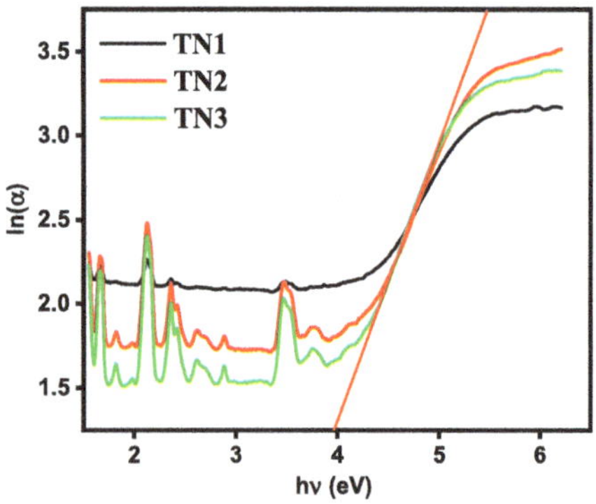

Fig. 6. Variation of ln(α) versus (hν) for TN glasses.

For both direct and indirect allowed transition in the suggested TN glasses, Fig. 5 shows the connection between $(\alpha h\nu)^2$ and $(\alpha h\nu)^{0.5}$ with regard to (hν). To calculate values of $E_{g\ .dir}^{Tauc\acute{s}}$ and $E_{g\ .indir}^{Tauc\acute{s}}$ were extrapolated the straight line that passes over many points on the graphs at the locations where $(\alpha h\nu)^2 = 0$ and $(\alpha h\nu)^{0.5} = 0$, respectively. In Table 3, values of $E_{g\ .dir}^{Tauc\acute{s}}$ and $E_{g\ .indir}^{Tauc\acute{s}}$ were provided. It was shown that the optical band gaps exhibited during direct transition had bigger diameters compared to those found during indirect transition. Moreover, it was observed that there was a considerable degree of similarity between the energy band gaps in each transition, indicating that the addition of Nd_2O_3 to glass samples had no effect on the optical band gaps [23].

Under low photon energy conditions, the absorption coefficient (α) is governed by Urbach's empirical rule, as expressed in Eq. (4),

$$\alpha = \alpha_0 \exp\left(\frac{h\nu}{E_u}\right) \quad (4)$$

where E_u and α_o stand for Urbach's energy and a constant, respectively. In low-crystalline or disordered materials, These numbers are often used to measure the magnitude of the band tail that generates from localized states within the band gap. One may rephrase empirical Eq. (4) as,

$$\ln \alpha = \ln \alpha_0 + \left(\frac{h\nu}{E_u}\right) \quad (5)$$

Therefore, the Urbach's energy (E_u) of Urbach may be determined by computing the gradient of the linear correlation between ln(α) and (hν). Figure 6 depicts correlation between the slope of the linear function (α) and the energy of photons (hν) for TN glasses, as shown in Table 3. The data suggest that the glasses produced have the lowest Urbach's energy, suggesting the existence of very stable and consistent glasses with few flaws and a decrease in disorder within the glass samples that were made [24].

3.3 ASF Method, $\left(E_g^{ASF}\right)$

A quantitative representation of the α(ν) as a function of photon wavelength (λ) has been proposed by Alarcon and colleagues [25] and Souri and Tahan [26]. The wavelength cutoff values for Planck's constant, light velocity, and the optical energy gap are denoted as h, c, and λ_g respectively [27]. The ASF approach, as indicated in Equation [22], allows the optical energy band gap to be determined only from the absorbance measurements, without taking the glass sample's thickness into account. Furthermore, optical energy band gap within the ASF model can be readily determined by using Eq. (6) and calculating λ_g

$$E_{gap}^{ASF} = \frac{hc}{\lambda_g} = 1239.83 \times \lambda_g^{-1} \quad (6)$$

Figures 7 and 8 illustrate the variations in both direct and indirect transitions made possible by the proposed TN glasses [27]. Using extrapolation, the values of $E_{g.dir}^{ASF}$ and $E_{g.indir}^{ASF}$ were determined by identifying the straight area that passes over a significant number of points on the generated plots at the points where $(A/\lambda)^2 = 0$ and $(A/\lambda)^{1/2} = 0$ respectively. In **Table 3**, the values of $E_{g.dir}^{ASF}$ and $E_{g.indir}^{ASF}$ are shown. Both Tauc's and ASF's approaches yielded very consistent outcomes [25].

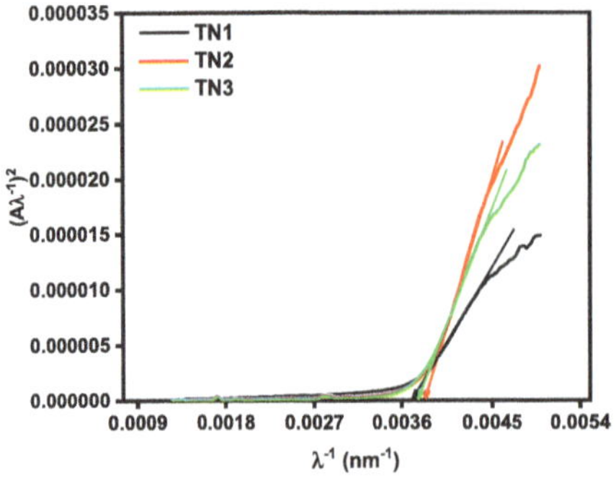

Fig. 7. Variation of $(A/\lambda)^2$ vs λ^{-1} for Nd^{3+} NPs doped TN glasses.

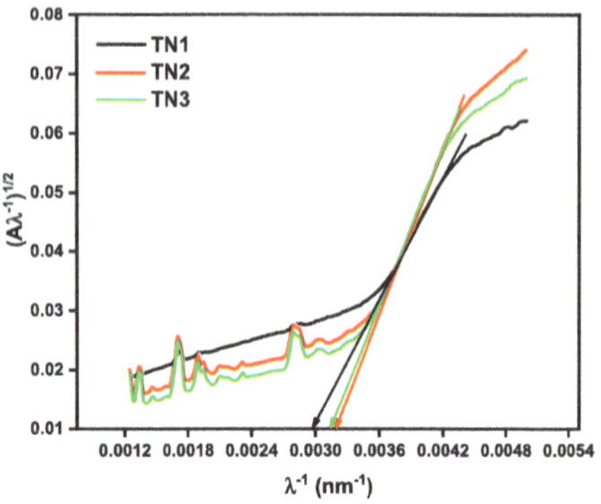

Fig. 8. Variation of (A/λ)^1/2 vs λ^−1 for Nd3+ NPs doped TN glasses.

3.4 ASF Method, $\left(E_g^{DASF}\right)$

The optical energy band gaps of the synthesised glasses were determined by DASF model. According to equation [22] where λ_g is the optical energy gap E_g^{DASF} calculate by,

$$E_g^{DASF} = \frac{hc}{\lambda_g} = 1239.83 \times \lambda_g^{-1} \tag{7}$$

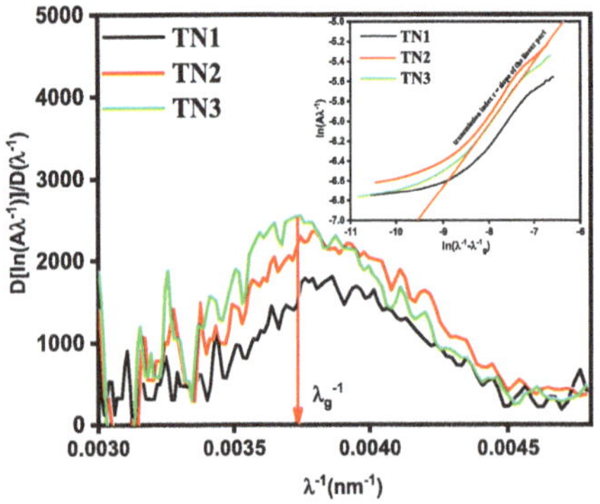

Fig. 9. Plot of d{ln[A(λ)/λ]}/d(1/λ) (inset ln $(A\lambda^{-1}$ vs $\ln(\lambda^{-1}$-$\lambda_g{}^{-1}))$ on (λ^{-1}) for Nd^{3+} NPs doped TN glasses. (DASF)

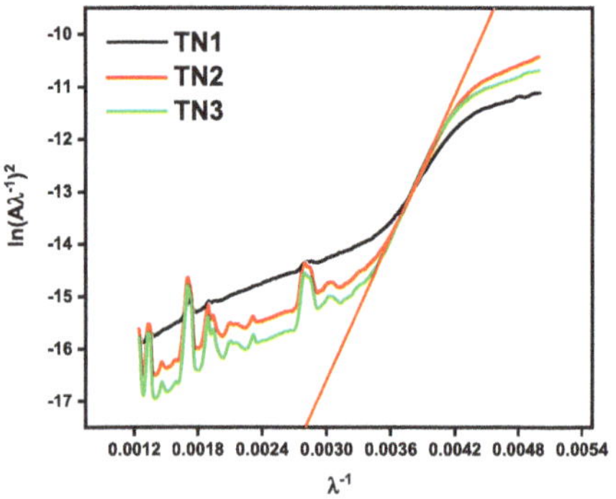

Fig. 10. $DASF_{urb}$ plot of $\ln(A\lambda^{-1})^2$ vs λ^{-1} for TN glasses.

Figure 9 illustrates the relationship between the derivative of TN glasses. The computed values of E_g^{DASF} were derived using Eq. (7) and reported in Table 3. The findings obtained from the evaluation of the three tests Tauc's, ASF, and DASF show a significant correlation with each other. The plot's linear section's slope, represented as ln((A(λ))/λ), may be compared to $\ln(1/\lambda - 1/\lambda_g)$ to get the r-value, as shown in the inset picture [28]. This enables for the determination of the precise characteristics of the charge carrier optical transition. Table 3 lists 'r' as the slope of the fitted linear part. It is approximately equal to 0.5 for all samples, indicating a direct transition. Figure 10 displays the dependence of $\ln(A\lambda^{-1})^2$ on λ^{-1} and shows the $DASF_{urb}$. The data show a strong correlation with Tauc's Urbach [29].

Table 3. Optical bandgaps, Refractive indices (n), Urbach energy (E_u) and Charge carrier optical transition (r) of present TN glasses.

Glasses	Tauc's					ASF				DASF			
	E_{dir}^{Tauc}	n	E_{indir}^{Tauc}	n	E_{urb}^{Tauc}	E_{dir}^{ASF}	n	E_{indir}^{ASF}	n	E_{Otp}^{DASF}	n	E_{urb}^{DASF}	r
TN1	4.580	2.066	3.840	2.201	0.903	4.605	2.062	4.104	2.150	4.488	2.082	0.385	0.61
TN2	4.763	2.037	3.943	2.181	0.665	4.762	2.037	3.930	2.184	4.636	2.057	0.295	0.65
TN3	4.652	2.054	3.804	2.209	0.604	4.680	2.050	3.905	2.188	4.562	2.069	0.269	0.63
Ref	[30]	–	–	–	–	–	–	[31]	[31]	–	–	–	[31]

This study included the calculation of the refractive index factor (n) for all types of glasses. Through the use of Eq. [12], We used Tauc's, ASF, and DASF techniques to compute the refractive index for permissible transitions that were both direct and indirect. The samples presented in Table 3 have their 'n' values derived from the current study. An inverse relationship was observed among the refractive indices and the optical energy band gaps of the analysed materials. Additionally, all of the glasses showed high n values, suggesting that they might be useful materials for optical and photoelectronic filter systems [27].

4 Conclusion

The major objective of this work is to analyse the physical and optical characteristics of borotellurite glasses that include Nd^{3+} nanoparticles. Presented below are the many characteristics of the outcome:

1. XRD verified the amorphous behaviour of the TN glasses, while Fourier Transform Infrared (FTIR) investigations established the molecular composition and structure of the sample.
2. Our analysis revealed that the density of the glass rose from 3.12 to 3.34 gm/cm^3, the molar volume (V_m) declined from 41.45 to 38.70 cm^3/mol, and the OPD increased from 58.37 to 64.59 gm/cm^3.
3. In Tauc's approach, the indirect optical energy band gap ranges from 3.80 eV to 3.94 eV, but in the ASF method and DASF, it varies from 3.90 eV to 4.10 eV and 4.48 eV to 4.63 eV.
4. The slope of the fitted linear segment 'r' is around 0.5 for all samples, suggesting a precise and direct transition.

5. The refractive index of TN3 glass was around 2.21. The verification of this discovery confirms that the glasses developed possess the capability to serve as efficient materials for photoelectronic and optical filter devices.

Acknowledgments. The authors like gratefully acknowledge the UGC-DAE-CSR located in Kolkata for their provision of characterization facilities.

References

1. Siva Rama Krishna Reddy, K., et al.: Opt. Mater. (Amst). **79**, 21–32 (2018). https://doi.org/10.1016/j.optmat.2018.03.005
2. Ravi, O., Reddy, C.M., Reddy, B.S., Raju, B.D.P.: Opt. Commun. **312**, 263–268 (2014)
3. Mahraz, Z.A.S., Sahar, M.R., Ghoshal, S.K., Dousti, M.R., Amjad, R.J.: Mater. Lett. **112**, 136–138 (2013)
4. Alzahrani, J.S., Alothman, M.A., Eke, C., Al-Ghamdi, H., Aloraini, D.A., Al-Buriahi, M.S.: Comput. Mater. Sci. **196**, 110566 (2021)
5. Jha, A., et al.: Prog. Mater. Sci. **57**, 1426–1491 (2012)
6. Ramachari, D., Rama Moorthy, L., Jayasankar, C.K.: Infrared Phys. Technol. **67**, 555–559 (2014). https://doi.org/10.1016/j.infrared.2014.09.020
7. Azlan, M.N., Halimah, M.K.: Res. Phys. **11**, 58–64 (2018). https://doi.org/10.1016/j.rinp.2018.08.017
8. Walia, T., Singh, K.: J. Non Cryst. Solids **564**, 120812 (2021). https://doi.org/10.1016/j.jnoncrysol.2021.120812
9. Dehelean, A., Rada, S., Popa, A., Culea, E.: J. Mol. Struct. **1036**, 203–208 (2013)
10. Al-Ani, S.K.J., Al-Rawi, S.S., Jassim, A.H., Al-Hilli, H.A.: Iraqi J. Appl. Phys. **2**, 23–25 (2006)
11. Madhu, A., Eraiah, B., Manasa, P., Srinatha, N.: Opt. Mater. (Amst). **75**, 357–366 (2018). https://doi.org/10.1016/j.optmat.2017.10.037
12. Saritha, D., Salagram, M., Bhikshamaiah, G.: Ser. Mater. Sci. Eng. **2**, 12057 (2009)
13. Yahya, N., et al.: Malaysian J. Anal. Sci. **22**, 296–302 (2018)
14. Kaur, A., Khanna, A., Aleksandrov, L.I.: J. Non Cryst. Solids **476**, 67–74 (2017). https://doi.org/10.1016/j.jnoncrysol.2017.09.025
15. Kreidl, N.J.: Glas. Berichte **62**, 213–218 (1989)
16. Duffy, J.A.: J. Non Cryst. Solids **297**, 275–284 (2002)
17. Rao, A.S., Ahammed, Y.N., Reddy, R.R., Rao, T.V.R.: Opt. Mater. (Amst). **10**, 245–252 (1998). https://doi.org/10.1016/S0925-3467(97)00055-4
18. Babu, M.R., Madhusudhana Rao, N., Mohan Babu, A.: Luminescence **33**, 71–78 (2018)
19. Mariselvam, K., Kumar, R.A., Manasa, P.: Infrared Phys. Technol. **91**, 18–26 (2018)
20. Siva Rama Krishna Reddy, K., Swapna, K., Mahamuda, S., Venkateswarlu, M., Rao, A.S., Vijaya Prakash, G.: Opt. Mater. (Amst.) **85**, 200–210 (2018). https://doi.org/10.1016/j.optmat.2018.08.057
21. Bhatia, B., Meena, S.L., Parihar, V., Poonia, M.: NJGC **05**, 44–52 (2015)
22. Rao, A.S., et al.: Phys. B Condens. Matter **404**, 3717–3721 (2009)
23. Sreedhar, V.B., Doddoji, R., Kumar, K.K., Reddy, V.R.M.: J. Non Cryst. Solids **553**, 120521 (2021)
24. Halimah, M.K., Awshah, A.A., Hamza, A.M., Chan, K.T., Umar, S.A., Alazoumi, S.H.: J. Mater. Sci. Mater. Electron. **31**, 3785–3794 (2020)

25. Escobar-Alarcón, L., Arrieta, A., Camps, E., Muhl, S., Rodil, S., Vigueras-Santiago, E.: Appl. Surf. Sci. **254**, 412–415 (2007). https://doi.org/10.1016/j.apsusc.2007.07.052
26. Souri, D., Tahan, Z.E.: Appl. Phys. B **119**, 273–279 (2015)
27. Rammah, Y.S., Mahmoud, K.A., Sayyed, M.I., El-Agawany, F.I., El-Mallawany, R.: J. Non Cryst. Solids **534**, 119944 (2020). https://doi.org/10.1016/j.jnoncrysol.2020.119944
28. Souri, D., Khezripour, A.R., Molaei, M., Karimipour, M.: Curr. Appl. Phys. **17**, 41–46 (2017). https://doi.org/10.1016/j.cap.2016.10.008
29. Souri, D., Sarfehjou, M., Khezripour, A.R.: J. Mater. Sci. Mater. Electron. **29**, 3411–3422 (2018). https://doi.org/10.1007/s10854-017-8276-5
30. Yamusa, Y.A., Hussin, R., Shamsuri, W.N.W., Tanko, Y.A., Jupri, S.A.: Optik **164**, 324–334 (2018)
31. Bulus, I., Hussin, R., Ghoshal, S.K., Tamuri, A.R., Danmallam, I.M., Yamusa, Y.A.: J. Non Cryst. SolidsCryst. Solids **534**, 119949 (2020). https://doi.org/10.1016/j.jnoncrysol.2020.119949

A Comparative Study of Phase and Optical Analysis of Nd^{3+}:Y_2O_3 and Sm^{3+}:Y_2O_3

S. Surya[1], S. Gokul Raj[1(✉)], and G. Ramesh Kumar[2]

[1] Department of Physics, Pondicherry University, Kalapet, Puducherry 605014, India
gokulraj@pondiuni.ac.in

[2] Department of Science and Humanities, University College of Engineering Arni, Anna University Chennai, Thatchur, Arni 632326, India

Abstract. The present investigation focuses on rare earth doped yttrium oxide (Y_2O_3). Neodymium (Nd), and Samarium (Sm) were used as dopant in the host yttrium oxide in the nano crystalline form. Simple coprecipitation technique was adopted for the preparation of nano crystalline, rare earth doped yttria. The dopant was maintained at 5% for both Nd^{3+} and Sm^{3+}. The synthesized samples were the calcinated at 750 °C to obtain highly crystalline Nd:Y_2O_3 and Sm:Y_2O_3. Powder X-ray diffraction analysis reveals that the synthesized sample sizes were found 10–30 nm and the particle morphology was studied using SEM. Strain analysis was calculated using Williamson–Hall (W–H plot). Optical properties was investigated using UV–Vis spectroscopy and the obtained results shows that the synthesized nanocrystalline Nd:Y_2O_3 and Sm:Y_2O have high band gap energy.

1 Introduction

The thermodynamically stable compound yttrium oxide commonly known as yttria (Y_2O_3) is a luminescent material which has high application in the field of Bio-imaging as a contrast agent [1]. Yttrium oxide have notable properties such as well-defined crystal and ionic structures, thermal stability, high dielectric constant, low phonon energy, stable chemical properties, etc. [2]. The nano particles of these rare-earth (RE) ions doped metal oxides have huge impact in the application of optical devices, luminescent biosensors, bioimaging probe and radiation detectors [3]. Nowadays due to the non-toxic and up-conversion property the RE doped yttrium oxide nanoparticles are being used as the substitute to CdSe quantum dots [4]. In recent decade researchers show wide interest on investigating the properties of Nd^{3+} doped Y_2O_3 nanoparticles which finds potential applications in optical and biological fields. Among all the established deep biological media materials that has high luminescent thermometry, these Nd^{3+}:Y_2O_3 is ruling over the area of this research because of its legacy in physical, chemical, optical and thermal properties, low toxicity, high reproducibility, sensitivity, and high damage threshold [5–7]. Also, they have strong absorption and emission lines in the range of 600–650 nm and 1000 to 1350 nm and it suggest that this interval of wavelength leads to the biological window (BW I and II) which mean it has interaction with the biological tissue [8]. The other rare earth dopant used for comparison was samarium in which

P. Joseph D. et al. (Eds.): IEMDST 2024, SPPHY 567, pp. 96–101, 2025.
https://doi.org/10.1007/978-981-96-5863-3_12

Sm^{3+} ions are doped with yttria (Sm^{3+}:Y_2O_3) so as to examine the optical property. This rare earth was choose as a dopant because in the biological window it has a peculiar characteristic of orange-red emission [9]. In this present work the phase analysis using XRD and the optical properties using UV-Vis analysis was performed, as both are good optically active material and the observed results were compared.

2 Materials and Methods

To synthesize the Nd:Y_2O_3 (NYO) and Sm:Y_2O_3 (SYO) nanoparticles, co-Precipitation method was adopted. On using deionised water as solvent, the aqueous solutions of $Y(NO_3)_3{\cdot}6H_2O$ and 5% of dopant material $Nd(NO_3)_3{\cdot}6H_2O$ had been mixed in the stoichiometric ratio. This precursor solution was stirred for 6 h to attain the homogeneity. As the pH is adjusted and maintained by adding few drops of NH_4OH, the solution starts to get precipitate at this point as metal hydroxide. The precipitate was then filtered and dried at 90 °C for 3 h. The obtained fine powders of dried Nd:Y_2O_3 was annealed at 750 °C. Similarly to synthesize Sm:Y_2O_3 (SYO) nanoparticles, Samarium was used as dopant. Now the stoichiometric aqueous solutions of $Y(NO_3)_3{\cdot}6H_2O$ and 5% of dopant material $Sm(NO_3)_3{\cdot}6H_2O$ was taken and got dissolved in the deionised water. The homogeneity of the mother solution was achieved by stirring it for 6 h. The pH got balanced by adding few drops of NH_4OH and at an instance the mother solution starts to get precipitate. This precipitate is usually metal hydroxide. Then it should be filtered, dried and annealed to form the expected crystalline metal oxide. The Sm:Y_2O_3 (SYO) sample was annealed at 750 °C.

The phase formation and crystallinity of the synthesized nano powders were confirmed from the PXRD pattern obtained from the Powder X-Ray Diffractometer with Cu-Kα radiation of wavelength 1.5406 nm in the range of 10° to 80°at a scan rate of 2°/min with a step size of 0.01. This data gives the structural details. The spectrophotometer shows the optical spectra analysis which includes the absorption spectrum in the range of 200–700 nm and the optical band gap energy was plotted using Tauc's plot. SEM analysis gives the data about the surface morphology on a scale of 20 μm.

3 Results and Discussion

3.1 XRD Analysis

The XRD patterns and intense peak shifts of the synthesized Nd:Y_2O_3 and Sm:Y_2O_3 which was annealed at 750 °C was shown in Fig. 1(a and b) respectively. The observed patterns are in good match with the JCPDS card number 41-1105. The analysis and comparison were made with the XRD patterns of NYO and SYO samples.

The single-phase cubic crystal system with Ia3 (206) as space group was confirmed [10]. The lattice constant, cell volume and crystallite size were explained in Table 1. The crystallite size was calculated using Debye Scherrer formula $D = K\lambda/\beta\cos\theta$, the average crystallite size was 9–15 nm. In Fig. 1b, the intense peak got shifted to lower diffraction angle which shows that there are structural modifications and the incorporation of Nd^{3+} and Sm^{3+} ions into Y_2O_3 [11]. From Table 1 the lattice constant for the NYO and SYO

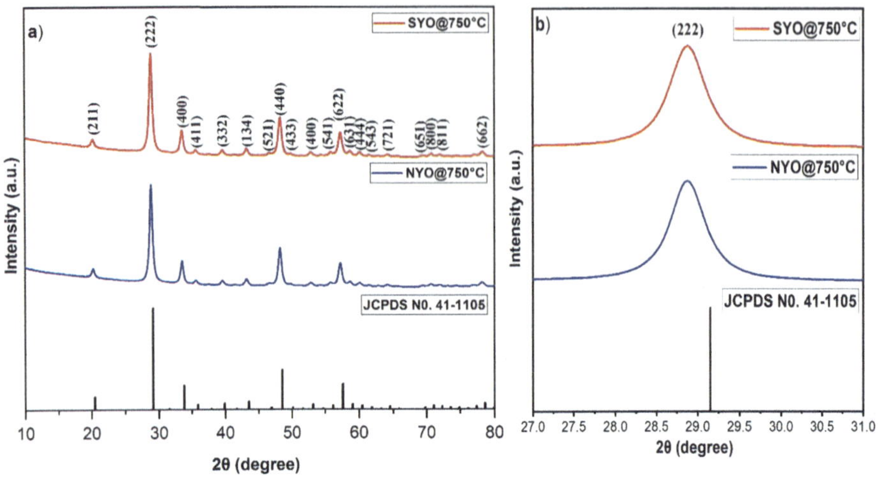

Fig. 1. (a) XRD pattern and (b) Peak shifts of NYO and SYO.

is higher than the lattice constant of reported pure Y_2O_3, and no secondary peaks of Nd and Sm are observed which is the indication of integration of Nd and Sm into the crystal lattice of Y_2O_3 [12]. The micro-strain (ε) of NYO and SYO was calculated from the Williamson–Hall (W–H) plot shown in Fig. 2a and 2b respectively. The micro-strain was found to be 8.68413×10^{-4} for NYO and 6.76327×10^{-4} for SYO.

Table 1. Comparison of lattice parameters, cell volume and crystallite size of NYO and SYO.

Sample code	Lattice Constant (Å)	Cell Volume	Crystallite size (nm)
NYO	10.697	1224.012	11 ± 1
SYO	10.697	1224.012	13 ± 1

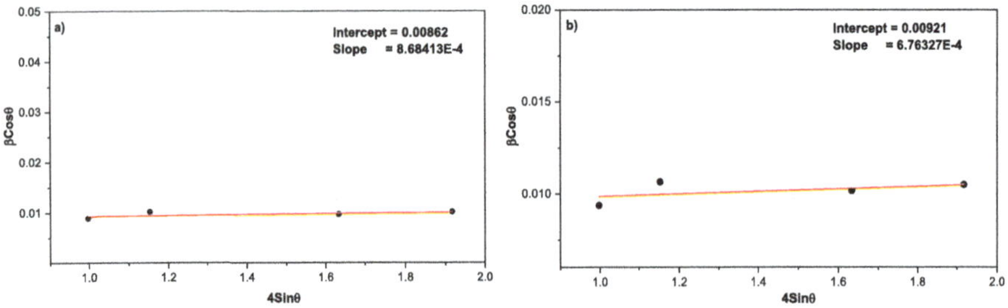

Fig. 2. W–H plot of a) NYO and b) SYO samples.

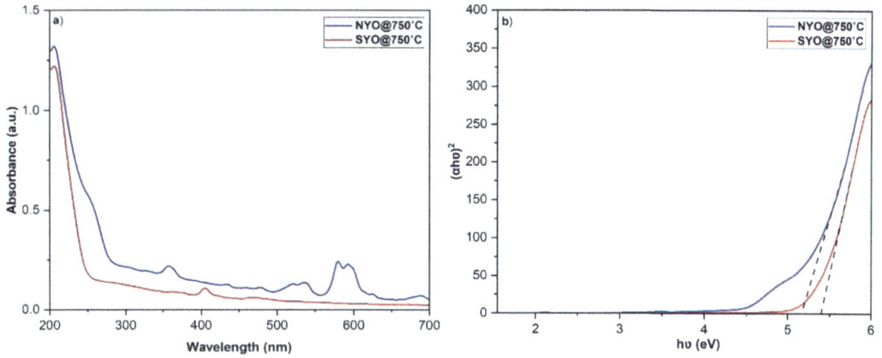

Fig. 3. a) Absorption spectrum and b) Tauc's plot of NYO and SYO.

3.2 UV–Visible Spectroscopy

Figure 3a and 3b is the absorption spectrum and Tauc' plot respectively. The absorption peaks are at 357, 580 and 592 nm for NYO and 404 nm for SYO. For both the samples the band edge absorption is at 245 nm and intense absorption peak in recorded below 250 nm. This peak contributes to the transition of Y_2O_3 from ground state by the creation of valance positions [13]. In the case of SYO no prominent peaks are found in the visible region. Figure 3b is the optical band gap energy arrived from Tauc's Plot from the relation between incident photon and the absorption coefficient equation $(\alpha h\nu)^2 = A(h\nu - E_g)$ plotted between (hv) vs $(\alpha h\nu)^2$. The energy band gap (Eg) for NYO and SYO is 5.1 eV and 5.4 eV respectively which is in good agreement with the reported observations. The reduction in energy band gap is because of the expansion in the strain that got induced in the lattice [14].

3.3 SEM Analysis

Scanning electron microscope (SEM) analysis was carried out to determine the surface morphology of the samarium and neodymium doped yttrium oxide. The samples were doped at 5% and was calcinated to 750 °C which was used for further analysis. ZEISS FE-SEM instrument fitted with EDAX was used measured and the images obtained neodymium and samarium doped yttria is shown in Fig. 4(a and b). It can be seen that the particles have agglomerated shapes when the magnification is at 20 mm. Moreover, in the image it can be seen that there are some spherical shaped agglomerated particles besides the larger particles.

Fig. 4. SEM images of a) NYO and b) SYO.

4 Conclusion

The nanosized yttria (Y_2O_3) doped with Nd^{3+} and Sm^{3+} was synthesized using co-precipitation method and annealed at 750 °C and the comparison between the two rare earth elements (REE) doped with yttrium oxide was analysed. The crystal structure and the single-phase formation of the samples was confirmed using the XRD analysis and the acquired average crystallite size was noted to be 10–30 nm. The UV–Vis analysis gives the optical properties, it shows that yttria with both the dopants makes the nanoparticles to have good NLO property and their energy band gaps were discussed. SEM images state the surface morphology. Further investigations are under progress and these preliminary investigations reveals that both the dopants show negligibly good outcomes for luminescent applications.

References

1. Chiechio, R.M., et al.: Chemosensors **11**(1), 20 (2022)
2. Arghavan, M.M., Sabouri-Dodaran, A.A., Ghamsari, M.S.: Heliyon **10**, 4 (2024). https://doi.org/10.1016/j.heliyon.2024.e25864
3. Gomes, M.A., Brandão-Silva, A.C., Avila, J.F., Alencar, M.A., Rodrigues, J.J., Jr., Macedo, Z.S.: J. Lumin. **200**, 43–49 (2018). https://doi.org/10.1016/j.jlumin.2018.04.004
4. Srinivasan, R., Yogamalar, R., Vinu, A., Ariga, K., Bose, A.C.: J. Nanosci. Nanotechnol. **9**(11), 6747–6752 (2009). https://doi.org/10.1166/jnn.2009.1467
5. Cesaria, M., et al.: Communication and Energy Conversion, pp. 273–284. Springer Netherlands (2015)
6. Saad, A.M., et al.: Braz. J. Phys. **53**(4), 108 (2023)
7. Saravanan, T., Raj, S.G., Chandar, N.R.K., Jayavel, R.: J. Nanosci. Nanotechnol. **15**(6), 4353–4357 (2015)
8. Laia, A.S., et al.: Opt. Mater. **142**, 114126 (2023). https://doi.org/10.1016/j.optmat.2023.114126
9. Devi, H.F., Singh, T.D.: Opt. Commun. **439**, 34–37 (2019). https://doi.org/10.1016/j.optcom.2019.01.045
10. Kumar, S., et al.: J. Mol. Struct. **1302**, 137463 (2024). https://doi.org/10.1016/j.molstruc.2023.137463
11. Zhang, L., Pan, W.: J. Am. Ceram. Soc. **98**(10), 3326–3331 (2015)

12. Goswami, B., Rai, N., Ahlawat, R.: J. Alloy. Compd. **730**, 450–457 (2018). https://doi.org/10.1016/j.jallcom.2017.09.269
13. Kruk, A.: Materials **13**(21), 4928 (2020). https://doi.org/10.3390/ma13214928
14. Sukul, P.P., Mahata, M.K., Kumar, K.: J. Lumin. **185**, 92–98 (2017). https://doi.org/10.1016/j.jlumin.2017.01.008

Phase Optimization and Structural Analysis of Mechanochemically Synthesized Tin Sulphide Nanoparticles

Bovaz Livingstone Baby[1], D. Paul Joseph[2], Agnes Ditto[3,4], D. Bharathi Mohan[3], and Benjamin Hudson Baby[2,3](✉)

[1] Department of Physics, Catholicate College, Pathanamthitta, Kerala 689645, India
[2] Department of Physics, National Institute of Technology, Warangal, Telangana 506004, India
benjaminhudsonbaby@gmail.com
[3] Department of Physics, Pondicherry University, Puducherry 605014, India
[4] Department of Materials Science and Engineering, Indian Institute of Technology, Kanpur 208016, India

Abstract. This report is on the synthesis of sulphur rich SnS phase using an inexpensive non-vacuum based mechanochemical method in which Sn and S powder are mechanically ground with subsequent heat treatment in oven. The optimized heating rate and temperature for the stabilization of SnS phase without any oxide phases is investigated using Raman and XRD analyses. Chemical structure analysis revealed the formation of sulphur rich phase of SnS without the formation of additional oxide phases for sample annealed at a low heating rate of 75 °C/hr. The absence of any XRD peak corresponds to pure tin and sulphur indicated the complete sulphurization of Sn atoms and the indexed peaks show the resultant SnS phase to be in orthorhombic crystal structure. Crystallite size and strain of the sample are calculated using Williamson-Hall analysis using uniform deformation model (UDM), uniform stress deformation model (USDM), uniform deformation energy density model (UDEDM). Surface morphology, elemental composition analysis and band gap calculation revealed the stabilization of sulphur rich spherical shaped SnS nanoparticles of 300 nm size having 1.38 eV of band gap.

1 Introduction

Tin (II) sulphide, a promising thin film solar cell absorber layer material is explored intensely due to its environmental abundance and acceptability as well as its ability of optoelectrical properties similar to other prominent absorber materials such as CIGS, CZTS and CdTe. SnS is exhibiting a band gap in the range of 1.3 to 1.7 eV which is near to the optimal value of 1.5 eV, high absorption coefficient ($\alpha > 10^4$ cm^{-1}), optimal carrier concentration ($\rho = 10^{15}$ to 10^{18} cm^{-3}) and has a theoretical efficiency of 24% [1–3].

Towards any application of large scale synthesis of the nanoparticles is an essential step and mechanical alloying using high energy ball milling is one of the easiest way. However, it is very challenging to avoid oxidation during these milling process. L. Yang

P. Joseph D. et al. (Eds.): IEMDST 2024, SPPHY 567, pp. 102–108, 2025.
https://doi.org/10.1007/978-981-96-5863-3_13

et al. reported the ball milling synthesis of SnS for lithium ion battery applications and observed the co-existence of Sn, S and various tin sulphide phases [4]. Here, we followed a non-vacuum based hot air oven assisted mechanochemical method for the synthesis of SnS and investigated the optimized heating rate for the complete sulphurization of Sn without any oxide phases.

2 Materials and Methods

Using an agate mortar and pestle, high pure Sn and S powder (1:1) were mixed for 30 min. The mixture was then heat treated at 240 °C for 15 min in a hot air oven, with heating rates of 75 and 90 °C per hour. An indication of the production of tin sulphide phases was the formation of a dark brown powder which was achieved after five cycles of hand grinding followed by heat treatment. X-ray diffraction (XRD) (Panalytical, X'Pert PRO) analysis was used to investigate the crystal structure and confocal Raman Spectrometer (Reinshaw inVia Raman Microscope, UK) with an excitation wavelength of 532 nm was used to investigate the chemical structure of the stabilized samples. SEM and EDAX using VEGA 3 and TESCAN were utilized to explore the morphology and elemental composition respectively. The UV-Vis-NIR spectrophotometer (Shimadzu UV-3600 Plus) with diffuse reflectance mode was used to study the optical properties.

3 Results and Discussion

The phase purity of the prepared sample devoid of any secondary phases such as SnS_2, Sn_2S_3, SnO, and SnO_2 is an important factor to consider SnS as an effective absorber layer in thin-film based solar cells. Confocal Raman spectrometer is used to examine the phase purity of the obtained samples. Raman spectra revealed the occurrence of mixed phases of Sn_2S_3 (154 and 186 cm^{-1}), SnO_2 (247, 438, and 473 cm^{-1}) and SnS (219 cm^{-1}) for the sample heated at a rate of 90 °C/hr (Fig. 1(a)) [6–8]. On the other hand, SnO_2 phase is completely suppressed with a slightly low rate of 75 °C/hr indicating the complete sulfurization of Sn without any oxide phases. The occurrence of the low formation energy Sn_2S_3 along with tin oxide phase is owed to the inhomogeneous sulphurization of Sn resulting from a high heating rate of 90 °C/hr. Similarly, increasing the Sn to S initial precursor weight ratio can help to prevent the resultant Raman peak of sulphur-rich SnS_2 phase at 315 cm^{-1} [5, 9].

Figure 1(b) shows the XRD pattern of the samples in the range of 20° to 60°. All the diffraction peaks are indexed after comparing with the JCPDS file No. 00-039-0354 for SnS, 01-088-0287 for SnO_2 and 01-072-0031 for Sn_2S_3 phases. XRD analysis reveals the formation of SnO_2 (110, 002 and 012) and Sn_2S_3 (220 and 121) as secondary phases for powders treated at 90 °C/hr and SnS_2 (100) as a secondary phase for powders treated at 75 °C/hr along with orthorhombic SnS as observed from Raman analysis. The absence of any diffraction peaks corresponding to elemental Sn and S confirmed the complete sulphurization of the precursor. The lattice constants are calculated using Eq. (1), Dislocation density (δ), which is related to the crystal growth defects can be calculated Eq. (4). The change in the FWHM values can be correlated to crystallite size

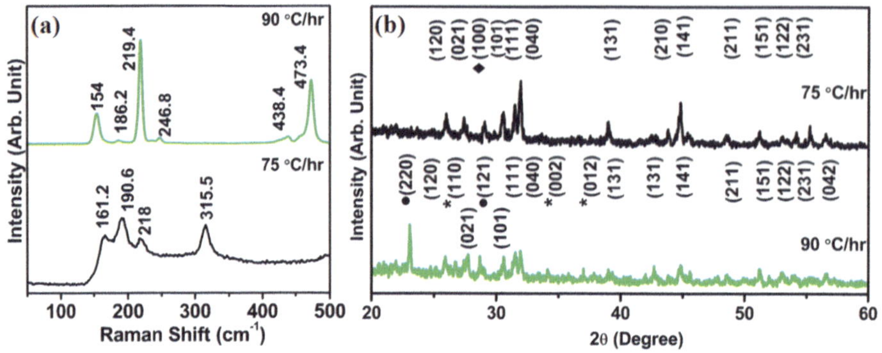

Fig. 1. (a) Raman spectra of samples in the range of 50–500 cm^{-1} with an excitation of 532 nm wavelength and (b) XRD pattern of the powders in the two theta range of 20°–60°. '◆' correspond to SnS_2, '●' correspond to Sn_2S_3 and '*' correspond to SnO_2 phases.

(D) and strain (ε) by making use of Eqs. (2) and (3) respectively and the calculated values are given in Table 1.

$$\frac{1}{d^2} = \frac{h^2}{a^2} + \frac{k^2}{b^2} + \frac{l^2}{c^2} \tag{1}$$

$$D = \frac{k\lambda}{\beta \cos\theta} \tag{2}$$

$$\varepsilon = \frac{\beta \cos\theta}{4} \tag{3}$$

$$\delta = \frac{1}{D^2} \tag{4}$$

The crystallite size and strain can also be calculated using Williamson-Hall equation by following three different models such as uniform deformation model (UDM), uniform stress deformation model (USDM) and uniform deformation energy density model (UDEDM). Williamson-Hall equation is given by Eq. 5, in which strain is considered as uniform in all crystallographic directions [10–14]. In UDM, graph drawn with 4sinθ along the x-axis and βcosθ along y-axis and the slope and y intercept values shows the average crystallite size (D) and micro-strain (ε) respectively as shown in Fig. 2.

$$\beta \cos\theta = \frac{k\lambda}{D} + 4\varepsilon \sin\theta \tag{5}$$

In the uniform stress deformation model (USDM), the lattice deformation stress (σ) is considered as isotropic and micro-strain (ε) and energy density are anisotropic [11–14]. In USDM, Williamson-Hall equation is modified with Hooke's law (σ = Yε) as,

$$\frac{\beta \cos\theta}{\lambda} = \frac{k}{D} + \frac{4\sigma \sin\theta}{\lambda Y_{hkl}} \tag{6}$$

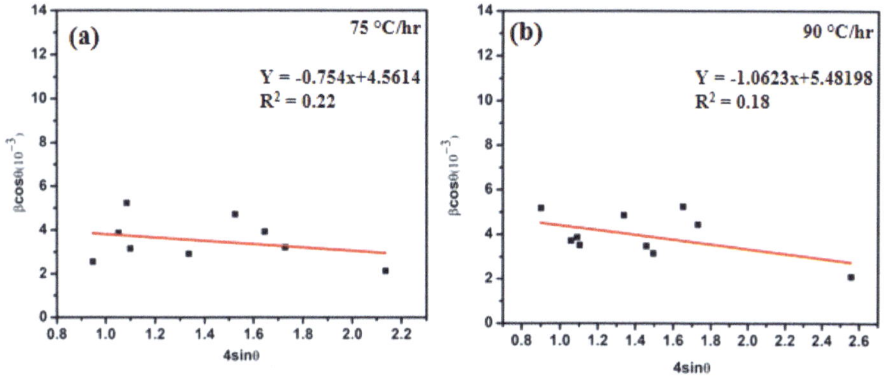

Fig. 2. The UDM plot between (β cosθ) vs (4 sinθ) for samples prepared with heating rate of (a) 75 °C/hr and (b) 90 °C/hr.

where Y_{hkl} is young's modulus and for an orthorhombic strucutre, Y is linked to their elastic compliance S_{ij} as [1, 11, 15],

$$\frac{1}{E} = S_{11}l_1^4 + (2S_{12} + S_{66})l_1^2l_2^2 + S_{22}l_2^4 + (2S_{23} + S_{44})l_2^2l_3^2 + S_{33}l_3^4 + (2S_{13} + S_{55})l_1^2l_3^2 \tag{7}$$

where l_1, l_2 and l_3 are direction cosines, $S_{11} = 13.549 \times 10^{-12}$, $S_{12} = -0.129 \times 10^{-12}$, $S_{13} = -6.5665 \times 10^{-12}$, $S_{33} = 15.8052 \times 10^{-12}$, $S_{44} = 27.5710 \times 10^{-12}$ and $S_{66} = 25.9538 \times 10^{-12}$ (TPa^{-1}) [1]. Accordingly, the y-intercept and slope of the plot drawn between $4\sin\theta/Y_{hkl}$ vs $\beta\cos\theta$ are used to compute σ and crystallite size (Fig. 3). Further the micro-strain (ε) can be calculated from the young's modulus of orthorhombic SnS ($Y_{hkl} = 105.3$ GPa) [11, 16].

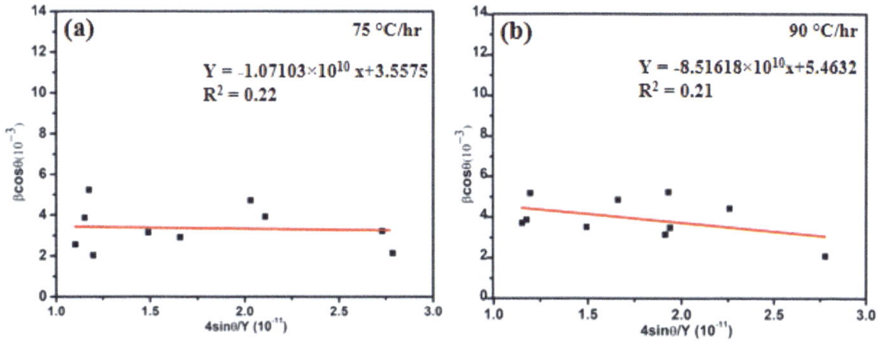

Fig. 3. The (β cosθ) vs (4 sinθ/Y) USDM plot for samples prepared with heating rate of (a) 75 °C/hr and (b) 90 °C/hr.

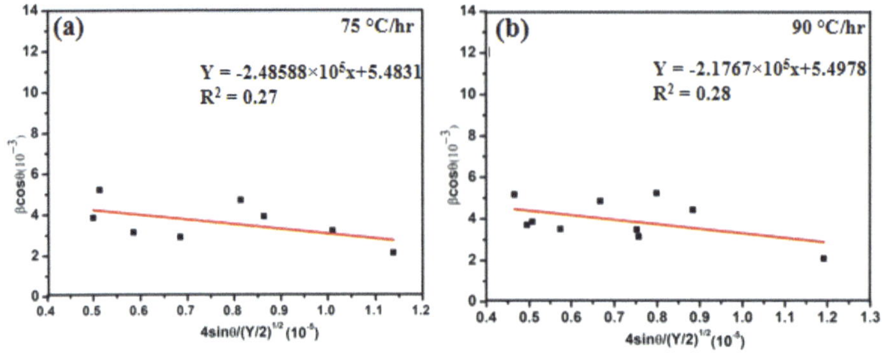

Fig. 4. The (β cosθ) vs (4 sinθ/(Y/2)$^{1/2}$) UDEDM plot for sample synthesized at a heating rate of (a) 75 °C/hr and (b) 90 °C/hr.

In UDEDM, energy density (u) is considered to be isotropic whereas the deformation stress is considered as anisotropic. For an elastic system that obeys Hooke's law, the energy density u can be calculated as $u = \varepsilon^2 Y_{hkl}/2$ [1, 11]. Then Eq. 5 can be modified as

$$\frac{\beta\cos\theta}{\lambda} = \frac{k}{D} + \frac{4\sin\theta}{\lambda}\left(\frac{2u}{Y_{hkl}}\right)^{1/2} \tag{8}$$

The graph drawn with $4\sin\theta/(Y_{hkl}/2)^{1/2}$ along the x-axis and βcosθ along the y axis and the crystallite size and energy density are calculated from the y intercept and slope as presented in Fig. 4. The calculated values of crystallite size and strain from three different models are given in Table 1 which is in good agreement with each other concluding that these three models can be used for the crystallite size and strain calculation. Whereas, the crystallite size and strain estimated using Eqs. 2 and 3 are higher than that obtained from W-H plot. The negative slope obtained from all three models indicate the formation of compressive strain [17]. Strain analysis shows a slight decrease in the micro-strain values upon reducing the heating rate.

Table 1. Crystallite size (D) and strain (ε) of SnS nanoparticles based on W-H analysis.

Sample code	Lattice parameter			V (nm^3)	$\delta \times 10^{-4}$ (nm^{-2})	D (nm)	$\varepsilon \times 10^{-3}$	UDM		USDM		UDEDM	
	a (nm)	b (nm)	c (nm)					D (nm)	$\varepsilon \times 10^{-3}$	D (nm)	$\varepsilon \times 10^{-3}$	D (nm)	$\varepsilon \times 10^{-3}$
00-039-0354	**0.433**	**1.119**	**0.398**	**0.193**	–	–	–	–	–	–	–	–	–
90 °C/hr	0.433	1.112	0.399	0.192	8.65	30	1.01	25	1.06	25	0.81	25	0.95
75 °C/hr	0.434	1.121	0.399	0.194	8.16	35	0.91	34	0.75	33	1.00	30	1.08

The surface morphology of the sample synthesized at a heating rate of 75 °C/hr is investigated using SEM analysis as shown in Fig. 5(a). SEM images show the formation of near spherical shaped SnS nanoparticles around 300 nm size. Figure 5(b) and (c)

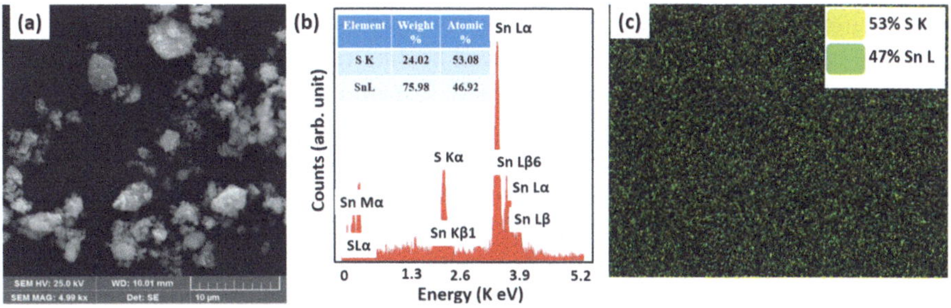

Fig. 5. (a) SEM image, (b) point EDAX spectra and (c) EDAX mapping analyses of sample synthesized at a heating rate of 75 °C/hr.

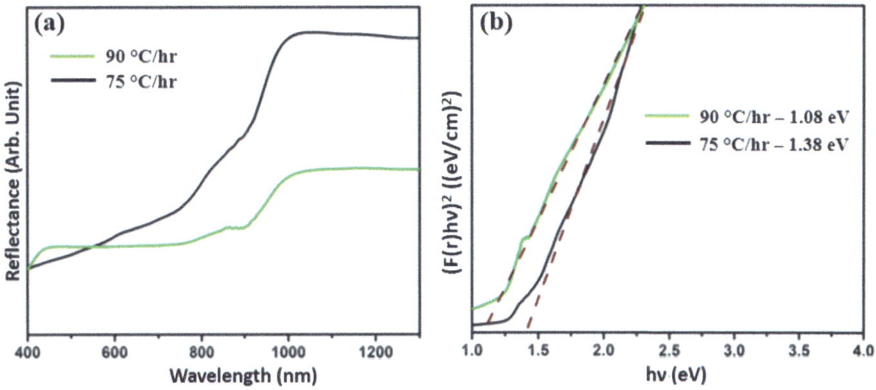

Fig. 6. (a) Reflectance spectra and (b) plot between $(F(r)h\nu)^2$ vs hν indicating the direct energy band gap value of the samples.

shows the EDAX point and mapping analyses of sample which reveals the stabilization of mild S rich SnS nanoparticles with S/Sn ratio as 1.13 which is very well matching with the Raman and XRD analyses. Figure 6(a) shows the reflectance spectra of both the samples in the wavelength range of 400–1300 nm. Figure 6(b) shows the direct band gap energy estimation using a plot between $(F(R)h\nu)^2$ vs hν, where F(R) is the Kubelka-Munk (K-M) function calculated using Eq. (9) [18–20]. Band gap calculation indicates a blue shift from 1.08 to 1.38 eV by decreasing the heating rate from 90 to 75 °C/hr.

$$F(R) = \frac{(1-R)^2}{2R} \tag{9}$$

4 Conclusion

This report investigated the optimized heating rate and temperature treatment for the complete sulphurization of Sn powder to result into tin sulphide phase utilizing non-vacuum based temperature assisted mechanochemical reaction. Raman analysis revealed the formation of oxide phases for the powders annealed at a higher rate of 90 °C/hr whereas it is completely absent for a reduced heating rate of 75 °C/hr. The crystallite size and strain was studied by Williamson Hall method by following three different models: uniform deformation (UDM), uniform stress deformation (USDM) and uniform deformation energy density (UDEDM). SEM analysis revealed the formation of near spherical shaped nanoparticles of size about 300 nm. EDAX analysis shows S/Sn ratio of 1.13 for sample prepared at a heating rate as 75 °C/hr which is very well corroborating the XRD and Raman analyses. S/Sn ratio can be further tuned to the optimum value by varying the initially taken Sn and S precursor weight ratio. Calculations showed sample with tin sulphide phase do have a direct energy band gap of 1.38 eV.

Acknowledgements. The author BHB thank NIT Warangal for granting institute post-doctoral fellowship (NITW/Admin-III/2023/4625). The authors thank Dr. D. Selvakumar, Scientist-F, DEBEL (DRDO), Bangalore for EDAX measurements.

References

1. Baby, B.H., Mohan, D.B.: Sol. Energy **174**, 373–385 (2018)
2. Baby, B.H., Mohan, D.B.: Mater. Chem. Phys. **192**, 317–329 (2017)
3. Loferski, J.J.: J. Appl. Phys. **27**, 777–784 (1956)
4. Yang, L., Qing, X.H., Ping, T.J.: Acta Phys. Chim. Sin. **25**, 365–370 (2009)
5. Baby, B.H., Philipson, A., Mohan, D.B.: Optik **240**, 166848 (2021)
6. Chandrasekhar, H.R., et al.: Phys. Rev. B **15**, 2177–2183 (1977)
7. Chandrasekhar, H.R., Mead, D.G.: Phys. Rev. B **19**, 932–937 (1979)
8. Dieguez, A., Rodriguez, A.R., Vila, A., Morante, J.R.: J. Appl. Phys. **90**, 1550–1557 (2001)
9. Smith, A.J., Meek, P.E., Liang, W.Y.: J. Phys. C Solid State Phys. **10**, 1321–1333 (1977)
10. Williamson, G.K., Hall, W.H.: Acta Metall. **1**, 22–31 (1953)
11. Kafashan, H., et al.: Phys. Status Solidi A **213**, 1302–1308 (2016)
12. Aly, K.A., Khalil, N.M., Algamal, Y., Saleem, Q.M.A.: Mater. Chem. Phys. **193**, 182–188 (2017)
13. Zak, A.K., Majid, W.H.A., Abrishami, M.E., Yousefi, R.: Solid State Sci. **13**, 251–256 (2011)
14. Bindu, P., Thomas, S.: J. Theor. Appl. Phys. **8**, 123–134 (2014)
15. Wen, Y., Wang, L., Liu, H., Song, L.: Crystals **7**, 39 (2017). https://doi.org/10.3390/cryst7020039
16. Koc, H., et al.: Phys. Status Solidi C **6**, 651–658 (2015)
17. Murugesan, C., Chandrasekaran, G.: RSC Adv. **5**, 73714–73725 (2015)
18. Kubelka, P., Munk, F., Tech, Z.: Phys. **12**, 593 (1931)
19. Kubelka, P.: J. Opt. Soc. Am. **38**, 448 (1948)
20. Lopez, R., Gomez, R.: J. Sol-Gel Sci. Technol. **61**, 1–7 (2012)

Structural Analysis of Ag Doped SnS Nanorods Prepared by CTAB Assisted Solvothermal Method

Agnes Ditto[1,2], D. Paul Joseph[3], Bovaz Livingstone Baby[4], D. Bharathi Mohan[1], and Benjamin Hudson Baby[1,3](✉)

[1] Department of Physics, Pondicherry University, Puducherry 605014, India
benjaminhudsonbaby@gmail.com
[2] Department of Materials Science and Engineering, Indian Institute of Technology, Kanpur 208016, India
[3] Department of Physics, National Institute of Technology, Warangal, Telangana 506004, India
[4] Department of Physics, Catholicate College, Pathanamthitta, Kerala 689645, India

Abstract. Ag doped SnS nanoparticles were prepared by CTAB assisted solvothermal method. XRD analysis of drop casted films (as deposited) revealed the presence of orthorhombic phase pure SnS nanoparticles. Surface morphology analysis revealed the growth of nanorods with 150 nm in length and 30 nm in width. The crystallite size and micro-strain due to the effect of doping is calculated from the observed broadening of the XRD peaks using Williamson-Hall equation by following three models such as uniform deformation model (UDM), Uniform stress deformation model (USDM) and uniform deformation energy density model (UDEDM). The difference in the calculated strain values by three different models may be due to the anisotropic nature of SnS. The observed strain value can be related with the intrinsic defects present in the sample and also revealed the polycrystalline nature of the prepared nanoparticles.

1 Introduction

Solar cells have paved an alternative pathway for the mankind whose energy demands keep rising every day. The second generation photovoltaics focuses on minimizing the material consumption, by using materials of high absorption coefficient, like CZTS, CIGS, CdTe and GaAs and these have achieved efficiencies comparable to that of conventional silicon solar cells. Even though they exhibit good efficiency, the scarcity of constituent elements and toxicity of Cd and Te raises practical issues in using them for commercial solar cells [1, 2].

Tin(II) Sulfide, with its excellent physical, structural and electrical properties, is considered as a suitable candidate for replacing the existing absorber layer materials in thin film photovoltaics. SnS, a p type material due to Sn vacancies, has absorption coefficient greater than 10^5 cm^{-1} and having 1.38 eV as the direct band gap [3]. Its constituent elements are nontoxic and abundant and has a chemical structure simpler than the complex quaternary compounds like CIGS and CZTS. In contradiction to the

P. Joseph D. et al. (Eds.): IEMDST 2024, SPPHY 567, pp. 109–116, 2025.
https://doi.org/10.1007/978-981-96-5863-3_14

theoretical efficiency of 24%, as predicted by Prince Lofferski, the higher efficiency achieved so far for SnS layer based solar cell is mere 4.36% and this fall in efficiency is mainly because of intrinsic defects, phase impurities and poor device fabrication [4, 5].

Here we are reporting the preparation of undoped and silver doped SnS nanoparticles (NPs) by following CTAB (Cethyl trimethyl ammonium bromide) assisted solvothermal method, where CTAB is used as the cationic surfactant. The crystal structure, the lattice spacing and the unit cell parameters are calculated from the obtained X-ray diffraction pattern. Strain and crystallite size of the prepared NPs were calculated by WH method by following three models, namely Uniform Deformation Model (UDM), Uniform stress deformation model (USDM) and Uniform Deformation Energy Density Model (UDEDM) [6, 7].

2 Materials and Methods

Initially for the synthesize of tin sulphide NPs, separate solutions of $SnCl_2$ (0.1M), Na_2S (1.2M) and CTAB (0.2 M) were made in 10 ml of EG (Ethylene Glycol). Then Na_2S and CTAB solutions were mixed into $SnCl_2$ solution under stirring at room temperature. To prepare Ag doped SnS NPs, x M (x: 0.02–0.08) of $AgNO_3$, is taken and dissolved in 10 ml of EG separately. Correspondingly the molarity of Sn precursor, $SnCl_2$ is changed to (0.1−x) M. Then Na_2S, $AgNO_3$ and Na_2S solutions were added into $SnCl_2$ at room temperature. Further, the final solution after mixed for 10 min was moved to a teflon linen autoclave and carried out heat treatment at 150 °C for 90 min in an oven. After 90 min, the solution was allowed to cooled down to the room temperature and peptized several times in methanol to remove traces of EG and other impurities. Then few microns thick film were deposited by following simple drop casting method. Further, the structural analysis of the prepared samples were carried out using XRD (Panalytical, X'Pert PRO) and the surface morphology were studied using HR-TEM (Libra 200 TEM) analysis.

3 Results and Discussion

The XRD pattern of as deposited films as shown in Fig. 1 are indexed after compared with standard JCPDS file No [00-039-0354]. The crystallite size (D) and micro-strain (ε) of the prepared NPs are calculated from the XRD peak broadening by following Williamson-Hall method using uniform deformation model (UDM), uniform stress deformation model (USDM) and uniform deformation energy density model (UDEDM). Williamson-Hall relation is given by [8],

$$\beta \cos\theta = \frac{k\lambda}{D} + 4\varepsilon \sin\theta \quad (1)$$

The uniform deformation model (UDM) as given in Eq. (1), it is assumed that strain is uniform in all crystallographic directions [9–11]. In W-H plot, average crystallite size (D) and micro-strain (ε) are calculated form the y intercept and slope of the plot between 4sinθ (x-axis) and βcosθ (y-axis) as shown in Fig. 2.

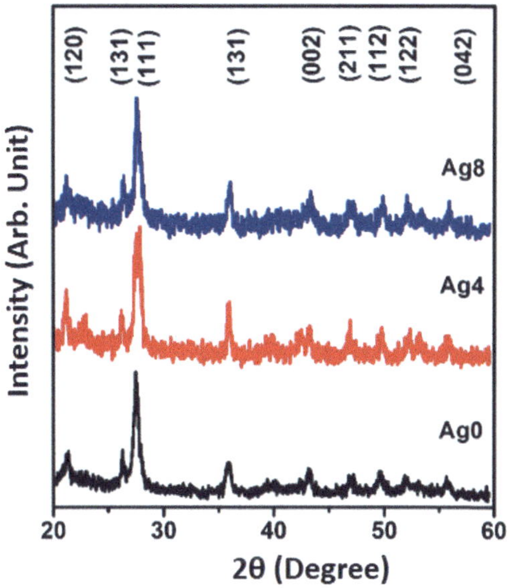

Fig. 1. XRD patterns of Ag 0, 4 and 8% doped SnS samples.

In the uniform stress deformation model (USDM), the lattice deformation stress (σ) is considered to be isotropic and Williamson-Hall relation is re-written using anisotropic strain (ε) [9]. In this approach Eq. (1) is modified based on the Hooke's law (σ = Yε) as,

$$\frac{\beta\cos\theta}{\lambda} = \frac{k}{D} + \frac{4\sigma\sin\theta}{\lambda Y_{hkl}} \tag{2}$$

For an orthorhombic structure, young's modulus (Y_{khl}) is linked to their elastic compliance S_{ij} as [12]

$$\frac{1}{E} = S_{11}l_1^4 + (2S_{12} + S_{66})l_1^2l_2^2 + S_{22}l_2^4 + (2S_{23} + S_{44})l_2^2l_3^2 + S_{33}l_3^4 + (2S_{13} + S_{55})l_1^2l_3^2 \tag{3}$$

where l_1, l_2 and l_3 are direction cosines. The elastic compliance of SnS ($S_{11} = 13.549 \times 10^{-12}$, $S_{12} = -0.129 \times 10^{-12}$, $S_{13} = -6.5665 \times 10^{-12}$, $S_{33} = 15.8052 \times 10^{-12}$, $S_{44} = 27.5710 \times 10^{-12}$ and $S_{66} = 25.9538 \times 10^{-12}$ (TPa^{-1})) can be calculated from the elastic constants using the following relations [13],

$$S_{11} = S_{22}, S_{23} = S_{31} = S_{13}, S_{44} = S_{55}, S_{66} = 2(S_{11} - S_{12})$$

$$S_{11} = \frac{C_{11} - C_{13}^2/C_{33}}{C_{11} - C_{12}}S \quad S_{12} = -\frac{C_{12} - C_{13}^2/C_{33}}{C_{11} - C_{12}}S \tag{4}$$

$$S_{13} = -\frac{C_{13}}{C_{33}}S \quad S_{33} = \frac{C_{11} + C_{12}}{C_{33}}S$$

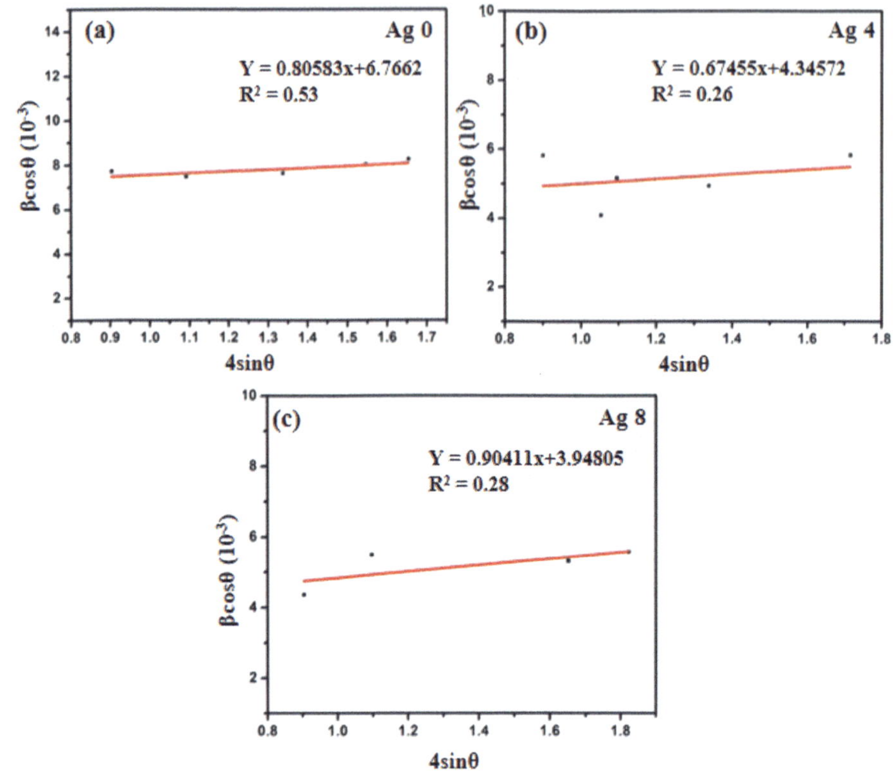

Fig. 2. The (β cosθ) vs (4 sinθ) UDM plot of (a) Ag 0, (b) Ag 4 and (c) Ag 8% doped SnS samples.

$$S_{44} = \frac{1}{C_{44}}S \quad S_{66} = \frac{1}{C_{66}}$$

$$S = \frac{1}{C_{11} + C_{12} - {}^{2C_{13}^2}\!/_{C_{33}}}$$

where C_{11}, C_{12}, C_{13}, C_{33}, C_{44} and C_{66} are the elastic constants of SnS with values 99.34, 26.23, 52.17, 106.62, 36.27 and 38.53 GPa respectively [13, 14]. The uniform deformation stress (σ) and crystallite size is calculated from the slope and y-intercept of the graph drawn between $4\sin\theta/Y_{hkl}$ and $\beta\cos\theta$ as shown in Fig. 3 and the strain (ε) can be estimated using Hooke's law ($Y_{hkl} = 105.3$ GPa) [7, 14].

The energy density of a lattice can be estimated using uniform deformation energy density model (UDEDM). In USDM model, we are considering the crystal as homogenous and isotropic in nature, whereas in many cases this is not justified. For an elastic systems that obeys Hooke's law, the energy density u can be calculated as $u = \varepsilon^2 Y_{hkl}/2$ [9, 10]. Then Eq. 2 can be modified as

$$\frac{\beta\cos\theta}{\lambda} = \frac{k}{D} + \frac{4\sin\theta}{\lambda}\left(\frac{2u}{Y_{hkl}}\right)^{1/2} \tag{5}$$

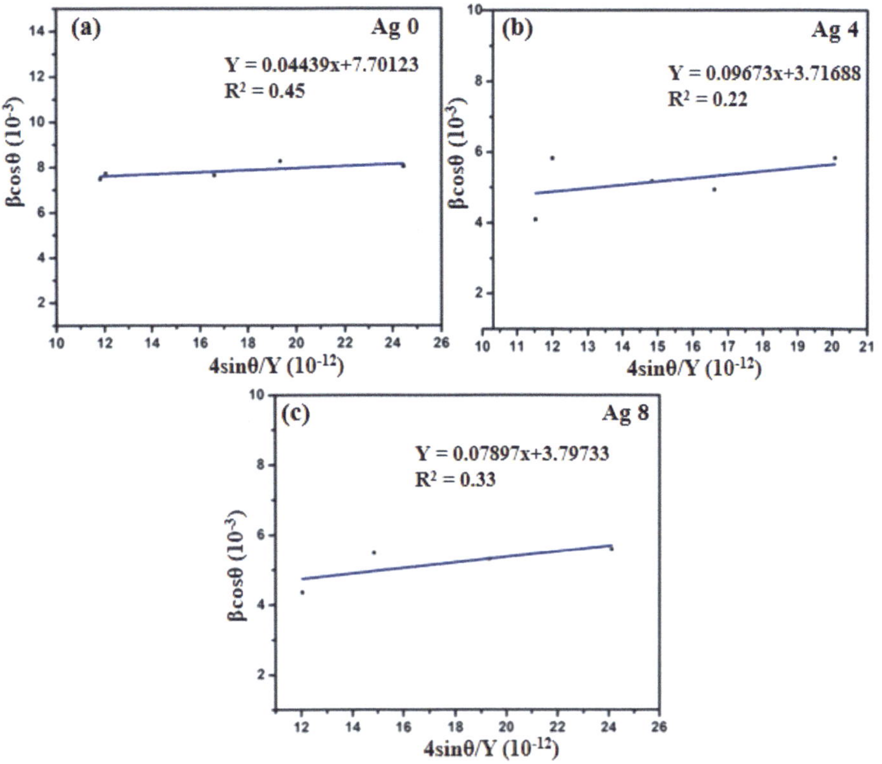

Fig. 3. The (β cosθ) vs (4 sinθ/Y_{hkl}) USDM plot of (a) Ag 0, (b) Ag 4 and (c) Ag 8% doped SnS samples.

The crystallite size and energy density (u) is obtained from the slop and y intercept of the plot between $4\sin\theta/(Y_{hkl}/2)^{1/2}$ vs βcosθ (Fig. 4). The obtained parameters are given in Table 1. From UDEDM model, crystallite size is increased from 21 to 36 nm due to the effect of doping. Micro-strain, which arises as a result of various intrinsic defects in the polycrystalline sample, is also increased by increasing the Ag doping concentration from 0 to 8%. The average crystallite size and micro-strain values calculated from three different models are in good agreement with each other indicating that these three models are acceptable for the calculation of crystallite size and strain. The surface morphology of Ag doped SnS is explored using HR-TEM analysis as presented in Fig. 5 which revealed the growth of highly homogenous SnS nanorods with 150 nm in length and 30 nm in width with an aspect ratio of 5.

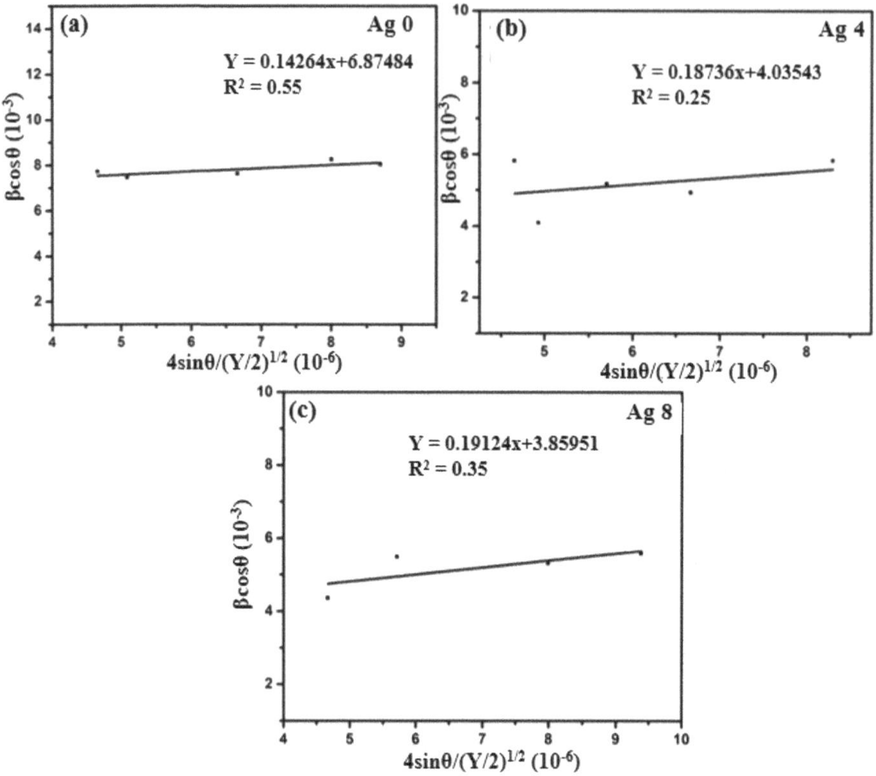

Fig. 4. The (β cosθ) vs (4 sinθ/$(Y_{hkl}/2)^{1/2}$) UDEDM plot of (a) Ag 0, (b) Ag 4 and (c) Ag 8% doped SnS samples.

Table 1. Crystallite size (D), micro-strain (ε), Stress (σ), and energy density (u) of SnS nanoparticles calculated using Williamson-Hall procedure.

Sample ID	UDM		USDM			UDEDM			
	D (nm)	ε (× 10^{-3})	D (nm)	Ε (× 10^{-3})	Σ (Mpa)	D (nm)	ε (× 10^{-3})	σ (Mpa)	U (KJm^{-3})
Ag 0	20	0.8093	21	0.5889	44	21	0.716	58	20
Ag 4	32	0.674	37	1.190	97	34	0.920	75	35
Ag 8	35	0.9041	37	1.018	79	36	0.971	75	36

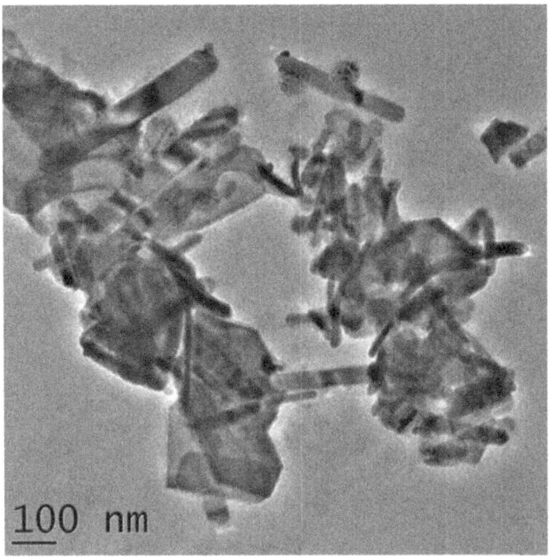

Fig. 5. HR-TEM image showing morphology of Ag (4%) doped SnS NPs.

4 Conclusion

Ag doped SnS NPs were synthesized by CTAB assisted solvothermal technique, by varying the Ag doping concentration from 0 to 8%. Structural properties of the prepared samples were studied using X-ray diffraction. XRD analysis reveals the formation of orthorhombic Ag doped SnS nanoparticles. The peak broadening effect due to crystallite size and lattice strain was studied by Williamson Hall method by following three different models: uniform deformation (UDM), uniform stress deformation (USDM) and uniform deformation energy density (UDEDM). W-H analysis shows an increase in strain value by increasing the Ag doping concentration. The calculated crystallite size and strain values from three models are in good matching to each other concluding that these three models are highly preferable for crystal structure analysis. HR-TEM analysis shows the formation of highly homogenous nanorods with an aspect ratio of 5.

Acknowledgements. The author BHB sincerely thank NIT Warangal for offering institute post-doctoral fellowship (NITW/Admin-III/2023/4625). The authors thank STIC-CUSAT, Kochi for providing access to use their HR-TEM facility.

References

1. Baby, B.H., Mohan, D.B.: Sol. Energy **174**, 373–385 (2018)
2. Park, H.H., et al.: Prog. Photovol. **23**, 901–908 (2015)
3. Baby, B.H., Mohan, D.B.: Mater. Chem. Phys. **192**, 317–329 (2017)
4. Baby, B.H., Mohan, D.B.: Appl. Sur. Sci. **423**, 1111–1123 (2017)
5. Malone, B.D., Gali, A., Kaxiras, E.: Phys. Chem. Chem. Phys. **16**, 26176–26183 (2014)
6. Kafashan, H.: Mater. Res. Express **5**, 046417 (2018)
7. Mote, V.D., Purushotham, Y., Dole, B.N.: J. Theor. Appl. Phys. **6**, 6 (2012)

8. Williamson, G.K., Hall, W.H.: Acta Metall. **1**, 22–31 (1953)
9. Aly, K.A., Khalil, N.M., Algamal, Y., Saleem, Q.M.A.: Mater. Chem. Phys. **193**, 182–188 (2017)
10. Zak, A.K., Majid, W.H.A., Abrishami, M.E., Yousefi, R.: Solid State Sci. **13**, 251–256 (2011)
11. Bindu, P., Thomas, S.: J. Theor. Appl. Phys. **8**, 123–134 (2014)
12. Wen, Y., Wang, L., Liu, H., Song, L.: Crystals **7**, 39 (2017). https://doi.org/10.3390/cryst7020039
13. Lubarda, V., Chen, M.: J. Mech. Mater. Struct. **3**, 153–171 (2008)
14. Koc, H., Simsek, S., Palaz, S., Oltulu, O., et al.: Phys. Status Solidi C **6**, 651–658 (2015)

Engineering Luminescence in $SrZrO_3$: Eu^{3+} Particles: Synthesis and Optimization for Enhanced Structural and Optical Properties

Aswin Raj Irudayaraj and Rajaboopathi Mani(✉)

Key Laboratory of Emergent Materials, Department of Physics and Nanotechnology, SRM Institute of Science and Technology, Kattankulathur, Tamil Nadu 603203, India
rajaboom@srmist.edu.in

Abstract. Lanthanide doping via crystal site engineering approach offers a viable strategy for tuning the emission wavelengths in luminescence applications. Nevertheless, a weak emission from low coordination sites and the complexities of heavily doped structures presents significant challenges. Inorganic materials with suitable coordination sites can enhance luminescence efficiency. In this study, Eu^{3+} doped $SrZrO_3$ inorganic particles with a perovskite structure were synthesized using hydrothermal method. Structural, morphological, and optical properties were characterized using X-ray diffraction, scanning electron microscopy, and photoluminescence spectroscopy. The XRD confirmed the orthorhombic phase of $SrZrO_3$ while the SEM revealed the formation of spherical particles. Under UV excitation at 293 nm, the $SrZrO_3$:Eu^{3+} particles emitted reddish luminescence, attributed to the electric dipole transition $^5D_0$ to $^7F_2$ dominated at 606 nm. The excessive doping of Eu further enhanced the red emission. The CIE color coordinates (x = 0.638, and y = 0.362) confirmed the reddish hue, with the respective CIE diagram plotted. These findings show notable luminescent properties, indicating the potential for applications in lighting and field emission display.

1 Introduction

$SrZrO_3$, an alkaline earth zirconate, is part of the $A^{2+}B^{4+}O_3$ family of inorganic oxides characterized typical perovskite structure [1, 2] and a high melting point of 2700 °C, making it ideal for high-temperature applications [3, 4]. Additionally, its enhanced proton conductivity at high temperatures makes it suitable material for electrochemical devices [5]. The physical and chemical properties of $SrZrO_3$ can be significantly improved by doping, which induces luminescent center through host-dopant interactions. $SrZrO_3$ perovskites have been prepared by various synthetic routes including solid-state reaction [6], co-precipitation [7], and sol-gel [8] methods. However, the traditional solid-state synthesis route requires extended calcination time and high temperature exceeding 1500 K, often resulting in coarse, inhomogeneous sample with a non-uniform size distribution. To obtain efficient luminescence characteristics, the resulting phosphors must exhibit fine particle size, spherical morphology, narrow size distribution and minimal agglomeration. Thus, selecting an appropriate synthesis method is critical for ensuring these favorable characteristics in luminescent materials.

P. Joseph D. et al. (Eds.): IEMDST 2024, SPPHY 567, pp. 117–122, 2025.
https://doi.org/10.1007/978-981-96-5863-3_15

The europium ion displays interesting photoluminescent properties in different oxidation states with Eu^{3+} producing distinct red ($^5D_0$ to $^7F_2$) or reddish-orange ($^5D_0$ to $^7F_1$) emissions, as observed in various Eu^{3+} doped host matrices such as aluminates [9], borates [10], and phosphates [11]. Achieving optimal red emission from Eu^{3+} requires selecting an appropriate host material to improve the emission intensity. In this study, Eu^{3+} was doped into the $SrZrO_3$ perovskite host structure using hydrothermal method, with heavy doping to induce intense red emission. We employed crystal site engineering [12] technique to tune the emission wavelength. The $SrZrO_3$ perovskite provides multi-coordination sites, with strontium and zirconium atoms coordinated with eight and six oxygen atoms respectively resulted in effective site-specific doping of Eu^{3+} within the host lattice. Our investigation focused on the crystal structure of the heavily doped sample, its morphology, and the photoluminescent properties of Eu^{3+} in the $SrZrO_3$ matrix, aiming to optimize the material emission characteristics.

2 Materials and methods

The $SrZrO_3$: Eu^{3+} particles were prepared via hydrothermal method. The precursors strontium nitrate ($Sr(NO_3)_2$), zirconium nitrate hydrate ($ZrO(NO_3)_2 \cdot xH_2O$), europium (III) nitrate hexahydrate ($Eu(NO_3)_3 \cdot 6H_2O$), and sodium hydroxide (NaOH) were purchased from SRL Pvt. Ltd. A 1:2 molar ratio of strontium nitrate and zirconium nitrate hydrate was dissolved in NaOH solution and stirred for 1 h at 80 °C using a hotplate magnetic stirrer. The resulting solutions were transferred into a 100 mL Teflon-lined autoclave, sealed, and placed inside a hot air oven at 180 °C for 24 h. After naturally cooled down to room temperature, the samples were washed multiple times with deionized water and ethanol via centrifugation to remove the impurities. The washed samples were dried at 60 °C for 6 h and ground into fine powder. Subsequent calcination was carried out at 1000 °C for 5 h in a muffle furnace, followed by cooling to room temperature. The resultant samples were a white powder, and bright in color for further characterization.

3 Characterization Techniques

The crystalline structures of the samples were examined by X-ray diffraction (XRD) on a PANalytical XRD spectrometer with the monochromatic Cu Kα radiation ($\lambda = 1.5406$ Å). Tscan rate was performed over the range (2Θ) of 10° to 90°. The structure and morphology were studied on Thermo Fisher Scientific Apreo S High Resolution Scanning Electron Microscope (HRSEM). The photoluminescence emission spectra were recorded using Edinburgh FLS100 Fluorescence Spectrophotometer. All the measurements were conducted at room temperature.

4 Results and Discussion

4.1 XRD Analysis

The Eu^{3+} doped $SrZrO_3$ particles, obtained in powder form, were analyzed using XRD as presented in Fig. 1. The major diffraction peaks are well indexed to the standard XRD pattern of $SrZrO_3$ (JCPDS No. 44-0161), confirming the orthorhombic crystal structure with the *Pnma* space group. However additional phase peaks corresponding to the $Eu_2Zr_2O_7$ phase (JCPDS No. 01-078-1292) were observed indicating the formation of secondary phase with peaks at 2Θ values of 28.41°, 29.91°, 34.66°, 49.86°, and 59.26° due to the heavy doping of Eu^{3+} in the $SrZrO_3$ host. This secondary phase likely arises because the similar ionic radii of Eu^{3+} with Sr^{2+} and Zr^{4+}, allow partially substitution of Eu^{3+} at both Sr and Zr lattice sites. Prior to calcination, impurities such as ZrO_2 and $SrCO_3$ may present. However, calcination at 1000 °C facilitate the formation of $SrZrO_3$ phase, minimizing the impurities. The major phase peaks at 2Θ values of 21.68°, 30.77°, 44.08°, 54.75°, 64.10°, 72.77°, 81.08° and 89.20° correspond to the planes (101), (121), (202), (042), (242), (323), (440), and (244) respectively.

The crystallite size of Eu^{3+} doped $SrZrO_3$ is estimated to be 29.79 nm calculated using the Debye-Scherrer equation as follows,

$$D = \frac{k\lambda}{\beta \cos\theta} \quad (1)$$

where D is the crystallite size (nm), k Scherrer's constant (0.9), λ wavelength of the X-ray beam ($\lambda = 1.5406$ Å), β symbolizes the full width at half maximum, and θ Bragg's diffraction angle.

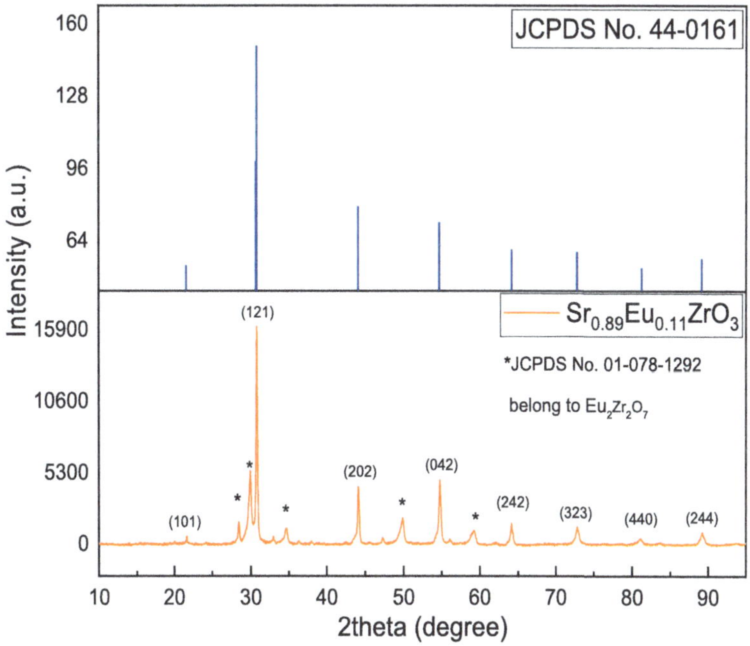

Fig. 1. XRD pattern of $Sr_{0.89}Eu_{0.11}ZrO_3$ sample calcined at 1000 °C for 5h.

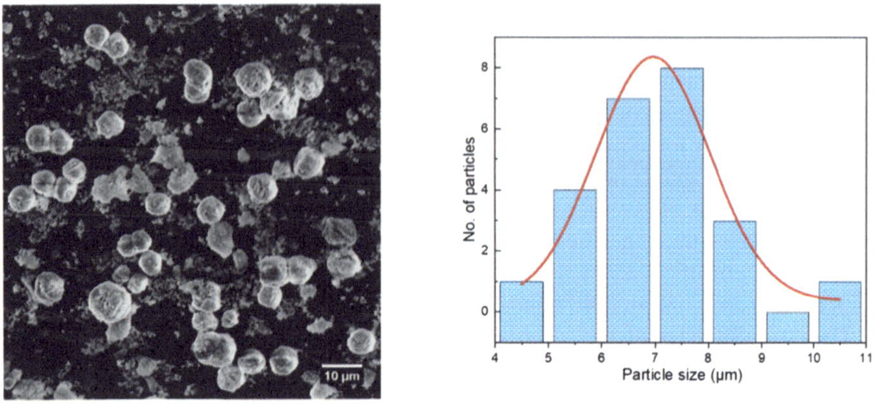

Fig. 2. SEM image of $Sr_{0.89}Eu_{0.11}ZrO_3$ sample synthesized via the hydrothermal method. The particles were calcined at 1000 °C and the particle size distribution in histogram.

4.2 SEM Analysis

The morphology of Eu^{3+} doped $SrZrO_3$ was studied using the SEM revealing a uniform distribution of predominantly spherical particles as shown in Fig. 2. The average particle size is about 6.9 μm in diameter. The particle size distribution is indicated in a histogram. Calcination at 1000 °C temperature resulted in increase in the particle size as high temperature promote the growth. Remarkably, there are uniformly distributed well-formed particles in the morphology of the sample with no agglomerations.

4.3 PL Spectrum Analysis and CIE Diagram

The photoluminescence emission spectra of hydrothermally synthesized Eu^{3+} doped $SrZrO_3$ sample were recorded under UV excitation at 293 nm with the results presented in Fig. 3. The sharp and major emission lines were observed at 590, 606, and 712 nm corresponding to $^5D_0$ ^7F_J ($J = 1, 2, 3$) transitions of Eu^{3+} ions, respectively. The emission peak at 606 nm ($^5D_0$ to $^7F_2$) is attributed to the hypersensitive electric dipole transition [13] which is affected by local symmetry of the Eu^{3+} ions. While the comparatively low-intensity emission peak at 590 nm ($^5D_0$ to $^7F_1$) arises from the magnetic dipole transition which is less affected by the site symmetry. The 606 nm emission is highly intense compared to 590 nm, leading to prominent red emission. The $^5D_0$ to $^7F_2$ transition of Eu^{3+} is highly sensitive to the local environment making it a useful structural probe. Given the similar ionic radii of Sr^{2+} (0.118 nm) and Eu^{3+} (0.095 nm), as well as Zr^{4+} (0.072 nm) and Eu^{3+} (0.095 nm), Eu^{3+} ions likely substitute at both Sr^{2+} and Zr^{4+} sites within the $SrZrO_3$ lattice [14]. It can be noted that a broad emission located in the high-energy spectral region, peaked at 468 nm is related to the trap emission of the host lattice $SrZrO_3$.

The CIE (Commission Internationale de l'Eclairage) coordinates of the $Sr_{0.89}Eu_{0.11}ZrO_3$ sample are represented in Fig. 4. The color performance of luminescent emission was evaluated following standard procedures. The values of the x and y coordinates 0.638 and 0.362 respectively, excited at 293 nm, confirm that Eu^{3+} doped

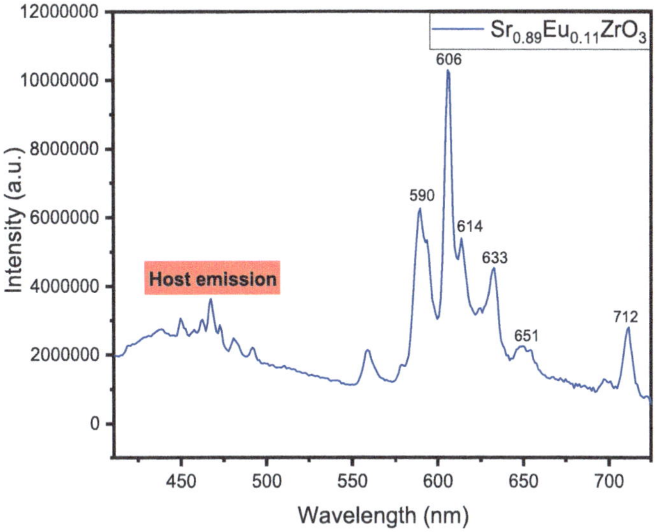

Fig. 3. Photoluminescence emission spectra of $Sr_{0.89}Eu_{0.11}ZrO_3$ sample synthesized via the hydrothermal method, calcined at 1000 °C.

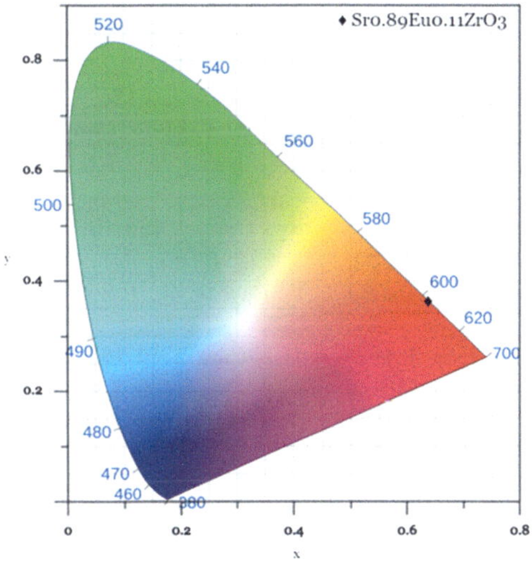

Fig. 4. CIE chromaticity coordinates of $Sr_{0.89}Eu_{0.11}ZrO_3$ sample, calculated from the emission in Fig. 3.

$SrZrO_3$ sample exhibits red luminescence. These coordinate values align well with the emission spectra, indicating the effective tuning of luminescence properties through Eu^{3+} doping in the $SrZrO_3$ matrix, and confirming the strong red-emitting capability.

5 Conclusion

In summary, the Eu^{3+} doped $SrZrO_3$ phosphor was successfully synthesized via hydrothermal method. The XRD analysis confirmed the orthorhombic phase with an additional phase attributed to the heavy doping of Eu^{3+} in the $SrZrO_3$ host lattice. The uniformly formed spherical-shaped particles were confirmed from the SEM images. The luminescence emission spectra under UV excitation revealed red emission due to electric dipole transition $^5D_0$ to $^7F_2$ for Eu^{3+} doped $SrZrO_3$ phosphor. These results suggest that the Eu^{3+}-doped $SrZrO_3$ phosphor holds promise as an efficient red luminescent material, with potential applications in lighting and field emission displays.

Acknowledgments. The authors gratefully acknowledge the NRC Research Facility and the SCIF at SRMIST for providing the necessary instrumentation support. The author A. I also extend his sincere thanks to SRMIST for the research fellowship support.

References

1. Jia, W., Xu, W., Rivera, I., Pérez, A., Fernández, F.: Solid State Commun. **126**, 153–157 (2003)
2. Lu, Z., Chen, L., Tang, Y., Li, Y.: J. Alloys Compd. **387**, L1–L4 (2005)
3. Slonimskaya, E.A., Belyakov, A.V.: Glass Ceram. **58**, 54–56 (2001)
4. Terki, R., Bertrand, G., Aourag, H., Coddet, C.: J. Alloys Compd. **456**, 508–513 (2008)
5. Davies, R.A., Islam, M.S., Gale, J.D.: Solid State Ionics **126**, 323–335 (1999)
6. Popescu, B., Enache, S., Ghica, C., Valeanu, M.: J. Alloys Compd. **509**, 6395–6399 (2011)
7. Potdar, H.S., Deshpande, S.B., Patil, A.J., Deshpande, A.S., Khollam, Y.B., Date, S.K.: Mater. Chem. Phys. **65**, 178–185 (2000)
8. Zhang, H., Fu, X., Niu, S., Xin, Q.: J. Alloys Compd. **459**, 103–106 (2008)
9. Satilmis, S.U., et al.: Opt. Mater. **34**, 1921–1925 (2012)
10. Naidu, S.A., Boudin, S., Varadaraju, U.V., Raveau, B.: J. Solid State Chem. **190**, 186–190 (2012)
11. Zhang, X.M., Li, W.L., Jang, K.H., Seo, H.J.: Curr. Appl. Phys. **12**, 299–302 (2012)
12. Mani, R., Gupta, S.K., Ghosh, P.S., Jiang, H.: Chem. Eur. J. **24**, 16149–16159 (2018)
13. Mani, R., Jiang, H., Gupta, S.K., Li, Z., Duan, X.: Inorg. Chem.. Chem. **57**, 935–950 (2018)
14. Huang, J., et al.: J. Alloys Compd. **487**, L5–L7 (2009)

Dipicolonic Acid (DPA) Capped Metal and Metal Oxide Dipicolonic Nanoparticles as Novel Nanocides

R. Amrutha and K. Jagajjianani Rao(✉)

Department of Biotechnology, Vel Tech Rangarajan Dr. Sagunthala R&D Institute of Science and Technology, Chennai 600062, India
drjagajjananirao@veltech.edu.in

Abstract. Metal and metal oxide nanoparticles capped with dipicolinic acid (DPA) represent a novel class of nanocides with considerable potential across multiple domains which ensure their distinct qualities and uses. The synthesis techniques used to create these nanoparticles, emphasize how important DPA is for maintaining and improving the stability of metal and metal oxide nanoparticles. The features that enable DPA-capped nanoparticles to effectively combat various diseases and contaminants are discussed, with particular attention paid to their structural, chemical, and physical qualities. The adaptability of DPA-based nanocides in treating cancer, preventing bacterial infections, and reducing environmental pollution is studied in the application section. Their potential to specifically target and establish cells makes a very intriguing use in medical and ecological cleanup settings. The article also discusses how employing DPA-capped nanoparticles may affect the environment, highlighting the necessity of carefully evaluating their toxicity and biodegradability to guarantee sustainable and safe procedures. The scope and prospects for DPA-based nanocides are finally explored, highlighting existing issues and possible directions for further study and advancement. The swift progress in nanotechnology and materials science implies that DPA-capped nanoparticles will persist in developing, providing inventive resolutions to urgent worldwide health and environmental concerns. In light of this, the abstract emphasizes how crucial it is to continue studying to fully grasp the potential of DPA-based nanocides in influencing medical and environmental protection in the future.

1 Introduction

DPA (dipicolinic acid) has chelating properties and consists of a pyridine ring and two carboxylic acid groups. Planar conformation facilitates coordination by enabling stronger ionic interactions with metal ions [1]. Flow diagram (Fig. 1) showing the formation of DPA in spore structure with exhibiting application posses through which DPA us formed. The DPA is engineered into a new class of nanocides that improve targeted infection and the biocompatibility of external applications. These nanocides are under careful evaluation for their safety in the environment and public health. The chemical reaction leading to the biosynthesis of DPA is expressed in Fig. 2 followed by the

P. Joseph D. et al. (Eds.): IEMDST 2024, SPPHY 567, pp. 123–131, 2025.
https://doi.org/10.1007/978-981-96-5863-3_16

reaction inside the spore cell with the formation of calcium dipicolinate Fig. 3. DPA-capped metal and metal oxide nanoparticles offer a cutting-edge solution by combining nanotechnology adaptability with DPA's inherent qualities. Dipicolinic acid (DPA) is a typical intracellular spore molecule and its biosynthesis in sporulating cells is involved in multiple enzymatic reactions It starts with L-aspartate, which is phosphorylated by aspartate kinase to give rise to the intermediate of L-aspartyl phosphate in the cytosol through the reaction of its formation from phosphorylation with adenosine triphosphate (ATP). And then L-aspartate-semialdehyde is produced by the reduction reaction of aspartate semialdehyde dehydrogenase along with NADPH. The next step involves the conversion of L-aspartate semialdehyde to dihydrodipicolonic acid (DHDP) by dihydrodipicolinate synthase. Lastly, DHDP is oxidized to DPA by dihydrodipicolinate reductase inside the spore core, requiring NAD + as a cofactor. The resulting DPA binds with calcium or magnesium ions in the spore cell, producing calcium dipicolinate, a resistant structure that protects spores from extreme conditions [2].

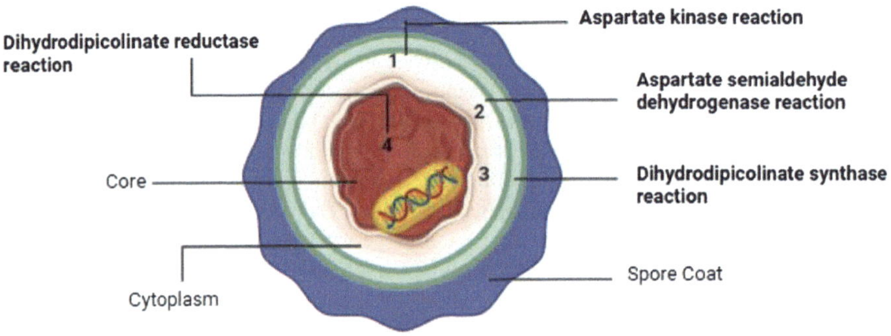

Fig. 1. Process of spore formation.

1. **L-aspartate + ATP → L-aspartyl phosphate + ADP**

   ↓

2. **L- Aspartyl phosphate + NADPH + H⁺ → L-aspartate semialdehyde + NADP⁺**

   ↓

3. **L- aspartate semialdehyde (DHDP) —Dihydrodipicolinate synthase(DHDPS)→ Dihydrodipicolinate acid**

   ↓

4. **Dihydrodipicolinate + NAD⁺ —Dihydrodipicolinate reductase→ Dipicolinic acid (DPA) + NADH + H⁺**

Fig. 2. Biosynthesis of DPA

2 DPA-Capped Metal and Metal Oxide Nanoparticles

The biocompatibility, reactivity, and the stability of different metal and metal oxide nanoparticles can be enhanced by the use of dipicolonic acid (DPA) as a capping agent. There are varied applications of easygoing nanoparticles such as silver, gold, copper oxide, titanium dioxide, and zinc oxide, when combined with DPA. DPA aids in stabilizing silver nanoparticles, improving their dispersion for antimicrobial coatings and wound dressings. Gold nanoparticles benefit from DPA by enhancing their colloidal stability, particularly useful for biomedical applications. Copper oxide nanoparticles show improved functionality for water purification when stabilized by DPA. Some of the commonly used nanoparticles in sunscreens and self-cleaning products are titanium dioxide and zinc oxide, their stability and dispersibility can be increased by the use of DPA. Nanoparticles enhance their efficacy depending on their size range which is stated in Table 1.

Fig. 3. Process and benefits of spore formation.

The use of nonmetal spores in synthesis processes is advantageous due to their simplicity, speed, and eco-friendliness. Studies have shown that silver nanoparticles exhibit enhanced proficiency not only against *B. stratosphericus* but also have an effective potential against organisms like *E. coli, K. pneumonia, P. aeruginosa*, and even *S. aureus*. Wide range use of the nanoparticles has made it an all-rounder in varied fields as exhibited in Table 2. The size differences, cost efficiency, and eco-friendly nature of spore-extracted DPA make it suitable for diverse applications in industry and medicine [8]. Some of the damage-causing fungal cankers such as *B. dothidea, R. solani, A. alternata, and V. pyri's*

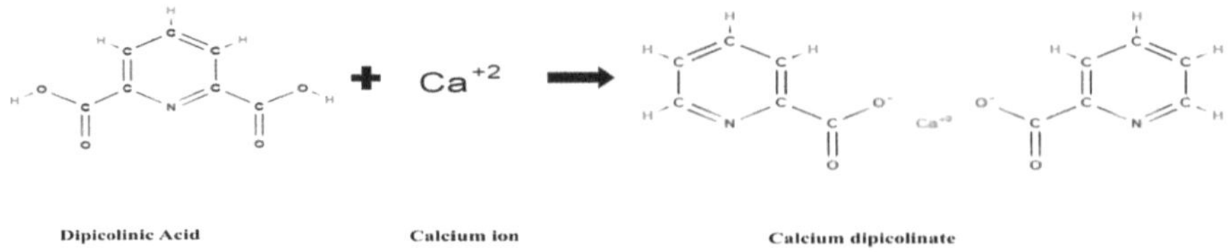

Fig. 4. Formation of Ca dipicolinate in the spores.

Table 1. Synthesis strategies and characterization

Synthesis and Approach	Characterization method	Particle Size (nm)	Type of metal/metal oxide	Ref.
Microwave-Assisted Synthesis	Disk diffusion assays and zone of inhibition tests and time-kill assays	6.9–7.3	Silver (Ag), Copper (Cu) Oxide (CuO)	[3]
Microbial Reduction	MTT assay, and LDH assay	10–20	Silver (Ag) and Gold (Au)	[4]
Turkevich Method	ICP-MS and EDS	15–30	Gold (Au)	[5]
Polyol Process	FTIR and Raman spectroscopy	10–50	Copper (Cu), Silver (Ag), Nickel (Ni), Cobalt (Co), Lead (Pb)	[6]
Radiation-Based Synthesis	TEM and SEM, and also DLS, XRD, and Brunauer–Emmett–Teller (BET) analysis	5–10	Gold (Au), Silver (Ag), Copper (Cu) Oxide (CuO), Titanium Dioxide (TiO_2), Zinc Oxide (ZnO)	[7]

actions can also be inhibited by the use of DPA. It prevents chitin formation in filamentous fungi and affects cellular structures, demonstrating potential for wider applications [9].

The use of dipicolonic acid (DPA) capped nanoparticles in the environment can have size-dependent toxicity for nitrifying bacteria. The chelating action of Ca towards DPA leading to the formation of Calcium dipicolinate (Fig. 4) is one good reason for combining DPA with nanoparticles [10]. These nanoparticles can cause oxidative stress and toxicity through reactive oxygen species (ROS) pathways, potentially harming environmental microorganisms and causing an adverse effect on the ecosystem. Moreover, the release of DPA-capped nanoparticles can lead to hazards such as bioaccumulation and ecological imbalance [11]. On the other hand, nanoscale materials like nanoscale zero-valent iron (nZVI) and nanosized titanium dioxide (TiO_2) offer effective solutions for environmental cleanup by removing pollutants from soil, sediment, and water. Additionally, the antibacterial characteristics of silver nanoparticles and the versatility of polymer-based nanomaterials in removing metallic contaminants demonstrate their potential for environmental remediation [12]. To ensure the prolonged use of nanoparticles in various applications while minimizing potential environmental harm, it is crucial to implement thorough risk assessment and regulation.

Table 2. Application in diverse fields.

Sectors	DPA capped metals	Application	Ref.
Medical and healthcare	Silver (Ag)	Antimicrobial coatings for textiles and surfaces to prevent infections- Use in wound dressings and bandages for accelerated wound healing and prevention of infection	[13]
	Gold (Au)	Diagnostic tools and contrast agents for medical imaging- Drug delivery systems for targeted therapy	[14]
	Copper (Cu)/ Copper Oxide(CuO)	Antimicrobial agents for combating hospital-acquired infections- Environmental decontamination and disinfection	[15]
	Titanium Dioxide (TiO_2)	UV protection in cosmetics and sunscreens- Self-cleaning surfaces in healthcare facilities	[16]
	Zinc oxide (ZnO)	Skin protectants and moisturizers for dermatological applications- Antimicrobial coatings for medical devices and implants	[17]
Agriculture	Silver (Ag)	Enhanced seed germination and plant growth by inhibiting fungal and bacterial pathogens	[18]
	Gold (Au)	Improved nutrient uptake and photosynthesis efficiency, leading to increased crop yield	[19]
	Copper (Cu)/ Copper Oxide(CuO)	Natural pest control and disease management, reducing reliance on synthetic pesticides	[20]

(continued)

Table 2. (*continued*)

Sectors	DPA capped metals	Application	Ref.
	Titanium Dioxide (TiO_2)	UV protection for crops, enhancing resilience against environmental stressors like extreme weather conditions	[21]
	Zinc oxide (ZnO)	Boosting nutrient absorption in plants, contributing to healthier growth and higher yields	[22]
Environment	Silver (Ag)	Wound healing and anti-inflammatory treatment-Anti-inflammatory treatment of psoriasis lesions	[23]
	Gold (Au)	Wound healing and antibacterial properties-Topical psoriasis treatment	[24]
	Copper (Cu)/ Copper Oxide(CuO)	Antibacterial and antioxidant properties for numerous biomedical applications	[25]
	Titanium Dioxide (TiO_2)	Cosmetics and dermato-cosmetic applications	[26]
	Zinc oxide (ZnO)	UV protection in sunscreen creams	[27]
Other Applications	Silver (Ag)	Electrically conductive textiles- Antibacterial properties in textiles	[28]
	Gold (Au)	Not explicitly mentioned in the provided sources for applications in Surface Coatings and Textiles when combined with DPA	[29]
	Copper (Cu)/ Copper Oxide(CuO)	Not explicitly mentioned in the provided sources for applications in Surface Coatings and Textiles when combined with DPA	[30]

(*continued*)

Table 2. (*continued*)

Sectors	DPA capped metals	Application	Ref.
	Titanium Dioxide (TiO_2)	Water repellence, dirt repellence, self-cleaning, UV protection in textiles	[31]
	Zinc oxide (ZnO)	Stiffness, abrasion resistance, self-cleaning, antibacterial properties, and UV protection in textiles	[32]

3 Conclusion, Scope and Future Perspectives of DPA Based Nanocides

The potential of metal and metal oxide nanoparticles capped with dipicolinic acid (DPA) as innovative nanocides is significant in enhancing antibacterial characteristics for various applications [33]. Some of their major contributions can also be included in the field of medical device coatings, wound dressing, developing antimicrobial fabrics in medical and healthcare industries, the promising power of the nanoparticle can also be used to combat antibiotic-resistant bacteria in case of environmental remediation [34]. Additionally, DPA exhibits antifungal properties and demonstrates stability in penetrating pear bark, reducing canker symptoms in Cuigan pear plants by 90%. A novel antifungal drug called DPA shows promise for treating canker disease caused by *V. pyri*. Silver nanoparticles capped with dipicolonic acid from cocoa bean extract demonstrate antimicrobial properties against drug-resistant plant and human bacteria, as well as some fungal species [35]. Infection control, antimicrobial and larvicidal activity, wound healing as well as anticoagulant properties as some among the varied biomedical applications that these nanoparticles hold [36]. Further research on advanced DPA-capped nanoparticles combining metals for broad-spectrum antibacterial properties is recommended. The safety and biocompatibility of these nanoparticles for medical and environmental use must be confirmed through extensive investigations [37]. Regulatory processes must be navigated to approve DPA-capped nanoparticles as novel nanocides, requiring comprehensive preclinical and clinical studies.

References

1. https://www.sciencedirect.com/topics/nursing-and-health-professions/dipicolinic-acid (2024)
2. Briffa, S.M., et al.: RSC Adv. **7**, 3894–3906 (2017). https://doi.org/10.1039/C6RA25064E
3. Mitev, D.P., Townsend, A.T., Paull, B., Nesterenko, P.N.: Diam. Relat. Mater. **48**, 37–46 (2014). https://doi.org/10.1016/j.diamond.2014.06.007
4. Swamy, M.K., Rudramurthy, G.R., Patra, J.K., Sinniah, U.R.: Microb. Biotechnol. Appl. Food and Pharmacol. **2**, 395–434 (2018). https://doi.org/10.1007/978-981-10-7140-9_19
5. Hou, Z., et al.: J Nanobiotechnol. 88 (2019). https://doi.org/10.1186/s12951-019-0522-y

6. Ghosh, I.D., Ghatak, S., Mahali, K., Bhattacharyya, S.S., Saha, B.: Tenside Surfactants Detergents **53**, 332–346 (2016). https://doi.org/10.3139/113.110440
7. Ma, Y.-Y., Li, Y., Huang, H.-X., Qian, D.-J.: Colloids and Surfaces A: Physicochemical and Engineering Aspects, 664 (2023). https://doi.org/10.1016/j.colsurfa.2023.131132
8. Hosseini-Abari, G., Emtiazi, S.M.: Ghasemi. World J. Microbiol. Biotechnol. **29**, 2359–2364 (2013). https://doi.org/10.1007/s11274-013-1403-4
9. Rai, M., Yadav, A., Gade, A.: Biotechnol. Adv. **27**, 76–83 (2009). https://doi.org/10.1016/j.biotechadv.2008.09.002
10. Shao, Z., et al.: Plant Biotechnol J. **18**, 141–154 (2020). https://doi.org/10.1111/pbi.13182
11. Yin, Z., et al.: New Phytologist **208**, 1202–1216 (2015). https://doi.org/10.1111/nph.13544
12. Luu, V.-T., Moon, H.Y., Yoo, S.J., Choo, J.H., Thak, E.J., Kang, H.A.: Appl. Microbiol. Biotechnol. **103**, 2277–2293 (2019). https://doi.org/10.1007/s00253-019-09614-4
13. Hosseini-Abari, G.E., Lee, S.-H., Kim, B.-G., Kim, J.-H.: Appl. Biochem Biotechnol. **174**, 270–282 (2014). https://doi.org/10.1007/s12010-014-1055-3
14. Sadiq, Z., Safiabadi Tali, S.H., Hajimiri, H., Al-Kassawneh, M., Jahanshahi-Anbuhi, S.: Critical Reviews in Analytical Chemistry, 1–36 (2023). https://doi.org/10.1080/10408347.2022.2162331
15. Sabei, F.Y.: J. Nanopart. Res. **25**, 250 (2023). https://doi.org/10.1007/s11051-023-05881-5
16. Raja, G., Cao, S., Kim, D.-H., Kim, T.-J.: Mater. Sci. Eng., C **107**, 110303 (2020). https://doi.org/10.1016/j.msec.2019.110303
17. Jamnongkan, T., Sukumaran, S.K., Sugimoto, M., Hara, T., Takatsuka, Y., Koyama, K.: J. Polym. Eng. **35**, 575–586 (2015). https://doi.org/10.1515/polyeng-2014-0319
18. Paladini, F., Pollini, M.: Materials 12 (2019). https://doi.org/10.3390/ma12162540
19. Arora, K.: Nanotechnology, Food Security and Water Treatment 1–52 (2018). https://doi.org/10.1007/978-3-319-70166-0_1
20. Chugh, B., Poddar, D., Singh, A., Yadav, P., Thakur, S., et al.: Nanosensors for Smart Agriculture, 117–146 (2022). https://doi.org/10.1016/B978-0-12-824554-5.00009-4
21. Sun, D.-S., et al.: Nanomaterials 6 (2016). https://doi.org/10.3390/nano6120237
22. Daccak, D., et al.: Plants 11 (2022). https://doi.org/10.3390/plants11111399
23. Mohammadi, S., Khayatian, G.: Spectrochim. Acta Part A Mol. Biomol. Spectrosc. **148**, 405–411 (2015). https://doi.org/10.1016/j.saa.2015.03.127
24. Kuo, Y.L., et al.: Nanomedicine **11**, 2497–2510 (2016). https://doi.org/10.2217/nnm-2016-0232
25. Alali, A., Hosseini-Abari, A., Bahrami, M., Mehr, Y.: Materials 16 (2023). https://doi.org/10.3390/ma16134670
26. Zhao, T.: (2024). https://kops.uni-konstanz.de/handle/123456789/32254
27. Mohapatro, U., Mishra, L., Mishra, M., Mohapatra, S.: Anal. Chem. **96**, 8630–8640 (2024). https://doi.org/10.1021/acs.analchem.4c00708
28. Saleem, H., Zaidi, S.J.: Materials (Basel). **13**, 5134 (2020). https://doi.org/10.3390/ma13225134
29. Alhashmi Alamer, F., Beyari, R.F.: Nanomaterials (Basel). **12**, 1609 (2022). https://doi.org/10.3390/nano12091609
30. Radetić, M., Marković, D.: Cellulose **26**, 8971–8991 (2019). https://doi.org/10.1007/s10570-019-02714-4
31. Caldas, J.S., Carneiro, E., Carvalho, S.: JTEFT **10**, 49–53 (2024). https://doi.org/10.15406/jteft.2024.10.00365
32. Vakayil, R., Muruganantham, S., Kabeerdass, N., Rajendran, M., Ramasamy, S., Alahmadi, T.A., et al.: Process Biochem. **111**, 1–8 (2021). https://doi.org/10.1016/j.procbio.2021.08.024
33. Barros, C.H.N., Fulaz, S., Stanisic, D., Tasic, L.: Antibiotics (Basel). **7**, 69 (2018). https://doi.org/10.3390/antibiotics7030069

34. Chavali, M.S., Nikolova, M.P.: SN applied sciences. **6**, 607 (2019). https://doi.org/10.1007/S42452-019-0592-3
35. Guerra, F.D., Attia, M.F., Whitehead, D.C., Alexis, F.: Molecules 23 (2018). https://doi.org/10.3390/molecules23071760
36. Song, X.G., et al.: Front. Microbiol. 11 (2020). https://doi.org/10.3389/fmicb.2020.00958
37. Seabra, A.B., Durán, N.: Metals 5 (2015). https://doi.org/10.3390/met5020934

Spray-Coated Fe_2O_3 Thin Film for Photocatalytic Degradation of Methylene Blue Dye

Gouranga Maharana[1], K. T. Hari Govind[2], Yuvashree Jayavelu[1], Meghana Pavithran[2], Alista Mariya Abraham[3], Kathyayani Nerella[4], Benjamin Hudson Baby[1], Kovendhan Manavalan[4], and D. Paul Joseph[1](✉)

[1] Department of Physics, National Institute of Technology, Warangal, Telangana 506004, India
paul@nitw.ac.in

[2] Department of Physics, Amrita School of Physical Science, Amrita Vishwa Vidyapeetham, Coimbatore 641112, India

[3] International School of Photonics, CUSAT, Kalamassery, Ernakulam, Kochi, Kerala 682022, India

[4] Department of Physics and Nanotechnology, SRM Institute of Science and Technology, Kattankulathur, Tamil Nadu 603203, India

Abstract. Oxides of metals are potential components for photodegradation applications due to its biocompatibility, high stability, capability to generate charge carriers, increased lifetime, favourable optoelectronic and charge transport aspects. Iron oxide is abundant, shows high absorbance and good chemical stability which influences its photodegradation aspects. Iron oxide thin films have been coated using ferric trichloride precursor by facile spray pyrolysis method. Structural property of iron oxide is studied using X-ray diffraction that ensures the formation of single phase Fe_2O_3 film with highly textured growth along (021) and (400) planes. Surface morphology investigated by the field emission scanning electron microscopy shows a rice grain shaped morphology of the deposited film. The band gap of Fe_2O_3 thin film estimated using UV-Vis spectroscopy is found to be around 2.59 eV. Photocatalysis of methylene blue demonstrated with Fe_2O_3 thin film showed degradation efficiency of 88.48% after 180 min irradiation under white LED light.

1 Introduction

Providing and ensuring clean water for the whole of the ecosystem is the biggest global problem of the twenty-first century. Global population growth has made it necessary for certain industries to grow, such as the textile and pharmaceutical sectors, where organic dyes are needed to modify a fabric's colour and antibiotics are used for healing [1]. Dye effluents are extremely hazardous and non-biodegradable, and they have a significant detrimental impact on the environment. Methylene blue (MB) is a very harmful and carcinogenic derivative of phenothiazine used in textile dyeing [2]. Water contaminants like methylene blue (MB), malachite green (MG), amoxicillin, diclofenac, ciprofloxacin,

P. Joseph D. et al. (Eds.): IEMDST 2024, SPPHY 567, pp. 132–139, 2025.
https://doi.org/10.1007/978-981-96-5863-3_17

sulfamethoxazole (SMX), and sulfisoxazole (SSX) have been removed using a variety of techniques, including floatation, ion-exchange, chlorination, adsorption, electrochemical, and photocatalytic disintegration [3–5]. Most of these techniques are pricy, need delicate equipment, and result in non-reusable items. Researchers settle on photocatalysis over the other two techniques because it can yield reproducible outcomes, is easy to perform, and has been demonstrated to disintegrate complex pollutants [6, 7].

Oxides of metals (such as Iron oxide, Zinc oxide, and Titanium oxide) have served as an efficient photocatalyst to degrade numerous pollutants from wastewater [8, 9]. In recognition of their photocatalytic traits, TiO_2 and ZnO are the most widely modified and explored semiconductors. One of the main challenges in utilizing these photocatalysts is its large band gap, which affects photocatalytic performance in visible light yet it has good chemical stability and cost-effectiveness. Hence, it is imperative to develop materials that can do both; absorbing the visible light region and minimizing electron-hole recombination. The potential of ferric oxides (Fe_2O_3) to absorb light up to 600 nm with a band gap of 2.1 to 2.3 eV, its excellent stability in an oxidative medium, its availability, and its cost-effectiveness of manufacturing make it an intriguing material for application in photocatalysis [10]. Considering α-Fe_2O_3, an oxide semiconductor with a low band gap, it has been the subject of intense research in recent years for photo-assisted electrolysis of water [11, 12]. Although α-Fe_2O_3's photochemical and electrochemical properties have been extensively studied, relatively little is known about its photocatalytic activity in thin film form.

Nanoparticles have a number of drawbacks as a photocatalyst, including low efficiency when exposed to light, stability problems, limited solar sensitivity, and the possibility of secondary contamination [13]. In the case of particulate films, environmental factors or reaction by products may cause these catalysts to deteriorate (peel off) over time, which may have an impact on their reactivity [14]. In contrast to powder photocatalysts, thin film photocatalysts perform better in a number of aspects, such as reactivity, stability, and overall efficiency. Thin film's unique features, which include notable activity at smaller material quantities, fit very nicely with the contemporary photocatalytic system's objectives of sustainability and efficiency. Their structural advantages render them ideal for practical uses in environmental cleaning and hydrogen production through improved photocatalytic activity. The main factor for this improved performance is the close interparticle connection found in thin films, which promotes better light and reactant interaction [15]. Thin film photocatalysts can capture light more efficiently and with comparatively less material on their surface. The dense and robust nature of the thin film enhances its suitability for photocatalytic reactions by facilitating increased contact between light and reactants.

The current effort aims to degrade organic pollutants with the least amount of power and catalyst possible. This work reports on the detailed characterization of the phase, morphology, texture, and optoelectronic aspects of Fe_2O_3 films coated by the facile spray pyrolysis process. In order to assess the photocatalytic performance of the produced Fe_2O_3 thin films, methylene blue (MB), a cationic dye, was exposed to visible white LED light for three hours and is explored.

2 Materials and Methods

The glass substrates were first cleaned using a mild detergent and then ultrasonicated in distilled water (DI water) followed by acetone, isopropyl alcohol, and ethanol for ten minutes. The process of chemical spray pyrolysis involves spraying a solution onto a substrate that has been heated, and the aerosol mixture reacts to result in oxide thin films. The bare glass is ultrasonically cleaned and dried prior to the deposition of iron oxide thin film. Ferric chloride as a precursor and DI water as a solvent were used to prepare the precursor solution. A 100 ml solution with a concentration of 0.3 M was made in total. With the assistance of a peristaltic pump, the precursor mixture loaded in a reservoir attached to the spray nozzle (with an inner diameter $\cong$ 0.3 mm) was sprayed over the heated substrate maintained at a distance of 32 cm. The optimum temperature for the substrate to evaporate and break down the chloride salts of metal so that they can be oxidized in the air and coat the glass substrate at 450 $\pm$ 5 °C. The mixture was sprayed using moisture-filtered compressed air at 30 kg/cm^2 pressure. The precursor mixture was sprayed over the glass substrate with a total number of 400 sprays with a duration of 0.5 s ON time and 30 s intervals.

The deposited films were subjected to structural characterization with the help of Rigaku SmartLab multifunctional X-ray diffractometer (XRD) (Cu-Kα radiation, λ = 1.5406 Å). The JSM-IT 800 Field Emission Scanning Electron Microscope (FESEM) was utilized to examine the surface morphology of films that were sprayed onto a glass substrate. UV-Visible spectrometer, (Analytik Jena, SPECORD 210 PLUS) was made use of to assess the optical parameters. A stylus profiler (DektakXT, Bruker) was used to determine the film thickness.

The photodegradation of an aqueous solution of methylene blue (MB) was performed to assess the photocatalytic activity of Fe_2O_3 thin film under visible white LED light (200 W). The photocatalytic activity was performed in a 25 ml Borosilicate beaker containing MB dye solution with a volume of 6 ml and a concentration of 5 mg/L. The electrode of size 4 cm^2 is kept immersed in the solution maintaining a distance of 12 cm from the LED light. At room temperature, the methylene blue mixture's pH was found to be 6.07 (neutral). In order to prevent pollutants in the air and undesired light interruptions, the photodegradation process was performed at room temperature for three hours in a closed cabinet. Prior to the experiment, to attain an equilibrium of adsorption and desorption between the pollutant and photocatalyst, the solutions were kept in darkness for a duration of 30 min. The absorption plot of the pristine and degrading MB dye mixture in the presence of a thin film catalyst was recorded every 30 min for a total of 180 min. Using an Analytik Jena, SPECORD 210 PLUS UV-Vis spectrometer, the absorbance of the suspension was measured.

3 Results and Discussion

The deposited iron oxide thin film was subjected to X-ray diffraction analysis to understand the details of its structural characteristics. The film's XRD patterns (Fig. 1a) reveal a good agreement with the JCPDS file # 96–210–8028, indicating a match with the space group C12/c1 of the monoclinic crystal structure. The distinctive diffraction peaks for

Fe_2O_3 were determined to be (111), (−311), (021), (400), (311), and (420) in Fig. 1a. These values correspond to the following 2θ values: 33.15, 35.62, 40.85, 49.44, 54.05, and 62.41°, respectively. The diffraction planes of the sample are narrower, based on the XRD patterns, implying higher crystallinity. The successful formation of single phase Fe_2O_3 is confirmed by the absence of peaks belonging to FeO, Fe_3O_4, and other secondary phases. The sample appears to have higher intensity and textured growth along the (111) plane than the other diffraction planes, according to the XRD spectra in the initial analysis. According to JCPDS file #96–210–8028, the (111) plane is the 100% peak and is observed in a thin film XRD pattern. The diffraction patterns indicate that the 2nd highest intensity peak (−311) is low in intensity. The ratio of intensity of different peaks in the XRD data of a material's thin film often does not match completely with the standard JCPDS file, that generally arises from its powder form, based on the deposition parameters. This is mostly because of how the deposition parameters affect the oriented growth of the crystallites while their positions are orientated across a substrate. The following calculation of Eq. (1) can be used to determine the texture coefficient [16], which indicates the textured growth of the coated films and provides insight into this.

$$TC_{hkl} = \frac{I_{hkl}/I_{0hkl}}{\frac{1}{N}\sum_{N=1}^{N} I_{hkl}/I_{0hkl}} \tag{1}$$

where I_{hkl} is the intensity measured experimentally, I_{0hkl} is the standard intensity as per the JCPDS file, N is the total count of diffraction peaks and TC_{hkl} is the texture coefficient of a particular (hkl) plane. A favored growth along a specific plane direction is indicated by the deviation from the unit value of TC_{hkl} and greater than 2 is interpreted as highly textured. The estimated TC_{hkl} parameters show that the growth of the (021) and (400) planes is highly textured, with TC_{hkl} values of 2.46 and 1.02, respectively. The spray deposited film's average crystallite size was determined to be 74.42 Å using the Debye-Scherrer equation [17].

FESEM analysis was performed to study the surface morphology of the spray-coated Fe_2O_3 thin film. Figure 1b illustrates the dense and homogenous distribution of crystallites with different sizes and some overgrown sizes that make up the surface morphology. The films appear denser due to the rice-grain-shaped particles present on the surface of the thin film. Several micro aggregates cover the film's surface, combining to form a granular and dense surface. Furthermore, tiny grains filled in the spaces between the aggregates to compact the thin layer's surface. Stylus profiler measurements reveal that the spray-coated Fe_2O_3 thin film has a thickness of 930 nm. The solvent's surface tension, vapor pressure, and viscosity significantly influence the film's thickness [18]. The UV-Vis spectrometer was used to record the optical absorbance plot in the 300 nm–1000 nm range in order to analyze the optical properties of the film. The spray-deposited Fe_2O_3 thin film's absorbance spectra in the UV–Vis range demonstrate that the curve to exhibit a strong absorption in the UV–A region (wavelength range: 300–420 nm) (Fig. 1c). Similar findings have been published by Mahadik et al. for iron oxide absorption spectra [19]. Iron oxide is a typical transition-metal oxide in II-VI semiconductors, exhibiting high electron-phonon binding and electron-electron correlation, leading to subtle electronic structures and an abundance of optical characteristics. The optical band gap value of the coated Fe_2O_3 film is determined through the Tauc plot (Fig. 1d), which establishes

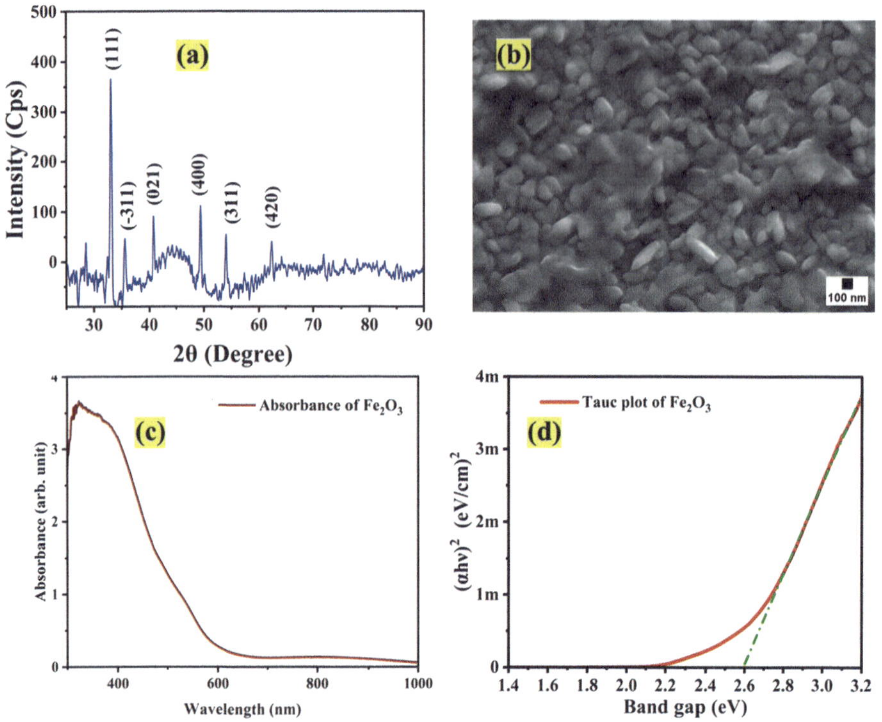

Fig. 1. (a) XRD pattern, (b) FESEM micrograph, (c) absorbance spectra, and (d) Tauc plot for the spray-deposited Fe_2O_3 thin film.

a relationship between the absorption coefficient (α) and incident photon energy (hυ). Estimating the band gap can be done by extrapolating the fitted linear part of the curve to the energy axis, as shown in Fig. 1d. Equation (2) is used to estimate the absorption coefficient (α) [20].

$$\alpha = \frac{1}{l}[\ln 10 \times A] \tag{2}$$

where absorbance and film thickness are denoted by A and l, respectively. The estimated value of the direct band gap from the Tauc plot is found to be 2.59 eV which falls in the visible spectrum. Iron (III) oxide, also known as hematite, has a bandgap value of approximately 2.0 to 2.2 eV. Depending on the particular phase of Fe_2O_3 and the measurement conditions, this value may vary slightly [21]. The electronic band structure of Fe_2O_3 is altered as a result of interatomic forces and electronic interactions. Increased bandgap results from pushing the band edge states further apart through enhanced interactions have been reported [22].

Methylene blue dye (MB dye) is considered as a test sample for Fe_2O_3 thin film's potential to degrade organic pollutants when exposed to visible white LED light. An experimental configuration with a 200 W LED lamp was utilized to carry out the photocatalytic activity as described in the materials and methods section. The intensity of MB's

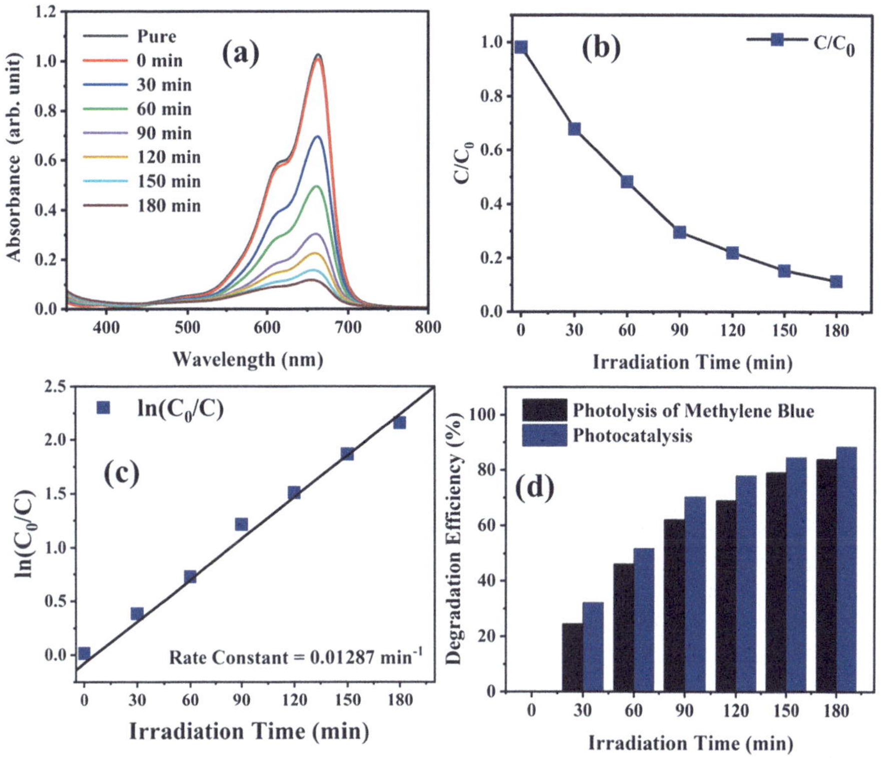

Fig. 2. (a) The absorbance spectra with varying irradiation times, (b) variation in MB dye concentration with illumination times, (c) pseudo-1st order kinetics with illumination time, and (e) disintegration efficiency in comparison with photolysis of MB dye for the iron oxide film under white LED visible light irradiation.

distinctive absorption peak at 664 nm gradually dropped, and after roughly 180 min, it drastically decreased (Fig. 2a). The MB dye solution's colour drastically decreased, indicating a breakdown in the dye's chromophoric structure. The photodegradation of the MB dye in the presence of Fe_2O_3 photocatalyst is shown in Fig. 2b. The photocatalytic disintegration rate of MB dye is represented using a simple Langmuir–Hinshelwood (L–H) kinetic model (Eq. 3), which was plotted against time, t, as a graph of $\ln(C/C_0)$ [23] as shown in Fig. 2c.

$$-ln\frac{C_0}{C} = kt \tag{3}$$

where k is the 1st order reaction rate constant and C_0 and C are the concentrations of MB in solution at time 0 and t, respectively. The obtained value of rate constant (k) is found to be 0.012 min^{-1}, which is a little higher than the rate of photolysis (0.010 min^{-1}). The degradation efficiency (η) of the photocatalyst is plotted in comparison with photolysis in Fig. 2d. The following formula (Eq. 4) was used to determine the percentage of MB dye degradation [24]:

$$\eta = \left(1 - \frac{C}{C_0}\right) \times 100\% \tag{4}$$

The degradation efficiency is found to be 88.48% (higher than photolysis: 83.9%) after 180 min for the Fe_2O_3 thin film, indicates the significant participation of electron-hole pair generated from the photocatalyst's surface. The electron-hole pairs are formed when Fe_2O_3 is excited with light at a wavelength (< 480 nm) larger than the band-gap energy. This charge-separated pair of electrons and holes reacts with the dissolved oxygen (O_2) and water (H_2O) in the mixture, respectively, resulting in redox processes that create superoxide and hydroxyl radical ions, which contribute to the disintegration of methylene blue dye.

4 Conclusion

Fe_2O_3 thin film has been successfully deposited by a facile spray pyrolysis process. According to the XRD analysis, single-phase Fe_2O_3, with highly textured growth in the (021) and (400) plane directions are formed. The thin film's surface morphology, as analysed by FESEM, reveals the formation of rice-grain-shaped particles. The iron oxide thin film's optical absorbance exhibits a significant absorption in the UV–A region with a 2.59 eV band gap. The photocatalytic activity for the thin film is observed to be quite significant with a disintegration efficiency of ~88% upon white visible LED light illumination. Even though photolysis and iron oxide thin film have modestly higher efficiency values, doping and surface modification can improve the efficiency of the latter towards a more cost-effective and sustainable catalyst for the disintegration of organic MB dye.

References

1. Ahmed, M.J.K., Ahmaruzzaman, M., Bordoloi, M.H.: RSC Adv. **5**, 74645–74655 (2015). https://doi.org/10.1039/c5ra13970h
2. Appavu, B., Thiripuranthagan, S., Ranganathan, S., Erusappan, E., Kannan, K.: Ecotoxicol. Environ. Saf. **151**, 118–126 (2018). https://doi.org/10.1016/j.ecoenv.2018.01.008
3. Maharana, G., Muniramaiah, R., Yuvashree, J., Mandal, D., et al.: Surfaces and Interfaces. **42**, 103413 (2023). https://doi.org/10.1016/j.surfin.2023.103413
4. Jayavelu, Y., Maharana, G., Rajender, G., Muniramaiah, R., et al.: Chemosphere **351**, 141128 (2024). https://doi.org/10.1016/j.chemosphere.2024.141128
5. Mezyen, M., El Fidha, G., Bitri, N., Harrathi, F., Ly, I., Llobet, E.: RSC Adv. **13**, 31151–31166 (2023). https://doi.org/10.1039/d3ra05424a
6. Hojjati-Najafabadi, A., Mansoorianfar, M., Liang, T., Shahin, K., Karimi-Maleh, H.: Sci. Total. Environ. **824**, 153844 (2022). https://doi.org/10.1016/j.scitotenv.2022.153844
7. González-González, R.B., et al.: J. Hazard. Mater. **423**, 127145 (2022). https://doi.org/10.1016/j.jhazmat.2021.127145
8. Akinola, P.O., et al.: Heliyon 6 (2020). https://doi.org/10.1016/j.heliyon.2020.e04610
9. Elango, G., Roopan, S.M.: J. Photochem. Photobiol. B Biol. **155**, 34–38 (2016). https://doi.org/10.1016/j.jphotobiol.2015.12.010

10. Mulmudi, H.K., et al.: Electrochem. Commun. **13**, 951–954 (2011). https://doi.org/10.1016/j.elecom.2011.06.008
11. Kennedy, J.H., Frese, K.W.: J. Electrochem. Soc. **125**, 709–714 (1978). https://doi.org/10.1149/1.2131532/xml
12. Kennedy, J.H., Anderman, M.: J. Electrochem. Soc. **130**, 848–852 (1983). https://doi.org/10.1149/1.2119833/xml
13. Feliczak-Guzik, A.: Materials (Basel). 16 (2023). https://doi.org/10.3390/ma16010193
14. Li, X., et al.: Chem Catal. **2**, 1315–1345 (2022). https://doi.org/10.1016/j.checat.2022.04.007
15. Méndez, F.J., et al.: Mater. Chem. Phys. **305**, 127925 (2023). https://doi.org/10.1016/j.matchemphys.2023.127925
16. Muniramaiah, R., Reddy, N.P., Santhosh, R., Maharana, G., Fernandes, J.M., et al.: Phys. Status Solidi **220**, 2200703 (2023). https://doi.org/10.1002/pssa.202200703
17. Joseph, D.P., Renugambal, P., Saravanan, M., Raja, S.P., Venkateswaran, C.: **517**, 6129–6136 (2009). https://doi.org/10.1016/j.tsf.2009.04.047
18. Strawbridge, I., James, P.F.: J. Non Cryst. Solids **86**, 381–393 (1986). https://doi.org/10.1016/0022-3093(86)90026-8
19. Mahadik, M.A., Shinde, S.S., Rajpure, K.Y., Bhosale, C.H.: Mater. Res. Bull. **48**, 4058–4065 (2013). https://doi.org/10.1016/j.materresbull.2013.06.031
20. Baby, B.H., Mohan, D.B.: Sol. Energy **189**, 207–218 (2019). https://doi.org/10.1016/j.solener.2019.07.059
21. Bhuvaneswari, K., et al.: Mater. Lett. 276 (2020). https://doi.org/10.1016/j.matlet.2020.128173
22. Tahir, D., et al.: ACS Omega **6**, 28334–28346 (2021). https://doi.org/10.1021/acsomega.1c04526/asset/images/large/ao1c04526_0009.jpeg
23. Paul, D.R., Sharma, R., Nehra, S.P., Sharma, A.: RSC Adv. **9**, 15381–15391 (2019). https://doi.org/10.1039/c9ra02201e
24. Kerkez, O., Boz, I., Ozge Kerkez, Boz, I.: Reac Kinet Mech Cat. **110**, 543–557 (2013). https://doi.org/10.1007/s11144-013-0616-8

Fabrication and Characterization of $Cu_9Fe_9S_{16}$ Thin Films Using Home-Made Automated SILAR System

C. P. Deepak[1,2], R. Dinesh[1,2], V. Vishnu Narayanan[1,2], K. S. Rajni[1,2(✉)], and S. Shyamlal[3]

[1] Department of Sciences Amrita School of Physical Sciences, Amrita Vishwa Vidyapeetham, Coimbatore, India
ks_rajani@cb.amrita.edu

[2] AMRITA Lab Amrita School of Engineering, Amrita Vishwa Vidyapeetham, Coimbatore, India

[3] Amrita Biosensor Research Lab, Amrita School of Engineering Coimbatore, Amrita Vishwa Vidyapeetham, Coimbatore 641112, India

Abstract. The SILAR method is a quick, affordable, and efficient way to create semiconducting thin films. However, when done by hand, producing the thin film with SILAR is laborious and time-consuming, and there is a high chance of human error. A completely automated SILAR unit prototype was developed using an open-source Arduino platform, and the parameters, such as the dipping time, cycle count, number of beakers, and substrate size, can all be altered as per the user needs. Using this automated SILAR unit, the $Cu_9Fe_9S_{16}$ (a derivative of $CuFeS_2$) thin films are fabricated and their structural, compositional, morphological, optical, and functional group properties are studied. The structural analysis indicated the formation of $Cu_9Fe_9S_{16}$ with a grain size ranging from 11 nm to 21 nm. The micro-Raman analysis confirmed the formation of $Cu_9Fe_9S_{16}$, and the morphological and elemental analysis of the thin films were also studied. The FTIR analysis indicated the formation of metal-sulfur bonds in the 561 cm^{-1} range. The optical analysis of the prepared thin film samples indicated that the $Cu_9Fe_9S_{16}$ has a direct band gap in the range of 3.35 eV to 3.75 eV due to the quantum confinement effect.

1 Introduction

A thin film is a material coating in the nanometer to micrometer range applied to a surface for a specific purpose. Thin films offer a significant benefit over bulk or thick coatings due to their high surface-to-volume ratio, allowing them to be efficient when surface reactions are required [1]. For the thin-film synthesis process, many approaches have been used. Roll coatings, spray coatings, dip coatings, and other coating methods, such as CVD and PVD, are examples [2]. Compared to other processes, the dip-coating approach is less expensive and can cover huge substrates, making it appropriate for industrial-scale applications. Successive ionic layer adsorption and reaction (SILAR) is a dip-coating method. One of the dip-coating processes is successive ionic layer adsorption

P. Joseph D. et al. (Eds.): IEMDST 2024, SPPHY 567, pp. 140–148, 2025.
https://doi.org/10.1007/978-981-96-5863-3_18

and reaction (SILAR) [3]. SILAR employs a layer-by-layer deposition process in which ionic species sequentially adsorb and react on the substrate. It is reasonably cheap and can create various materials such as sulfides, composites, oxides, and chalcogenides.

In one cycle of the SILAR process, the cleaned substrate first goes into a cationic solution, then a wash solution, then a cationic solution, and another wash. Depending on the compound we want to develop, we alter the number of cationic and anionic solutions. The parameters that influence the SILAR method are the pH of the anionic and cationic solution, dipping time, number of cycles, the concentration of the ions in the solution, solvent, and nature of the substrate [3].

When done by hand, the SILAR process (Fig. 1) is laborious and time-consuming, and there is a high chance of human error. Although there are instruments for the process available on the market, these are relatively inexpensive. So, a prototype was designed and developed that is economical, Efficient, and fully automated. Then, using this home-made prototype, the $Cu_9Fe_9S_{16}$ thin films were fabricated on glass plates, and their structural, compositional, morphological, and functional group properties were analyzed. The Mooihoekite nanoparticles have a tetragonal structure with a = b = 10.585 Å and c = 5.383 Å with space group P-42 m [4]. The studies on $Cu_9Fe_9S_{16}$ indicate that they have a significant negative Seeback coefficient of 100 to 140 and are suitable for thermoelectric applications [5, 6].

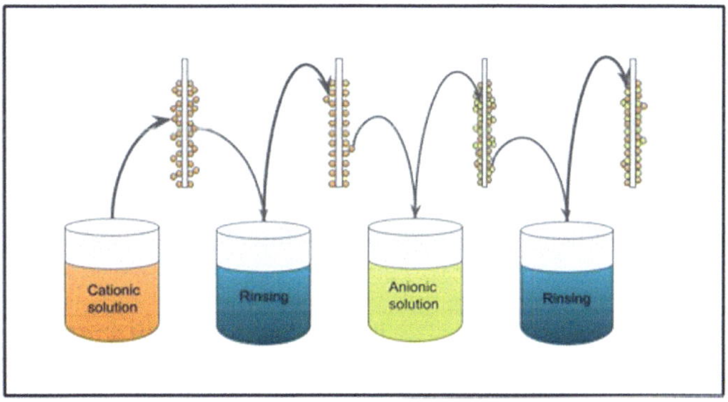

Fig. 1. A four-step SILAR process that involves one cationic and one anionic solution.

2 Materials and Methods

For the development of the SILAR prototype, an Arduino UNO was used along with two stepper motors were employed, which were controlled by L298 modules. The body of the SILAR unit was fabricated using acrylic sheets. Insulated copper wires and jumper wires were used for the connection. A 2 V, 2A power supply was provided for the device to function. For the fabrication of $Cu_9Fe_9S_{16}$ thin films, analytical grade $Cu(NO_3)_2.3H_2O$, $FeSO_4.7H_2O$ and Na_2SxH_2O. The solvent is DDI water. The anionic solution consists of Cu $(NO_3)_2.3H_2O$ and $FeSO_4.7H_2O$ dissolved in DDI water, and the cationic solution was Na_2S. xH_2O. The stoichiometric ratio of precursors was 1:1:2.

The glass plates/FTO were cleaned using soap water first and then ultrasonicated in ethanol and acetone. Finally, the glass plates are washed with DDI water and dried. The adsorption and reaction time was 30 s, while the rinsing time was 10 s (Fig. 1). Five samples were prepared with 40, 50, 60, 70 and 80 cycles using the prototype developed, of which 50,60, and 70 cycles produced the best results and are denoted as D350, D360, and D370, respectively, which we further studied for their structural, optical, morphological, and functional group properties.

3 Development of SILAR Prototype

The project's fundamental idea is that the coating substrate will be retained on the vertical axis while the beakers that hold the solutions will sit in the disc. The vertical axis will move similarly to the hand, which will do the dipping procedure. The disc will rotate, presenting the alternate solution sequentially by the procedure (Fig. 2).

One of the unique features of the prototype is the platform that holds the beakers (Disc). The prototype is built in such a way that the Disc is variable. Hence, according to the user's requirements, the number and size of beakers that the device can hold can be varied. Usually, one cationic and one anionic solution will be used for the SILAR process. As a result, the initial suggestion is to make two discs: one that can hold four beakers and another that can hold eight beakers.

On the curved surface of the disc, button magnets were inserted along the line of beaker holders. This is done to stop the motor's rotation at the proper points so that the dipping processes can occur. The hall sensor switch was used to find the magnet, hence finding the position of the beaker. Two rods are attached perpendicular to the base. One of which is threaded and attached to a bipolar stepper motor. A threaded T-nut and a linear ball bearing attach the threaded rod and smooth road with a platform. The platform parallel to the base will hold the substrate for the dipping. This type of alignment will

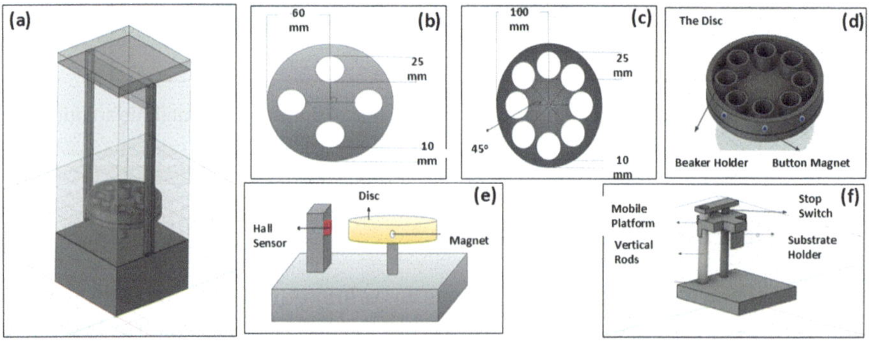

Fig. 2. (a) A basic model that was prepared for the development, (b) Top view of the disc that can hold 4 beakers of base 23 mm, placed at right angles, (c) Top view of the disc that can hold 8 beakers of base 23 mm, placed at right angles, (d) The disc which acts as a platform to hold the beakers, (e) Placement of hall sensor switch, (f) Design for the vertical platform which holds the substrate.

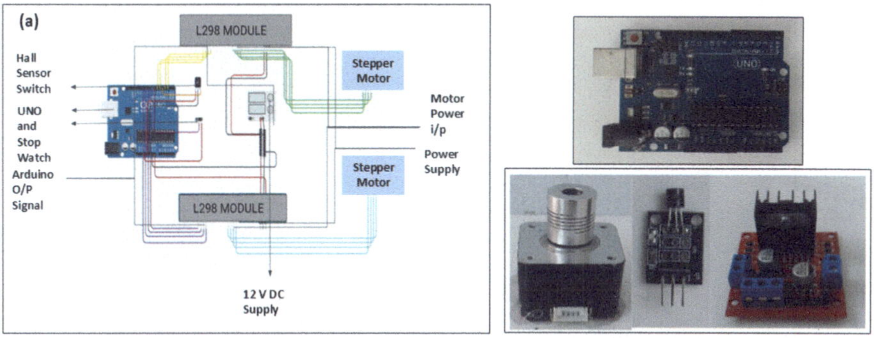

Fig. 3. (a) Circuit Diagram, (b) Arduino UNO, (c) Image of components, which include a stepper motor, hall sensor switch, and stepper motor control board.

enable vertical movement of the substrate. The movement will assist in substrate dipping and rising. As the solutions are kept in the disc, the dipping takes place sequentially.

Arduino was used as the central processing unit. Arduino is an open-source platform for electronics that makes working with microcontrollers easy. It takes digital and analog inputs, processes them, and gives signals to output devices as per the user's instruction. Arduino is an extendible open-source cross-platform which is inexpensive and simple to use. For the prototype development, Arduino Uno was used, which is a microcontroller-based board that uses an ATmega328p microcontroller, which is widely used for automation and robotic applications [7]. To identify the locations of beakers, a hall sensor was utilized to detect magnets. The location of the substrate that needs to be used for coating is determined by an end stop affixed to the end of the vertical rods. One bipolar stepper motor is utilized for the disc's rotation, and another is used for the substrate's vertical movement. To precisely instruct stepper motors, there are two stepper motor control boards (Fig. 3).

Stepper motors connected the threaded rod and disc. The end stop was attached to the end of the vertical rod, and the hall sensor switch was maintained, as shown in Fig. 2f. Both devices received power from an Arduino UNO. The sensors' sensor terminals were also connected to the Arduino. Stepper motor control modules, powered by 12 V and 2 A, were connected to stepper motors. One of these modules also provided 5 V for the Arduino to operate. To code, an Arduino IDE was employed. To prepare different thin films according to user requirements, the number of cycles and dipping times in each solution were left flexible in the code. The images of prototype are as shown in Fig. 4.

Fig. 4. Picture of the prototype.

4 Results and Discussion

The X-ray diffraction (XRD) analysis is done to analyze the structural properties of the prepared thin films. The XRD analysis confirmed the formation of $Cu_9Fe_9S_{16}$ thin films, and the diffraction data matched well with the JCPDS card No 25–0286 (Fig. 5a). $CuFe_2S_3$, another derivative of $CuFeS_2$ and Cu_2S, was also formed along with $Cu_9Fe_9S_{16}$. The crystallinity of Cu_2S increased with the increase in the number of SILAR cycles, which is due to its low energy of the formation [8]. The microstructural properties of the thin films were calculated using the Scherrer formula [9], as shown in Table 1. The particle size varied from 11 nm to 21 nm, and the lattice parameters were similar to those reported earlier [5]. The reduction in the grain size is due to the synthesis process at room temperature and low-temperature annealing.

Table 1. Microstructural analysis of the prepared samples.

Sample	Grain Size (D) nm	Dislocation Density (10^{-15})	Micro-strain
D350	11	8.26	0.0091
D360	21	2.26	0.0071
D370	15	4.44	0.0121

The micro-Raman analysis indicated the formation of CuS and FeS bonds (Fig. 5b). The 270 cm^{-1} and 474 cm^{-1} peaks represent the formation of FeS and CuS bonds, respectively [10]. There is no previous data about the micro-Raman analysis of $Cu_9Fe_9S_{16}$. The FTIR analysis of the thin films had a broad peak at 591 cm^{-1} and a narrow peak at 2333 cm^{-1} (Fig. 5c). The peak at 591 cm^{-1} indicates the presence of metal sulfur bonds, and the broadening is due to the resonance vibrations of sulfur-sulfur bonds. In contrast, the peak at 2333 cm^{-1} indicates the CO_2 adsorption by the samples from the atmosphere [11–13].

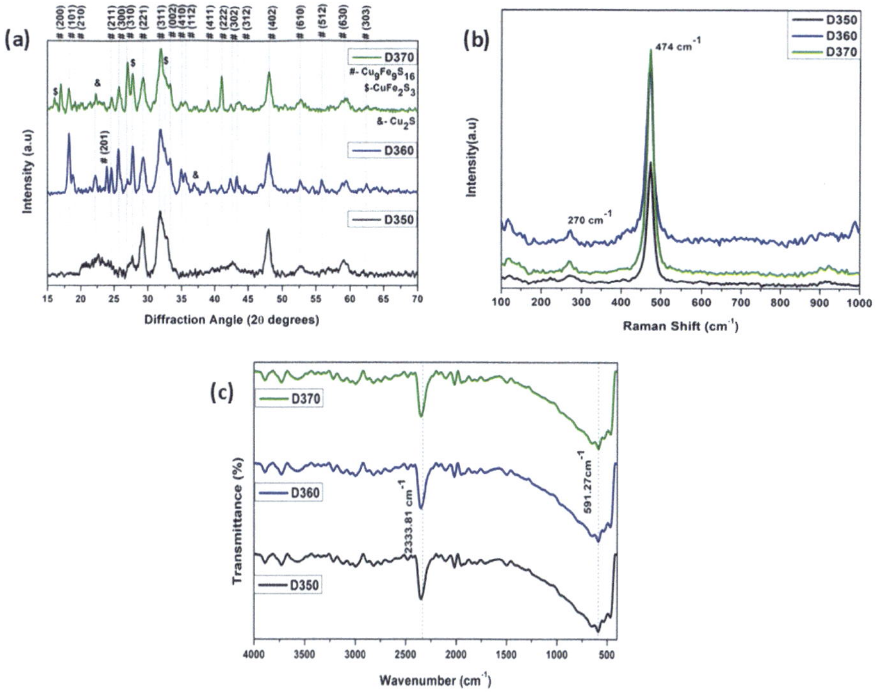

Fig. 5. (a) X-ray Diffractogram, (b) micro-Raman analysis, (c) FTIR analysis of the samples.

FESEM analysis indicated the formation of spherical grains with an average size of 25 nm for the sample D360. This agrees with the grain size calculated with the Scherrer equation. The EDS analysis confirmed the presence of Copper, Iron, and Sulfur; no other impurities were detected (Fig. 6 a,b). The optical absorption properties of the material are analyzed (Fig. 6c), and the band gap is calculated using the Taue plot method [14]. The samples had a direct band gap of 3.35 to 3.75 eV (Fig. 6d). There is no previous report about the optical properties of $Cu_9Fe_9S_{16}$ yet. Using the band gap value, the refractive indices of all three thin films were calculated using three methods as follows [15].

a. ***Moss relation***

$$n^4 \times Eg = 95\,\text{eV} \tag{1}$$

b. ***Herve-Vandamme relation***

$$n^2 = 1 + \frac{A}{(Eg + B)^2} \tag{2}$$

where A is the hydrogen ionization energy (13.6 eV) and B is a constant of value 3.47 eV and is assumed to be the difference between UV resonance energy and band gap energy.

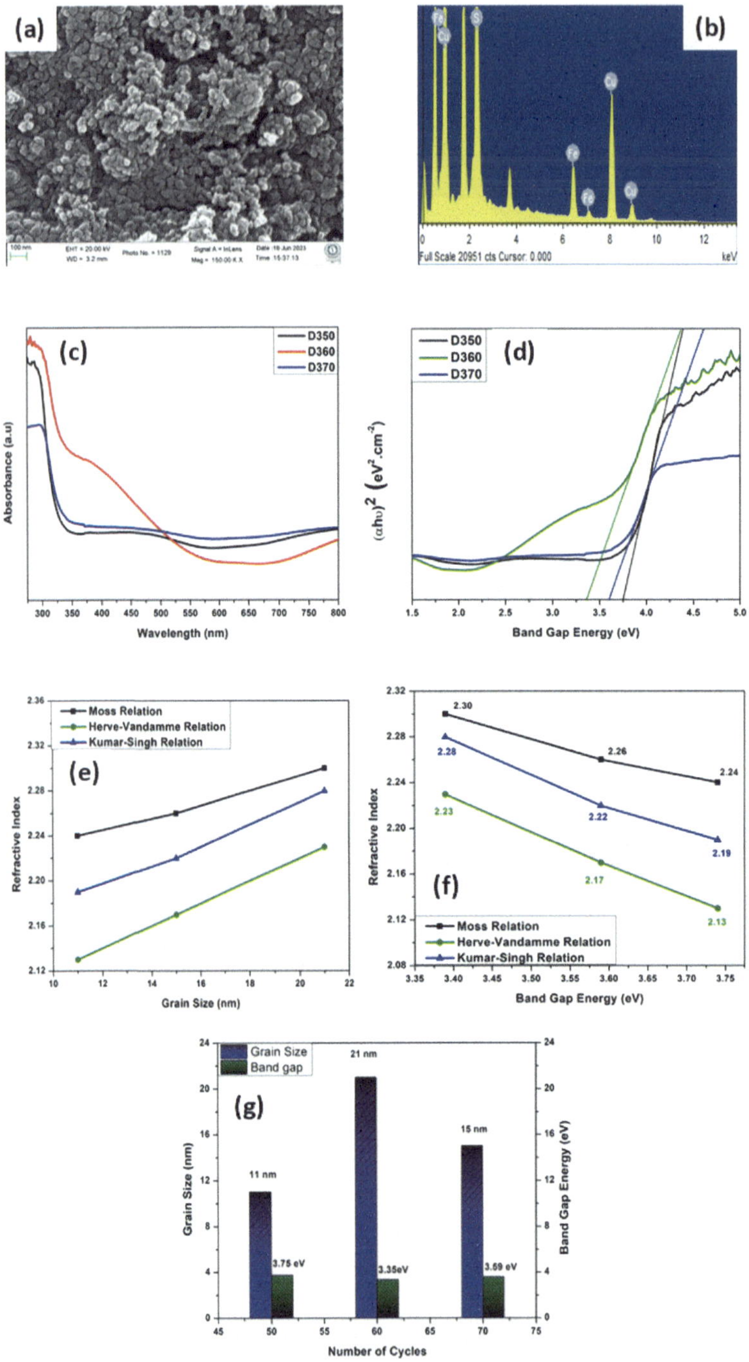

Fig. 6. (a) FESEM, (b) EDS, (c) UV-Vis absorption spectrum, (d) Taue plot, (e) relation between band gap and refractive index, (f) relation between grain size and refractive index, (g) relation between the number of cycles, grains size, and band gap energy.

c. ***Kumar and Singh's relation***

$$n = KEg^C \tag{3}$$

K and C are constants with values of 3.3668 and −0.32234, respectively.

The calculated values were found to be in close proximity to each other. The correlation between band gap energy, refractive indices, grain size, and number of cycles is shown in Fig. 6(e-g).

5 Conclusion

The developed prototype will help with completely automated and easy thin-film deposition that is also inexpensive. The technology will assist in preventing human error and in resolving tedious, lengthy tasks with a high possibility of error. The adjustable disc will also allow for using substrates of various sizes, and the number of anionic and cationic solutions can also be varied. Additionally, we can alter the dipping duration in each solution. The $Cu_9Fe_9S_{16}$ is a derivative of $CuFeS_2$ (chalcopyrite) and is not appropriately studied except for its thermoelectric properties. The structural analysis indicated the formation of $Cu_9Fe_9S_{16}$ with a grain size ranging from 11 nm to 21 nm. The micro-Raman analysis confirmed the formation of $Cu_9Fe_9S_{16}$, and the morphological and elemental analysis of the thin films were also studied. The FTIR analysis indicated the formation of metal-sulfur bonds in the 561 cm^{-1} range. The optical analysis of the prepared thin film samples indicated that the $Cu_9Fe_9S_{16}$ has a direct band gap in the range of 3.35 eV to 3.75 eV due to the quantum confinement effect.

Acknowledgments. The authors sincerely acknowledge Dr. S Mahadevan, Principal, of Amrita School of Physical Sciences, Coimbatore, for his constant support. Dr. Senthil Kumar, CoE – AMGT, Assistant Professor, Department of Chemical Engineering and Material Science, Amrita Vishwa Vidyapeetham, Coimbatore.

References

1. Moharrami, N., Bull, S.J.: Thin Solid Films **572**, 189–199 (2014)
2. Butt, M.A.: Coatings **12**(8), 1115 (2022)
3. Ratnayake, S.P., et al.: Small **17**(49), 2101666 (2021)
4. Sidharth, B., Aslam, S.S., Narayanan, V.V., Rajni, K.S.: Emergent Materials, 1–14 (2024)
5. Tsujii, N., Mori, T.: MRS Online Proc. Libr. **1680**, 1–6 (2014)
6. Qiu, X., et al.: Zeitschrift für anorganische und allgemeine Chemie **646**(14). https://doi.org/10.1002/zaac.202000017
7. Kumar, R.H., Roopa, A.U., Sathiya, D.P.: Int. J. Innov. Res. Electr. Electron. Instrum. Control Eng **3**(4), 27–29 (2015)
8. Kravchenko, T.A.: New Data on Minerals **48**, 80 (2013)
9. Kiruthiga, G., Raguram, T., Rajni, K. S., Selvakumar, P., Nandhakumar, E.: Journal of Materials Science: Materials in Electronics **32**, 22780–22791 (2021)
10. Chandra, U., Sharma, P., Parthasarathy, G.: Chemical Geology **284**(3–4), 211–216 (2011)

11. Ranjani, P.R., Anjana, P.M., Rakhi, R.B.: Journal of Energy Storage **33**, 102063 (2021)
12. Smyrnioti, M., Tampaxis, C., Steriotis, T., Ioannides, T.: Catal. Today **357**, 495–502 (2020)
13. Waluś, E., Manecki, M., Cios, G.: Materials **13**(19), 4440 (2020)
14. Raguram, T., Rajni, K.S.: J. Sol-Gel. Sci. Technol. **93**, 202–213 (2020)
15. Mohan, L., et al.: Journal of Inorganic and Organometallic Polymers and Materials **30**(7), 2626–2632 (2020)

Effect of Annealing Rate and Annealing Temperature on Structural Properties of PZT Thin Films

Karthika P. Moni[1], P. N. Vinod[2], and Maneesh Chandran[1](✉)

[1] Department of Physics, National Institute of Technology Calicut, Kozhikode, Kerala 673601, India
maneesh@nitc.ac.in

[2] Naval Physical and Oceanographic Laboratory, Kochi, Kerala 682021, India

Abstract. Lead zirconate titanate (PZT) thin films are of significant interest in various technological applications, including MEMS sensors/actuators and non-volatile memories, owing to their exceptional piezoelectric and ferroelectric properties. In this work, PZT [$Pb(Zr_{0.52}Ti_{0.48})O_3$] thin films are deposited on $Pt/Ti/SiO_2/Si$ substrates using the sol-gel method. The deposited films are annealed at different temperatures (600, 650 and 700 °C) and heating rates (2, 10 °C/min) to realize the perovskite phase formation. We systematically investigated the effect of annealing temperature as well as the rate of annealing on the structural properties of PZT thin films, using x-ray diffraction and Raman spectroscopy. Our results indicate the importance of the annealing rate as well as the annealing temperature for the formation of perovskite phase.

1 Introduction

Lead zirconate titanate (PZT) films are used in wide range of applications like sonar devices, energy harvesting, resonators, acoustic transducers, biomedical applications, non-volatile memory etc., due to their exceptional ferroelectric properties. PZT thin film has a significant role in the field of MEMS technology [1–5]. In general, PZT thin films are deposited on several substrates including silicon, stainless steel, FTO glass, 2D mica, ITO, quartz [6–10] etc. However, direct deposition of PZT on silicon is not feasible due to the lattice mismatch. Therefore, PZT has been deposited on silicon substrates using interlayers such as platinum, RuO_2, (La, Sr)CoO_3, etc. The widely used interlayer is $Pt/Ti/SiO_2$, with Pt acting as the bottom electrode, Ti functioning as the adhesion layer, and SiO_2 acting as an unintentional insulating layer. As mentioned, PZT thin film can be coated deposited using different techniques and the sol-gel method is one of the most widely used methods for PZT deposition due to its great control over stoichiometry, cost effectiveness, etc. [7].

In general, as-deposited PZT thin films are amorphous in nature, and multiple heat treatments are required to realize a phase pure perovskite PZT thin films. Several studies have been reported related to the effect of annealing temperature on the PZT thin films. Hu *et al.* studied the effect of heating rate by comparing conventional furnace annealing

P. Joseph D. et al. (Eds.): IEMDST 2024, SPPHY 567, pp. 149–155, 2025.
https://doi.org/10.1007/978-981-96-5863-3_19

(CFA) and rapid thermal annealing (RTA). In the case of CFA, the transformation from the amorphous phase to perovskite phase is through an intermediate metastable pyrochlore phase. On the other hand, when PZT films are annealed using RTA, the amorphous phase directly transforms to perovskite phase [11]. Zhu *et al.* reported the effect of heating rate in RTA for sol-gel derived PZT thin films. It was found out that the complete perovskite formation occurs at a lower temperature when it is annealed at higher heating rates [12]. Dang *et al.* theoretically studied the effect of RTA on perovskite phase formation. The report asserted two main reasons for the formation of intermediate pyrochlore phase. The perovskite phase is denser than the pyrochlore phase. It is easier to form less dense phase when thin films in amorphous phase are annealed. Excess lead also promotes the formation of pyrochlore phase as an intermediate phase [13]. Huang *et al.* reported that during low temperature annealing, the crystallization is controlled by the phase transformation from pyrochlore to perovskite phase rather than a nucleation controlled process [14]. On the contrary, Velu *et al.* reported lead deficient pyrochlore phase appears when thin films are annealed at higher temperature for a longer duration [15].

In this work, we studied the effects of annealing temperature and rate of annealing on crystallization and perovskite phase formation of PZT thin films deposited on platinized silicon wafer using the sol-gel method.

2 Materials and Methods

The precursor materials used for the preparation of precursor solution were lead acetate trihydrate, zirconium oxynitrate hydrate and titanium n-butoxide. The lead acetate trihydrate ($(CH_3COO)_2Pb{\cdot}3H_2O$, Sigma Aldrich) and zirconium oxynitrate hydrate ($ZrO(NO_3)_2{\cdot}xH_2O$, Sigma Aldrich) were mixed in ethylene glycol and refluxed at 125 °C for 3 h. To compensate lead loss, an excess of 10% of lead precursor was added to the solution. The titanium n-butoxide ($Ti(OC_4H_9)_4$, Alfa Aesar) was added to ethylene glycol and stirred for 2 h. The obtained solutions were then mixed and stirred for 3 h to get uniform mixture. The substrate used for depositing PZT thin film was Pt (111)/Ti/SiO_2/Si (100). The thickness of Pt, Ti, SiO_2, Si were 200 nm, 20 nm, 200 nm and 375 μm, respectively. The schematic of the PZT thin film deposited on platinized Si wafer is shown in Fig. 1. The platinum coated silicon substrate was cleaned by sonicating the wafer in both acetone and isopropyl alcohol for 5 min and then rinsed using deionised water. The wafer was then dried using a hot plate.

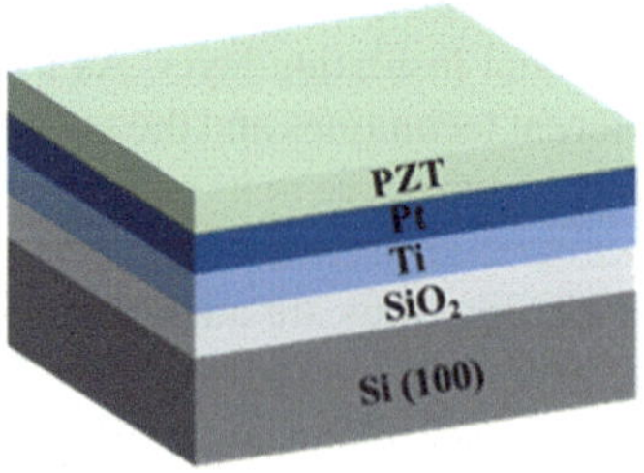

Fig. 1. Schematic of PZT/Pt/Ti/SiO_2/Si.

The precursor solution was spin coated on the platinized silicon substrate at 3000 rpm for 30 s using a spin coater (WS-650Mz-23NPPB, Laurell Technologies Corporation). The coated substrate was dried using a hot plate at 150 °C for 5 min. The process was repeated multiple times to achieve an optimum thickness for the film. The deposited thin film was annealed at different temperatures (600, 650 and 700 °C) at two different annealing rates (2 and 10 °C/min) for 1h using a muffle furnace (L 9/12, Nabertherm). Multistep annealing was also carried out for studying the effect of intermediate holding step. The thin film was annealed at 700 °C at a heating rate of 2 °C/min for 1 h with an intermediate holding at 650 °C for 1 h. For brevity, obtained samples are labelled as PZT-600_2, PZT-600_10, PZT-650_2, PZT-650_10, PZT-700_2 and PZT-700_10, PZT-650_700_2 respectively. To understand the effect of controlled cooling thin film annealed at 650 °C was allowed natural cooling with an intermediate holding at 600 °C for 10 min. The sample was labelled as PZT-650_600_2.

The crystallinity and phase purity of the prepared samples were studied using x-ray diffraction (XRD, SmartLab SE Rigaku) with Cu- Kα source. The Raman spectra of the samples were recorded using Horiba LabRam HR Evolution ($\lambda = 532$ nm) Raman spectrometer.

3 Results and Discussion

The XRD patterns of the prepared samples are shown in Fig. 2. The obtained XRD peaks obtained match with the standard ICDD (00–033-0784), which confirmed the formation of $Pb(Zr_{0.52}Ti_{0.48})O_3$ with tetragonal crystal structure. By comparing all the XRD patterns recorded, we found that phase pure perovskite PZT thin film was obtained when the sample was annealed at 650 °C at a heating rate of 2 °C/min. The coexistence of pyrochlore and perovskite phases at the lowest temperature of 600 °C were expected. When the thin films are annealed using CFA the phase transformation follows the route of amorphous to pyrochlore and then to perovskite phase [11]. The complete transformation of pyrochlore to perovskite phase happened when the thin film was annealed at 650 °C at a heating rate of 2 °C/min. However, the presence of an undesired pyrochlore phase can be observed in XRD patterns of thin film annealed at 650 °C at a heating rate of 10 °C/min. This clearly showed the importance of the annealing rate for the formation of phase pure perovskite PZT. The thin films annealed at 700 °C at both 2 and 10 °C/min heating rates showed the presence of pyrochlore phase. It can be asserted that lead deficient pyrochlore phase formed at a higher annealing temperature and longer annealing duration [15]. The thin film annealed at 700 °C at heating rate of 2 °C/min with an intermediate holding at 650 °C for 1 h also showed the presence of pyrochlore phase. This indicates that the pyrochlore phase can be again formed even after the thin film achieved phase purity at 650 °C. The formation of perovskite phase in PZT thin film can be further ascertained by recording high temperature XRD at different temperatures. The thin film annealed at 650 °C at heating rate of 2 °C/min with controlled cooling by holding it at 600 °C for 10 min also shows phase purity.

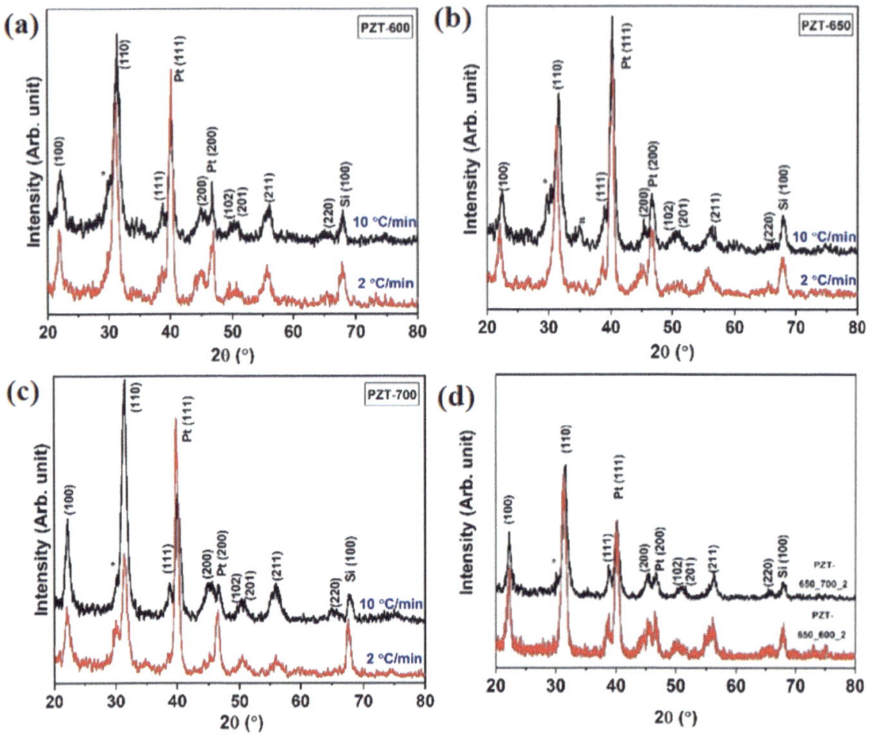

Fig. 2. XRD patterns of PZT thin films annealed at a) 600 °C, b) 650 °C, and c) 700 °C at two different heating rates. XRD pattern of film annealed at 700 °C with an intermediate holding step at 650 °C for 1 h and film annealed at 650 °C and controlled cooling with intermediate holding at 600 °C for 10 min is shown in (d).

The preferred orientation of the PZT thin film is along (110) direction. It is observed that the full width at half maximum (FWHM) of the XRD peaks, corresponding to (110) decreases with an increase in the annealing temperature. As the annealing temperature is increased, the thermal energy provided enables crystal grains within the PZT thin film to grow larger, thereby exhibit large crystallite size.

Raman spectra of the PZT thin films are shown in Fig. 3, which further ascertain at the formation of perovskite PZT. The Raman modes of PZT-650_2, which is a phase pure perovskite sample, are 123 cm^{-1} (A_1(1 TO)), 204 cm^{-1} (A_1(1 LO)), 281 cm^{-1} (B_1 + E), 335 cm^{-1} (A_1(2 TO)), 437 cm^{-1} (E(2LO)), 522 cm^{-1} (E (3 TO)), 574 cm^{-1} (E (3 TO)), 606 cm^{-1} (A_1(3 TO)) and 740 cm^{-1} (A_1(3 LO)). The Raman peaks of the other samples are tabulated in Table 1.

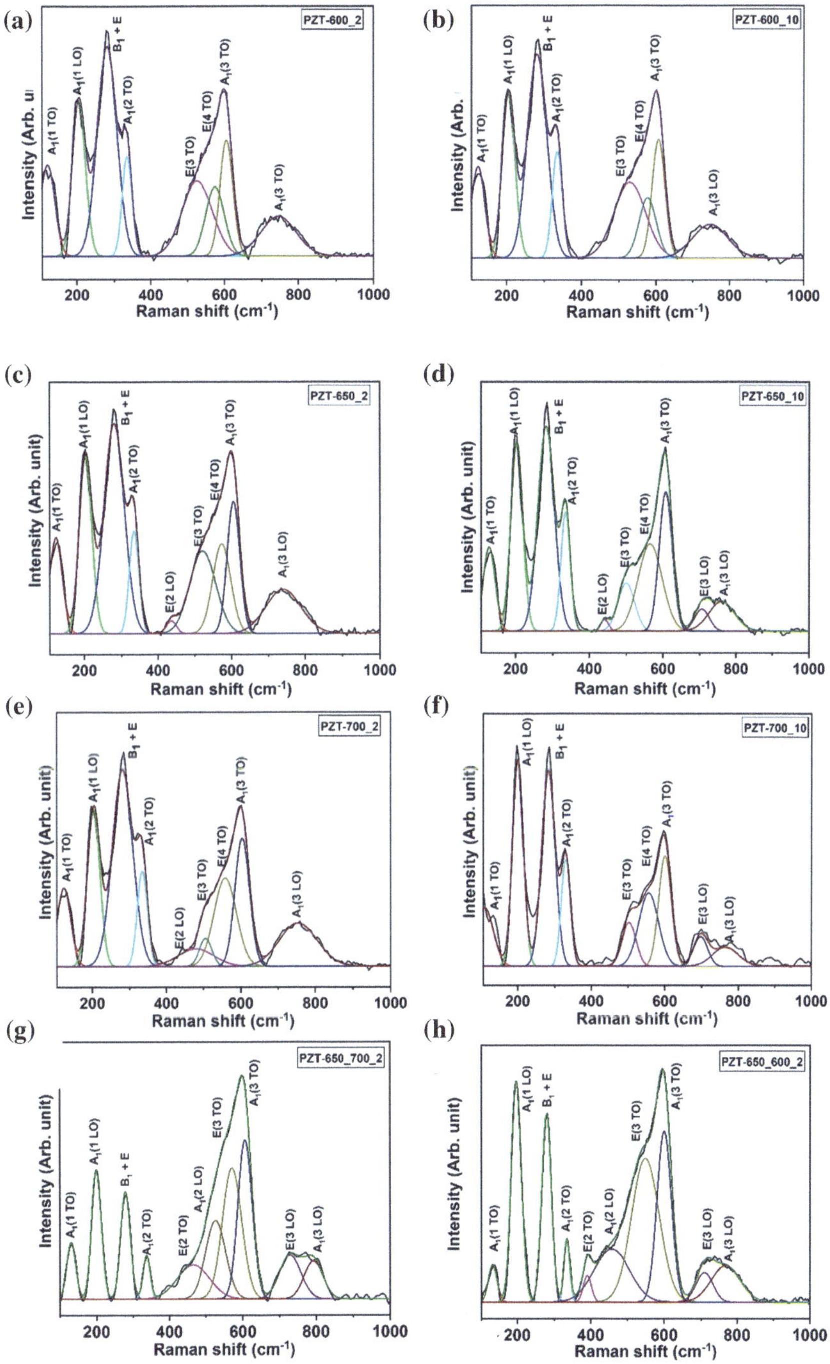

Fig. 3. Raman spectra of PZT thin films annealed at a) and b) 600°C, c) and d) 650 °C and e) and f) 700 °C at a heating rate of 2 °C/min and 10 °C/min, respectively. Raman spectra of PZT thin film annealed at 700 °C with an intermediate holding step at 650 °C for 1 h is shown in (g). Raman spectra of PZT thin film annealed at 650 °C and controlled cooling with intermediate holding at 600 °C for 10 min is shown in (h).

Table 1. Raman modes of the prepared PZT thin films annealed at different temperature at annealing rates of 2 and 10 °C/min.

Raman modes	Raman shift (cm^{-1})							
	PZT-600_2	PZT-600_10	PZT-650_2	PZT-650_10	PZT-700_2	PZT-700_10	Ref [16]	Ref [17]
A_1(1 TO)	120	123	123	128	125	109	133	140
A_1(1 LO)	206	206	204	204	204	201	192	-
B_1 + E	282	282	281	283	282	283	287	273
A_1(2 TO)	336	335	335	335	335	330	349	326
E(2LO)	-	-	437	442	480	-	400	466
E (3 TO)	527	530	522	500	506	503	-	535
E (4 TO)	575	578	574	565	558	556	-	-
A_1(3 TO)	606	607	606	608	604	600	592	603
E (3 LO)	-	-	-	-	-	697	678	702
A_1(3 LO)	746	748	740	760	754	765	843	750

4 Conclusion

PZT thin films were synthesized using the sol-gel method and analyzed using XRD and Raman spectroscopy. The effects of annealing temperature as well as the rate of annealing on the structural properties of PZT thin films were systematically analyzed by annealing the films at different temperatures (600, 650, and 700 °C) at different heating rates (2 and 10 °C/min). Intermediate holding steps were introduced in the annealing process to investigate their effect on the realization of phase pure PZT. XRD analysis revealed that phase pure perovskite PZT thin films were formed under optimum conditions, i.e., when the films were annealed at 650 °C at an annealing rate of 2 °C/min. Further investigations are needed to explore the role of intermediate holding steps in the annealing process to realize phase formation.

Acknowledgments. The authors acknowledge the Department of Physics, National Institute of Technology Calicut, for providing XRD facility (FIST SR/FST/PSI-113/2019). The authors also acknowledge the Centre for Materials Characterization, National Institute of Technology Calicut, for Raman spectroscopy measurements.

References

1. Shoghi, A.S., Abdizadeh, H., Golobostanfard, M.R.: Procedia Materials Science **11**, 386–390 (2015). https://doi.org/10.1016/j.mspro.2015.11.136
2. Rodríguez-Aranda, M.C., Calderón-Piñar, F., Mayén-Mondragón, R., Yáñez-Limón, J.M.: J. Mater. Sci. Mater. Electron. **26**, 3486–3492 (2015). https://doi.org/10.1007/s10854-015-2859-9
3. Gueye, G.L., Gergaud, P., Renault, O., Defay, E., Barrett, N.: Applied Surface Science **363**, 21–28 (2016). https://doi.org/10.1016/j.apsusc.2015.11.118
4. Prasad Bag, S., Her, J.-L., Pan, T.-M.: Materials Science in Semiconductor Processing **64**, 24–28 (2017). https://doi.org/10.1016/j.mssp.2017.03.007

5. Schwartz, R.W.: Chem. Mater. **9**, 2325–2340 (1997). https://doi.org/10.1021/cm970286f
6. Muralt, P., et al.: IEEE Trans. Ultrason. Ferroelect. Freq. Contr. **52**(12), 2276–2288 (2005). https://doi.org/10.1109/TUFFC.2005.1563270
7. Lu, G., Dong, H., Chen, J., Cheng, J.: J. Sol-Gel Sci. Technol. **82**, 530–535 (2017). https://doi.org/10.1007/s10971-017-4311-5
8. Shoghi, H.A., Shakeri, A., Golobostanfard, M.R.: J. Sol-Gel Sci. Technol. **93**, 623–632 (2020). https://doi.org/10.1007/s10971-019-05121-0
9. Liu, S., Zou, D., Yu, X., Wang, Z., Yang, Z., Appl, A.C.S.: Mater. Interfaces **12**, 54991–54999 (2020). https://doi.org/10.1021/acsami.0c16973
10. Monga, S., Tomar, S., Vilarinho, P.M., Singh, A.: Materials Today: Proceedings **36**, 616–620 (2021). https://doi.org/10.1016/j.matpr.2020.03.175
11. Hu, H., Peng, C.J., Krupanidhi, S.B.: Thin Solid Films **223**, 327–333 (1993). https://doi.org/10.1016/0040-6090(93)90540-6
12. Zhu, M.W., Wang, Z.J., Chen, Y.N., Kobayashi, T., Maeda, R.: Thin Solid Films **540**, 73–78 (2013). https://doi.org/10.1016/j.tsf.2013.06.014
13. Dang, E.K.F., Gooding, R.J.: Phys. Rev. Lett. **74**, 3848–3851 (1995). https://doi.org/10.1103/PhysRevLett.74.3848
14. Huang, Z., Zhang, Q., Whatmore, R.W.: J. Appl. Phys. **86**, 1662–1669 (1999). https://doi.org/10.1063/1.370944
15. Velu, G., Remiens, D., Thierry, B.: J. Eur. Ceram. Soc. **17**, 1749–1755 (1997). https://doi.org/10.1016/S0955-2219(97)00031-9
16. Chandran, M., Tiwari, B., Kumaran, C.R., Samji, S.K., Bhattacharya, S.S., Rao, M.S.R.: J. Phys. D Appl. Phys. **45**, 202001 (2012). https://doi.org/10.1088/0022-3727/45/20/202001
17. Rath, M., Kumar, D., Rao, M.S.R.: Appl. Phys. Lett. **113**, 032904 (2018). https://doi.org/10.1063/1.5035450

Spray Coated CuO Thin Film for Photocatalytic Degradation of Methyl Violet Dye

Yuvashree Jayavelu[1], Meghana Pavithran[2], Gouranga Maharana[1], K. T. Hari Govind[2], Alista Mariya Abraham[3], Kathyayani Nerella[4], Benjamin Hudson Baby[1], M. Kovendhan[4], and D. Paul Joseph[1](✉)

[1] Department of Physics, National Institute of Technology, Warangal, Telangana 506004, India
paul@nitw.ac.in

[2] Department of Physics, Amrita Vishwa Vidyapeetham, Coimbatore, Tamil Nadu 641112, India

[3] International School of Photonics, CUSAT, Kalamassery, Ernakulam, Kochi, Kerala 682022, India

[4] Department of Physics and Nanotechnology, SRM Institute of Science and Technology, Kattankulathur, Chennai, Tamil Nadu 603203, India

Abstract. Oxides of transition metals are widely explored for their optoelectronic properties; among them Copper oxide (CuO) is a wide bandgap semiconductor exploited for its photoconductive and photochemical properties. CuO thin film was prepared over glass substrate by the aerosol spray pyrolysis technique using only cupric nitrate trihydrate (Cu $(NO_3).3H_2O$). Structural analysis and phase purity of CuO film is explored by XRD and the surface topography are studied using FE-SEM. The optical band gap of 2.67 eV for CuO film is estimated using Tauc plot. Photocatalytic degradation of methyl violet organic dye is demonstrated using CuO thin film photocatalyst with an efficiency of 76.7% under white light LED for 180 min of illumination.

1 Introduction

Nanomaterials are widely exploited due to their wide range of significance in chemical, biological, and environmental applications. Metal oxide nanoparticles like TiO_2, ZnO, CuO, MnO_2, SnO_2, CdS, WO_3 etc., have extensive application due to their innate properties of being chemically stable and tunability of physio-chemical properties [1, 2]. Their properties and morphology is engineered for suitable applications in various fields including sensors, solar cells, electrocatalysis, photocatalysis, biological applications [3–6]. Among them, copper oxide nanoparticles with bandgap (E_g) of 1.2 to 2.5 eV as a p-type semiconductor makes it significant for solar energy conversion, batteries, humidity sensors, PN junction diodes, photocatalyst, antimicrobial, and antifouling and other biological applications [7, 8]. Monoclinic CuO crystal structure has large surface area, thermal conductivity, electrochemical redox potential and stability making them versatile for applications [7]. Different synthesis techniques including wet chemistry route, alkoxide-based preparation, hydrothermal technique, sol–gel, micro-emulsions,

P. Joseph D. et al. (Eds.): IEMDST 2024, SPPHY 567, pp. 156–160, 2025.
https://doi.org/10.1007/978-981-96-5863-3_20

aerosol spray pyrolysis etc., have been explored for the synthesis of the nanomaterials. The comparatively high precursor cost of these methods restricted their extensive usage [7–11]. In the field of Photocatalysis, CuO with lower bandgap can generate maximum reactive oxygen species (ROS) for visible and near infrared region photocatalysis [7]. Among the water purification techniques, photocatalytic degradation is extensively used to degrade pollutants in a cost effective and large-scale accessibility manner [2, 5]. Predominantly powder photocatalysts are used for laboratory scale photodegradation due to their cost effectiveness and ease of synthesis. Thin film photocatalysts can overcome self-contamination, in particular spray deposited films can overcome poor particulate adherence, by reducing peeling off of layers and easier cyclic usage [2, 12, 13]. This study focuses on facile, cost-effective immobilized copper oxide photocatalyst and demonstrates spray deposited immobilized CuO film for photocatalytic degradation of methyl violet (MV) organic dye in the visible light region by using the cost-effective white LED light.

2 Materials Used

All the procured precursor chemicals were of analytical grade and used as it is for spray deposition and photocatalytic activity without any further purification treatments. Methyl Violet (MV) dye was procured from Sigma-Aldrich. Deionized water (DI) was used for preparation of all the solutions for all the experiments.

3 Spray Deposition Process

CuO thin film is deposited on glass substrate (50 mm x 100 mm x 1 mm) by spray pyrolysis technique at a substrate temperature of 380 ± 5 °C. Initially, 0.5 M cupric nitrate trihydrate (Cu $(NO_3).3H_2O$) were allowed to dissolve in 100 ml of DI water. Then, the precursor solution was loaded in the reservoir and was sprayed on to the heated substrate at a distance of 30 cm at 30 kg/ cm^2 pressure. A total of 600 sprays were performed with 0.5 s spray ON and 30 s OFF durations. The fabricated CuO film is cooled down to room temperature and explored for photocatalysis.

4 Results and Discussion

Phase purity of CuO thin film investigated by XRD matched with the JCPDS file #96-901-4581 with space group C12/c1 of monoclinic structure (Fig. 1a). The highest intensity peak at (111) match with the 100% peak of the JCPDS file #96-901-4581. The surface topography confirms the formed film is dense and uniform. The morphology is granular with an average particle size of 134 nm (Fig. 1b). The film was dark in color with only a transmittance of 45% at 590 nm (Fig. 1c). The film has low transmittance in the visible region and comparatively higher transmittance in the IR region. The optical bandgap calculated from Tauc plot is found to be 2.67 eV (Fig. 1d). The film thickness measured using stylus profiler is 140 nm for this spray deposited film.

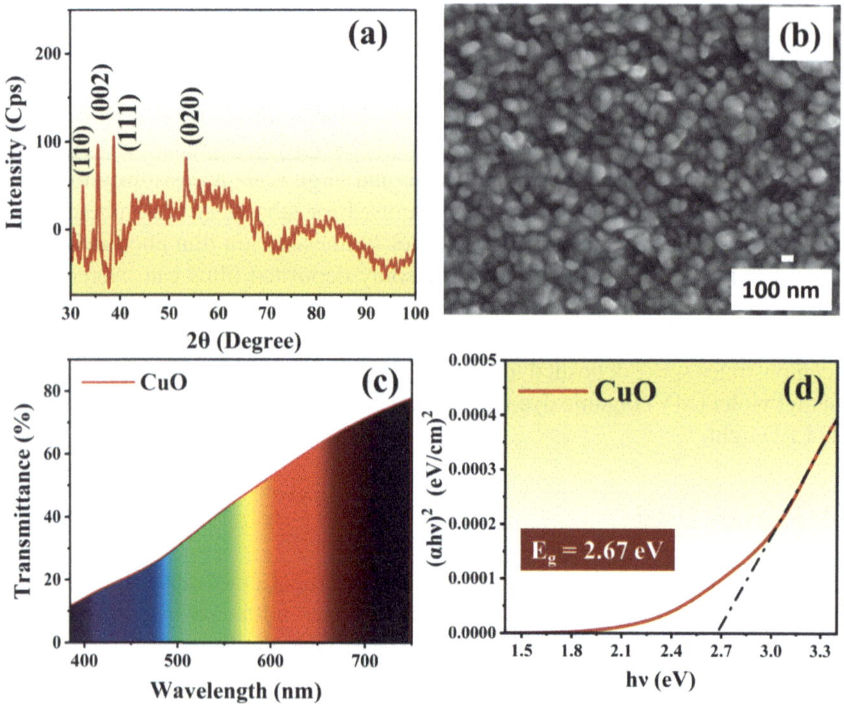

Fig. 1. (a) XRD pattern, (b) FE-SEM image showing granular morphology, (c) transmittance curve in the visible region and (d) optical bandgap calculated from tauc plot of the spray deposited CuO thin film.

The photodegradation of MV dye with maximum absorbance at 590 nm in the visible region is used as the model pollutant under white LED light of 200 W. CuO film (2 x 2 cm) is immersed in 5 ppm concentration of 6 ml volume of MV dye to demonstrate visible light photodegradation. Initially 30 min of adsorption-desorption equilibrium was carried out in dark for the pollutant MV dye to be effectively adsorbed on the film's surface. The absorbance spectra of photodegradation of CuO thin film is represented in Fig. 2a with respect to their irradiation time of 180 min with a step size of 30 min. The gradual decrease in concentration of MV dye is represented with respect to C/C_0 vs irradiation time plot (Fig. 2b). The photodegradation rate constant of 0.006 min^{-1} for 180 min (Fig. 2c) with bar graph representing degradation efficiency of photolysis (75. 83%) and photocatalysis (76.70%) of MV, indicating only a small improvement (Fig. 2d). Initial degradation efficiency was high with respect to photolysis, but it decreased with time which may be due to the pacification of the film's surface due to the degraded components settling on the surface. The photocatalytic degradation efficiency (η) of the CuO thin film is calculated using the relation [4],

$$\eta = \left[\frac{C_0 - C}{C_0}\right] \times 100\% \quad (1)$$

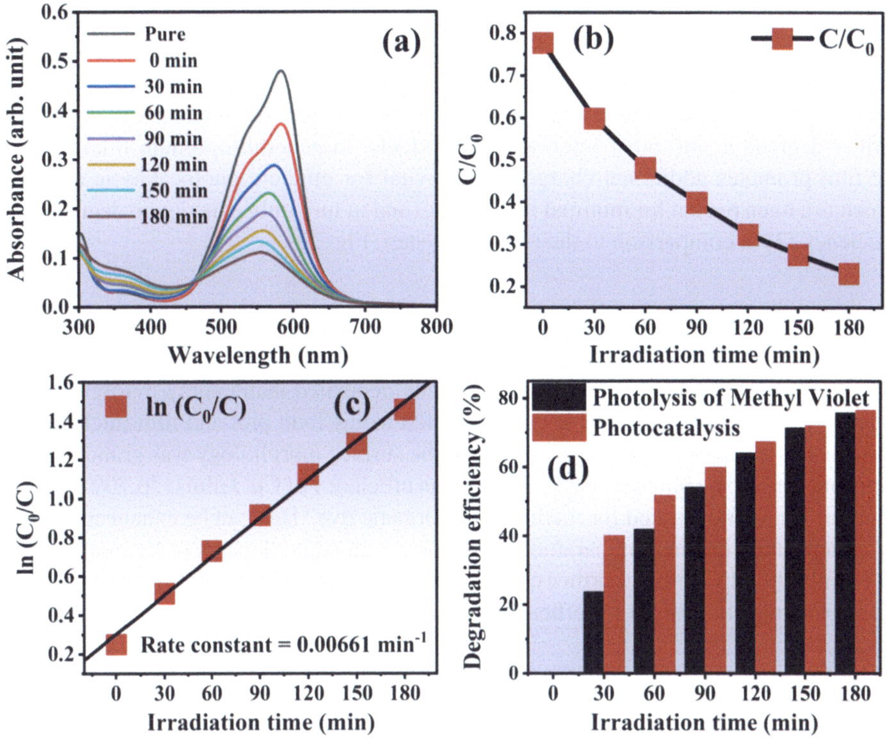

Fig. 2. (a) Absorbance spectra, (b) (C/C_0) Vs irradiation time, (c) $\ln(C_0/C)$ Vs irradiation time (d) bar graph representing degradation efficiency for CuO thin film in comparison with photolysis of methyl violet dye.

where C_0 is initial concentration, C is concentration of the pollutants at varying duration of irradiation.

The following plausible photocatalytic degradation mechanism of MV dye is proposed below,

$$CuO + h\upsilon \rightarrow e^- + h^+ \tag{2}$$

$$h^+ + H_2O \rightarrow {}^{\cdot}OH + H^+ \tag{3}$$

$$O_2 + e^- \rightarrow {}^{\cdot}O_2^- \tag{4}$$

$$MV + h\upsilon \rightarrow MV* \tag{5}$$

$$MV * + {}^{\cdot}O_2^- \text{or} {}^{\cdot}OH \rightarrow \text{Degraded Products} \tag{6}$$

The preliminary step includes adsorption of the dye onto the CuO thin film surface during the adsorption-desorption equilibrium. As the white LED light strikes the MV adsorbed CuO film, the electron–hole (e^-–h^+) pairs are generated [2]. These photogenerated holes react with H_2O molecules producing $^{\cdot}OH$ radicals [2, 4]. Similarly, the electrons generated react with oxygen and instigate highly reactive $^{\cdot}O_2^-$ radicals [4]. These highly reactive radicals react with methyl violet dye adsorbed on CuO film and further degrade it into other intermediate products. In general, optimum thickness of the film promotes additional charge carriers vital for efficient photocatalytic activity, which is a main reason for minimal film contact and in turn only a moderate degradation efficiency [3] in comparison to the photolysis step (Fig. 2d).

5 Conclusion

CuO thin film of monoclinic structure has been deposited using spray pyrolysis technique. Optical bandgap of 2.67 eV is calculated using Tauc plot and film thickness of 140 nm was measured using stylus profiler. The surface morphology was granular, uniform and dense. The photocatalytic degradation efficiency of CuO films (76.70%) within 180 min was demonstrated for methyl violet organic dye. This can be enhanced further by either tuning the coating parameters, altering with some dopants or by maintaining an optimal film thickness. Further exploration in this direction by tuning several other parameters may be fruitful for efficient and cost-effective photocatalysts.

References

1. Huang, X., Zhao, G., Wang, G., Irvine, J.T.S.: Chem. Sci. **9**, 3623–3637 (2018)
2. Maharana, G., Muniramaiah, R., Yuvashree, J., Mandal, D., et al.: Surf. Interfaces **42**, 103413 (2023)
3. Behnam, M.A., Emami, F., Sobhani, Z., Dehghanian, A.R.: Iran. J. Basic Med. Sci. **21**, 1133 (2018)
4. Jayavelu, Y., Maharana, G., Joseph, D.P., Divyadharshini, S., et al.: Electrochim. Acta. 144777 (2024)
5. Jayavelu, Y., Maharana, G., Rajender, G., Muniramaiah, R., Divyadharshini, S., et al.: Chemosphere **351**, 141128 (2024)
6. Reddy, N.P., Santhosh, R., Thogiti, S., Muniramaiah, R., Joseph, D.P., Murali, B.: J Mater Chem C Mater **10**, 1403–1413 (2022)
7. Diachenko, O., et al.: Coatings 11 (2021)
8. Aktar, A., Ahmmed, S., Hossain, J., Ismail, A.: ACS Omega **5**, 25125–25134 (2020)
9. Mahana, D., Mauraya, A.K., Pal, P., Singh, P., Muthusamy, S.K.: Mater. Res. Bull. **145**, 111567 (2022)
10. Menazea, A.A., Mostafa, A.M.: J. Environ. Chem. Eng. **8**, 104104 (2020)
11. Ahmed, A.M., Abdalla, E.M., Shaban, M.: The Journal of Physical Chemistry C **124**, 22347–22356 (2020)
12. Huang, A., Cao, L., Chen, J., Spiess, F.-J., Suib, S.L., et al.: J. Catal. **188**, 40–47 (1999)
13. Fernandes, J.M., Muniramaiah, R., Maharana, G., Janbandhu, S.Y., et al.: J. Mater. Sci.: Mater. Electron. **34**, 2131 (2023)

Layered N Doped-CQDs/ZnO NRs Heterostructure Thin Films for Improved UV Photodetector Performance

Lavanya Thyda, Koppula Naresh, Joel K. Joseph, S. Suneetha, and Kuppusamy Thangaraju(✉)

Organic Optoelectronic Device Laboratory, Department of Physics, National Institute of Technology, Warangal, Telangana 506004, India
ktr@nitw.ac.in

Abstract. The layered heterostructure of Nitrogen-doped carbon quantum dots (N doped-CQDs) spin-coated onto the hydrothermally grown ZnO nanorods surface (N doped-CQDs/ZnO NRs) was prepared and characterized for their potential application in the UV photodetection. Photoluminescence study exhibits the decreased intensity of near-band-emission (400 nm) of layered N doped-CQDs/ZnO NRs heterostructure due to the energy transfer from N doped-CQDs to surface states of ZnO nanorods and decreased the recombination of charge carriers. N doped-CQDs/ZnO NRs heterostructure showed the improved photocurrent density of 105.7 μA/cm^2, higher responsivity (105.3 mA/W) and EQE (35.8%) at 1 V [UV (365 nm) of 1 mW/cm^2] when compared to that (42.2 μA/cm^2, 41.98 mA/W and 14.2% at 1 V) of bare ZnO nanorod based device.

1 Introduction

Over the past decade, the UV photodetectors (UV PDs) have been of most important devices for various applications such as environmental monitoring, fire-detection, bioimaging, camera sensors, military, and industrial applications [1]. Until recently the conventional Si- based photodetector devices had widely been used as a prominent device for UV photodetection but have some limitations such as requiring filters for the blocking the infrared and limited absorption of visible photons [1, 2]. The broad bandgap semiconductors such as TiO_2, SnO_2, GaN, SiC, and ZnO have been made available and became common choices for UV photodetection since they show the significant ultraviolet (UV) absorption and efficient performances. Among the various semiconducting materials, the ZnO has become a promising material for efficient UV detection because of its superior carrier mobility, wideband gap (3.3 eV), easy synthesis process. ZnO nanostructures exhibited the excellent UV detection performance owing to their huge surface to volume ratio and high electron mobility [3]. Because of their high surface area, the ZnO nanostructures showed more adsorption and desorption of oxygen atoms on the surface [4]. Carbon quantum dots (CQDs) has recently been explored to the extent possible in the optoelectronic devices owing to its good thermal stability, easy synthesis, increased electrical conductivity and excellent absorption of UV and visible photons

P. Joseph D. et al. (Eds.): IEMDST 2024, SPPHY 567, pp. 161–167, 2025.
https://doi.org/10.1007/978-981-96-5863-3_21

[5–7]. The decoration of CQDs onto the ZnO surface has formed the active interface through non-covalent bonding which efficiently enhance the separation of charges and transfer electrons from CQDs to the conduction band (CB) of ZnO due to the excellent alignment of energy levels, resulting in the improved performances of photodetectors [5–9].

In the present work we deposited N doped-CQDs layer onto the hydrothermally grown ZnO nanorods surface (Layered structure of N doped-CQDs/ZnO NRs) by using the spin coating method. The prepared layered N doped-CQDs/ZnO NRs heterostructure thin films were characterized for their optical and electrical properties in order to understand their effective use in the UV photodetectors. Al metal contacts are deposited onto the layered N doped- CQDs/ZnO NRs heterostructure to get the active device area of 0.08 cm^2. The fabricated layered N doped-CQDs/ZnO nanorod heterostructure film-based UV PDs were characterized to estimate the current density (J - V), photoresponse (I - t), responsivity, EQE, and detectivity.

2 Materials and Methods

2.1 Hydrothermally Grown ZnO Nanorods

The ZnO seed layer preparation was followed as-such reported in our earlier work published elsewhere [8, 10]: zinc acetate dihydrate (0.5 M) (Sigma Aldrich; 99.9%) dissolved in 15 mL of 2-methoxy ethanol and being added by stabilizer monoethanolamine (MEA). The above precursor was stirred for 2 h at 65 °C and spin-coated at 2500 rpm for 30 s onto the cleaned glass substrates in order to get the ZnO seed layer. It was then annealed for 10 min at 180 °C and this procedure was repeating for 5 coatings to get the optimal thickness (~200 nm) with uniformity. The prepared seed layer film was annealed for 2 h at 350 °C to improve the film quality. The ZnO NRs were grown onto the prepared ZnO seed layer by the hydrothermal process using the precursor solution [0.05 M of zinc nitrate dehydrate and 0.05 M of hexamethyl tetra amine (HMTA) dissolved in 100 mL of Deionized water (DI) and stirred for 1 h at normal temperature. Hydrothermal reaction took place for 4 h at 95 °C for the growth of NRs. The grown ZnO NRs were post-annealed for 2 h at 400 °C to improve the crystalline properties [named as film (**a**) [7]].

2.2 Deposition of N Doped-CQDs Layer onto ZnO Nanorods Thin Films

The synthesis of Nitrogen-doped carbon quantum dots (N doped-CQDs) using the citric acid and urea as the sources for carbon and nitrogen, respectively using the hydrothermal method (reacted at 180 °C for 8 h). Then, 0.8 mg of N doped-CQDs powder [26.66 $\times$ 10^{-3}%(w/v)] was dissolved into 3 mL of 2 -methoxy ethanol and stabilizer mono ethanolamine at 1:1 ratio. The resultant solution was sonicated for 30 min, then stirred at 65 °C for 16 h [5, 8]. Finally, the obtained solution was used to spin-coat onto the hydrothermally grown ZnO nanorods at 2000 rpm. The N doped-CQDs/ZnO NRs heterostructure film obtained by spin-coating (single-coating using 0.12 mL of precursor solution) of N doped-CQDs layer onto the surface of ZnO NRs film is named as film

b and the heterostructure film obtained by spin-coating (triple-coating of 0.36 mL of precursor solution) named as film **c**. All the prepared heterostructure thin films were post-annealed at 110 °C for 1 h to evaporate the residue solvents.

2.3 UV Photodetectors Device Fabrication

The layered N doped-CQDs/ZnO NRs heterostructure thin films-based UV Photodetectors device fabricated by depositing the Al metal electrodes with 120 nm thickness on the N doped-CQDs coated ZnO NRs heterostructure film under the vacuum of 5.5×10^{-5} mbar using the thermal evaporation method so as to get the active device area of 0.08 cm^2. The pristine ZnO nanorods based device is named as device **A**, the film **b**-based device is named as device **B** and that of film c is named as device **C**.

2.4 Instrumentation

The PL characteristics of layered N doped-CQDs/ZnO heterostructure films (**a** – **c**) were carried out by the fluorescence spectrometer (Horiba Jobin-Yvon, Flourolog-3 model). The current versus voltage (I – V) and photoresponse (I-t) measurements of devices (**A** – **C**) were obtained using source measure unit (SMU) [Keithley (2450 model)] under the exposure of UV light (UV spot-light source: 365 nm, Hamamatsu Japan, L9588-06A model and UV light intensity of 1 mW/cm^2).

3 Results and Discussion

The PL spectra of layered N doped-CQDs/ZnO NRs heterostructure films were acquired by exciting them at 325 nm as in the Fig. 1. The prepared ZnO NRs film (**a**) reveals the near-band-edge (NBE) emission at 402 nm, attributed to the free exciton recombination. The widely observed broad defect level emission peaked 563 nm ascertain the native defects (interstitials, vacancies) of ZnO. There is no much change in the peak position of near-band-edge emission of layered N doped-CQDs/ZnO NRs heterostructure films (**b** and **c**) but emission intensity got decreased as compared with film (**a**), ascribing to the N doped-CQDs adsorbed onto the ZnO nanorods surface, subsequently creating the depletion layer at the interface, and reducing the charge recombination [5, 6]. It is also observed that the slight increase in the intensity of defect emission in the film (**b**) and (**c**) when compared with film (**a**) can affirm the N doped-CQDs adsorbed on the surface of ZnO NRs and facilitating that energy transfer from N doped-CQDs to defect levels of ZnO.

The layered N doped-CQDS/ZnO NRs heterostructure film- based devices **A** - **C** exhibited the ohmic behaviour of current density versus voltage (J - V) under the dark and UV illumination of 365 nm with light intensity (1 mW/cm^2) as reveals in the Fig. 2. Device **A** shows the current densities of 1.55/1.47 μA/cm^2 at (8/−8 V) and 0.2/0.19 μA/cm^2 at (1/−1 V)] under dark condition, whereas when illuminated with the UV light of 1 mW/cm^2, it revealed the increased current density of 346.9/246 μA/cm^2 at (8/−8 V) and 42.2/40.7 μA/cm^2 at (1/−1 V) (Fig. 2(i)) due to increased photogenerated charge carries and efficient transport and collection of charge carriers [5, 6]. The device

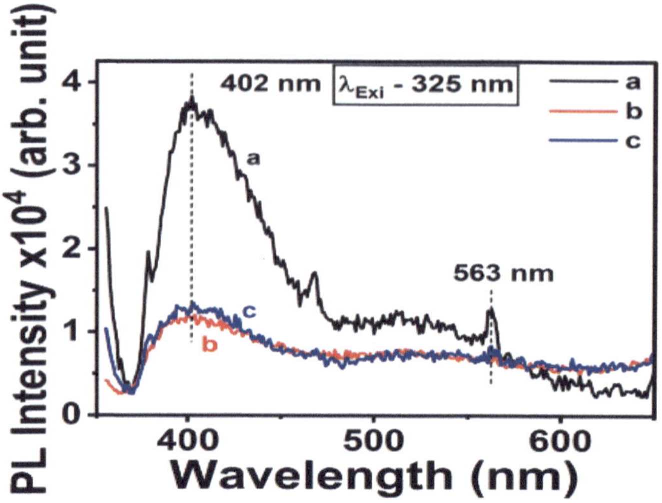

Fig. 1. The Photoluminescence spectra of layered N doped-CQDs/ZnO NRs heterostructure films (a - c) excited by 325 nm.

B exhibited the higher dark current density of 4/7.9 μA/cm^2 at (8/−8 V) and 0.47/0.46 μA/cm^2 at (1/−1 V)] compared with device **A**, attributed to the enhanced crystalline properties of grown ZnO NRs. Under the illumination of UV light, the device **B** reveals the higher photocurrent density of 494/461.4 μA/cm^2 at (8/−8 V) and105.7/89 μA/cm^2 at (1/−1 V), attributed to the fact that photogenerated electrons are transferred from N doped-CQDs to conduction band (CB) of ZnO and eventually collected by the electrodes. The device **C** exhibited the reduced dark current density [0.52/0.47 μA/cm^2 at (8/−8 V) and 0.07/0.06 μA/cm^2 at (1/−1 V)] than that of devices **A** and **B**, corroborated to the enhanced interfacial potential due to the thick layer of N doped-CQDs forming the surface of ZnO nanorods [5, 6]. It reveals the less photocurrent density [334.4/127.2 μA/cm^2 at (8/−8 V) and 48.7/32.2 μA/cm^2 at (1/−1 V)] as compared with device **A** and **B**, determining the increased energy barrier at the N doped-CQDs/ZnO interface which hinders the charge transfer to ZnO and reducing the charge conducting path in the ZnO NRs.

The photoresponse measurements [time dependent current (I - t)] of devices (**A** - **C**) as given in Fig. 2(**ii**) are acquired for ON/OFF (switching) ratio of 30 s upon the 365 nm UV light with intensity (1 mW/cm^2) at 1 V. Device **A** revealed good current stability and reproducibility with rise-time (the duration required for the photocurrent to rise from 10% to 90%) of 14.3 s and fall-time (the essential time for current to fall from 90% to 10%) of 19.4 s at voltage of 1 V. It has been reported that (i) in the normal ambient under dark, when the oxygen molecules are adsorbed onto the ZnO nanorod's surface by acquiring the free electrons and forming anions [O_2 (gas) + $e^- \rightarrow O_2^-$], making the high potential gradient close to the interface and resulting the reduced dark current and (ii) upon the exposure of UV, the photoexcited holes move to the surface of ZnO and desorb the adsorbed oxygen molecules [$h^+ + O_2^- \rightarrow O_2$] [11]. Then, photogenerated free electrons subsequently produce photocurrent in the device **A**. The holes are captured

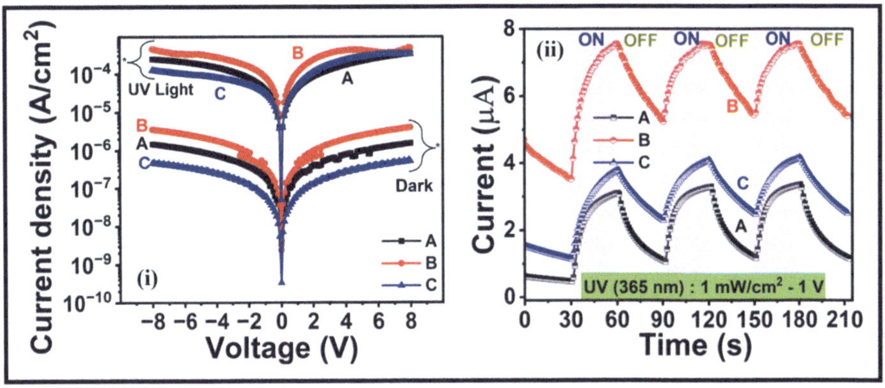

Fig. 2. (i) The current density of layered N doped-CQDs/ZnO NRs heterostructure device (A - C) under dark and under the 365 nm UV light with intensity (1 mW/cm^2), and (ii) I-t photoresponse of layered N doped-CQDs/ZnO NRs heterostructure devices (A - C) at bias 1 V under the 365 nm UV light with intensity (1 mW/cm^2).

at the ZnO surface states generating the photo persistentcurrent when the UV light is switched OFF. The device **B** reveals the enhanced photoresponse current of 7.4 µA due to the effective charge transfer between the N doped-CQDs and ZnO nanorod surface [5–7]. The rise-time of 17 s and fall-time of 23 s are slightly higher than that of device A because of the slow adsorption and desorption process due to the occurrence of adsorbed N doped-CQDs. The device **C** exhibits the slightly less photoresponse current of 3.7 µA with rise-time 19 s and fall-time 23 s as compared with that of the device B, due to the excess amount of N doped-CQDs are covered the surface of ZnO NRs (effective covering) and it is obstructing UV light from entering into the ZnO which results into decreased photocurrent [6]. The Responsivity (R), EQE (η) and Detectivity (D) of devices (**A – C**) are determined using the following equations and shown in Fig. 3. $R = J_{ph}/P$, $EQE = 1240J_{ph}/\lambda P$, and $D = (2qJ_{dark})^{1/2}/R$, where J_{ph} is the photocurrent density, P - incident light, q - charge of electron, J_{dark} - dark current density and λ - wavelength of UV light.

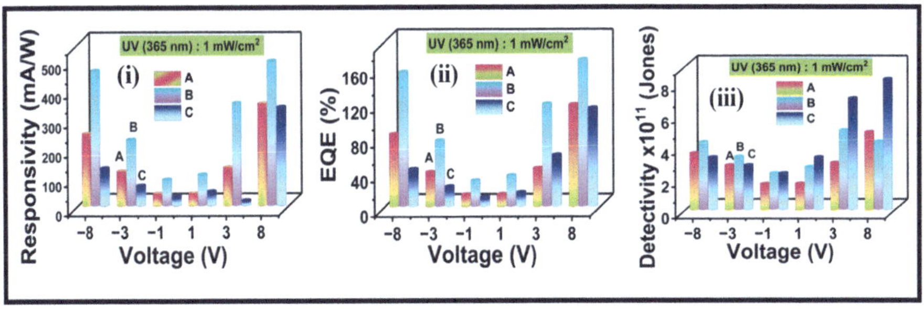

Fig. 3. (i) Responsivity (R), (ii) EQE (η) and (iii) Detectivity (D) of layered N-CQDs/ZnO nanorods heterostructure based devices (A- C) under the 365 nm UV light with the intensity (1 mW/cm^2).

The device **A** reveals the responsivity [345.4/245.3 mA/W at (8/−8 V) and 42/40.5 mA/W at (1/−1 V)] at different voltages (Fig. 3(i)). The device **B** exhibits the enhanced responsivity [490/458 mA/W at (8/−8 V) and 105.3/88.5 mA/W at (1/−1 V)] which is 2.7 times greater as compared with device A, ascribing the efficient charge transfer from N doped-CQDs to ZnO nanorods and resulting in the increased photo-generated charge carriers. The increased content of N-CQDs onto the ZnO nanorods surface reduces the responsivity [334.4/127.3 mA/W at (8/−8 V), 48/32.2 mA/W at (1/−1 V) for device **C**], attributed to the increased energy barrier at the interface of N doped-CQDs/ZnO NRs due to effective coverage of N doped-CQDs on the ZnO NRs and resulting in the reduced current conducting path in the film [5, 6]. The device **A** exhibits the EQE of 117.3/83.3 mA/W at (8/−8 V) and 14.2/88.5 mA/W at (1/−1 V) (Fig. 3(**ii**)). The 2.7 times higher EQE of 169.8/155.5% at (8/−8 V) and 35.8/30% at (1/−1 V) was observed for the device **B**. Device **C** shows the reduced EQE [113.6/43.2% at (8/−8 V), 16.5/10.9% at (1/−1 V)] than that of device **B**. It is also observed that more than 100% efficiency is observed at higher voltages (±8 V) due to the multiple photons are absorbed onto the nanorods surface and/or efficient charge transfer from N doped-CQDs to ZnO NRs [12, 13]. The Fig. 3(**iii**) reveals the obtained detectivity [4.89/3.6 × 10^{11} (Jones) at (8/−8 V) and 1.65/1.63 × 10^{11} (Jones) at (1/−1 V)] for device **A**. The device **B** reveals the detectivity of 4.28/4.25 × 10^{11} Jones at (8/−8 V) and 2.69/2.28 × 10^{11} Jones at (1/−1 V) which are slightly enhanced as compared to that of device **A**. At the higher voltage (8 V), the detectivity has slightly decreases this may be attributed to the nonlinear behaviour of electrodes. Enhanced detectivity of 8.17/3.29 × 10^{11} Jone at (8/−8 V) and 3.629/2.28 × 10^{11} Jones at (1/−1 V) in device **C** is related to the decreased dark current density.

The layered N doped-CQDs/ZnO NRs heterostructure thin films-based devices exhibit the significantly enhanced performance (2.7 times higher) as compared with bare ZnO NRs based device at a significantly low voltage of 1 V. This may be due to the fact decorated N doped-CQDs not only absorb the UV light but also effectively transfer the photogenerated charge carriers to the surface states of ZnO NRs, resulting in the increased UV photodetector performances. The prepared layered N doped-CQDs/ZnO heterostructure films will be advantageous in the optoelectronic devices particularly in the UV PDs.

4 Conclusion

The layered N doped-CQDs/ZnO NRs heterostructure films showed the reduced intensity of near-band-edge PL emission, ascertaining the N doped-CQDs adsorbed onto the ZnO NRs surface and possible energy transfer from N doped-CQDs to ZnO surface energy states. The layered N doped-CQDs/ZnO NRs heterostructure thin films-based UV photodetectors exhibit the higher photoresponsivity (105.3 mA/W) and EQE (16.5%) which are 2.7 times greater and the detectivity 2.5 times greater compared with bare ZnO NRs based device at a low bias of 1 V, ascertaining that the decorated N doped-CQDs not only absorb the UV light but also effectively transfer the photogenerated charge carriers to the surface states of ZnO NRs, resulting in the increased UV photodetector performances. These results are beneficial for the fabrication of low-cost UV PDs for commercial applications.

Acknowledgements. The authors acknowledge that the Scheme for promotion of academic and research collaboration (SPARC) project (Phase-II) (project ID: P2268) for the financial support. The authors acknowledged to the Department of Physics and Central Instrumentation Research Facility (CRIF) for providing the experimental and characterization facilities.

References

1. Lam, K.-T., et al.: Nanoscale Research Letters **12**, 1–7 (2017). https://doi.org/10.1186/s11671-016-1818-6
2. Jin, Y., Wang, J., Sun, B., Blakesley, J.C., Greenham, N.C.: Nanoletters **8**, 1649–1653 (2008). https://doi.org/10.1021/nl0803702
3. Jun, J.H., Seong, H., Cho, K., Moon, B.-M., Kim, S.: Ceramics International **35**, 2797–2801 (2009). https://doi.org/10.1016/j.ceramint.2009.03.032
4. Li, Q.H., Gao, T., Wang, Y.G., Wang, T.H.: Appl. Phys. Lett. **86**, 1–3 (2005). https://doi.org/10.1063/1.1883711
5. Thyda, L., et al.: Applied Surface Science **611**, 155664 (2023). https://doi.org/10.1016/j.apsusc.2022.155664
6. Thyda, L., et al.: Thin Solid Films **790**, 140221 (2024). https://doi.org/10.1016/j.tsf.2024.140221
7. Tran, M.H., Park, T., Hur, J.: Appl. Surf. Sci. **539**, 14822 (2021). https://doi.org/10.1016/j.apsusc.2020.148222
8. Thyda, L., et al.: J. Mater Sci: Mater Electron **35**, 64 (2024). https://doi.org/10.1007/s10854-023-11768-x
9. Liu, D., et al.: Nanoscale Research Letters **13**, 261 (2018). https://doi.org/10.1186/s11671-018-2672-5
10. Dasi, G., et al.: Thin Solid Films **710**, 138265 (2020). https://doi.org/10.1016/j.tsf.2020.138265
11. Das, S.N., et al.: Applied Physics Letters **97**, 022103 (2010). https://doi.org/10.1063/1.3464287
12. Dhar, S., Majumder, T., Mondaljumder, S.P.: ACS Appl. Mater. Interfaces **8**, 31822–31831 (2016). https://doi.org/10.1021/acsami.6b09766
13. Goswami, L., Aggarwal, N., Verma, R., Bishnoi, S., Husale, S., Gupta, R.P.G.: ACS Appl. Mater. Interfaces **12**, 47038–47047 (2020). https://doi.org/10.1021/acsami.0c14246

Carbon Quantum Dots Incorporated Li-Doped ZnO (CQDs/LZO) Hybrid Thin Films for Improved UV Photodetection

Joel K. Joseph, Lavanya Thyda, Koppula Naresh, S. Suneetha, and Kuppusamy Thangaraju(✉)

Organic Optoelectronic Devices Laboratory, Department of Physics, National Institute of Technology, Warangal, Telangana 506004, India
ktr@nitw.ac.in

Abstract. Organic-inorganic hybrid photodetector devices exploit the advantages of both such as low ionization potential and low toxicity of the organic molecules and high electron affinity and mobility of inorganic semiconductors. The incorporation of carbon quantum dots (CQDs) in the lithium doped ZnO (LZO) thin film has exhibited enhanced UV absorption and higher transmittance (89%) due to the quantum confinement effect. It has also improved the UV Photodetection performances [responsivity (57 mA/W), EQE (20%) and photocurrent (2.1 μA) at 5 V] when in comparison with LZO based devices, attributed to the efficient energy transfer between LZO and CQDs through the heterojunction. The CQDs/LZO hybrid thin films showed the photo-response (2.1 μA) at 3 V which is 3 times higher than that (0.7 μA) of LZO films. The slightly increased band gap of the hybrid thin film may be attributed to adsorption of CQDs on to the surface of the LZO and controlling the particle size (quantum confinement effect).

1 Introduction

Organic/inorganic hybrid thin films have recently emerged as promising materials for the optoelectronic applications. The long-term stabilities and high mobilities of the carriers inside inorganic materials, together with the levels of cost-effectiveness and flexibilities of organic semiconductor materials, make them the optimum alternatives for their application in hybrid systems. The ZnO has gained much interest among other metal oxides such as TiO_2, MgO, V_2O_5, and so on due to its immense potential use in the optoelectronic devices owing to its long-term stability, higher transmittance, good UV-visible light absorption, and solution processability [1]. Lithium doped ZnO (LZO) has widely been used as an effective material in the transparent conductive oxides (TCOs) and UV photodetectors [2]. Li atoms possess p-type characteristics, helps to reduce the concentration of zinc vacancies which are the defects that can negatively impact the material's electronic properties and also contribute in altering the optical characteristic. This reduction in defects concentration contributes to better crystallinity and hence better photdetection responses. Without appreciable change in the work function and valence band edge, the improved conductivity is achieved upon doping of lithium ions (intercalated) in the ZnO lattice. Because of their exceptional thermal stability, unique broad

P. Joseph D. et al. (Eds.): IEMDST 2024, SPPHY 567, pp. 168–173, 2025.
https://doi.org/10.1007/978-981-96-5863-3_22

emission, minimal cytotoxicity, and adjustable fluorescence emission, carbon quantum dots (CQDs) have demonstrated great potential for use in optoelectronic applications in recent decades [3]. CQDs have an unique ability to broaden the light absorption spectrum of ZnO, allowing it to capture not only UV light but also visible light. This is achieved through the introduction of additional energy levels within the bandgap, which facilitates the absorption of lower-energy photons. The CQDs can also exhibit up-conversion luminescence, further enhancing the light harvesting capabilities of the composite material. The zero-dimensional confinement of CQDs facilitates inter-sub-band transitions between the quantized energy levels, enabling broadband photodetection from UV to near-infrared. The CQDs can also improve the stability of ZnO-based photodetectors under ambient conditions. The organic components in CQD composites help to mitigate degradation effects commonly associated with pure inorganic materials, thereby enhancing device longevity and reliability. Deep-UV photoluminescence and photodetection are possible using CQDs because of its absorption peak at ~250 nm, which is obtained from the π–π^* transitions in the sp_2 framework. The type-I heterojunction formed between CQDs and ZnO allows for the fast response and extended broadband photodetection. CQDs-sensitized ZnO photodetectors show faster response times compared to bare ZnO [4]. This research work focusses on improving the photodetection capabilities of LZO thin films with the incorporation of CQDs.

2 Materials and Methods

Zinc acetate dihydrate served as the zinc supply material. Both the solvent and the stabilizer, 2-methoxyethanol and monoethanolamine (MEA), were employed in the process, with them mixed in a 1:1 ratio to create 0.5 M of ZnO sol-gel precursor solution.[3] Lithium chloride (LiCl) was added into the ZnO precursor solution in three different weight percentages (2%, 4% and 6%) named as solutions L2ZO, L4ZO, and L6ZO, respectively. These solutions were then filtered with a syringe filter of 0.22 μm and was left undisturbed for 24 h for gelation [5]. Carbon quantum dots (CQDs) were prepared by bottom-up approach using citric acid and urea. The CQDs precursors were dispersed in 100 mL of DI water and was made to undergo 4 h of hydrolysis. The resultant brown-black solution was heated at 100° to evaporate the residue solvent. The solute particles were rinsed using ethanol for 3 times to remove the leftover solvent and unreacted precursors. 3 mg of the prepared CQDs was dispersed in 3 mL of 2-methoxyethanol solvent. 1 mL of CQDs dispersed solution was added to 3 mL of each LZO solutions (L2ZO, L4ZO, and L6ZO) to get three more solutions named as CL2ZO, CL4ZO, and CL6ZO respectively. All the 6 solutions were then spincoated at 2000 rpm onto the glass substrates and post-annealed at 350 °C for 2 h under ambient atmosphere. The optical studies of six spin-coated thin films such as L2ZO, L4ZO, L6ZO, CL2ZO, CL4ZO, and CL6ZO were characterized using UV–Visible Spectrophotometer (Agilent Carry 5000). The Al electrodes with active device area of 0.16 cm^2 were deposited onto the above six thin films by thermal evaporation process under the vacuum of 5×10^{-6} mbar for UV photodetection. The photoresponse characteristics of the above deposited thin films were carried out using a keithley source meter (SMU Keithley 2450 model) and UV spot-light source of wavelength 365 nm (3 mW/cm^2).

3 Results and Discussion

The solution processed (spin-coated) lithium (2%, 4%, and 6%) doped ZnO (LZO) thin films were named L2ZO, L4ZO, and L6ZO, respectively, whereas the CQDs (0.37 M) incorporated lithium (2%, 4%, and 6%) doped ZnO (CQDs/LZO) thin films were named CL2ZO, CL4ZO, and CL6ZO. The transmittance spectra of CQDs/LZO hybrid thin films are shown in the Fig. 1(i).

The hybrid thin films showed the hightest average transmittance (89.3%) in the visible region when compared to that (84.9%) of LZO thin films, attributed to the improved film quality of LZO films upon incorporation of CQDs due to the quantum confinement effect. It also exhibited increased transmittance in the UV region (310 nm – 365 nm) as indicated by arrow in the Fig. 1(i). These results depict the fact that the LZO film quality is improved upon CQDs incorporation. The direct bandgap is determined from the Tauc plot using the relation: $(\alpha h\upsilon)^2 = \beta\,(h\upsilon - E_g)$ where α is the optical absorption coefficient, hν is the energy of incident photon, E_g is the bandgap energy and β is constant. It is observed that the band gap values are slightly increased (4.0 eV) upon incorporation of CQDs in LZO film from that (3.9 eV) of bare LZO film, ascertaining the adsorption of CQDs onto the surface of LZO and control the particle size (improved film quality). A slight shift in the band gap of hybrid thin film when compared to that of bare LZO reveal effect of confinement of electrons (quantum confinement effect) and quantization of energy states in the LZO for increasing CQDs content.

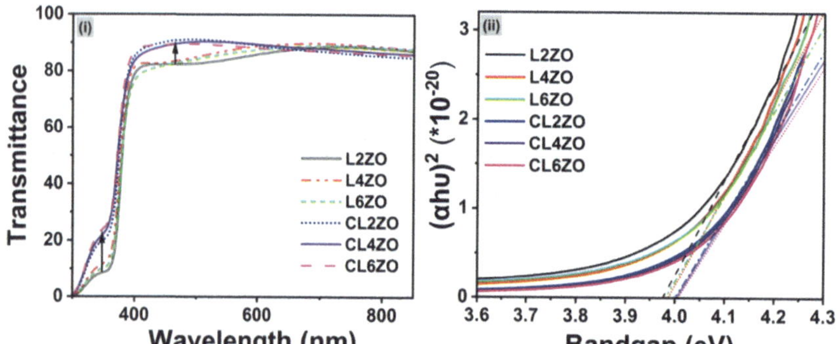

Fig. 1. (i) Transmittance of the films and (ii) Bandgap of the films found using Tauc relation.

The Al electrodes with an active device area of 0.16 cm^2 were deposited by Thermal evaporation method on the above three LZO thin films (L2ZO, L4ZO and L6ZO) to carry out UV photodetection measurements. The linear J-V characteristics of LZO based photodetector devices ascribe the ohmic nature of the metal contacts. LZO devices (L2ZO, L4ZO and L6ZO) excibited dark current densities of 6.4×10^{-5}, 36×10^{-5}, and 4.4×10^{-5} A/m^2, respectively at 3 V. Upon UV light illumination of power 3 mW/cm^2, the photocurrent densities of the LZO devices got improved for the same bias voltage (3 V) as observed from Fig. 2. Oxygen (gas) molecules are adsorbed onto the surface of ZnO by becoming oxygen ions (anions). This leaves a large potential

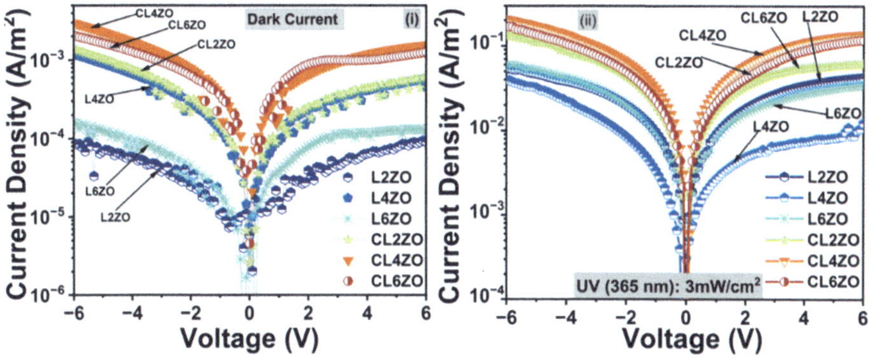

Fig. 2. The J-Vcharacteristics of the UV photodetector devices under (i) dark condition and (ii) UV light illumination (365 nm and 3 mW/cm^2).

gradient near the surface, resulting in a lower dark current. When UV illumination (365 nm at 3 mW/cm^2) is on, electron-hole pairs are generated in ZnO. Photo-generated holes diffuse to the surface via the potential gradient to desorb the oxygen molecules and/or become stuck near the surface, lowering the depletion width near the surface and thereby improving photocurrent. L4ZO exhibited the highest photocurrent of 0.75 μA as compared to L2ZO (0.38 μA) and L6ZO (0.44 μA) for an applied voltage of 5V under the same UV light illumination (365 nm at 3 mW/cm^2). As Li concentration increases, Li ions replace Zn in the ZnO lattice and increasing ionic conductivity by lowering the activation energy for ionic transport. The reduction of Zn vacancies improve the material's electrochemical properties. The addition of Li as a shallow acceptor improves charge transfer while reducing electron-hole recombination, increasing the accessible charge carriers and hence enhancing overall conductivity. At elevated doping levels, a phenomenon known as concentration quenching occurs due to which the increased number of dopant atoms leads to non-radiative recombination pathways [6]. Beyond a specific threshold (6%), excessive Li can cause defects such as interstitial Li atoms that operate as electron donors, counteracting p-type activity and resulting in decreased performance and diminishing luminescence. Agglomeration of dopants can be another reason for the reduction in the device performance.

The CQDs incorporated LZO (CQDs/LZO) hybrid thin film-based UV photodetectors were fabricated by depositing Al metal electrodes (active device area of 0.16 cm^2) onto the hybrid composite films such as CL2ZO, CL4ZO, and CL6ZO. The devices yield a slightly improved dark current [5.6, 12 and 9.6 × 10^{-4} A/m^2 at 3 V)] when compared to that obtained from LZO based devices. It has been observed that CQDs get adsorbed (chemisorbed) onto the ZnO crystallites in the bulk of the hybrid film, supporting a previous report published elsewhere. By manipulating the size at the nanoscale, CQDs improve the hybrid film quality and result in a slightly improved dark current density.[5] Upon the UV light illumination (365 nm at 3 mW/cm^2), it is distinctly observed from Fig. 3 that the light current density [0.067, 0.10, 0.077 A/m^2 at 3 V] of the devices based on hybrid films CL2ZO, CL4ZO, and CL6ZO, shows a significant increase as compared to that of respective LZO films. The photocurrent response of CL4ZO showed higher

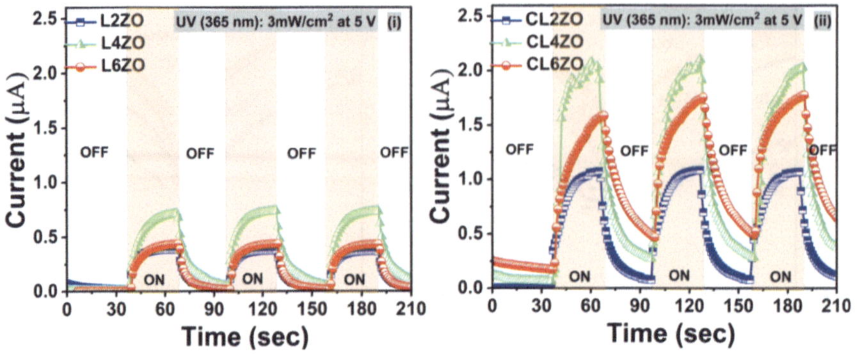

Fig. 3. Represents the UV photo-response curves (I-V) of the (i) LZO and (ii) CQDs/LZO hybrid thin film devices for an applied bias of 5 V.

values (2.1 μA) over that obtained for CL2ZO (1.1 μA) and CL6ZO (1.76 μA). The chemisorbed CQDs absorb the UV light energy creating e-h pairs. The heterojunction formed between CQDs and ZnO creates an interface that promotes efficient charge separation. When the UV light excites, the electrons in the CQDs can be transferred to the conduction band of LZO, effectively reducing recombination rates of electron-hole pairs [5]. This mechanism enhances the overall photocurrent generation in the device.

The CQDs/LZO hybrid film (CL2ZO, CL4ZO, and CL6ZO) based devices exhibited the responsivity [37, 57 and 45 mA/W at 5 V] and EQE (η) [12, 20 and 16% at 5 V] which are found to be significantly higher than the respective LZO based devices (responsivity of 15, 10 and 17 mA/W and EQE of 3.4, 1.7 and 3.6% for L2ZO, L4ZO and L6ZO respectively). The incorporation of CQDs in the Li-doped ZnO have shown significant improvements in responsivity and EQE values of UV photodetection when compared to that of LZO films, even though detectivity values are lower as compared to LZO devices. The incorporation of CQDs may have introduced additional noise sources, facilitating the non-radiative recombination of charge carriers, overshadowing the signal generated by the photodetector and hence lowering the detectivity [7, 8]. This increment in responsivity and EQE is attributed to the increased number of photogenerated charge carriers injected from CQDs into LZO, which enhances the photocurrent under UV illumination [7]. These results show that the heterojunction between LZO and CQDs aids the process of charge seperation, contributing to increased photocurrent. Figure 4 shows the plots of responsivity, detectivity and external quantum efficiency of the devices.

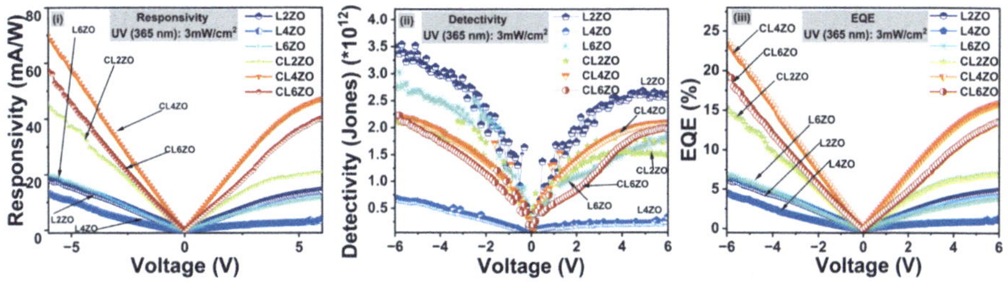

Fig. 4. Responsivity (i), Detectivity (ii), and Exteral Quantum Efficiency (EQE) (iii) of the LZO and CQDs incorporated LZO (CQDs/LZO) hybrid thin films.

4 Conclusion

The incorporation of carbon quantum dots forming heterojunction between inorganic and organic interface, improved the UV photodetector performance to a significant level (3 times) over the Li-doped ZnO films, revealing its potential in the optoelectronic industry. The CQDs/LZO hybrid thin films deposited by sol-gel spin-coating process exhibit the higher transparency of 87% at 550 nm over the 82% obtained for LZO films. The optical bandgap of CQDs/LZO hybrid films found to be slightly shifted towards the higher energy region, revealing the effect of confinement of electrons (quantum confinement effect) and quantization of energy states in the ZnO upon increasing CQDs content. The significant improvement in the responsivity (57 mA/W) and EQE (20%) of CQDs/LZO hybrid films make them useful for commercial UV photodetector applications.

Acknowledgements. The authors acknowledge the financial support from SPARC project (Phase – II, project ID: P2268). The authors thank The Dept. of Physics and CRIF, NIT Warangal for providing the experimental and characterisation facilities.

References

1. Chen, Z., Wang, J., Wu, H., Yang, J., et al.: Nature Communications **13**, 4387 (2022)
2. Wang, J., Pan, H., Xiaoyun, X., et al.: ACS Appl. Mater. Interfaces **14**, 12450–12460 (2022)
3. Thyda, L., Dasi, G., Abdul Azeez, M.S., Naresh, K., et al.: Appl Surf Sci **611**, 155664 (2023)
4. Zhou, Y.H., Zhang, Z.B., Xu, P., et al.: Nanoscale Research Letters **14**, 364 (2019)
5. Thyda, L., et al.: Thin Solid Films **790**,, 140221 (2024)
6. Dasi, G., Ramarajan, R., Joseph, D.P., Vijayakumar, S., et al.: Thin Solid Films **710**, 138265 (2020)
7. Huy, V.P.H., Bark, C.W.: RSC Adv. **14**, 12360–12371 (2024)
8. Jing-Jing, X., Yi-Ning, L., Tao, F.-F., Liang, P.-F., et al.: ACS Omega **8**, 7845–7857 (2023)

Needle-Like Nanostructures and Mixed Storage Behaviour of Co_3O_4 Electrode Materials for Improved Supercapacitor Applications

Koppula Naresh[1], Vinay Pratap Singh Rajawat[1], Thyda Lavanya[1], Joel K. Joseph[1], S. Suneetha[1], Murali Mohan Seepana[2], and Kuppusamy Thangaraju[1](✉)

[1] Organic Optoelectronic Device Laboratory, Department of Physics, National Institute of Technology, Warangal, Telangana 506004, India
ktr@nitw.ac.in

[2] Depaterment of Chemical Engineering, National Institute of Technology, Warangal, Telangana 506004, India

Abstract. The excellent electrochemical properties, structural and morphological advantages of cobalt oxide (Co_3O_4) have gained a lot of attention in the energy storage devices. The hydrothermally synthesised Co_3O_4 characterized using the XRD study exhibits the crystalline structure, phase purity. FESEM analysis reveals the formation of Co_3O_4 nanoneedle-like structures. This research study explored the capacitive- and morphology-dependant storage behavior of the Co_3O_4. The mixed capacitive behavior (capacitive and diffusion-controlled) and needle-like morphological structures of the synthesized Co_3O_4 electrode material enhances the supercapacitor performances, exhibiting the high specific capacitance of 2397 F g^{-1}.

1 Introduction

Depletion of natural resources such as fossil fuels is causing harm to the ecosystem. It is need of the hour to find an alternate appropriate and long-lasting energy storage technology to meet the demand for renewable energy. In order to store the renewable energy, suitable energy storage technology highly demanding one [1]. Batteries and Supercapacitors are the promising energy storage devices for the future applications. In this contest, supercapacitors are currently being explored owing to their wide attention in the energy storage field [2]. Supercapacitors became most popular in the recent years owing to their promising energy storage abilities due to the fast charge-discharge rates and long cycle life as compared to that of batteries [3]. Based on the charge-storage mechanism, supercapacitors are divided into two categories: (i) Electric double layer capacitor (EDLC) and (ii) Pseudocapacitor based on faradaic effect. The transition metal oxides are found to be better pseudocapacitive electrode materials for supercapacitor applications due to the various oxidation states, different nanostructures, and high conductivity, ascribed to enhanced electrochemical properties. Metal oxides such as MnO_2, RuO_2, Fe_2O_3, NiO and V_2O_5, Co_3O_4 etc. have high specific capacitances. Among these, Co_3O_4 is having

P. Joseph D. et al. (Eds.): IEMDST 2024, SPPHY 567, pp. 174–180, 2025.
https://doi.org/10.1007/978-981-96-5863-3_23

excellent theoretical and practical specific capacitances, cyclic stability, different nanostructures, electrochemical and structural properties, making it a promising electrode material for supercapacitor applications [4, 5]. Co_3O_4 is the pseudocapacitive electrode material and exhibits different oxidation states. In this study, we prepared Co_3O_4 by using hydrothermal method and characterized to study their structural and morphological properties. Further, the Co_3O_4 sample is used as an electrode in the supercapacitors and results are discussed in detail.

2 Experimental Details

Analytical-grade cobalt nitrate hexahydrate ($Co(NO_3)_2{\cdot}6H_2O$) and urea ($CO(NH_2)_2$) were used as-such without any further purification as Co source and binder, respectively. 40 mL of deionized (DI) water consisting of $Co(NO_3)_2{\cdot}6H_2O$ (50 mM) and urea (100 mM) was swirled for an hour to get the uniformly mixed precursor solution. The glycine (200 mM) the templating agent was added to 20 mL of DI water and stirred for 1h. It was then drop-wise added to the above source precursor and stirred for another 3 h. The final precursor (60 mL) was transferred to a Teflon-lined autoclave (volume of 120 mL) and a nickel foam (NF) (substrate with the area of 2 cm $\times$ 4 cm) was immersed in the precursor solution. It was then placed in an oven at 120 °C for 2 h to perform the hydrothermal reaction. After the reaction, the Co_3O_4 coated NF substrate was removed from the autoclave and washed with DI water for several times to remove the impurities. It was then dried in an oven at 70 °C for 6 h and finally calcinated at 300 °C for 2 h in the muffle furnace to promote the formation of nanostructures on the NF surface. It was then used for various characterizations such as XRD, FESEM, and cyclic voltammetry (CV) in order to study its suitability to be used as an electrode in the supercapacitor.

3 Results and Discussion

3.1 Structural Study

Figure 1 depicts the X-ray diffraction (XRD) pattern of hydrothermally synthesized Co_3O_4 coated onto the NF substrate. The diffraction peaks obtained at the 2-theta values of 30.96°, 36.6°, 59.34°, 65.24° are assigned corresponding to the (hkl) planes of (220), (311), (511), (440), respectively, showing the face-centered cubic spinel structure of Co_3O_4 consistent with JCPDS data (Card number of 42-1467) [6]. The dominant additional peaks observed at 44.35°, 51.74°, 76.32° are related to the (hkl) planes (111), (200), (220) of NF and are well matched with JCPDS data (Card number of 04-0850), confirming the crystalline structure, phase purity, and the formation of Co_3O_4 onto NF substrate during the hydrothermal process [7].

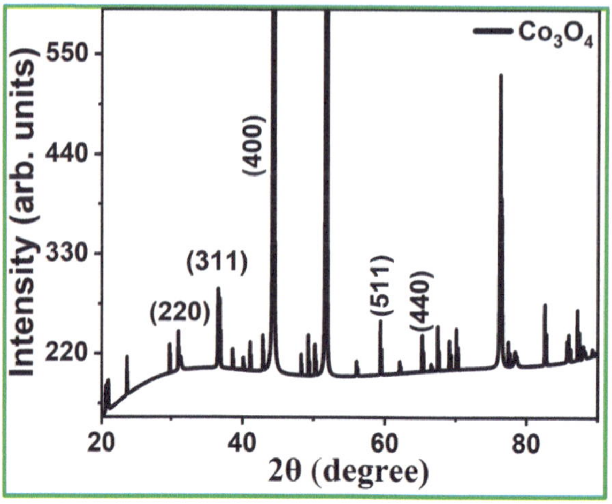

Fig. 1. XRD pattern of the hydrothermally coated Co_3O_4 onto NF substrate.

4 Morphological Studies

Figure 2 shows the FESEM image of the synthesised Co_3O_4 coated on NF substrate, exhibiting the well-distributed needle-like nanostructured morphology across the surface of NF substate. The nanoneedle-like structure is providing the additional active sites for ion adsorption and desorption. These additional electrochemical active sites elevate the reaction kinetics and promote the fast redox rates at the bulk as well as at the electrode/electrolyte interface [8]. These nanoneedles significantly increase the surface area (increased surface to volume ratio) at the edges, allowing better exposure of the material to the ion's diffusion. As a result, large surface-active sites of the Co_3O_4 is engaged-in

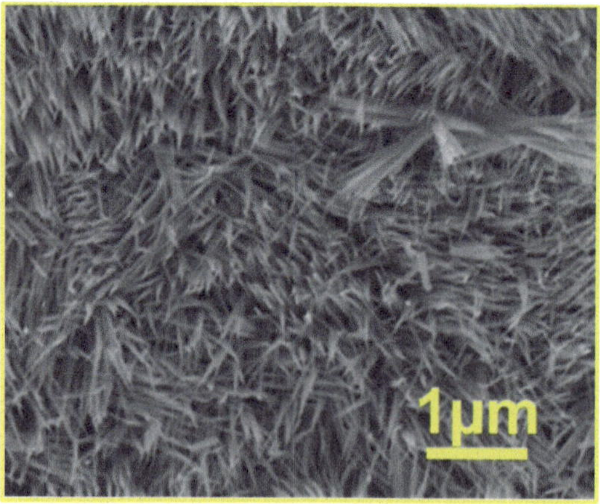

Fig. 2. FESEM image of hydrothermally coated Co_3O_4 on NF substrate.

for the reaction with ions in the electrolyte which increases the capacitance and results in the increased reaction efficiency and obviously electrochemical performance [9].

5 Electrochemical Studies

The prepared nanoneedle-like Co_3O_4 electrode was used as a working electrode (1 cm × 1 cm) and it was characterized using the three-electrode system with Pt, Ag/AgCl and 2 M KOH are the counter electrode, reference electrode and electrolyte, respectively. The nanoneedle-like Co_3O_4 electrode exhibited the distinct redox curves in the CV in the potential window from −0.1 to 0.6 V. While increasing the scan rate from 5 to 100 mVs^{-1}, the redox peaks shift towards more negative and positive potentials due the materials electrochemical reversibility as shown in the Fig. 3 (a) [10].

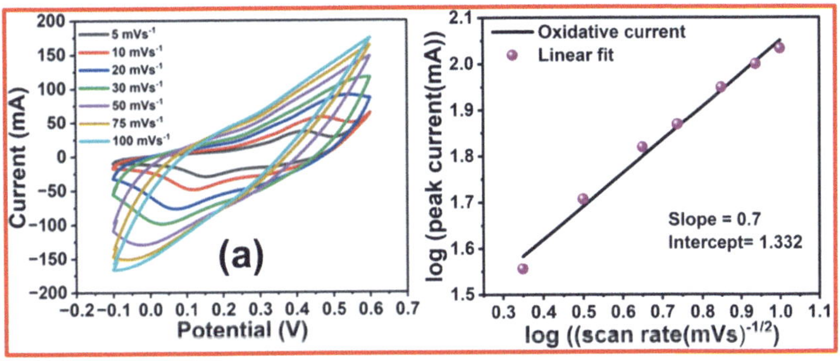

Fig. 3. a) CV curves of Co_3O_4, b) The oxidative peak current vs (scan rate)$^{1/2}$.

CV curves exhibit the non-linear curves, showing that the pseudocapacitive nature of the electrode material. For increasing scan rate (5–100 mVs^{-1}), shape of the CV curves remain same but peak current is increased due to the electrochemical reversibility [2]. Specific capacitance (C_s) of the Co_3O_4 nanoneedles is estimated using the following formula [11].

$$C_S = \frac{\int I(V)\mathrm{d}V}{m \times \Delta V \times \nu} \tag{1}$$

where $\int I(V)\mathrm{d}V$ – is the area under the CV curve, ΔV – potential window, and ν – scan rate. The Co_3O_4 nanoneedles have exhibited a high specific capacitance of 2397 F g^{-1} at 5 mV s^{-1} scan rate.

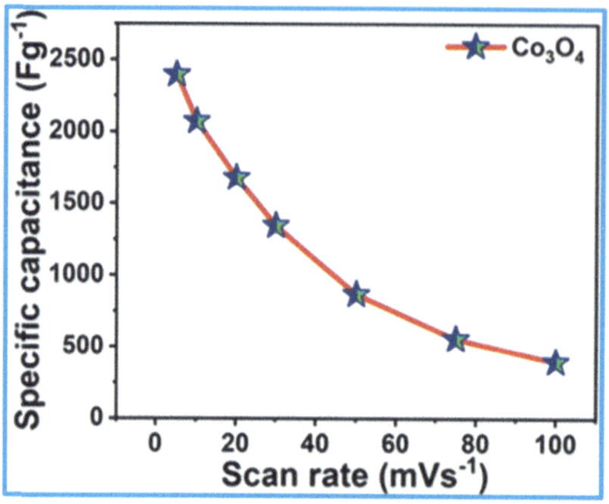

Fig. 4. Specific capacitance of Co_3O_4 at different scan rates.

Figure 4 depicts that the specific capacitance of Co_3O_4 electrode is dropped while increasing scan rate due to at higher scan rates the ions have less time to intercalate in to the electrode material [12]. For a better understanding of charge storage behavior of the Co_3O_4 electrode, the CV curves are analysed using power law. The Eqs. (1) and (2) indicate the current response with respect to the scan rate.

$$i(\nu) = a\nu^{b} \tag{2}$$

$$log[i(\nu)] = log(a) + blog(\nu) \tag{3}$$

where i – peak current (at fixed potential), ν – scan rate and a, b – linear fitting parameters [13].

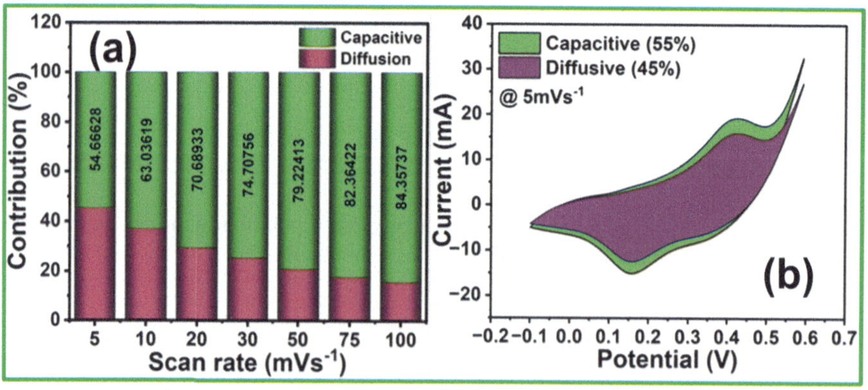

Fig. 5. Charge-storage contribution percentage (%, (a) At fixed potential and (b) At fixed scan rate.

The b-value from slope of linear fitting ranges from 0 to 1. When the b-value is close to 0.5, it indicates diffusion-controlled (redox storage behavior) dependence, whereas a b-value of 1 indicates capacitive-controlled (EDLC storage behavior) dependence. In this work, the b-value lies between 0.5 and 1 (specifically at 0.7), suggesting that the prepared Co_3O_4 exhibits mixed behaviour due to its nanoneedle structure.

CV curves are deconvoluted using Eqs. (4) and (5) to differentiate the dominant charge storage behaviour.

$$i(v) = k_1 v + k_2 v^{0.5} \tag{4}$$

$$\frac{i(\nu)}{\nu^{0.5}} = k_1 \nu^{0.5}_{+} k_2 \tag{5}$$

where $k_2 v^{0.5}$ is diffusion controlled, $k_1 v$ *is* capacitive controlled, k_1 and k_2 are the linear fitting constants [13].

Figure 5 (a) shows that the current response with different scan rates (5–100 mV s^{-1}) at fixed potential (0.45 V), exhibiting that the nanoneedles-like Co_3O_4 electrode possesses the mixed charge storage behaviour (redox and EDLC) and is slightly more capacitive-controlled (55%) than diffusion-controlled (45%) charge storage. Nanoneedles of Co_3O_4 electrode were uniformly distributed onto the surface of NF substate which leads to increasing the surface area of the electrode and creating more active sites for ions interaction. Figure 5 (b) depicts the redox current at fixed scan rate (5 mV s^{-1}) scanned over potential window, revealing the slight capacitive-controlled dominance over the diffusion-controlled nature. Nanoneedle-like structures increase the reaction surface at the electrode/electrolyte interface, resulting in the improved electrode performance [8].

6 Conclusion

The nanoneedle-like Co_3O_4 coated onto the NF substrate was successfully synthesised using in-situ hydrothermal technique. XRD studies confirmed the formation of nanoneedle-like Co_3O_4 with face-centered cubic spinel structure, whereas FESEM images support the uniform distribution of nanoneedle-like structures over entire surface of NF substrate. The exposure of increased surface area of nanoneedle-like Co_3O_4 enhances the ion interaction at the electrode edges, leading to the rapid redox reactions at the electrode/electrolyte interface. Electrochemical study revealed Co_3O_4 exhibits the non-rectangular CV curves, confirming the faradaic (redox) behavior and combined charge-storage nature and resulting in a high specific capacitance value. These findings demonstrate that the hydrothermally prepared nanoneedle-like Co_3O_4 will be an excellent electrode material for supercapacitor applications.

Acknowledgements. This work was supported by the Scheme for Promotion of Academic and Research Collaboration (SPARC) project (Phase-II) (project ID: P2268) for its financial assistance. In addition, the authors acknowledge the Department of Physics and Central Instrumentation Research Facility (CRIF) for providing the experimental and characterisation resources.

References

1. Kebede, A.A., Kalogiannis, T., Mierlo, J.V., et al.: Renew. Sustain. Energy Rev. **159**, 112213 (2022)
2. Guziejewski, D., Stojanov, L., Gulaboski, R., Mirceski, V.: J. Phys. Chem. C **126**, 5584–5591 (2022)
3. Zallouz, S., Réty, B., Vidal, L., Meins, J.M.L., et al.: ACS Appl. Nano Mater. **4**, 5022–5037 (2021)
4. Umar, A., Raut, S.D., Ibrahima, A.A., Algadi, H., et al.: Electrochim. Acta **389**, 138661 (2021)
5. Jayakumar, S., Santhosh, P.C., Mohideen, M.M., Radhamani, A.V.: A comprehensive review of metal oxides (RuO_2, $Co3O_4$, MnO_2 and NiO) for supercapacitor applications and global market trends. J. Alloys Compounds **976**, 173170 (2024)
6. Liang, H., Raitano, J.M., Zhang, L., Chan, S.-W.: Controlled synthesis of Co3O4 nanopolyhedrons and nanosheets at low temperature. Chem. Commun. **48**, 7569 (2009)
7. Swanson, T.: Natl. Bur. Stand. (U.S.) Circ. 539(**I**), 13 (1953)
8. Anjana, P.M., Sarath Kumar, S.R., Rakhi, R.B.: MnCo2O4 nanoneedles self-organized microstructures for supercapacitors. Mater. Today Commun. **28**, 102720 (2021)
9. Rajalakshmi, R., Remya, K.P., Viswanathan, C., et al.: Nanoscale Adv. **3**, 2887–2901 (2021)
10. Elgrishi, N., Rountree, K.J., McCarthy, B.D., et al.: J. Chem. Educ. **2018**(95), 197–206 (2018)
11. Ranjan, B., Kaur, D.: Small **20**, 2307723 (2024)
12. Vijayakumar, S., Nagamuthu, S., Muralidharan, G.: ACS Appl. Interfaces **5**(2), 188–219 (2013)
13. Balamurugan, R., Chandra Bose, A.: ACS Appl. Energy Mater. **7**(3), 974–985 (2024)

Fabrication of Different Red-Sensitive Cross-Linked Photopolymers for Holography

A. B. Sreebha[1(✉)] and V. P. Mahadevan Pillai[2,3]

[1] Department of Physics, Sree Sankara Vidyapeetom College, Valayanchirangara, Perumbavoor, Kerala 683556, India
sreebhaab@ssvcollege.ac.in

[2] Department of Optoelectronics, University of Kerala, Karyavattom, Thiruvanathapuram, Kerala 685581, India

[3] School of Physical Sciences, Amrita Viswa Vidyapeetham, Tamil Nadu, Coimbatore 641112, India

Abstract. Polyvinyl alcohol (PVA) based photopolymer with acrylamide (AA) monomer is extensively researched for holographic applications due to its high sensitivity, superior diffraction efficiency, and impressive dynamic range. Some PVA based red sensitive photopolymers with different dyes are developed for holographic applications. The effect of the cross-linker on the photopolymer composition has been examined. Holograms with about 75% diffraction efficiency are obtained. This material has broad potential for holographic applications and can be manufactured to commercial standards.

1 Introduction

The demand for advanced materials to record and store data is increasing rapidly. Volume holograms show significant promise for information storage, offering extreme density and rapid retrieving capabilities, as the recorded data is dispersed over a thick holographic recording material's volume [1, 2]. In addition to data storage, holographic materials can be utilized in a variety of applications, including non-destructive testing, solar holography, imaging, anti-counterfeiting, and advertising. Common recording media for holography include silver halides, dichromated gelatin, photopolymers and photoresists [1]. Among these, photopolymers are particularly noteworthy due to their versatility, ease of processing, relatively low cost, and excellent diffraction efficiencies [3].

Photopolymer recording materials for holography typically consist of a monomer, a photo-initiator, photo sensitizing dye and a binder that holds everything together [3, 4]. The binder serves to integrate all components of the photopolymer without participating directly in any photo induced chemical reactions [5, 6]. Monomers are tiny molecules that go through chemical reactions to form gratings through refractive index modulation [5]. The photo-initiator generates free radicals upon exposure to light, reacting with monomers and binders to create connections [5, 6]. The light that falls on sensitizing dye is being absorbed and subsequently photo-initiator initiates the polymerization of

P. Joseph D. et al. (Eds.): IEMDST 2024, SPPHY 567, pp. 181–187, 2025.
https://doi.org/10.1007/978-981-96-5863-3_24

monomers. Supplementary cross-linking monomers may be introduced to enhance the stability of the final hologram optically written in the photopolymer recording material [6].

The primary mechanism for the construction of hologram in these photopolymers is free radical polymerization, which happens through consecutive phases of initiation, propagation, and termination [5]. Free radicals are generated during the initiation phase [6], and during the propagation phase, the monomer molecules assemble and lengthen to create a long polymer chain. Polymerization is stopped during the termination phase to bring the process to an end. The distance between the monomers decreases as a result of covalent bonds forming between them once polymerization is finished. In photopolymer materials, this polymerization results in shrinkage, which modifies the refractive index and thickness [6].

This paper focuses on the synthesis of various red-sensitive photopolymers tailored for holographic applications, examining how cross-linkers can enhance the optical quality of the material.

2 Materials and Methods

In the present case, photopolymer is fabricated using Poly Vinyl Alcohol (PVA) as binding matrix, Acrylamide (AA) as monomer, Tri-ethanolamine (TEA) as photo-initiator, N,N' -methylene-bis-acrylamide (BAA) as cross-linking monomer and dyes such as Methylene Blue (MB), Grams Crystal Violet solution (for microscopy) and Nile Blue chloride as photo sensitizing dyes.

At the outset, four different sets of photopolymer solutions with 10% PVA, 0.300 molar AA, 0.250 molar TEA and sensitized with different values of MB dye concentration viz. 2×10^{-5}, 4×10^{-5}, 6×10^{-5} and 8×10^{-5} molar are prepared at normal laboratory conditions. The prepared photopolymer solutions is made to films with an average thickness of 200 μm on cleaned glass slides using gravity settling method and are designated as $P1_{M2}$, $P1_{M4}$, $P1_{M6}$, $P1_{M8}$.photopolymer films. First order holographic diffraction gratings are written in the fabricated films using transmission geometry and their DEs are studied.

In order to find the influence of cross-linker as a next phase, 0.020 molar BAA is also added to the photopolymer solution prepared with 10% PVA, 0.300 molar AA, 0.250 molar TEA and sensitized with different values of MB dye concentration viz. 2×10^{-5}, 4×10^{-5}, 6×10^{-5} and 8×10^{-5} molar respectively to fabricate four sets of P2 photopolymer samples viz: $P2_{M2}$, $P2_{M4}$, $P2_{M6}$, $P2_{M8}$. (The photopolymer samples fabricated without cross-linking monomer are designated as P1 samples and those fabricated with cross-linking monomer are designated as P2 samples.) For optimizing the concentration of cross-linker, four different sets of P2 films (sensitized with the most favourable concentration of MB) with various concentrations of BAA viz. 0.009, 0.020, 0.035 and 0.052 molar are fabricated with 10% PVA, 0.300 molar AA, 0.250 molar TEA. They are designated as $P2_{M4B0.009}$, $P2_{M4B0.02}$, $P2_{M4B0.035}$, and $P2_{M4B0.052}$ respectively.

Apart from MB doped photopolymer films, two different categories of photopolymer samples viz: $P2_{GCV}$ and $P2_{NBC}$ are fabricated respectively using 0.003 molar Grams Crystal Violet solution (GCV) and 0.003 molar Nile Blue chloride dye (NBC) dissolved in distilled water. Four sets of photopolymer films such as $P2_{GCV2}$, $P2_{GCV4}$, $P2_{GCV6}$, $P2_{GCV8}$ are fabricated by respectively doping 2×10^{-5} molar, 4×10^{-5} molar, 6×10^{-5} molar, 8×10^{-5} molar GCV dye and another four sets of photopolymer samples viz: $P2_{NBC2}$, $P2_{NBC4}$, $P2_{NBC6}$, $P2_{NBC8}$ are fabricated by respectively doping 2×10^{-5} molar, 4×10^{-5} molar, 6×10^{-5} molar, 8×10^{-5} molar NBC dye in the photopolymer solution prepared using 10% PVA, 0.300 molar AA, 0.250 molar TEA and optimum concentration of BAA. All the photopolymer films are made with an average thickness of 200 μm.

2.1 Recording and Reading of Holographic Gratings in the Photopolymer Films

Standard two beam transmission geometry is used to record first order transmission gratings in the fabricated photopolymer, as in Fig. 1(a) [4]. The laser beam is split into two using a beam splitter. These beams are expanded with beam expanders and then collimated. Both the collimated beams are set to an area of 1 cm^2 at the recording plate using suitable apertures and the path of two beams are made equal at the interfering region (photopolymer film). Holographic grating elements are recorded in the photopolymer samples using a red (He – Ne laser) laser radiation of wavelength 632.8 nm by keeping the beam ratio at 1:1 and at an inter-beam angle (angle between reference and object beams) of 30° and at different laser exposure energies.

First order diffraction efficiencies (DE) are measured with the schematic representation shown in Fig. 1(b). The laser beam (wavelength of 632.8 nm) of area 1 cm^2 is allowed to fall on the holographic grating positioned at Bragg's angle for analysing the first order diffracted beam. The power intensities of the first order diffracted beam and that of incident beam are measured using a power meter. The DE is calculated using the following equation.

$$DE = \left({}^{P_1}/_{P_i}\right) * 100 \tag{1}$$

where P_1 and P_i are the power intensities of first order diffracted beam and incident beam respectively [4].

2.2 Optimization of MB Dye and Cross-Linker Concentration in Photopolymer Samples

First order holographic transmission gratings are first recorded in the P1 photopolymer samples that have been sensitized with various MB concentrations. Recordings are done using the two beam transmission geometry at different laser exposure energies (viz: 50 mJ/cm^2, 100 mJ/cm^2, 150 mJ/cm^2 200 mJ/cm^2 and 250 mJ/cm^2). The first order DEs are measured with the recording wavelength and the dye concentration for which the highest DE obtained is found.

As a next phase, 0.020 molar BAA monomer cross-linker is also added to the above prepared P1 photopolymer with the optimum MB dye concentration. First order holographic transmission gratings are recorded and read as detailed before. These samples are designated as P2 samples. To find the optimum concentration of BAA cross-linker, four different concentrations of BAA viz. 0.009, 0.020, 0.035 and 0.052 molar are added

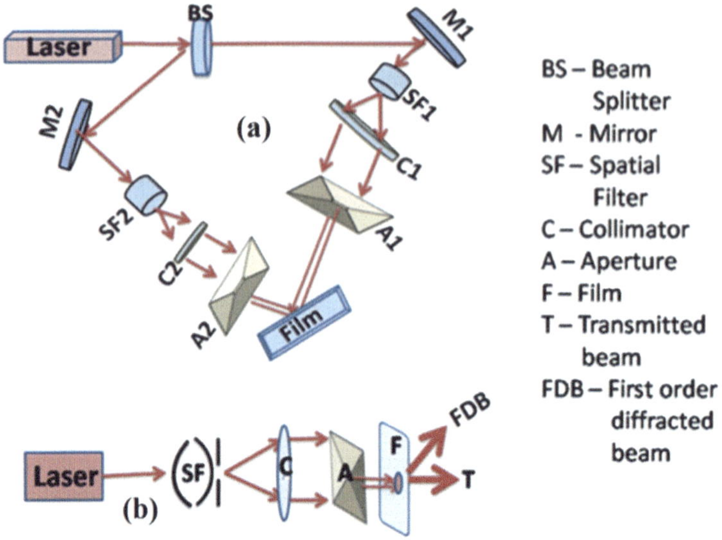

Fig. 1. (a) Experimental set up to record transmission holographic gratings, (b) Schematic diagram to measure first order diffraction efficiency of holograms.

separately to P2 photopolymer solution sensitized with the optimum MB concentration. These solutions are designated as $P2_{M4B0.009}$, $P2_{M4B0.02}$, $P2_{M4B0.035}$, and $P2_{M4B0.052}$ respectively.

2.3 Preparation of Photopolymer Doped with Other Red Sensitive Dyes

Two more sets of photopoymer solutions are again prepared with other red sensitive dyes such as grams crystal violet (GCV) and Nile blue chloride (NBC) with the same concentrations of PVA, AA, TEA and the optimum value of BAA taken for the preparation of P2 samples.

Four sets of GCV sensitized and four sets of NBC sensitized photopolymers with dye concentrations of 2×10^{-5}, 4×10^{-5}, 6×10^{-5} and 8×10^{-5} molar are prepared. Those samples are designated as $P2_{GCV2}$, $P2_{GCV4}$, $P2_{GCV6}$, $P2_{GCV8}$ and $P2_{NBC2}$, $P2_{NBC4}$, $P2_{NBC6}$, $P2_{NBC8}$ respectively. First order holographic transmission gratings are written in those photopolymer samples and their first order diffraction efficiencies are studied.

3 Results and Discussion

Figure 2(a) illustrates how DE of P1 photopolymer films changes with exposure energy. All the photopolymer samples under study show their maximum DE at laser exposure energy of 200 mJ/cm^2. Among the samples chosen, the highest DE (64.0%) is exhibited by $P1_{M4}$ photopolymer sample which is fabricated with a dye concentration of 4×10^{-5} molar MB dye. Hence, 4×10^{-5} molar is chosen as the most favourable MB dye concentration for the fabrication of P1 photopolymer.

Figure 2(a) shows that the highest DE is only achieved for a specific dye concentration. A decrease in DE is seen in the holographic transmission gratings recorded in the photopolymer samples with dye concentrations above the optimal value (in this case, 4 $\times 10^{-5}$ M). This reduction may be the result of scattering loss brought on by inadequate bleaching in the material with high dye concentrations [7]. In the same way, holographic transmission gratings recorded in photopolymer samples with dye concentrations below the optimal value show a decrease in DE. This might happen if the substance doesn't have enough dye molecules to produce enough free radicals for effective photopolymerization [8].

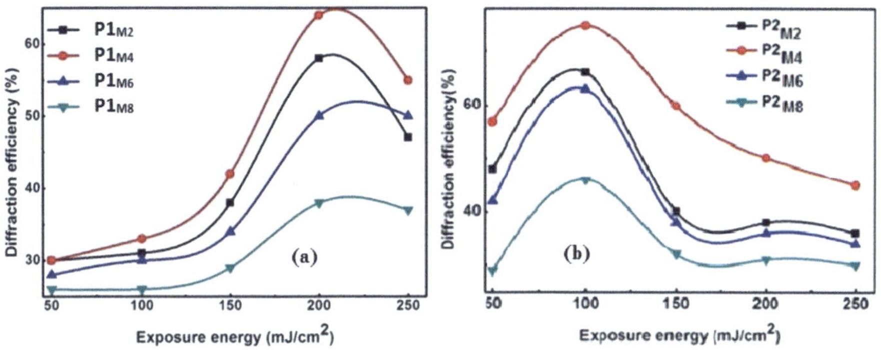

Fig. 2. (a) Plot showing how diffraction efficiency changes with exposure energy for P1 photopolymer samples doped with different MB concentrations (b) Plot showing how diffraction efficiency changes with exposure energy for P2 photopolymer samples doped with different MB concentrations.

Plot on the diffraction efficiency changes with exposure energy for the gratings recorded in P2 photopolymers is shown in Fig. 2(b). Holographic gratings recorded in all the photopolymer samples show the highest diffraction efficiency at laser exposure energy of 100 mJ/cm^2. The optimal value of MB dye concentration does not change by the incorporation of cross-linking monomer. Here, holograms written in the photopolymer films fabricated without monomer cross-linker (P2 films) also exhibit maximum DE at the concentration of 4 $\times 10^{-5}$ molar MB dye. But, the DEs of all holographic gratings are found to be enhanced by the incorporation of BAA. Furthermore, the samples show their maximum DEs at an exposure energy of 100mJ/cm^2 while the gratings written in the photopolymer samples fabricated without BAA display their greatest DEs at an exposure energy of 200 mJ/cm^2. Thus, the addition of monomer cross-linker not only enhances the diffraction efficiency of holographic gratings, but enhances their sensitivity to laser light also. The addition of cross-linker can improve the monomer polymerization reaction and enhance the photopolymer's refractive index modulation. Thus, adding a monomer cross-linker to the photopolymer composition improves both diffraction efficiency (DE) and sensitivity [7].

All the photopolymer samples fabricated for optimising BAA concentration ($P2_{M4B0.009}$, $P2_{M4B0.02}$, $P2_{M4B0.035}$, and $P2_{M4B0.052}$) exhibit highest diffraction efficiency at laser exposure energy of 100 mJ/cm^2 (Fig. 3(a)). Among the four sets of photopolymer samples, the highest DE (around 75% is exhibited by P2 $_{M4B0.02}$ photopolymer sample which is fabricated with a concentration of 0.020 molar cross-linker and hence, 0.020 molar BAA is chosen as the optimum cross-linker concentration for the fabricated photopolymer. When the concentration of BAA is less than the optimum value, it will be insufficient for getting stabilized holograms with good diffraction efficiency and when the concentration is higher than the optimum value, it may cause increment in cross-linking density and hence can retard mobility of monomers [9, 10]. High cross-linker concentration also makes the photopolymer films opaque and cause imperfect solubility of monomer and cross-linking monomer in a viscous binding matrix [9, 10]. Here, the optimum concentration of BAA enhances contrast in refractive index and consequently high diffraction efficiency.

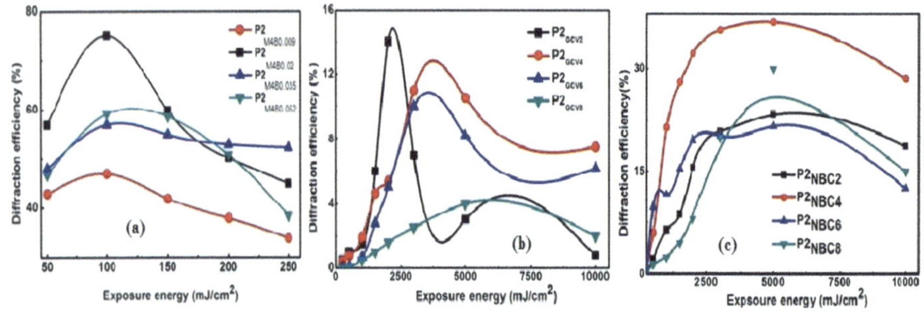

Fig. 3. (a) Plot showing how diffraction efficiency changes with exposure energy for P2 photopolymer samples doped with different cross-linker concentrations (b) Plot showing how diffraction efficiency changes with exposure energy for the gratings recorded in GCV sensitized photopolymer samples and (c) Plot showing how diffraction efficiency changes with exposure energy for the gratings recorded in NBC dye sensitized photopolymer samples.

Initial measurements in the gratings recorded in GCV and NBC dye sensitized photopolymer films reveal that the grating formation with the previously used laser exposure energies from 50 to 250 mJ/cm^2 is not sufficient to get an appreciable first order DE. Hence, gratings are recorded in those photopolymer films for higher exposure energies from 250 to 10,000 mJ/cm^2. Figure 3 (**b** and **c**) respectively show how diffraction efficiency changes with exposure energy for the gratings recorded in GCV and NBC sensitized photopolymer samples.

Compared with the gratings recorded in MB dye sensitized photopolymer films, gratings recorded in all GCV and NBC sensitized photopolymer films show very less first order DEs even with very high laser exposure energies. All these holographic gratings show appreciable values of DEs only at exposure energies above 2000 mJ/cm^2. It is also observed that as the dye concentration of GCV and NBC sensitized photopolymer samples increases, photosensitivity of the photopolymer decreases. All the photopolymer samples fabricated with NBC dye show their highest DE at the energy exposure of 5000 mJ/cm^2. However, the DEs estimated for the holographic gratings in NBC dye sensitized photopolymer films are better than those fabricated on GCV sensitized photopolymers.

4 Conclusion

Methylene Blue (MB) sensitized red-sensitive photopolymers are developed for holography under standard laboratory conditions. First-order diffraction gratings are successfully recorded in the fabricated photopolymer using transmission geometry. An enhancement in diffraction efficiency and photosensitivity of the holographic gratings is observed upon incorporating N,N'-methylenebisacrylamide (BAA) cross-linking monomer. Better diffraction efficiency is obtained when optimum concentration of cross-linker is added in the photopolymer solution. Grams Crystal Violet (GCV) and Nile Blue Chloride (NBC) sensitized red sensitive photopolymers are also prepared. Among the three different red sensitive photo-sensitizing dyes chosen for the fabrication of photopolymer, MB dye can be considered as the appropriate dye for the fabrication of photopolymer with better diffraction efficiency and sensitivity for holographic applications.

Acknowledgments. A. B. Sreebha is sincerely grateful for the financial assistance from the 'Back to Lab' Scheme of Kerala State Council for Science, Technology and Environment.

References

1. Vorzobova, N., Sokolov, P.: Polymers (Basel) (2020). https://doi.org/10.3390/polym11122020
2. Deepak, L.N., et al.: Sci. Reports (2016). https://doi.org/10.1038/srep26163
3. Heena, Sharma, K., Kumar, K., Das, B., Kumar, R.: Progress Org. Coat. (2024). https://doi.org/10.1016/j.porgcoat.2023.108108
4. Hima, V., Sreebha, A.B., Prasad, R., Tiwari, G., Mahadevan Pillai, V.P.: Materials Today: Proc. **57**, 2031–2035 (2020). https://doi.org/10.1016/j.matpr.2021.11.165
5. Kumar, S., Padmanapan Rao, V., Joseph, J.: Asian J. Phys. **24**, 1449–1464 (2015)
6. Sreebha, A.B., Mahadevan Pillai, V.P.: Thesis, University of Kerala (2018). http://hdl.handle.net/10603/364429
7. Guo, J., Gleeson, M.R., Sheridan, J.T.: Phys. Res. Int. (2012). https://doi.org/10.1155/2012/803439
8. Nimmi, K.P., Pramitha, V., Sreekumar, K., Kartha, C.S., Joseph, R.: J. Appl. Polym. Sci. **125**, 1238–1243 (2012). https://doi.org/10.1002/app.34963
9. Martin, S., Leclere, P., Renotte, Y.L., Lion, Y.F.: Opt. Eng. **33**, 3942–3946 (1994)
10. Srivastava, A., Sawant, S.M., Mainali, L., Bhatt, R., Gupta, A.K., Samui, A.B.: Polymers Adv. Technol. **23**(6), 967–974 (2012). https://doi.org/10.1002/pat.1999

Chemo-Green Synthesized NiO/CuO Nanocomposite: Photoluminescence and Antimicrobial Activity

Lakshmi Pavani Dukkipati[1(✉)], Madhavi Jetti[2], Rekha Rani Maddula[3], and P. Kalyan Kumar[4]

[1] Department of Applied Sciences, SIES Graduate School of Technology, Navi Mumbai 400706, India
lakshmid@sies.edu.in
[2] Malla Reddy College of Engineering, Hyderabad, Telangana 500014, India
[3] Gokaraju Lailavathi Engineering College, Hyderabad, Telangana 500090, India
[4] Koneru Lakshmaiah Education Foundation, Hyderabad, Telangana 500075, India

Abstract. A bimetallic mixed metal oxide (NiO/CuO) nanocomposite was synthesized, using both chemical and green methods, and investigated. The crystalline nature, phase purity, and size of the nanocomposite were investigated using the X-ray diffraction (XRD) technique. Scanning electron microscopy (SEM) and high-resolution transmission electron microscopy (HRTEM) examinations revealed about 30 nm size for the composite, which further aided in the deep physical and chemical analysis of the chemical and green route synthesized NiO/CuO composite. The studies also revealed aggregation and hexagonal spherical shape in the nanocomposite. The presence of the Cu, Ni, and O elements in the mixed valence state of the NiO/CuO nanocomposite was revealed using X-ray photoelectron spectroscopy (XPS) and energy dispersive X-ray spectroscopy (EDS) studies. Consequently, the nanocomposite's potential towards antibacterial and photoluminescence activities were investigated. The NiO/CuO (1 mg/ml) composite demonstrated antimicrobial properties against Escherichia coli, Staphylococcus aureus, and Coagulase-negative staphylococci. The NiO/CuO composite's ability to inhibit the growth of bacteria was confirmed by the investigation of reactive oxygen species generation. In addition, the PL emission spectra of the synthesized NiO/CuO nanocomposite revealed two visible emission peaks, demonstrating its relevant photoluminescence potential.

1 Introduction

Sustainability today is a 360-degree multidimensional deep concept, which is not possible to achieve without considering the minutest of organisms that affect both our lives and livelihood. The two most important parameters for a sustainable environment are water and microbes. A fine balance thrives sustainability and any adverse reaction hampers growth. Microbes survive everywhere including extreme conditions such as deserts, freezing water bodies, and equator. Amidst the interactions of microbes with all kinds of environment, nanotechnology has since long been used as a powerful tool not only

P. Joseph D. et al. (Eds.): IEMDST 2024, SPPHY 567, pp. 188–198, 2025.
https://doi.org/10.1007/978-981-96-5863-3_25

to combat diseases but also environmental pollution. To survive and sustain, everyone must have access to pure water. Moreover, water is also a crucial component of many vital industries. Recent years have worsened the water frontier situation due to continuous industrial growth [1]. This situation results in severe adverse effects, including environmental contamination, water shortages, and increased global temperatures. Primarily, the textile industry has been producing waste material with various chemical characteristics. Numerous colours found in waste are typically harmful and stable in the environment. The large-scale use of these harmful organic dyes destabilizes the environment and increases the risk of cancer. Environmental contaminants resulting from dyes and harmful bacteria are one of the most significant issues that needs immediate attention. Therefore, finding affordable and eco-friendly ways to remove these dangerous dyes from the environment and groundwater has become a significant concern.

New solutions in various domains, particularly the treatment of industrial dyes in wastewater, can be provided by advancements in science and technology along with the quick development and improvement in related disciplines. Absorption, natural processes, aeration, photo-oxidation, antioxidation and other processes are used in various physical, chemical, and biological approaches to remove contaminants. Photo-oxidation is one practical method of purification, which is also a financially sound choice [2]. Its quick photocatalytic oxidation reaction at low concentrations, absence of byproducts, and creation of undesirable products are some of the valid reasons for this. However, in more sophisticated oxidation processes, the hydroxyl radical plays a significant role and functions as a potent oxidizer that causes indiscriminate and efficient oxidation of colours.

Along with water related problems, infectious diseases are another significant challenge today. Growing bacterial infections from new emerging strains are a cause of huge concern. Consequently, antimicrobial agent analysis is crucial as microbes are the leading cause of illness and mortality worldwide. The primary goal of this study has been to identify significant medication resistance in widespread diseases. By their chemical makeup, antibiotic agents are divided into two categories: natural and artificial. Due to their unusual physical, electrical, optical, and chemical properties, metal oxide nanoparticles (NPs) have attracted considerable interest. Over the past ten years, researchers have investigated metal oxide NPs for catalysis, sensing, energy storage, optical activity, pharmaceutical applications, environmental remediation, and biological pursuits. NiO, CuO, SnO_2, TiO_2, and ZnO are some examples of inorganic semiconductors. Due to the quantum confinement effect, surface to volume ratio, and specific nanostructures, such as nanospheres, nanowires, and nanotubes, NP's display unique electrical and optical capabilities. NiO is well known for its extensive uses in photocatalysis and energy technology [3]. NiO nanoparticles synthesized, according to Ezhilarasi and Vijaya, for use in photocatalytic and antimicrobial processes, has been studied for investigations. The work by Duan et al. in the synthesis of NiO nanoparticles for the photocatalytic degradation of rhodamine B dye under ultraviolet rays, has also been studied extensively.

Fazlali and his colleagues produced NiO spherical nanoparticles and reported 81 per cent methyl orange degradation under UV light. Thus, it is accurate to state that the UV zone of electromagnetic radiation is the only one region where NiO exhibits photocatalytic activity. As a result, it requires additional consideration to become a strong contender towards visible or solar light response. NiO has undergone several modifications for this purpose. Examples involve doping other elements into NiO and creating composites of NiO and other metal oxides. Their bandgap energy is decreased by doping with other transition metals or their cations, making them less efficient candidates for visible light response. The greater rate of e^--h^+ pair recombination is one of the causes of this decreased efficiency. CuO is a transition metal oxide with a narrow band gap ($E_g = 1.2$ eV), making it a p-type semiconductor [4]. Due to the interaction of metal-metal and metal-oxygen-metal bonds in mixed metal oxide, the number of charge carriers, surface area, and charge transfer increase, which in turn improve its chemical and physical characteristics. As a result, mixed metal oxide has higher catalytic effect on the breakdown of amino perchlorate than a single metal oxide, according to Hosseini et al. The primary mixed metal oxide, NiO/CuO, has Cu and Ni as metal centres. The NiO/CuO NP active surface area, crystallinity, and particle size significantly impact the material's catalytic performance [5].

2 Materials and Methods

2.1 Materials

Nickel nitrate hexahydrate [Ni $(NO_3)_2{\cdot}6H_2O$], copper nitrate trihydrate [Cu $(NO_3)_2{\cdot}3H_2O$], sodium hydroxide [NaOH] were procured from Sigma Aldrich, India. Neem fruits of uniform size were procured from the local market.

2.2 Synthesis

2.2.1 Neem Fruit Extract Preparation

About 40 g of neem fruits (washed and dried) were added to 150 ml deionized water under continuous stirring and placed on a hot plate at 90 °C for 3 h. The fruit extract solution was subsequently cooled and filtered with filter paper to obtain pure neem fruit aqueous extract solution. The extract solution was used for the preparation of NiO/CuO nanocomposites.

2.2.2 Bio and Chemical Synthesis of the NiO/CuO Nanocomposite

The bi-metallic NiO/CuO nanocomposite was synthesized using the green synthesis method [6]. An aqueous solution of 0.01 M of nickel nitrate hexahydrate and copper nitrate trihydrate salt was prepared in a beaker. The mixture solution was stirred for 30 min, and the neem fruit extract was slowly added. The solution colour gradually changed from greenish blue to black. The mixture solution was maintained at 80 °C on magnetic stirrer until evaporation of the solution. The dried sample was calcined at 300 °C for 4 h, and finally, a black NiO/CuO nanocomposite was obtained [7]. For the chemical route, the same process was carried out using the NaOH solution instead of the neem extract solution. 1 M of NaOH solution was added to the mixture solution. The residue was washed with deionized water several times and dried at 80 °C for 14 h. The dried sample was calcined at 500 °C for 4 h. The obtained final NiO/CuO nanocomposite was black in colour [8, 9]. The end to end synthesis process is illustrated in Fig. 1.

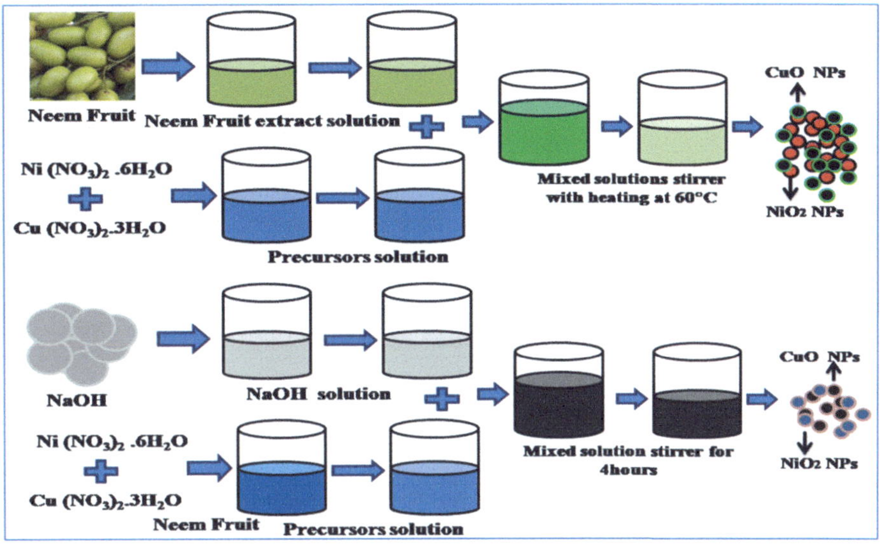

Fig. 1. Synthesis of NiO/CuO nanoparticles: chemical route (top) and green route (bottom).

2.2.3 Antibacterial Investigation

The effectiveness of the inhibitory capabilities against gram-positive and gram–negative bacteria is evaluated using the well diffusion method. The bacteria were swabbed over freshly prepared Mueller Hinton agar medium in petri plates. Six mm diameter wells containing 2 mg of the samples with different concentrations within were created and incubated at 37.2 °C for 24 h. Zones of inhibition (ZOI) were determined to have evolved around the samples from where the bacteria's inhibitory properties were accessible [10].

3 Results and Discussion

3.1 XRD Analysis

X-ray diffraction (XRD) analysis was used to study the crystallinity and phase of the composite. The as-synthesized nanocomposites demonstrated crystalline structures and distinct diffraction peaks. The crystal structure and phase purity of synthesized nanocomposite materials were evaluated, using XRD analysis as shown in Fig. 2. The observed peaks are assigned to cubic NiO (JCPDS no. 47-1049) and monoclinic CuO (JCPDS no.

45-0937). The average crystallite size determined for the chemical and green synthesis for the NiO/CuO composite was 17.23 nm and 15.24 nm, respectively. Previous works have revealed, the crystallite size to be about 24 nm for the NiO/CuO nanocomposite in the green route and about 30 nm in the chemical route using the Debye-Scherrer equation [11–15].

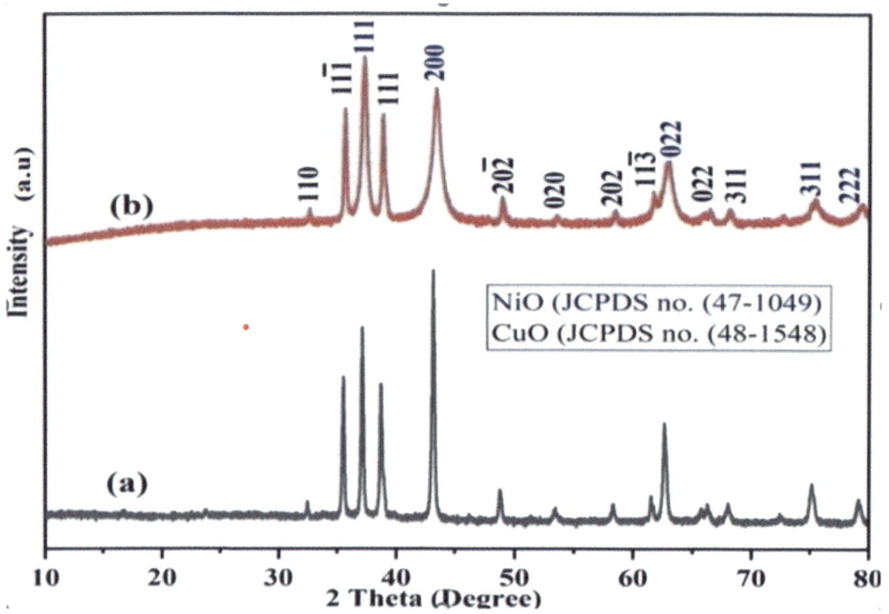

Fig. 2. XRD analysis of (a) chemically synthesized NiO/CuO nanocomposite and (b) green route synthesized NiO/CuO nanocomposite

3.2 FESEM

FESEM analysis provided morphological and topographical examinations. Figure 3 shows the microscopic surface morphology of the synthesized NiO/CuO nanocomposite using green and chemical routes as characterized by FESEM technology. The scan reveals the broad spherical structure of the composite, indicating copper oxide deposition on the nickel oxide surface. By observing FESEM images, we can say that the green method realized a clear spherical structure as compared to the chemical method [16].

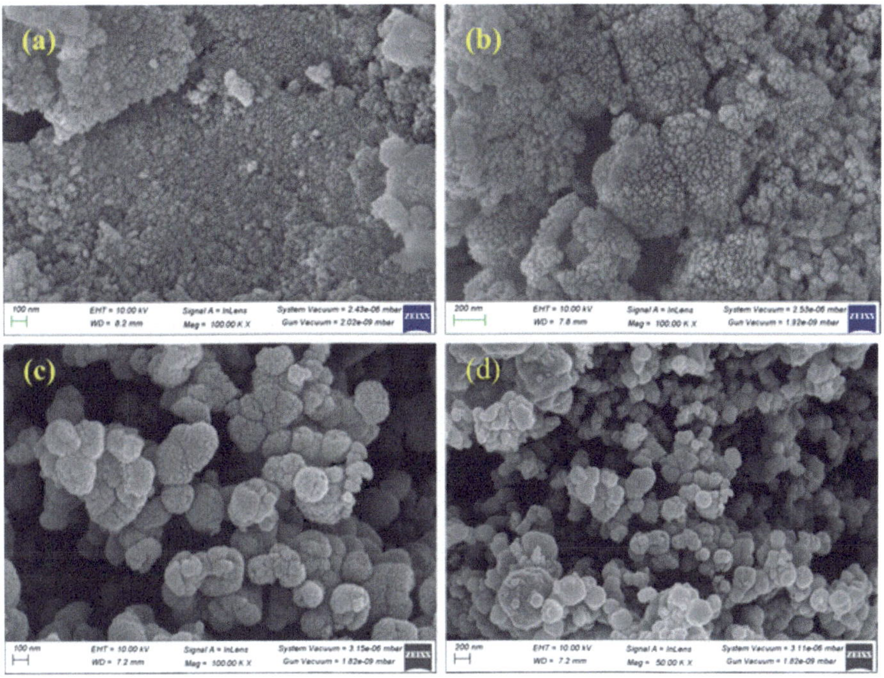

Fig. 3. FESEM images of (a) chemically synthesized NiO/CuO nanocomposite and (b) green route synthesized NiO/CuO nanocomposite

3.3 XPS

We have carried out the XPS studies of the NiO/CuO composite to find the oxidation states of the elements present in the sample [17, 18]. As seen from the Fig. 4(a), the full spectrums of XPS indicate the presence of Ni, Cu and O. We have further carried out high-resolution XPS studies for each element present in the sample. Figure 4(b) shows the XPS spectra of Cu 2p showing two peaks at 933.8 eV and 953.8 eV corresponding to the Cu 2p3/2 and Cu 2p1/2 core level components of CuO. The other peaks at 941.4 eV, 943.8 eV and at 962.0 eV are the satellite peaks of the Cu^{2+}. Figure 4(c) shows the XPS spectra of Ni 2p. In this spectrum, the 854.2 eV and 872.9 eV peaks are attributed to Ni $2p_{3/2}$ and Ni $2p_{1/2}$, respectively. The peaks at 861.4 eV and 879.5 eV are the satellite peaks of Ni^{2+}. The obtained results are consistent with the other available reports [1–5]. Figure 4(d) shows O 1s spectra indicating two peaks at 529.2 eV and 531.2 eV, which is characteristic of the presence of oxygen ions in the NiO/CuO composite [1–5].

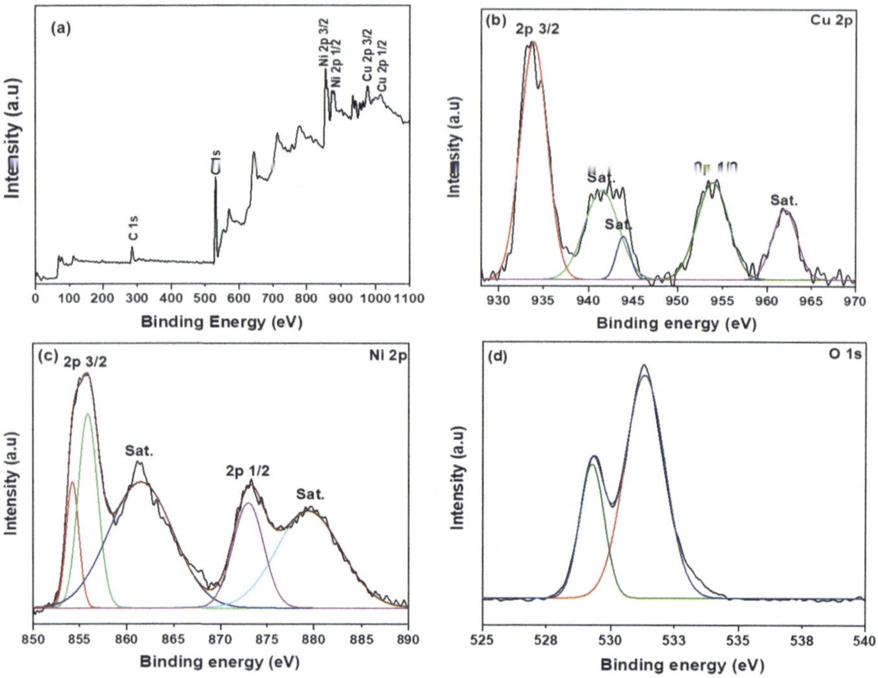

Fig. 4. XPS spectra of green synthesized NiO/CuO (a) wide spectrum (b) Cu 2p (c) Ni 2p and (d) O1s

3.4 HRTEM

Using HRTEM, the size of the synthesized NiO/CuO nanocomposite was determined. Figure 5 demonstrates a significant assembly of nanocomposites and particle aggregation with a hexagonal spherical shape [19]. HRTEM analysis revealed the NiO/CuO nanocomposite size at about 30 nm. Figure 5a and b, display the HRTEM picture of both the chemical and green synthesized NiO/CuO nanocomposites. The EDS spectrum (Fig. 5c and d) of the chemically synthesised NiO/CuO nanocomposite indicate Ni, Cu, and O peaks thus revealing their presence whereas the G-NiO/CuO revealed one additional elemental peak in the EDS spectrum. Cu content was observed to be higher than Ni in the C-NiO/CuO composite and Ni content was higher than Cu in the G-NiO/CuO composite, likely dominating the material's structural, morphological, thermal, optical, photoconductive, magnetic, and antibacterial properties, on which further investigation are in progress. Particle size distributions were studied using the dynamic light scattering (DLS) technique (Fig. 5). C-NiO/CuO and G-NiO/CuO particle size distribution (Figs. 5e and f), based on DLS analysis, are indicated, which complies well with FESEM and XRD results discussed in the study.

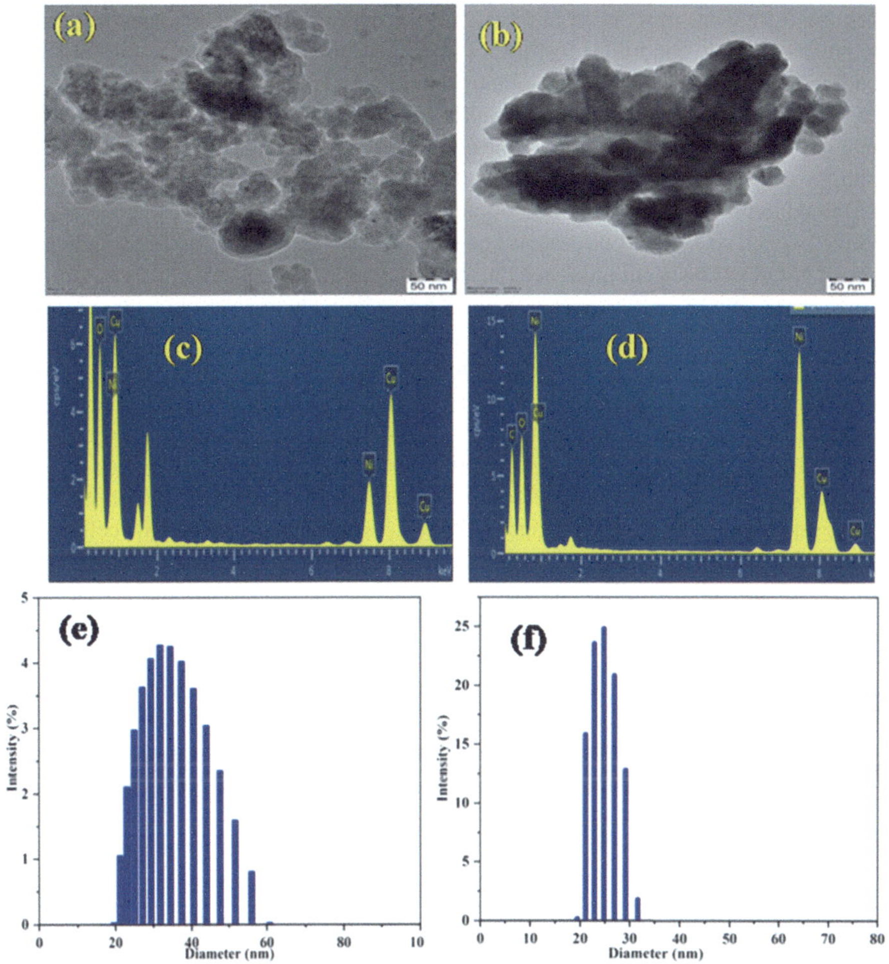

Fig. 5. TEM, EDS and DLS analysis of (a, c, e) chemically synthesized and (b, d, f) green synthesized NiO/CuO nanocomposites respectively.

3.5 Antibacterial Activity of Green and Chemically Synthesized NiO/CuO Nanocomposites

The antibacterial activity of NiO/CuO NCs was evaluated against E. coli, coagulase-negative staphylococci (CoNS), and S. aureus via the well diffusion method, discussed previously [20, 21]. Agar plates were incubated under suitable conditions depending upon the test microorganisms. 50, 75, 100, and 125 μl concentrations of the NiO/CuO nanocomposite were introduced into the well for investigations. The plates were incubated under suitable conditions for further investigations. The diameter values of corresponding zones of inhibition (ZOI) obtained are displayed in Fig. 5. It is interesting to note that, as compared to the chemical NiO/CuO nanocomposite, the green NiO/CuO nanocomposite significantly suppress the test bacteria, though E. coli bacteria were more suppressive than both S. aureus and CONS bacteria. The highest ZOI for E. coli was determined to be 16 mm, while for S. aureus it was 14 mm at 125 μl. The generation of reactive oxygen species (ROS), release of soluble ions, and charge on the active membrane of NiO/CuO NCs, all influence the antibacterial activity. From Fig. 5 the zones of inhibition of G-NiO/CuO(green route) and C-NiO/CuO (chemical route) were indicated to be 12, 13, 14, and 16 mm, 10, 11, 11, and 13 mm, and 9, 10, 11, and 14 mm and 9, 10, 11, and 12 mm, 10, 11, 12, and 12 mm, and 9, 10, 12, and 14 mm, respectively. The remarkable antibacterial inhibition activity realised might be due to i) breakdown, in which positively charged Ni^{2+} and Cu^{+} ions are generated. These ions were firmly connected to the negatively charged bacterial cell wall owing to the electrostatic force of attraction. ii) The production of reactive oxygen species (ROS) such as hydroxyl (OH*), superoxide ($O*_2$) radicals, and hydrogen peroxide (H_2O_2) enter the cell wall of the bacteria, oxidising coenzymes and inhibiting cell respiration, ultimately killing the microorganisms. The NiO/CuO nanocomposite's improved anti-bacterial activity may be attributed to its lower crystallite size. Smaller crystallites produce more NiO/CuO nanoparticles per unit volume, which increases the production of reactive oxygen species and effectively destroy harmful microorganism cells. The NiO/CuO nanocomposite is suitable for biological applications because of its ferromagnetic properties as targeted drug carriers. Further, the NiO/CuO composites can be possibly applied for antibacterial purpose in wastewater treatment systems.

3.6 Photoluminescence

Materials' luminescence characteristics can predict their crystal quality and purity [22]. The PL emission spectra of the synthesized NiO/CuO nanocomposites are shown in Fig. 6, reflecting two visible emission peaks at 412 nm and 496 nm under 300 nm excitation wavelengths. The visible emission peaks have been attributed to the oxygen vacancy defects, which have been noticed. Oxygen vacancies are the most prevalent defects in poly or nanocrystalline oxides as well as centres of radiation in the luminescence process. From the spectra, we observed that the green route synthesized NiO/CuO nanocomposite demonstrated higher intensity than the chemical route prepared NiO/CuO.

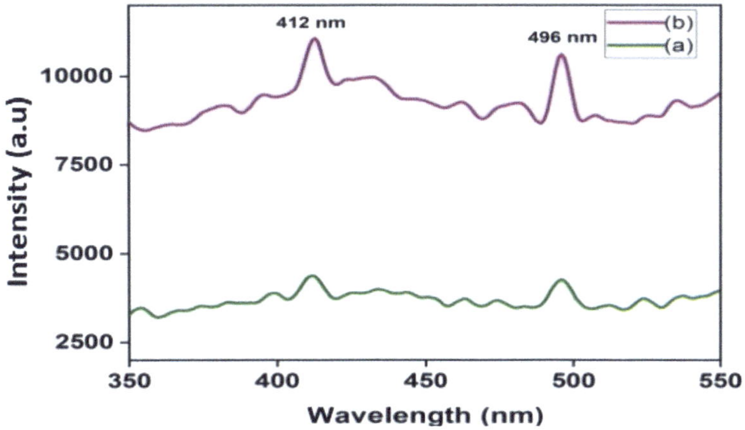

Fig. 6. PL analysis of chemically synthesized NiO/CuO nanocomposite (a) and green synthesized NiO/CuO nanocomposite (b)

4 Conclusion

NiO/CuO bimetallic oxide nanoparticles, with NaOH as the reducing agent and neem fruit extract solution as the reducing and capping agent, were successfully synthesized via chemical and green routes and investigated for photoluminescence and antimicrobial properties. The average crystallite size determined for the NiO/CuO composite was 17.23 nm and 15.24 nm, respectively, for the chemical and green synthesis, using XRD studies. FESEM studies reveal the spherical structure morphology of the samples. TEM analysis revealed the synthesized nanocomposite size at about 30 nm, in line with XRD and DLS studies. XPS and EDS analyses showcased the elemental and oxidation states of the nanocomposite. EDS analysis demonstrated a relative increase of elemental Ni and decrease of elemental Cu in the G-NiO/CuO nanocomposite. The PL analyses of composites reflect upon the greater intensity of emission in the lower visible region with green NC, well matched with the CIE graph. Concentrations of NiO/CuO NCs viz., 50, 75, 100, and 125 μl were prepared and largest ZOI for *E. coli* was determined to be 16 mm, while the largest ZOI for *S. aureus* was 14 mm at 125 μl. The study infers that the green synthesized (secondary metabolites available in neem plant) NiO/CuO composite shows significant activity against E. coli, S. aureus and Coagulase-negative staphylococci when compared with the chemically synthesized composite. The work realizes the composite's chemical identity, surface characteristics, and adaptable optical properties most suitable for bioimaging and theranostics.

Acknowledgments. We are grateful for the support and cooperation received from the Department of Physics, NIT Warangal, Department of Nanotechnology, JNTU Hyderabad, and SIES-Graduate school of Technology, Navi Mumbai for the study, research work, and facilities.

References

1. Tanihara, A., Kikuchi, K., Konno, H.: Inorg. Chem. Commun. **131**, 108782 (2021)
2. Sharma, S., Sharma, A., Chauhan, N.S., et al.: Inorg. Chem. Commun. **146**, 110093 (2022)
3. Daniel, P.S., Thangadurai, T., Indumathi, T., et al.: Inorg. Chem. Commun. **146**, 110077 (2022)
4. Nayak, R., Ali, F.A., Aswal, V.K., et al.: J. Mater. Res. Technol. **9**, 11045–11059 (2020)
5. Abo Zeid, E.F., Nassar, A.M., Hussein, M.A., et al.: J. Mater. Res. Technol. **9**(2), 1457–1467 (2020)
6. Balu, A.S., Balamurugan, A.R., Suganya, S., et al.: Inorg. Chem. Commun. **134**, 109073 (2021)
7. Ding, Y., Zhuang, Q., Guo, X., Li, H., Liang, C., Du, B., et al.: Appl. Surf. Sci. **612**, 155 (2023)
8. Das, D., Nath, B.C., Phukon, P., et al.: Colloids Surf. B Biointerf. **101**, 430–433 (2013)
9. Shanker, K., et al.: Colloids Surf. B Biointerf. **174**, 199–206 (2019)
10. Jones, N., Ray, B., Ranjit, K.T., Manna, A.C.: FEMS Microbiol. Lett. **279**, 71–76 (2008)
11. Sankar, S., Sharma, S.K., An, N., Lee, H., Kim, D.Y., et al.: Photocatalytic properties of Mn-doped NiO spherical nanoparticles synthesized from sol-gel method. Optik **127**(22), 10727–10734 (2016)
12. Anandan, K., Rajendran, V.: Mater. Sci. Semicond. Process. **14**, 43–47 (2011)
13. Motevalli, K., Zarghami, Z., Panahi-Kalamuei, M.: J. Mater. Sci.: Mater. Electron. **27**(5), 4794–4799 (2016). https://doi.org/10.1007/s10854-016-4360-5
14. Xu, J.F., et al.: J. Solid State Chem. **147**, 516–519 (1999)
15. Said, A.E.-A., et al.: Synthesis and characterization of nano CuO-NiO mixed oxides. Nanosci. Nanoeng. **2**(1), 17–28 (2014)
16. Rawat, M., Rawat, D.S.: ACS Sustain. Chem. Eng **8**, 13701–13712 (2020)
17. Bhuyan, T., Mishra, K., Khanuja, M., Prasad, R., Varma, A.: Mater. Sci. Semicond. Process. **32**, 55–61 (2015)
18. Elumalai, K., Velmurugan, S.: Appl. Surf. Sci. **345**, 329–336 (2015)
19. Krishnan, S.G., Nand, D., Sivadas, S., Alfarhan, A., Muthusamy, K.: J. King Saud Univ. – Sci. **34**, 101718 (2022)
20. Krishna, M.S., Joy, B.: Sundaresan. J. Food Sci. Technol. **52**, 3561–3570 (2015)
21. Ghasemzadeh, A., Jaafar, H.Z.E., Ashkani, S., Rahmat, A., et al.: BMC Complement. Alternat. Med. **16** (2016).
22. Saravanakumar, K., Shanmugam, S., Varukattu, N.B., MubarakAli, D., Kathiresan, K., Wang, M.H.J.: Photochem. Photobiol. B Biol. **190**, 103–109 (2019)

Bulk Growth of Highly Transparent Triglycine Sulphate (TGS) Single Crystals for IR Sensor Applications

I. Karthikeyan[1], S. Surya[1], S. Gokul Raj[1](✉), V. C. Bharath Sabarish[2], and G. Ramesh Kumar[3]

[1] Department of Physics, Pondicherry University, Kalapet, Puducherry 605014, India
gokulraj@pondiuni.ac.in

[2] Department of Science and Humanities, K.Ramakrishnan College of Technology (Autonomous), Kariyamanikam Road, Samayapuram, Trichy 621112, India

[3] Department of Science and Humanities, University College of Engineering Arni, Anna University Chennai, Thatchur, Arni 632326, India

Abstract. The focus of the present investigation is to grow bulk good quality single crystals for ferroelectric applications. Triglycine Sulphate (TGS) was crystallized by simple low temperature solution growth. Further for the cubic growth of TGS analysis slow evaporation methods adopted. The structural analysis of the grown single crystal was studied using Powder X-Ray Diffraction successive analysis, which confirms the crystal system of the grown single crystal is monoclinic. UV-Vis-NIR analysis confirms the transparency in the range 300–1100 nm with a lower cut-off at 241 nm. Room temperature dielectric ferroelectric properties was investigated to test in electrical properties of the grown crystals. These preliminary investigation reveals that TGS crystals as a potential room temperature ferroelectric material.

1 Introduction

Ferroelectric triglycine sulphate ($(NH_2CH_2COOH)_3.H_2SO_4$) (TGS) single crystals were synthesized and grown by many authors and were reported by slow evaporation, conventional slow cooling and also by unidirectional solution growth method. TGS was first reported by Matthias et al. in 1956 which exists two ferroelectric and paraelectric phases and belongs to the monoclinic system with non-centrosymmetric space group P21. It has second order phase transition at 49.3 °C and above which there is change of ferroelectric phase to centrosymmetric P21/m in the paraelectric phase. There were reports on the modification in the Curie temperature by the use of various dopants [1–4]. The fabrication and applications of pyroelectric and electro-optical devices, the curious area of interest is shown on ferroelectric single crystals due to their peculiar properties [5]. It is also one such good ferroelectric material which was widely used for the fabrication of IR detector [6]. Highly sensitive to IR radiation, soft ferroelectric properties along with optimum piezoelectric constants makes Triglycine sulphate (TGS) single crystal as a good crystalline material [7].

P. Joseph D. et al. (Eds.): IEMDST 2024, SPPHY 567, pp. 199–206, 2025.
https://doi.org/10.1007/978-981-96-5863-3_26

In the fabrication of infrared (IR) devices these TGS single crystal that are been grown in room temperature in different oriented crystals and were being used because of its low cost and low power consumption, also these single crystals can operate on wide range of temperature and frequency [8, 9]. TGS crystallizes in monoclinic structure with the curie temperature of 49 °C. This 49 °C low Curie temperature leads to depolarization which is a disadvantage. But this depolarization caused by low Curie temperature can be rectified by substituting an optically active molecule on the sites of glycine [10]. Single crystal of pure TGS crystal was grown and harvested in a duration of 7 days followed by multiple recrystallizations from an aqueous solution using one of the temperature lowering techniques, in general termed as (SEST) Slow Evaporation Solution Growth Technique. In our current investigation the reports are recorded on the preliminary investigation on powder XRD, UV-Vis-NIR spectroscopy and an attempt was made to grow bulk single crystals for IR detector application. Room temperature dielectric and ferroelectric properties was investigated to compare the electrical properties.

2 Experimental Procedure

2.1 Synthesis of TGS

Calculated quantities of glycine ($C_2H_5NO_2$) and sulphuric acid (H_2SO_4) are mixed in double distilled water in the required ratio and the solution were left in petri dish to get seed crystals of TGS which was obtained after a period of 7 days.

$$\text{Reaction Mechanism}: \; C_2H_5NO_2 + H_2SO_4 = (NH_2CH_2COOH)_3 \cdot H_2SO_4$$

One of the major parameters in growing a transparent single crystal is the solubility curve measurement of the mother solution. The solubility standardization of the mother solution determines the single nucleation and growth rate of the crystal [11]. Triglycine sulphate $(NH_2CH_2COOH)_3.H_2SO_4$ (TGS) solution was prepared as per the solubility curve of TGS and the amount of solute is given in Table 1. The solubility curve of TGS powder Vs Temperature is given in Fig. 1.

Table 1. Temperature Vs Solubility of TGS powders (gm/100 ml)

Temperature (°C)	Powdered TGS (gm/100 ml)
30	22
40	28
50	34
60	40
70	46
80	52
90	56

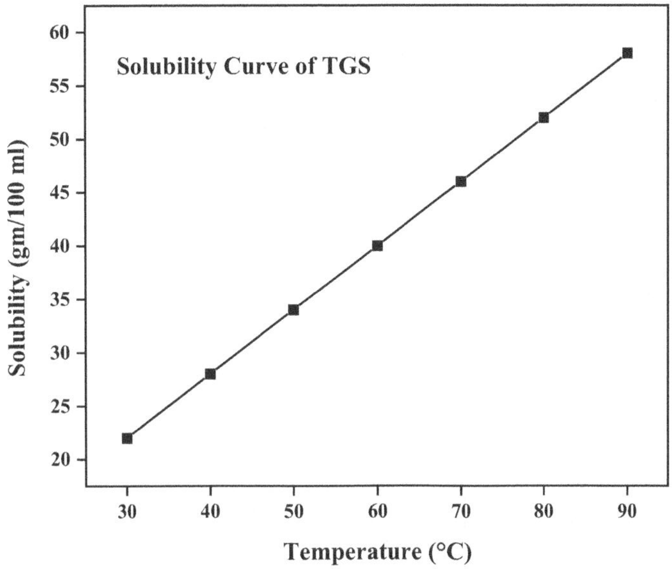

Fig. 1. Solubility curve of TGS powder in solvent Vs Temperature

According to solubility data, 22.5 gm solute per 100 ml of solvent is chosen for the preparation of solution. Using magnetic stirrer, the solution is stirred for 4 h at room temperature to get highly homogeneous and a transparent good saturated mother solution of TGS. Whatman filter paper was used to filter the prepared homogenous TGS solution and poured in to a beaker and a petri dish to set seed crystals of TGS. It was kept opened and to obtained solution evaporation was kept in a constant temperature bath at 36 °C in a dust free atmosphere. After 7 days the solution has been completely evaporated. To achieve good transparent bulk crystals, recrystallization process is carried out multiple times. During slow evaporation or slow cooling, crystals with inclusions and without inclusions can be observed. The purity and transparency of the crystals was enhanced by successive multiple recrystallization process. Figure 2 represents the pictorial steps involved in growth of TGS seed crystal. During the initial phase of seed crystal preparation, Multinucleated TGS appears and during successive recrystallization, single crystals were obtained from homogenous solution and this muti-nucleation formation during (a) First Crystallization and (b) Second recrystallization is shown in Fig. 3. Transparent single crystals were obtained in successive recrystallization from the mother solution were used for further analysis and this is shown in Fig. 4. As Grown TGS single crystals from slow evaporation method in the department of physics, Pondicherry University is shown in Fig. 5.

3 Steps Involved in the Bulk Growth of TGS from Solution Growth Technique

TGS crystal were grown in room temperature by slow evaporation solution growth technique. This technique involves the following steps:

1. Selection of solvent and solute.
2. Preparation of saturated solution and Filtering.
3. Solution cooling and slow evaporation.
4. To monitor the solution regularly and to observe the formation of growth of seed crystals.
5. Harvesting crystals, once they have reached to the desired size.
6. Drying and recrystallization for further characterization.

Fig. 2. Pictorial representation for the steps involved in growth of TGS seed crystal.

Fig. 3. Multinucleated TGS: A) First Crystallization B) Second Recrystallization.

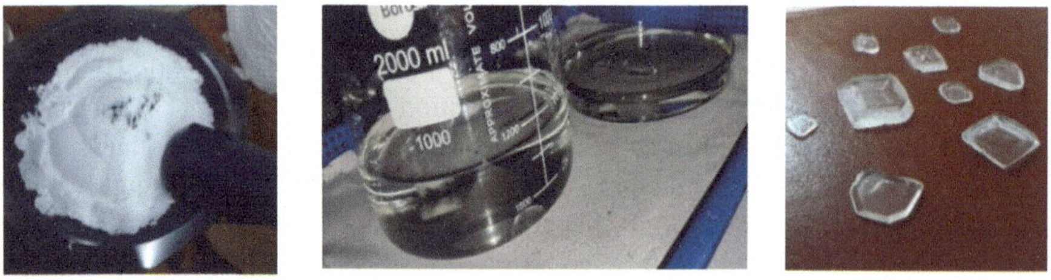

Fig. 4. Pictorial representation of successive recrystallize transparent crystal in mother solution.

Fig. 5. As grown TGS single crystals from slow evaporation method.

4 Results and Discussion

4.1 Powder X-Ray Diffraction (PXRD) Analysis

Samples of TGS were obtained from slow evaporation technique. These samples were the powdered and was further used for powder x-ray diffraction analysis. From powder x-ray analysis we could obtained the prominent planes and the XRD pattern is compare with standard JCPDS (14-0873) [12] pattern as shown in Fig. 6.

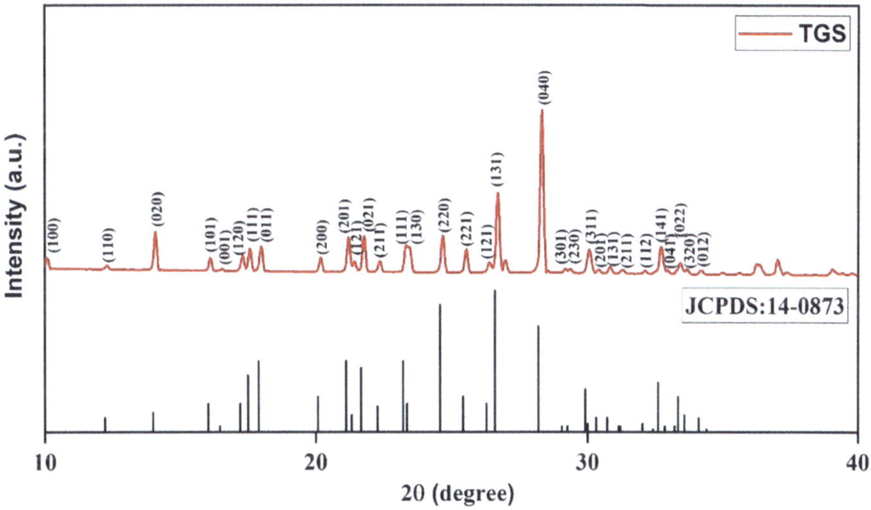

Fig. 6. PXRD pattern of TGS crystal

The Rigaku X-Ray Diffractometer was used to carry out the PXRD analysis which uses Cu Kα of wavelength (λ) 1.5406Å. The analysis was done on the 2θ range of 100–600 on a scan rate of 20/min. The synthesized TGS crystals crystallized in monoclinic structure with unit cell parameters a = 0.9417 nm, b = 1.2643 nm, c = 0.5735 nm; α = γ = 90° β ≠ 90° and this is in close agreement with that gate reported values and standard JCPDS pattern. Since the Bragg's peak for the diffraction angle which were obtained from the standard JCPDS pattern and compared with the obtained experimental values.

5 Optical Properties

5.1 UV-Vis-NIR Spectroscopy

Some of the optical parameters used to enquire the Non-Linear Optical (NLO) behaviour of the single crystal were absorbance, transmittance and energy band gap (Eg). Using the UV spectrophotometer these behaviours were studied [13, 14]. The absorption spectrum was shown in Fig. 7, which has the lower cutoff wavelength at 241nm and strong absorption peak was found below 240 nm, this strong absorption peak and lower cutoff wavelength of the absorption spectrum corresponds to the state of π-π* electronic transition [15]. There is no absorption peaks found in the visible region between 300–1100 nm which denotes the wide transparency and this is found to be the major requirement for a single crystal to be optically active [16].

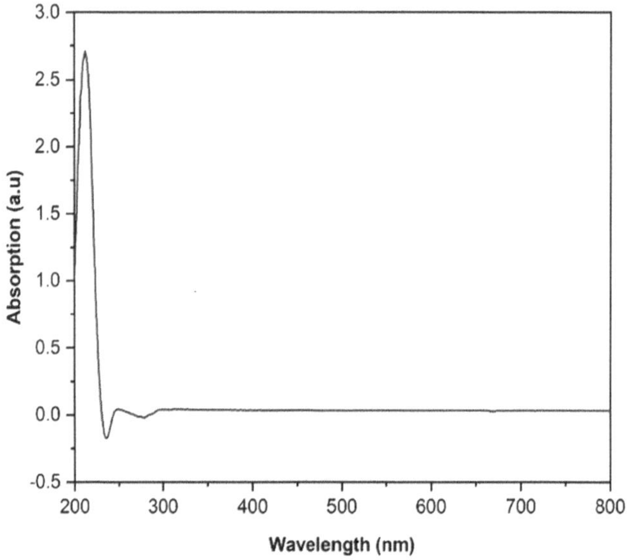

Fig. 7. Absorption spectrum

6 Optical Energy Band Gap

The optical band gap energy (Eg) was found from Tauc's Plot using the absorption coefficient (α) derived from the absorption spectrum. The Tauc's plot was plotted between hυ vs $(\alpha h\upsilon)^2$ using the formula $h\upsilon\alpha = A(h\upsilon - E_g)^2$ [17, 18]. The optical band gap energy (E_g) graph was shown in Fig. 8. The TGS single crystal have high optical band gap energy which was calculated to be 5.2 eV.

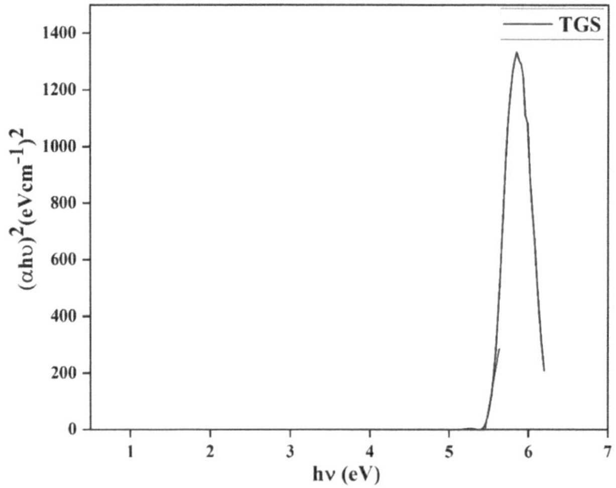

Fig. 8. Tauc's Plot of the single crystal (hυ vs $(\alpha h\upsilon)^2$)

7 Conclusion

The pure and transparent TGS single crystal was grown using slow evaporation solution growth technique (SEST) in a duration of 7 days followed by multiple recrystallizations. The length of the as grown crystal is 2.9 cm. The PXRD analysis shows that structural detailing of TGS single crystal and the XRD pattern is in good agreement with the JCPDS. The crystal system is found to be monoclinic and the lattice parameters were calculated. The UV-Vis-NIR analysis shows the absorption spectrum which have no absorption peak in the visible region, stating that the material is optically active. From Tauc's plot the energy band gap was found to be 5.2 eV. Optical results infer that the crystal also have NLO properties due to its wide transparency observed from the absorption spectrum.

References

1. Bhatt, V.P., Patel, R.M.: **53**(3) 633–635 (1981)
2. Pandian, M.S., Verma, S., Karuppasamy, P., Padmanabhan, V., et al.: J. Cryst. Growth **546**, 125793 (2020)
3. Trybus, M., Szlachta, A.: Infrared Phys. Technol. **123**, 104211 (2022)
4. Matthias, B.T., Miller, C.E., Remeika, J.P.: Physi. Rev. **104**(3), 849–850 (1956)
5. Deepthi, P.R., Shanthi, J.: RSC Adv. **6**(40), 33686–33694 (2016)
6. Zhao, X., Sun, C., Si, Y., Liu, M., Xue, D.: Mod. Phys. Lett. B **23**, 3809–3815 (2009)
7. Mahendra, K., Kumar, H.K.T., Udayashankar, N.K.: Appl. Phys. A **125**, 1–15 (2019)
8. Deepthi, P.R., Shanthi, J.: Growth and characterization of l-alanine doped TGS crystal. Int. J. Eng. Res. **3**(6) (2014)
9. Yarmarkin, V.K., Shul'man, S.G., Pankova, G.A., Zaĭtseva, N.V., Lemanov, V.V.: Phys. Solid State **50**(5), 937–940 (2008)
10. Alexandru, H.V., Mindru, C., Bacşei, R., Ganea, C.P.: J. Optoelectron. Adv. Mater. **16**, 457–462 (2014)

11. Andriyevsky, B., Esser, N., Patryn, A., Cobet, C., Ciepluch-Trojanek, W., Romanyuk, M.: Materials **373**, 328–333 (2006)
12. Balamurugan, N., Lenin, M., Bhagavannarayana, G., Ramasamy, P.: J. Exp. Ind. Crystallogr. **42**, 151–156 (2007)
13. Batra, A.K., Mathur, S.C., Mansingh, A.: 399–405 (1983)
14. Fletcher, S.R., Keve, E.T., Skapski, A.C.: Ferroelectrics **14**(1), 775–787 (1976)
15. Hilczer, B., Balanicka, S.: Physica Status Solidi (a) **19,** 717–723
16. Horcas, I., Fernández, R., Gomez-Rodriguez, J.M., Colchero, J.W.S.X., Gómez-Herrero, J.W.S.X.M., Baro, A.M.: Rev. Sci. Instrum. **78**, 013705 (2007)
17. Hoshino, S., Okaya, Y., Pepinsky, R.: Phys. Rev. **115**, 323 (1959)
18. Ikeda, T., Tanaka, Y., Toyoda, H.: Japanese J. Appl. Phys. **1**, 13 (1962)

Molecular Docking and Trace Elements Analysis in *Senna occidentalis*, an Anti-asthmatic Medicinal Herb

S. Jyothsna[1,2], T. Lavanya[1], D. Sammaiah[3], Mamidala Estari[4], Joel K. Joseph[1], S. Suneetha[1], Kopula Naresh[1], and Kuppusamy Thangaraju[1](✉)

[1] Organic Optoelectronics Lab, Department of Physics, National Institute of Technology, Warangal 506004, India
ktr@nitw.ac.in

[2] Department of Physics, University College of Engineering and Technology for Women, Kakatiya University, Warangal 506009, India

[3] Department of Botany, Kakatiya Government College-Kakatiya University, Warangal 506009, India

[4] Department of Zoology, Kakatiya University, Warangal, Telangana, India

Abstract. The docking investigation is conducted on five plant based substances which shall produce useful lead molecules in asthma medication. Crystal structure of target enzyme (PDE4) was obtained from the RCSB Protein Database, whereas that of phytochemical molecules (Apigenin, Betain, Stachydrine, Trigonelline and Anthraquinone) were obtained from the Pub chem database. Using a molecular modeling software programme called Auto-Dock Tools, the successful molecules were used to dock to the active regions of enzymes. Our findings show that Apigenin and Anthraquinone which are both natural components of the fruits *Senna occidentalis* (*S. occidentalis*) performed exceptionally well in the docking research, having low binding energies of −5.84 and −5.23 kcal/mol, respectively. This suggests their potential application for the asthma-inhibiting molecular interactions. Furthermore EDXRF technique is used to estimate the 13 elements and their concentrations in *S. occidentalis* fruits. The presence of K, Fe, Mn, Cu, Zn and Se indicates their promising role in the administration and management of asthma.

1 Introduction

Asthma is a major pulmonary disease that causes inflammation and narrowing of the airway. It is associated with some significant alterations in the concentrations of numerous distinct cells including eosinophils, mast cells, lymphocytes, cytokines, and certain allergic chemical compounds [1]. Asthma is a complex pulmonary disorder which can occur in people of all ages across the globe [2]. Asthma affects over 300 million people worldwide with an increase to 100 million by 2025 [3]. Bronchodilator medications including albuterol, terbutaline, theophylline, and cromolyn sodium can be used to treat asthma [4]. Because of numerous side effects and financial implications, current asthma treatments are not sufficiently effective. As a result, individuals with asthma are looking

P. Joseph D. et al. (Eds.): IEMDST 2024, SPPHY 567, pp. 207–215, 2025.
https://doi.org/10.1007/978-981-96-5863-3_27

for alternative medications that have no harmful side effects. Treatment and management of asthma without any side effects is still a challenge for medical field. Plants have been a prime source of healing since prehistoric times [5]. The Indian ayurveda system mentioned that the use of plants and herbs in several ailments including common cold to asthma [6]. Numerous studies have shown that herbal species are an excellent source of traditional medicines with the best potential for controlling and preventing various forms of asthma [6, 7]. Researchers have become increasingly determined over time to separate and characterize the bioactive components from these plants. *S. occidentalis* is a well-known plant with potential anti-inflammatory, immunomodulatory, smooth muscle relaxant, and anti-allergic properties. The folkloric use of this valuable herb in the treatment of a variety of ailments including asthma, has garnered substantial attention [8].

In traditional medicine, *S. occidentalis* has been used as a remedy for various respiratory ailments including asthma. S. *occidentalis* contains a variety of bioactive substances (Flavonoids, alkaloids, phenolic compounds and other secondary metabolites) which were thought to contribute to its therapeutic benefits. Specific chemicals in the plant may interact with the numerous cellular and molecular targets involved in these processes. Several reports that laid claims to fruit of *S. occidentalis* possessing anti-inflammatory, bronchodilatory, anti-oxidant and immune modulatory properties have been documented [9]. Computational research on the binding of phytochemicals from various medicinal plants with PDE4 Enzyme might be useful for learning in-depth information about their inhibitory mechanism with high precision. Different extracts produced from distinct *S. occidentalis* sections have been noted for their effectiveness in the management of asthma. Therefore, molecular docking was used to determine the molecular basis of contact and consequently the inhibiting ability of the phytochemicals extracted from this significant medicinal plant towards PDE4 Enzyme [10]. Additionally the elemental analysis was carried out in *S. occidentalis* fruit samples by EDXRF Technique.

The importance of trace elements to human health (both helpful and harmful) has come to light in recent years. Trace elements are important in inflammatory and may have an impact on asthma symptoms [11]. Critical trace elements including zinc, copper, and selenium are recognized to be crucial for redox equilibrium maintenance which is also necessary for the immune system to function properly. Changes in the condition of these minerals may lead to increased oxidative stress and inflammatory reactions [12]. Many studies have previously claimed that different herbs are used to treat asthma [13, 14] but only a few numbers of researchers reported the elemental quantification of herbal plants used to treat asthma by various analytical techniques [15, 16]. In the present study *S. occidentalis* was selected for assessment of elemental levels via EDXRF-tech. It focused on major, minor and trace element concentrations in *S. occidentalis* fruits commonly used against asthma disease by the Telangana tribal people. The primary goal of the investigation is to analyze the phytochemicals with molecular docking and evaluate trace elemental concentrations.

2 Materials and Methods

Senna occidentalis (*S. occidentalis*) fruits were collected from Warangal district, Telangana, India (Fig. 1 indicated the image of plant sample).

Fig. 1. Photograph of selected plant *S. occidentalis* (fruits).

The fruit samples were cleaned separately with double distilled water and dried in an oven at 60 °C before being finely pulverized with an agate mortar and pestle to reduce particle size and ensure homogeneity. The samples of fine powder were stored in butter paper at room temperature. Each powder sample weighted 150 mg and pellets were made for experimental investigation. Three pellets were prepared by each sample. The NIST 1515 apple leaf certified values were used to verify the accuracy of the present results [16].

2.1 Preparation of Target Enzyme

PDE4 is a cyclic nucleotide phosphodiesterase enzyme that plays a key role in regulating the levels of intracellular cyclic AMP (cAMP) in various cells including those in the respiratory system. It is a promising target for the management of respiratory conditions like chronic obstructive pulmonary disease (COPD) including asthma [17]. The study aims to investigate the anti-asthmatic effect of *S. occidentalis* phytochemicals by targeting phosphodiesterase type 4 (PDE4), a key enzyme involved in asthma pathophysiology. Utilizing the high-resolution structure of PDE4 allows for an in-depth analysis of ligand binding to its active site, offering insights into potential therapeutic effects. The crystal structure of PDE4 (PDB ID: 5K1I) was retrieved from RCSB protein database [18] (Fig. 2). Using a software programme called auto-dock, the target was prepared for molecular docking by removing the water molecules from the protein structure and adding hydrogen atoms. Kollman or Gasteiger charges were applied to the protein to show its electrostatic characteristics. At last, the protein architecture was built and stored in PDBQT form.

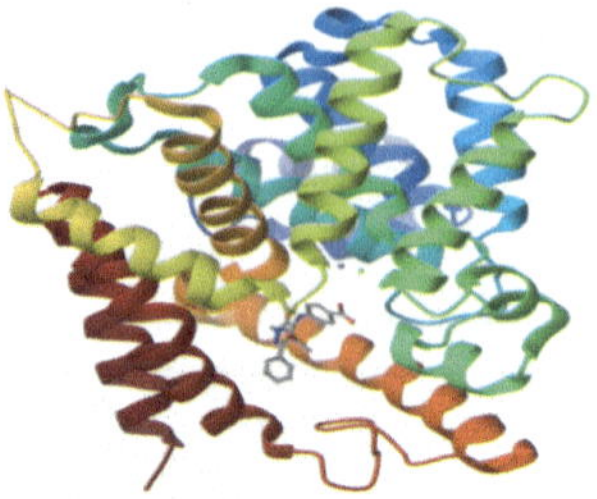

Fig. 2. 3D structure of Phosphodiesterase type 4 (PDE4) (PDB ID: 5K1I) with its inhibitor retrieved from RCSB protein database.

2.2 Phytochemicals Selected for Docking

The chemical structures of the chosen phytochemicals (Anthraquinone, Apigenin, Betain, Stachydrine, and Trigonelline) were downloaded from the Pub Chem database in the SDF (Fig. 3) and transformed by the Open Babel program into PDB format [19].

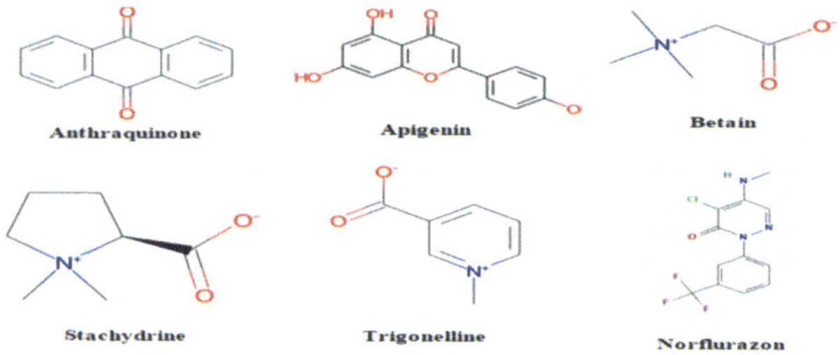

Fig. 3. Phytochemicals of *S. occidentals.*

2.3 Experimental Analysis of ED-XRF

There are several analytical spectroscopic methods including NAA, XRD, EDXRF, and PIXE which can be used to measure the trace elements concentration of medicinal plants [16, 20]. Better detection limits and more elements would be obtained using X-ray tubes as excitation sources instead of radionuclide sources. The current experiments are being carried out at the India at UGC-DAE and CSR trace element laboratory, Kolkata. Operating specifications for the Energy Dispersive X-ray Fluorescence (ED-XRF) spectrometer include an X-ray tube with a Rh anode as the X-ray source (50 kV voltage) up to 1 mA current, and Si (Li) detector with a resolution and emsp of 143 eV at 5.9 k eV and a detection limit of 1–10 ppm [16].

3 Results and Discussion

3.1 Molecular Docking Analysis

The results of the molecular docking simulations showed that Anthraquinone and Apigenin had the lowest binding energies (−5.84 kcal/mol and −5.23 kcal/mol, respectively) among the selected phytochemicals from *S. occidentalis*, indicating a stronger binding affinity to PDE4 compared to Betain, Stachydrine, and Trigonelline (Table 1). Furthermore, both Anthraquinone and Apigenin had binding energies comparable to the standard drug Pyridazinone (−5.19 kcal/mol), suggesting that they may have similar or even better potential as anti-asthmatic agents.

The number of hydrogen bonds and residues involved in forming hydrogen bonding can provide additional insights into the binding interactions of the selected phytochemicals with PDE4. Among the selected phytochemicals, Apigenin had the highest number of hydrogen bonds (2) with two hydrogen bond interactive residues (Asn321 and Gln369) followed by Anthraquinone and Stachydrine with one hydrogen bond each (Gln369 and

Table 1. Molecular docking interactions of phytochemicals from *S. occidentalis* with target enzyme PDE4

S.No	Compound	Binding Energy (kcal/mol)	Number of H_2 bond interactions	H_2 bond interactive residues	Hydrophobic bond interactive residues
1	Anthraquinone (PubChem CID: 6780)	−5.84	1	Gln369	Tyr159, Leu319, Ile336, Phe372, Met357, Ser368, Met337
2	Apigenin (PubChem CID: 5280443)	−5.23	2	Asn321, Gln369	Met357, Phe372, Leu319, Tyr159, Trp332, Tyr329, Thr333, Ile336, Ser368
3	Betain (PubChem CID: 11545)	−2.88	0	-	Lys367, Ile410, Ser364, Thr409, glu366
4	Stachydrine (PubChem CID: 115244)	−3.55	1	Lys367	Thr409, Glu366, Ile410, Lys367
5	Trigonelline (PubChem CID: 5570)	−3.22	0		Ala235, Asn231, Tyr223, Arg261, Phe238, Leu234, Ile265
6	Pyridazinone (33775) (Standard drug)	−5.19	2	Arg261, Gln242	Tyr223, Leu234, Ala235, Phe238, Lys239

Lys367 respectively). Betain and Trigonelline had no hydrogen bonds with the receptor. In comparison with standard drug, Pyridazinone also had two hydrogen bonds with two hydrogen bond interactive residues (Arg261 and Gln242), indicating a similar mode of binding as Apigenin.

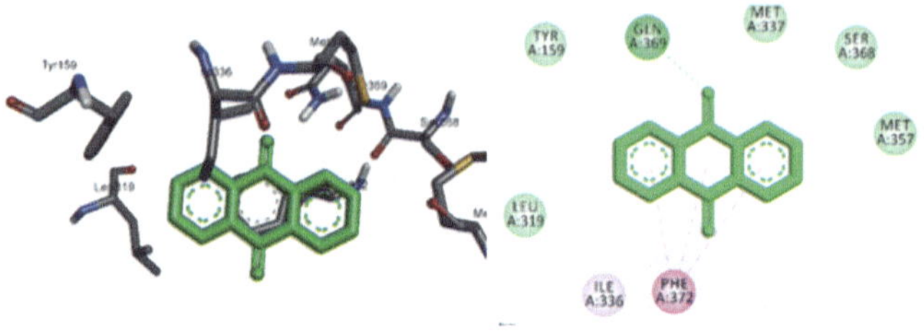

Fig. 4. Molecular Docking interactions (3D and 2D) of Anthraquinone with PDE4.

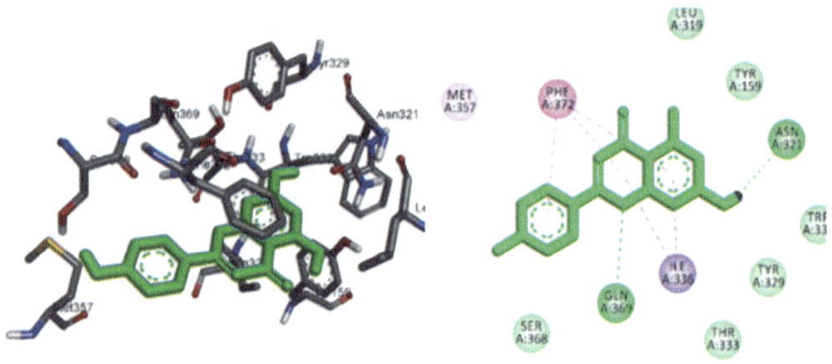

Fig. 5. Molecular Docking interactions (3D and 2D) of Apigenin with PDE4.

Anthraquinone and Apigenin had the lowest binding energies and the highest number of hydrophobic bond interactive residues including Tyr159, Leu319, Ile336, Phe372, Met357, Ser368, Met337, Met357, Phe372, Leu319, Tyr159, Trp332, Tyr329, Thr333, Ile336, and Ser368 (Figs. 4 and 5). Betain and Trigonelline had weaker or no binding interactions through hydrogen bonding or hydrophobic interactions while Stachydrine had moderate binding interactions with Thr409, Glu366, Ile410, and Lys367. The presence of hydrophobic bond interactive residues shared between Anthraquinone, Apigenin, and the standard drug Norflurazon (including Tyr223, Leu234, Ala235, Phe238, and Lys239) suggests that they may have comparable or even better anti-asthmatic activity than Pyridazinone (Table 1).

3.2 Experimental Analysis of ED-XRF

The elemental concentration of *S. occidentalis* fruits was estimated by ED-XRF spectrometer and total 13 elements that is P, Cl, S, K, Mn, Ca, Fe, Zn, Cu, Se, Rb, Br, and Sr were examined. The analyses of present results are shown in Table 2 and the Fig. 6 shows the EDXRF spectra of *S. occidentalis* fruit samples. The elements found at high concentrations are P, Cl, Fe, Cu, Zn, Se, Br, Rb, and Sr, whereas the elements S, K,

Ca and Mn were discovered at smaller concentrations with reference to the NIST 1515 apple leaf standard value. From the current study, the presence of K, Fe, Mn, Cu, Zn and Se were correlated with the anti-asthmatic curative properties of this fruit plant samples. According to standard values of NIST 1515 (16100 ppm), the mean K content was found to be high in S. *occidentalis* fruits (14138.2ppm). Inadequate K consumption may impair lung function, increase airway responsiveness, and limit flow but the relationships between these factors are complicated [16]. The mean amount of Mn was discovered to be lower in *S. occidentalis* fruits (38.04 ± 3.1 ppm) when compared with certified values of NIST 1515 (54 ppm). Inhaling of manganese, a crucial trace element can result in an inflammatory response in the lungs of humans. The average content of Fe was found higher in *S. occidentalis* fruits 93.6 ± 7.9 when compared to certified values of NIST 1515 (83 ppm). Fe is a necessary element for humans.

Table 2. The Average Elemental Concentrations in *S. occidentalis* ± SD and contrasted to certified reference material NIST 1515

Elements	*Senna occidentalis* (Fruits)	NIST 1515 apple leaf
P	3090.9 ± 1195.3	1590
S	1491.6 ± 384.8	1800
Cl	3546.9 ± 1025.0	579
K	14138.2 ± 2439.0	16100
Ca	5868.4 ± 860.3	15260
Mn	38.04 ± 3.1	54
Fe	93.6 ± 7.9	83
Cu	56.2 ± 8.7	5.64
Zn	46.2 ± 5.6	12.50
Se	0.40 ± 0.4	0.05
Br	3.34 ± 1.1	1.80
Rb	28.6 ± 2.8	10.20
Sr	42.1 ± 6.3	25

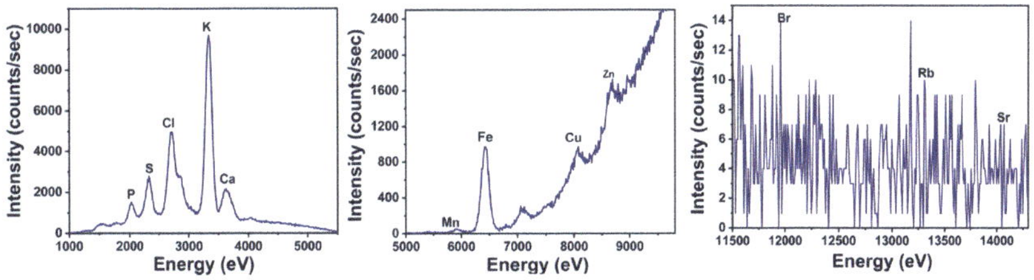

Fig. 6. EDXRF-spectrums of anti-asthmatic medicinal plant of *S. occidentalis* fruits.

Coarse Grain Modeling of ZnO Tetrapods and Generic DNA Hydrogels

Vrinda Garg, Mehul Mishra, Yuvashree Jayavelu, K. Satish Kumar, D. Paul Joseph, and Surya K. Ghosh(✉)

Department of Physics, National Institute of Technology, Warangal, Telangana 506004, India
skghosh@nitw.ac.in

Abstract. The unique structural aspects of tetrapods can be utilized to model a variety of systems ranging from generic ZnO structures to DNA hydrogels. In this study, we introduce two distinctly different coarse-grained models representing the basic building blocks. One is the 3D inert model representing a central core each connected to three arms via a harmonic bond mimicking the ZnO tetrapod structure. Second is 3D non-inert model of tetrapod whose features mimics three double-stranded DNA arms with single-stranded sticky ends, represented by bead chains and patchy particles, to accurately capture both the geometry and interactions of the complementary units. We did a comparative study of the dynamics and self-assembly of these two models. We study the dynamical properties of tetrapods using extensive Langevin dynamics simulation. The finite size effect and defined structure of tetrapods highlight an optimum surface deposition density. We quantify the dynamics by observing the mean squared displacement at distinct time scales compared to the relaxation time and diffusion constant of the centre of mass of each model of the tetrapods.

1 Introduction

Subjective to the incredible stability and porosity, the molecular and self-assembly phenomena particular to ZnO and DNA hydrogels can be studied using model systems with tetrapods as building blocks. Moreover, the functionalization can be introduced either over the entire surface or at the tips only. This amalgamation of the geometry, functionality has led to utilization of tetrapods as biosensors, electric components, and for lasing action [1].

In this study, we considered two scenarios of inert and non-inert model of tetrapod involving the study of their dynamics, under the influence of gravitational force. We have developed coarse-grained models representing particles with tetrahedral shape consisting of four arms interconnected via a central core. The tetrapods are incredibly stable due to their symmetric geometry, fully accessible (surface area of each arm), high porosity (self-organized structure), and uniform force distribution.

P. Joseph D. et al. (Eds.): IEMDST 2024, SPPHY 567, pp. 216–224, 2025.
https://doi.org/10.1007/978-981-96-5863-3_28

In general, 3D inert model of the tetrapod mimics ZnO properties while the 3D non-inert model with sticky patchy ends that mimics the DNA hydrogel. Tetrapod DNA hydrogel models [2–4, 6, 7] attracted considerable attention due to their uses in controlled drug delivery [8–10], tissue engineering, 3D cell cultures, biosensors [6], cell transplant therapy, and for biomedical applications [11]. The dynamics of tetrapods under gravity and until they settle down on the depositing surface has been observed for various properties. The dynamics of the center of mass of the tetrapod have been considered. Effectiveness of surface deposition provides a perspective of their optimal density. The time taken by the tetrapod to deposit and the effectiveness of the surface deposition elucidate the phenomenon meticulously. Mean Squared displacement (MSD (t)) assumes the power law form: $\langle \Delta r^2 (t) \rangle \sim t^\alpha$, where exponent α determines the nature of dynamics. MSD at distinct time scales compared to the relaxation time (τ_R) of the center of mass of tetrapod is considered. At smaller time scales, when time t is less as compared to the tetrapod relaxation time τ_R ($t < \tau_R$) the MSD scales with $\alpha = 1.5$. Whereas, at longer time scales τ_R ($t > \tau_R$), at higher concentrations, the interlocking and interaction among tetrapods leads to $\alpha = 1.7$. The diffusion constant D of the tetrapods at different times and various densities is observed.

2 Materials and Methods

1. 3D Inert Tetrapod Model

Bead spring model to represent a coarse-grained modeling of tetrapod has been implemented. Each arm is connected to a central core and is equally distributed around the center in 3D. All the neighboring beads [5] are linked by the harmonic spring using the harmonic potential of the form

$$V_{harmonic}(l) = k_r (l - l_0)^2 \tag{1}$$

where k_r is the spring constant, l_0 is the equilibrium bond length. To maintain the structural integrity of tetrapod structure and to allow only small fluctuations, k_r is set to a high value of 300 $k_B T$.

To mimic the actual building block we have symmetrically placed the four beads on vertices with the center bead, each with an angle of 109.5° from the center (see Fig. 1(a)).

The excluded volume interaction among the tetrapods is modeled using truncated Lennard Jones potential, also referred as Weeks-Chandler-Andersen (WCA) potential as

$$U_{LJ}(r_{ij}) = \begin{cases} 4\epsilon\left[\left(\frac{\sigma_{ij}}{r_{ij}}\right)^{12} - \left(\frac{\sigma_{ij}}{r_{ij}}^{6}\right)\right] + \varepsilon, & \text{for } r_{ij} \le 2^{\frac{1}{6}}\sigma_{ij} \\ 0, & \textit{otherwise}. \end{cases} \tag{2}$$

Here $r_{ij} = |r_i - r_j|$ represent the interparticle distance, $\sigma_{ij} = (\sigma_i + \sigma_j)/2$ is the overall size of the interaction domain of the interacting particle of size σ_i, σ_j and ϵ is strength of the interaction. To mimic a good solvent condition we have used repulsive aspect of the Lennard Jones potential by eliminating the attractive part.

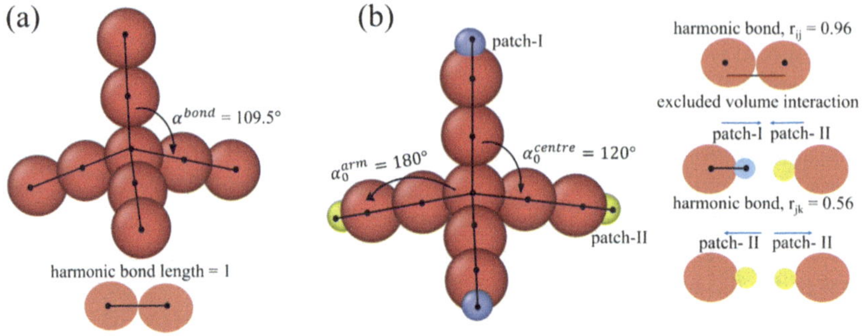

Fig. 1. Schematics of a bead spring model of tetrapods. (a) 3D inert model. Four beads at the vertices are connected to a central core with a harmonic bond of length = 1, spring constant = 300, interplanar angle = 109.5° (b) 3D non-inert model of tetrapod. Each arm end has a small patch connected to its end. Same patch has repulsive interaction between them while the opposite patches have attractive interaction between them. Bead-bead and bead-patch are connected via harmonic bond of length $r_{ij} = 0.96$ and $r_{jk} = 0.56$ respectively. We have given $\alpha_0^{arm} = 180°$ with each arm and the three arms are equally positioned around the central bead with minima at $\alpha_0^{center} = 120°$.

2. 3D Non-Inert Tetrapod Model: Sticky Patches

To incorporate the functionality we have added small patches at the end of each arm named as patch-I and patch-II. The same type of patches have repulsive interaction, while the opposite patches have attractive interaction between them (see Fig. 1(b)).

The excluded volume effect among the individual beads is modeled using truncated LJ potential, referred as WCA potential using Eq. (2). For our simulation, to avoid any overlap among the arms of the tetrapod we provide repulsive interaction with the help of $\epsilon = \epsilon_{LJ}, \sigma = \sigma_{LJ}, \mathrm{r_{cutoff}} = 1.12\ \sigma_{LJ}$ between two large beads. Moreover, for the small beads in the patch we have given ~$\epsilon = \epsilon_{LJ}, \sigma = 0.67\sigma_{LJ}, \mathrm{r_{cutoff}} = 0.67\ \sigma_{LJ}$ between the same patches having repulsive interaction and to avoid attractive interaction between different patches.

The attraction among the complementary patches is given by the full LJ potential truncated at a farther distance

$$\mathrm{U_{LJ}(r_{ij})} = \begin{cases} 4\epsilon\left[\left(\frac{\sigma_{ij}}{r_{ij}}\right)^{12} - \left(\frac{\sigma_{ij}}{r_{ij}}^{6}\right)\right] & \text{for } r_{ij} \le 5\sigma_{ij} \\ 0, & \textit{otherwise}. \end{cases} \tag{3}$$

For the interaction among patch-I and patch-II, $\epsilon = 4.5\epsilon_{LJ}, \sigma = 0.2\sigma_{LJ}$. All the neighboring large beads are linked by the harmonic spring via the harmonic potential using Eq. (1). The equilibrium bond length between the two large beads is set to be 0.96 σ_{LJ} and between the large bead and the patch is 0.56 σ_{LJ}. The angle constant is set by the harmonic potential

$$\mathrm{V_{angle}}(\theta) = K_\theta(\theta - \theta_0)^2, \tag{4}$$

where K_θ is the spring constant and θ_0 is the bond angle at equilibrium bond angle. K_θ is used for maintaining the bead-patch rigidity. We provide $\alpha_0^{arm} = 180°$ with each arm and the three arms are symmetrically arranged around the central bead, each positioned at a minima of $\alpha_0^{center} = 120°$.

We define our simulation parameters using energy, mass, and distance scales ϵ, m, σ corresponding to $t = \sqrt{m\sigma_{LJ}^2/\epsilon_{LJ}}$ and the subsequent time scales as f = ϵ/σ and force in ps and pN respectively.

Initially, tetrapods are arranged with equal spacing in various concentrations (number of tetrapods) at the top surface of the simulation box, of 60σ height. The bottom (depositing) surface is attractive with a 1.5 σ cut-off. Tetrapods falls under the influence of gravity with a force of 0.1 LJ units in the negative z-direction (see Fig. 2). We performed this coarse-grained simulation using LAMMPS simulation package. In this simulation, the dynamics of each bead is governed using the equation of motion and is administered by the Langevin relation which can be written as

$$m\frac{d^2r(t)}{dt^2} = -\xi\frac{dr(t)}{dt}(t) + F^R(t) - \nabla(U_{LJ} + U_{harmonic}) + mg, \tag{5}$$

where position is defined by r(t), mass by m, gravity as g, friction coefficient by ξ, and F^R refers to the random force pertaining to thermal fluctuation fulfilling the fluctuation-dissipation theorem,

$$F_i^R(t) = 0, \langle F_i^R(t) \cdot F_j^R(t') \rangle = 2dk_BT\xi\delta_{ij}\delta(t - t') \tag{6}$$

here D is friction coefficient defined as k_BT/ξ relating it to the diffusivity of the particle. Here dimension of the system is given by d, in the present model d = 3, Boltzmann constant as k_B, and temperature by T. In the simulation the dimensionless parameters are temperature T = 1.2 and friction ξ = 0.7 in LJ units. We have used Langevin thermostat to execute the simulations, to start with, a truly random initial configuration. Random initial configurations are giving the same results as that of the ordered initial configuration. The velocity Verlet algorithm is used to integrate the equation of motion at each step. After the equilibration period, tetrapods are allowed to move downward from the top surface. We averaged our data for 100 independent realizations, to obtain satisfactory statistics, with Δt = 0.001 as simulation time step.

3 Results and Discussion

3.1 Mean Squared Displacement

To understand the dynamics of tetrapods, we compute MSD as a function of time. The MSD for the tetrapod's center of mass is estimated using the relation

$$MSD(t) = \left\langle r^2(t) \right\rangle = \frac{1}{N}\sum_{i=1}^{N}\left\langle \left[r_i(t + t') - r_i(t)\right]^2 \right\rangle, \tag{7}$$

where total number of tetrapods is given by N, time difference as t′. MSD are calculated till deposition of tetrapods (center of mass) for various concentrations (number of tetrapods) have been analyzed. It has power law form: $\langle \Delta r^2(t) \rangle \sim t^\alpha$, where the exponent α is defined as the anomalous diffusion exponent.

For the inert model of tetrapod, we observe that at smaller time scales when the time t compared to the relaxation time τ_R is much lesser ($t < \tau_R$), it shows scaling of $\alpha = 1.5$. Intermediate time acts as a relaxation time for tetrapods which were initially put in a congested space and allows them to breathe in and occupy the provided space. At a later time, when the time t in comparison to the relaxation time is larger ($t > \tau_R$), the MSD curve shows a scaling of $\alpha = 1.7$ (Fig. 3(a)).

For the non-inert model, we observe similar trend in the MSD behavior. At smaller time scales when the time t compared to the relaxation time τ_R is much lesser, tetrapods are diffusing with a scaling of $\alpha = 1.5$ while after crossing the relaxation time, at higher times all the tetrapods are moving with a scaling of $\alpha = 1.7$ (Fig. 3(b)).

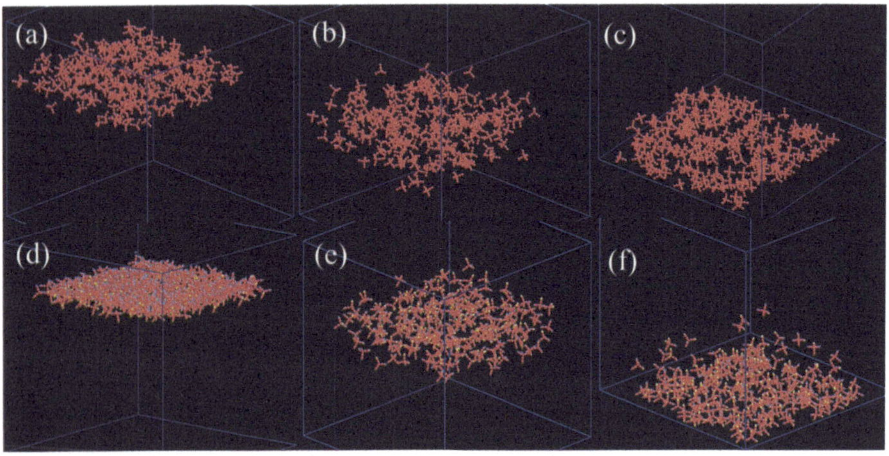

Fig. 2. First row represents the 3D inert model of tetrapods. Second row represent the 3D non inert model of tetrapods. Each row represents three stages of tetrapod deposition on the depositing surface. Initially, they are arranged at the top of the simulation box. At the second stage, tetrapopds falls under the influence of gravity from the top surface. Finally, tetrapods are allowed to settle down on the depositing surface. (a), (b), and (c) are the three stages of inert model of tetrapods. (d), (e), and (f) are the three stages of tetrapod deposition for the non-inert model of tetrapods.

3.2 Diffusion Constant

Diffusion constant D is defined from the MSD as,

$$\langle r^2(t)\rangle = 2dDt, \tag{8}$$

where $\langle r^2(t)\rangle$ is the mean squared displacement over time t, dimension of the system is given by d, and diffusion constant as D. For d = 3, D can be obtained as the rate of variation of MSD as,

$$D = \frac{\langle r^2(t)\rangle}{6t} \tag{9}$$

We observe that for shorter time the diffusion constant is less as the tetrapods are moving slower and diffusing through the medium slowly, but at intermediate times, as the increased density (number of tetrapods) kicks in due to interlocking and intertwining of the tetrapods, they will exhibit collective motion, resulting in a comparatively higher diffusion constant with increasing time. Furthermore, as the density increases further, an increase in the number of tetrapods leads to competition between the excluded volume and the entropy, leading to their lower diffusion. The trend of diffusion constant is similar for the inert and non-inert model of tetrapod (see Fig. 3(c) and (d)).

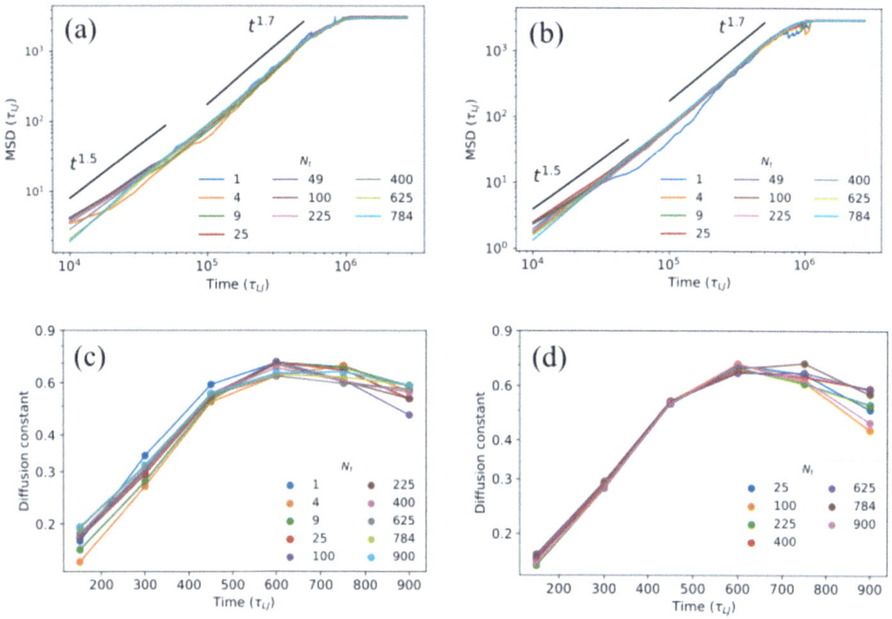

Fig. 3. Dynamics of the tetrapods falling from the top surface under the influence of gravity. (a) MSD of the inert tetrapod's center of mass with time for various initial concentrations. (b) MSD of the non-inert tetrapod's center of mass with time for various initial concentrations. At small time scales, MSD varies as ~$t^{1.5}$ with time whereas for the longer time tetrapods are showing faster diffusion with a scaling ~ $t^{1.7}$. (c) Diffusion constant as a function of simulation time for the inert tetrapods. (d) Diffusion constant as a function of time for the non-inert tetrapods for various initial concentration of tetrapods falling from the top surface under the effect of gravity.

3.3 Optimal Number for Effective Deposition: Entropy vs. Excluded Volume

In order to see the effectiveness of surface deposition we have plotted number of tetrapods deposited with the simulation time for various concentration (number of tetrapods). We observe that at low density (number of tetrapods), tetrapods exhibit random zig-zag motion, resulting in slower deposition. While at intermediate density, excluded volume restricts the tetrapods from direction movement, resulting in faster deposition. At high density, tetrapods are entangled to each other which does not allow them to relax and finite structure effect leads to compete them with each other for space, resulting in slower deposition. These fluctuations shows the existence of an optimum number of concentrations (number of tetrapods). We can see a strong competition on the higher concentration to occupy vacant spots on the depositing surface from the nature of this curve. There is a competition between the entropy and the excluded volume. The trend of the curve is showing similar behavior for inert and non-inert model of tetrapods (see Fig. 4(a) and (b)).

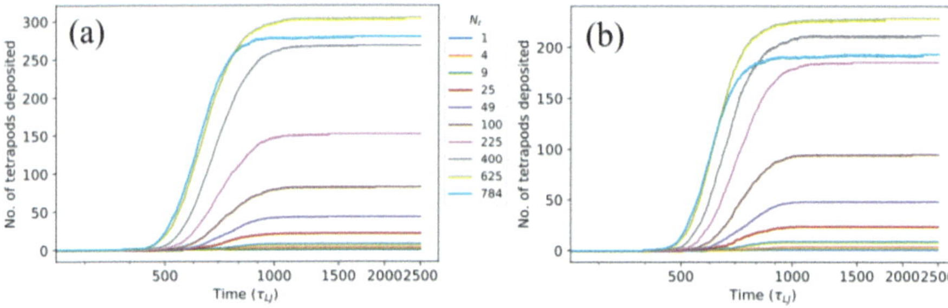

Fig. 4. (a) Number of inert tetrapods deposited on the depositing surface as a function of simulation time in LJ units for various initial concentrations of the tetrapods. (b) Number of non-inert tetrapods with time for various initial concentrations of the tetrapods.

3.4 Time of First Touching of Tetrapod on the Depositing Surface

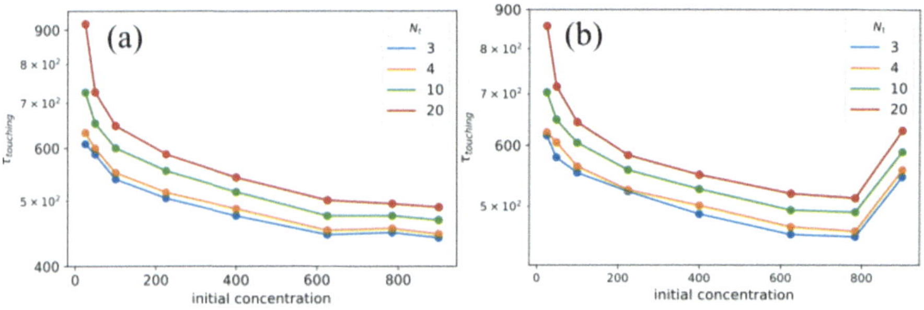

Fig. 5. Time taken for touching the depositing surface $(\tau_{touching})$ of a specific number of tetrapods with the initial concentration (number of tetrapods). (a) $\tau_{touching}$ of inert tetrapods falling from the top surface. (b) $\tau_{touching}$ of non-inert tetrapods as a function of their initial concentrations.

Now, we examine the time of first touching $(\tau_{touching})$ of tetrapods. It is the time when the tetrapods just touch the depositing surface. We observe $\tau_{touching}$ against the various concentration (number of tetrapods) for different number of tetrapods deposited (N_t). $\tau_{touching}$.

decreases with the increase in concentration (number of tetrapods). Time taken at higher initial concentrations is less than that at lower initial concentrations. This suggests an enhancement in deposition speed with increasing number of tetrapods. At higher concentrations, interactions between the tetrapods would be much higher because of the congested space; also, due to the optimal structure of tetrapods, there could be interlocking between the tetrapods, which would, in turn, drag the tetrapods together towards the depositing surface which increases the speed of deposition (see Fig. 5(a) and (b)). While for the case of non-inert model of tetrapod, at higher initial concentration (number of tetrapods), there is slight increase in the time of touching due to the sticky ends of the tetrapods, they are clustering together and competition for occupying the surface has increased which leads to delay in their deposition on the surface.

4 Conclusion

We have developed two distinct models of tetrapods: inert and non-inert models. We have done a comparative study for the dynamics and self-assembly of these two distinct models. A study on the deposition of tetrapods with time on the depositing surface reveals an optimum number. There is an interplay between the entropy and the excluded volume. Random zig-zag motion of tetrapods or their higher entropy at lower density results in a slower deposition. While at intermediate density, due to the excluded volume effect, which restricts the direction of movement of tetrapods, it leads to faster deposition. Subsequently, at the higher density, it has been observed that the tetrapods are entangled to each other which does not allow them to relax. The finite structure effect of tetrapods leads them to compete for space, resulting a slowing down of the deposition dynamics. MSD of tetrapods's center of mass exhibits that at smaller time scales, when t is less as compared to the relaxation time it shows a scaling of $t^{1.5}$ while when the time is greater than their relaxation time, scaling reaches to $t^{1.7}$ showing an enhanced diffusion at larger time scales. The diffusion constant, which tells about the rate of change of MSD, also reveals a higher diffusion with an increase in the time of deposition. Moreover, as the density increases further, an increase in the number of tetrapods leads to competition between the excluded volume and the entropy, resulting to their lower diffusion.

Acknowledgments. The author would like to thank MATRICS, SERB, DST (Project # MTR/2022/000833) for financial support. The author also acknowledges NIT Warangal for the research support.

References

1. Sinkovits, D.W., Luijten, E.: Nano Lett. **12**(4), 1743–1748 (2012). https://doi.org/10.1021/nl200961z
2. Naskar, S., Bhatia, D., Lin, S.T., Maiti, P.K.: Phys. Chem. Chem. Phys. **25**(11), 7847–7858 (2023). https://doi.org/10.1039/D2CP04683K
3. Araki, Y., Kobayashi, Y., Arai, N.: Mol. Syst. Des. Eng. **4**(6), 1095–1102 (2019). https://doi.org/10.1039/C9ME00112C

4. Xing, Z., Ness, C., Frenkel, D., Eiser, E.: Macromolecules **52**(2), 504–512 (2019). https://doi.org/10.1021/acs.macromol.8b01948
5. Garg, V., Mathew, R., Ibrahim, R., Singh, K., Ghosh, S.K.: Iscience **27**(4), 109348 (2024). https://doi.org/10.1016/j.isci.2024.109348
6. Murphy, K.A., Reiser, N., Choksy, D., Singer, C.E., Jaeger, H.M.: Granular Matter **18**(2), 26 (2016). https://doi.org/10.1007/s10035-015-0600-2
7. Ghosh, S.K., Kundu, T., Sain, A.: Phys. Rev. E—Stat. Nonlinear, Soft Matter. Phys. **86**(5), 051910 (2012). https://doi.org/10.1103/PhysRevE.86.051910
8. Maitra Roy, S., et al.: ACS Appl. Nano Mater. **6**(20), 18988–18998 (2023). https://doi.org/10.1021/acsanm.3c03481
9. Roy, S.M., et al.: J. Drug Deliv. Sci. Technol. **83**, 104408 (2023). https://doi.org/10.1016/j.jddst.2023.104408
10. Roy, S.M., Garg, V., Barman, S., Ghosh, C., Maity, A.R., Ghosh, S.: K. Front. Bioeng. Biotechnol. **9**, 785937 (2021). https://doi.org/10.3389/fbioe.2021.785937
11. Castañeda-Priego, R., Pica Ciamarra, M., Corezzi, S., Ghosh, S.K., Liu, Y.: Front. Phys. **12**, 1388370 (2024). https://doi.org/10.3389/fphy.2024.1388370

Control Experiment Attempts for Self-assembly of ZnO Tetrapods Under Gravity in Homogeneous Toluene Medium

Mehul Mishra, Yuvashree Jayavelu, Vrinda Garg, Surya K. Ghosh, and D. Paul Joseph(✉)

Department of Physics, National Institute of Technology, Warangal, Telangana 506004, India
paul@nitw.ac.in

Abstract. ZnO, a multifunctional material has viability for various applications in a specific nanostructured form like four legged tetrapods. In this work, ZnO tetrapods synthesized by a modified catalyst free thermal evaporation technique is attempted to test for plausible self-assembly under gravity in a homogenous toluene medium. Based on the theoretical modeling and analysis on the self-assembly of the coarse-grained bead-spring model of tetrapods, preliminary control experiments were designed and attempted using FESEM image analysis. These control experiments were carried out by taking 3 ml of toluene as a homogenous and moderately viscous medium. 10 mg of ZnO tetrapods was taken in three beakers with toluene and is ultrasonicated. Three different substrates have been considered for assembling the tetrapods on their surface including carbon tape, copper sheet and Indium Tin Oxide coated glass (ITO). All the substrates were placed horizontally into separate beakers and the ultrasonicated ZnO tetrapods were allowed to settle over them. After complete evaporation of toluene from the beakers, the substrates were carefully taken for image analysis. The FESEM images confirmed the settlement of the ZnO tetrapods. However, the number density of tetrapods standing upright was very negligible. Although the experimental results confirmed the formation of small number of interlocked structures and vertically aligned tetrapods, the experimental data deviated from the self-assembly structures from those predicted by theoretical simulations. These control experiments need to be further tuned to extend statistical properties such as ensemble average mean square displacement, diffusion constant, etc.

1 Introduction

Since 1974, after the term was coined by Norio Taniguchi, nanomaterials have seen a rapid rise in academia and in industry as an emerging technology [1]. This new era of ingenuity, combining many fields, including Physics, Chemistry, Biology and Engineering ushered unprecedented advances in medical, optics, telecommunications, instrumentation and sustainable technologies [2]. Nanomaterials refer to materials in the size range of 1–100 nm [1, 3]. Although this definition seems trivial at first, the change in the properties at this scale leads to various interesting optical, thermal, catalytic and diffusive properties [2].

P. Joseph D. et al. (Eds.): IEMDST 2024, SPPHY 567, pp. 225–232, 2025.
https://doi.org/10.1007/978-981-96-5863-3_29

In metal oxide nanoparticles the oxygen anions and the metal cations are bonded together while the size of the particles is in the range of a few tens to hundreds of nanometers [4]. These metal oxide nanoparticles can be engineered for a wide range of applications due to the properties like high ductility at increased temperature, unique catalytic behaviour, superparamagnetic, and nonlinear optical properties, making them most versatile materials for industrial applications [5, 6]. These materials can be utilized for miniaturization of components possibly by adopting a self-assembly process [2]. In particular, ZnO was found to be one of the most significant metal oxides with multifunctional properties with direct and wide bandgap of 3.4 eV [7, 8]. Its high surface to volume ratio and multitude of morphologies make it suitable for sensing and electronic applications. It also possesses defects, which changes its properties so as to make it much better suited for photoluminescence and photodetection applications [7, 9]. Morphology plays an important role in the utility of ZnO, both from physical and chemical viewpoints. ZnO can form nanotubes, nanoribbons, nano pellets, nanosheets, nanowires, nanobelts, tetrapods, flowers, snowflakes etc. [10].

Tetrapods consist of four arms interconnected together via a central core [9, 11]. Physical and mechanical properties of a material are not only dependent on the chemical structure but also by the self-assembled structure. The coarse-grained theoretical model is considered to simulate the physical properties of real time ZnO tetrapods effectively. Hence these models are used in theoretical analysis while ZnO tetrapods are used for the experimental analysis [11]. The manner these tetrapods are arranged is as important as the morphology itself since it can be utilized for various applications. The arrangement can occur through the process of self-assembly, however not much has been studied about this phenomenon with respect to tetrapods, in a free fall mode both theoretically and experimentally. Hence, in this work preliminary control experiments with ZnO tetrapods are given a try upon various substrate substances copper sheet, carbon tape and Indium Tin Oxide coated glass to look for self-assembly structures in a homogeneous and moderately viscous medium of toluene.

To rationalize the underlying physical mechanism of interplay between entropy and excluded volume of tetrapods, a 3D coarse-grained simulation model of tetrapod mimicking ZnO structural properties has been proposed. A bead-spring model [13] is used to assent the actual representation of tetrapods. Each arm is connected to the central core and is equally distributed around the center. All the neighboring large beads are connected by the harmonic potential of the form.

$$\mathrm{V_{harmonic}(l)} = \mathrm{k}\,(\mathrm{l} - \mathrm{l_0})^2 \tag{1}$$

where k is the spring constant, $\mathrm{l_0}$ is the equilibrium bond length. To maintain the structural integrity of the tetrapod structure and to allow only for small fluctuations, k is set to a higher value of 300 $\mathrm{K_BT}$. The equation of motion of the dynamics of each bead under the influence of gravitational force is governed by the Langevin equation which can be written as,

$$m\frac{d^2r(t)}{dt^2} = -\xi v(t) + F^R(t) - \nabla(U_{LJ} + U_{harmonic}) + mg, \tag{2}$$

where r is the position, m is the mass, v is the velocity of each bead, g is gravity, and ξ is the friction coefficient, and thermal fluctuation is considered by F^R which is the random forces.

Tetrapods are allowed to fall freely from the top surface under the effect of gravitational force. A study on the dynamical properties of the falling tetrapods has been done observed. Deposition of the tetrapods with time and different numbers of tetrapods provides a perspective of their optimal density. Mean squared displacement (MSD, (τ)) takes the power law form: $\langle \Delta r^2 (t) \rangle \sim t^{\alpha}$, where α is the anomalous diffusion exponent. MSD at distinct time scales compared to the relaxation time of the center of mass of tetrapod is also considered. The diffusion constant of the tetrapods at different times and densities are observed.

2 Materials and Methods

Due to practical difficulties like sprinkling the ZnO tetrapods in a free fall manner uniformly from a height, moderate deviations are considered like dispersing the ZnO tetrapods in a homogenous and moderately viscous toluene medium so as to maintain any situation nearly equivalent to the theoretical model.

Here, self-assembly of ZnO tetrapods upon three different substrates is performed by a control experiment designed preliminarily. Since no such experiment is reported in the literature, it is decided to use copper sheet, carbon tape and ITO coated glass substrate for this preliminary self-assembly experiment. A homogeneous liquid, toluene, of moderate viscosity is chosen as a medium to perform these preliminary control experiments. Based on literature reports, toluene, is used as solvent to benefit easy evaporation on a hotplate at a moderate temperature [12].

2.1 Control Experiments

This is a new attempt to study the self-assembly of ZnO tetrapods. Many techniques have been proposed to study self-assembly of superstructures but those are template assisted process and none of them are free fall of tetrapods on a flat surface. Therefore, based on the experiment by Nimai Mishra et al [12] the self-assembly was tried in toluene medium. There are a few of the many unavoidable deviations from the theoretical consideration and analysis. Due to these deviations, there may be differences in the self-assembly between the theoretical and experimental analysis. 10 mg of ZnO tetrapod powder was dispersed in toluene and is ultrasonicated for 10 min with the respective substrate kept horizontally at the bottom of the beaker (Fig. 1. Left). After ultrasonication, the contents were kept on a hotplate at 80 °C without any disturbance (Fig. 1 Middle) so as to allow toluene to evaporate gradually [12]. After 2 h, when toluene was evaporated completely (Fig. 1 Right), the substrates were carefully taken for further analysis under FE-SEM.

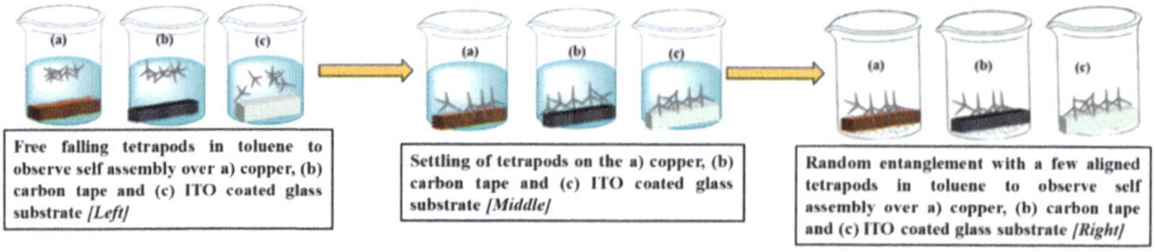

Fig. 1. Schematic representation of the expected output from the control experiment performed on three different substrates, (a) copper sheet, (b) carbon tape, and (c) ITO coated glass in toluene medium.

2.2 FE-SEM Characterization and Analysis

The FE-SEM (JEOL, JSM-IT800) measurements were performed for all the three different substrates at about 5-micron scale and then sampling was done for four different locations in the range 0.5 – 1.0 micron. These showed self-assembly like structures of the ZnO tetrapods very sparingly with mostly random alignment of the tetrapods and the entangled arrangement caused by the toppling of tetrapods due to the high number of nearest neighbor tetrapods.

3 Results and Discussion

The FESEM images of ZnO tetrapods taken on the three different substrates, are analysed for plausible self-assembly. A few of the ZnO tetrapods standing upright are encircled in yellow dotted line in the respective FE-SEM images (Figs. 2, 3 and 4) shown below.

1. ***Upon copper sheet***

2. ***Upon carbon tape***

3. ***Upon indium tin oxide coated glass***

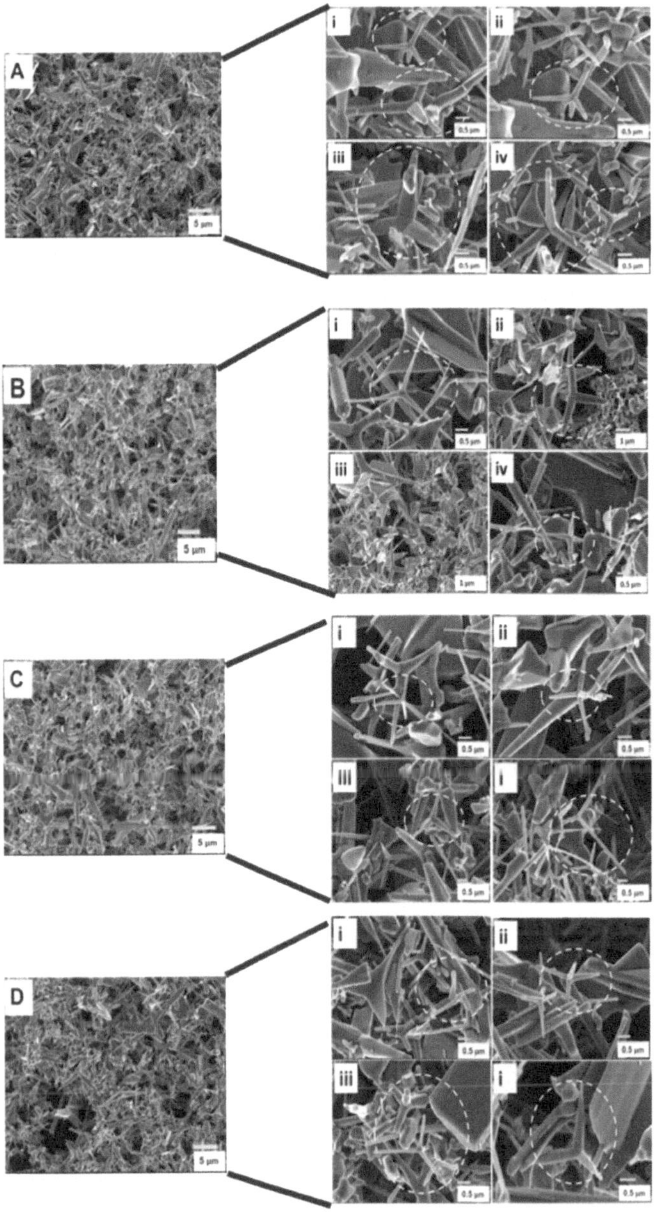

Fig. 2. FESEM images at different locations on the surface of the Cu sheet (at 5 μm scale) (A, B, C, D). Sampling at four different locations on each specific region A, B, C, D taken at different scale (i, ii, iii, iv). Few of the ZnO tetrapods aligned upright are encircled among the entangled and toppled tetrapod structures.

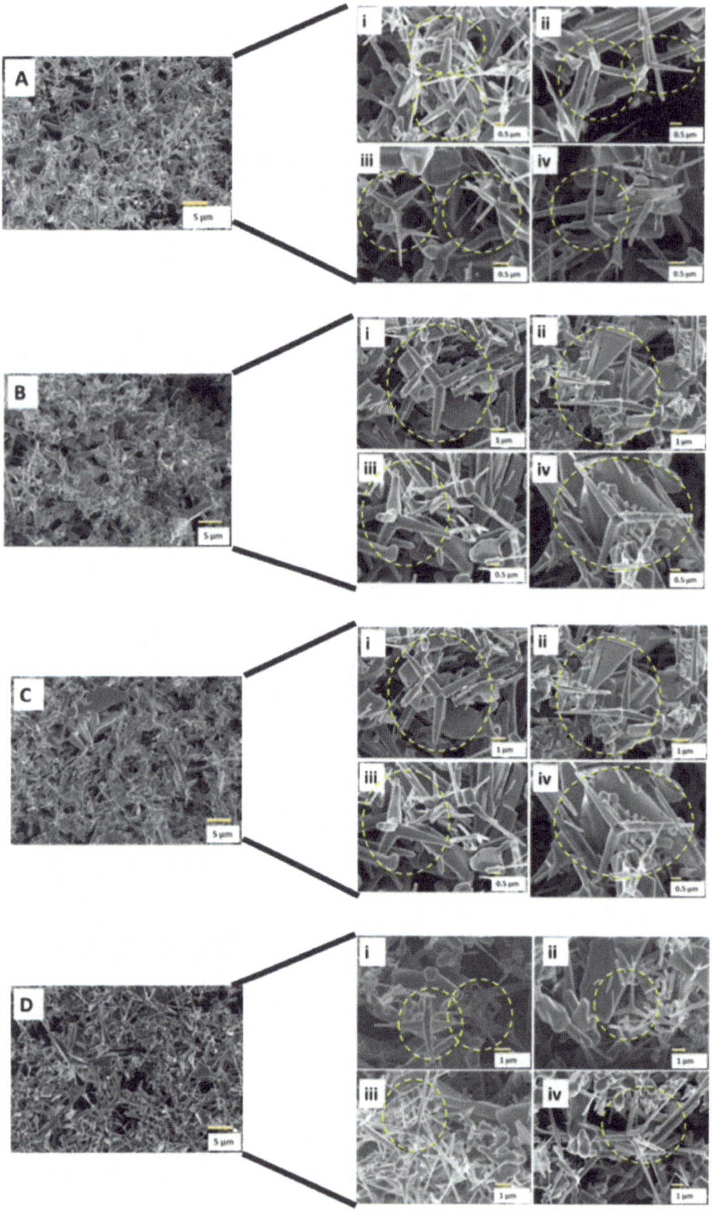

Fig. 3. FESEM images of ZnO tetrapods at different locations on the surface of the carbon tape (at 5 μm scale) (A, B, C, D). Sampling at four different locations on each specific region A, B, C, D taken at different scale (i, ii, iii, iv). Few of the ZnO tetrapods aligned upright are encircled among the entangled and toppled tetrapod structures.

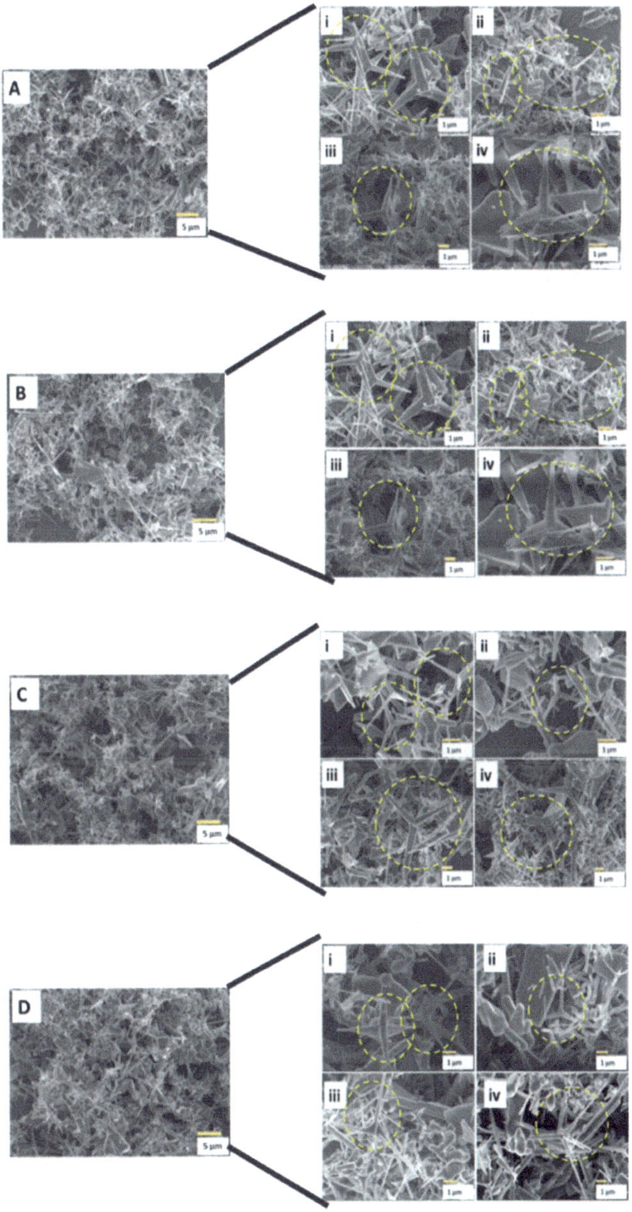

Fig. 4. FESEM images of ZnO tetrapods at different locations on the surface of the ITO glass (at 5 µm scale) (A, B, C, D). Sampling at four different locations on each specific region A, B, C, D taken at different scale (i, ii, iii, iv). Few of the ZnO tetrapods aligned upright are encircled among the entangled and toppled tetrapod structures.

4 Conclusion

Out of these three substrates it is inferred that ITO glass provided comparatively and marginally better results for studying self-assembly of ZnO tetrapods, since there were a comparatively greater number of unbroken tetrapods and less concentration of different morphologies. However careful deliberations need to be made about how to conduct further studies in the context of self-assembly under the action of Brownian motion and gravity, as no substantial correlation can be made between the theoretical model and the experimental data as of now. These preliminary experimental results show that more fine tuning of the control experiment is needed in terms of starting quantity of ZnO tetrapods and size uniformity to minimize entanglement and plausible nano scale texturing of the substrates to induce self-assembly and to minimize toppling.

References

1. Baig, N., Kammakakam, I., Falath, W.: Mater. Adv. **2**, 1821–1871 (2021). https://doi.org/10.1039/d0ma00807a
2. Mekuye, B., Abera, B.: Nano Select **4**(8), 486–501 (2023). https://doi.org/10.1002/nano.202300038
3. Baby, B.H., Shajan, N.T., Biju Balan, N., Joseph, D.P., Mohan, D.B., Mumthas, S.: Mater. Today Commun. **39,** 109041 (2024). https://doi.org/10.1016/j.mtcomm.2024.109041
4. Huang, X., Zhao, G., Wang, G., Irvine, J.T.S.: Chem. Sci. **9**, 3623–3637 (2018). https://doi.org/10.1039/c7sc03920d
5. Jayavelu, Y., et al.: Chemosphere **351**, 141128 (2024). https://doi.org/10.1016/j.chemosphere.2024.141128
6. Reddy, N.P., Santhosh, R., Thogiti, S., Muniramaiah, R., Joseph, D.P., Murali, B.: J. Mater. Chem. C Mater. **10**(4), 1403–1413 (2022). https://doi.org/10.1039/d1tc04584a
7. Klingshirn, C.: Chem. Phys. Chem. **8**(6), 782–803 (2007). https://doi.org/10.1002/cphc.200700002
8. Schmidt-Mende, L., MacManus-Driscoll, J.L.: Mater. Today **10**, 40–48 (2007). https://doi.org/10.1016/S1369-7021(07)70078-0
9. Mishra, Y.K., Adelung, R.: Mater. Today **21**(6), 631–651 (2018). https://doi.org/10.1016/j.mattod.2017.11.003
10. Raha, S., Ahmaruzzaman, M.: Nanoscale Adv. **4**, 1868–1925 (2022). https://doi.org/10.1039/d1na00880c
11. Yan, L., Uddin, A., Wang, H.: Nanomaterials Nanotechnol. **5**, 19 (2015). https://doi.org/10.5772/60939
12. Mishra, N., Wu, W.-Y., Srinivasan, B.M., Hariharaputran, R., Zhang, Y.-W., Chan, Y.: Continuous shape tuning of nanotetrapods: toward shape-mediated self-assembly. Chem. Mater. **28**(4), 1187–1195 (2016). https://doi.org/10.1021/acs.chemmater.5b04803
13. Garg, V., Mathew, R., Ibrahim, R., Singh, K., Ghosh, S.K.: Iscience **27**(4), 109348 (2024). https://doi.org/10.1016/j.isci.2024.109348

Effects of Confinement Shape on Binding Energy and Transition Energy of a D^0 Impurity in a Quantum Dot with Spin-Orbit Interactions

Anusha Kachu, Swapna Vemula, and Aalu Boda(✉)

Department of Physics, National Institute of Technology, Warangal, Telangana 506004, India
abdop@nitw.ac.in

Abstract. In this study, we investigate how the confinement shape affects the hydrogenic donor D^0 impurity, trapped within the quantum dot (QD). Considering the effects of Rashba and Dresselhaus spin-orbit interactions, along with an applied magnetic field, we analyze the system using the power exponential potential. We estimated the energies of the ground and first excited states using the variational approach. A systematic study of binding and transition energies was performed, exploring the effects of QD parameters and various shapes of the confinement potential. The results show transformative behaviour in response to variations in potential shape, QD characteristics, and coupling strength of the spin-orbit interaction. This study gives light on the fundamental physics of QDs and impurities, allowing for improvements in detector and sensor technology.

1 Introduction

The rapid expansion of nanotechnology has sparked interest in low-dimensional systems [1–5], particularly quantum wires (QWs), quantum wells (QWLs), and QDs, due to their distinct magnetic, electric, and optical properties [6–8]. QDs, which have spatially confined electrons, have exceptional properties that enable applications in optoelectronics, photonics, energy conversion, and biological imaging [9–13]. The confinement potential in QDs plays a critical role and is influenced by the QD's shape, applied voltages, and external fields. Early experimental results led researchers to propose a harmonic confinement potential in QDs, which led to extensive study of parabolic QDs [14–21]. Recent research indicates that QD confinement potentials exhibit non-parabolic and finite characteristics, deviating from traditional assumptions. The Gaussian potential model, initially proposed by Adamowsky et al. [22], has garnered significant recognition for its precise representation of QD characteristics. Following this, various confinement potentials have been explored, including the power-exponential (PE) potential introduced by Ciurla et al. [23], which closely approximates realistic QD confinement profiles. Researchers have probed the influence of confinement shapes on QD optical and electrical properties, such as Kwasniowski et al. [24] work on electron exchange interactions and Xie [25] research on the photoionization cross-section.

Building on the discussion of confinement potentials in QDs, Spin-orbit interaction emerges as a key factor in determining the transport characteristics of nano-scale systems, particularly about transport properties. SOI is a central focus in spintronics, with

P. Joseph D. et al. (Eds.): IEMDST 2024, SPPHY 567, pp. 233–241, 2025.
https://doi.org/10.1007/978-981-96-5863-3_30

two primary types: Rashba spin-orbit interaction (RSOI) [26], arising from structural inversion asymmetry, and Dresselhaus spin-orbit interaction (DSOI) [27], originating from bulk inversion asymmetry. While RSOI can be tuned by external factors such as gate voltage, DSOI depends on intrinsic QD parameters. Both types of SOI play a key role in the development of spintronic devices, including spin filters [28], spin transistors [29], and components for quantum computing [30], which rely on spin currents for enhanced performance. Understanding the effects of impurities in QDs is also essential, as achieving pristine nanostructures without defects remains an ongoing challenge in nano-technology. Early work by Bastard [31] laid the foundation for studying hydrogenic impurities in quantum wells using variational methods, and subsequent studies, including those by Sundqvist et al. [32], have explored how external potentials control impurity distributions in low-dimensional systems. A significant contribution by Xie [25] explored the photoionization cross section of impurities with power-exponential potential, while Kumar et al. [33] studied how the RSOI affects the ground state energy in GaAs QDs. Saini et al. [34] focused on understanding the interplay between spin-orbit interactions, confinement geometry, and impurity properties. More recent research by Kachu et al. [21, 35] examined the impact of confinement potential geometry on impurity behaviour in QD with RSOI and DSOI. This study examines the impact of confinement potential shape on ground state (GS) energy, first excited energy state (FES) energy, binding energy, and 1s to 2p transition energy. The manuscript is structured as follows: theoretical framework (Sects. 2), numerical results and discussion (Sect. 3), conclusion (Sect. 4), and references.

2 Theoretical Approach

This section consists of two parts: (1) Hamiltonian of a D^0 system, which outlines the theoretical framework, and (2) Variational Method, which describes the computational approach used to solve the ground and first excited state energies.

2.1 Hamiltonian of a D^0 System

The mathematical formulation for a donor D^0 impurity in a 2D power-exponential potential in a GaAs QD, incorporating magnetic field and spin-orbit interactions, is given by:

$$H_{D^0} = \left[\frac{1}{2m^*}\left(\boldsymbol{P} + \frac{e}{c}\boldsymbol{A}\right)^2 + H_{imp} + V_{Conf}(r)\right]I + H_R + H_D, \quad (1)$$

The first term in Eq. (1) represents the D^0 center in a QD under the effect of an applied magnetic field $\boldsymbol{B}$. Here, $\boldsymbol{P}$ denotes the canonically conjugate momentum operator, and $\boldsymbol{A}$ denotes the magnetic vector potential arising from the magnetic field, defined by $\boldsymbol{B} = \nabla \times \boldsymbol{A}$. The symmetric gauge transformation is employed, where $\boldsymbol{A} = B\left(\frac{-y}{2}, \frac{x}{2}, 0\right)$. The terms m^* and e correspond to the effective mass and the charge of an electron, respectively. The impurity Hamiltonian, $H_{imp} = -\frac{e^2}{\varepsilon r}$, represents the Coulomb potential energy between the impurity and the electron, where r is the position of an electron from the origin, where I denotes the order 2-unit matrix, and ϵ is the dielectric constant. The confinement potential $V_{Conf}(r)$ is modeled as a power-exponential potential, given by: $V_{Conf}(r) = -V_0 e^{-\left(\frac{r}{R}\right)^{\gamma}}$. In this expression, $r = \sqrt{x^2 + y^2}$, represents the position coordinate of the electron in a QD, R denotes the size of the QD, V_0 is the depth of the potential, and γ is a parameter that characterizes the shape of the confinement potential. We have categorized the potential shapes into three primary cases. The potential's shape varies with γ: triangular ($\gamma = 1$), Gaussian-like ($\gamma = 2$), and square ($\gamma \to \infty$), for finite V_0.

The Hamiltonians for Rashba and Dresselhaus SOIs are expressed as follows:

$$H_R = \frac{x_R}{\hbar}\left[\sigma_x\left(p_y + \frac{e}{c}A_y\right) - \sigma_y\left(p_x + \frac{e}{c}A_x\right)\right], \tag{2}$$

$$H_D = \frac{x_D}{\hbar}\left[\sigma_x\left(p_x + \frac{e}{c}A_x\right) - \sigma_y\left(p_y + \frac{e}{c}A_y\right)\right], \tag{3}$$

In Eqs. (2) and (3), x_R represents the RSOI coefficient and x_D represents the DSOI coefficient. The symbol $\hbar$ is the reduced Planck's constant and σ_x, and σ_y are the Pauli spin matrices. The explicit form of Eq. (1), incorporating the terms from Eqs. (2) and (3), is:

$$\begin{aligned} H_{D^0} = &\left[\frac{1}{2m^*}\left(P + \frac{e}{c}A\right)^2 - \frac{e^2}{\varepsilon r} - V_0 e^{-\left(\frac{r}{R}\right)^{\gamma}}\right] I + \frac{x_R}{\hbar}\left[\sigma_x\left(p_y + \frac{e}{c}A_y\right) - \sigma_y\left(p_x + \frac{e}{c}A_x\right)\right] \\ &+ \frac{x_D}{\hbar}\left[\sigma_x\left(p_x + \frac{e}{c}A_x\right) - \sigma_y\left(p_y + \frac{e}{c}A_y\right)\right] \end{aligned} \tag{4}$$

To address the SOI terms, we employed a unitary transformation. Specifically, we applied the transformation $UH_{D^0}U^{\dagger}$, where $U = e^{\varsigma}$, with the generator $\varsigma = i\left(\frac{m^*}{\hbar^2}\right)\left[x_R\left(y\sigma_x - x\sigma_y\right) + x_D\left(x\sigma_x - y\sigma_y\right)\right]$. Through this transformation, the Hamiltonian is diagonalized in the spin space up to quadratic terms in x_R and x_D, which represent the Rashba and Dresselhaus coupling strengths, respectively. As an outcome, the Hamiltonian is re-expressed as a polynomial expansion in terms of SOI coupling constants.

$$\begin{aligned} H_{D^0} = &\left[\frac{p^2}{2m^*} + \frac{1}{8}m^*\omega_c^2 r^2 - \frac{e^2}{\varepsilon r} - V_0 e^{-\left(\frac{r}{R}\right)^{\gamma}} - \frac{m^*}{\hbar^2}\left(x_R^2 + x_D^2\right)\right] I - \frac{m^*}{\hbar^3}\left(x_R^2 - x_D^2\right)\sigma_z L_z + \frac{\omega_c}{2}L_z \\ &- \frac{m^*}{2\hbar^3}\left(x_R^2 - x_D^2\right)\omega_c \sigma_z r^2. \end{aligned} \tag{5}$$

The $\frac{\omega_c}{2}L_z$ term reflects the contribution from the magnetic field (with cyclotron frequency, ω_c) to the orbital angular momentum L_z where $L_z = -i\hbar\frac{\partial}{\partial\phi}$. Finally, the term $\left(\frac{m^*}{2\hbar^3}\left(x_R^2 - x_D^2\right)\omega_c\sigma_z r^2\right)$ includes both the spin and magnetic field effects on the system. To simplify the Hamiltonian and eliminate dimensions, we used the energy and length units are chosen to be donor Rydberg $R_y^* = \frac{m^* e^4}{2\varepsilon^2\hbar^2} = 6\,\text{meV}$ and the effective Bohr radius $a_B^* = \frac{\varepsilon\hbar^2}{m^* e^2} = 9.8\,\text{nm}$, throughout this work. These units provide a convenient framework for scaling the system parameters in dimensionless form, facilitating easier interpretation of the results. The Hamiltonian Eq. (5) can be transformed into its nondimensional form as follows:

$$H_{D^0} = \left[-\nabla^2 + \frac{1}{16}\omega_c^2 r^2 - \frac{2}{r} - V_0 e^{-\left(\frac{r}{R}\right)^\gamma} - \left(\frac{x_R^2 + x_D^2}{2}\right)\right] I - \left(\frac{x_R^2 - x_D^2}{2}\right)\sigma_z L_z + \frac{\omega_c}{2} L_z - \left(\frac{x_R^2 - x_D^2}{4}\right)\omega_c \sigma_z r^2, \quad (6)$$

2.2 Variational Method

As mentioned in the abstract, the ground and first excited state energies of the D^0 system in a GaAs QD are determined using the Ritz variational method. In this method, the energy E_i for a given state i is calculated as:

$$E_i = \frac{\langle \psi_i | H_{D^0} | \psi_i \rangle}{\langle \psi_i || \psi_i \rangle}, \quad (7)$$

The ground an first excited state energies of the D^0 system is represented as E_i for $E_i(i = 1s)$ and $E_i(i = 2p)$, respectively. To compute these energies using Eq. (7), specific trial wavefunctions are proposed for different quantum states, which are given below as

$$\psi_{1s(r)} = B_1 e^{-ar^2 - br - im\phi}, \quad (8)$$

$$\psi_{2p}(r) = B_2 r \left(\delta e^{-ar^2 - br - im\phi}\right), \quad (9)$$

For both D^0 and one-electron systems, the GS wave function $\psi_{1s}(r)$ (Eq. 8, $m = 0$) has no nodes, while the FES wave function $\psi_{2p}(r)$) (Eq. 9, $m = -1$) has one node due to a linear r factor. Parameters a, b, δ are variational parameters, and B_1, B_2 are normalization constants. The binding energies for the D^0 center are

$$E_{B(D^0_{1s})} = E_{e^-_{1s}} - E_{D^0_{1s}}, \quad (10)$$

$$E_{B\left(D^0_{2p}\right)} = E_{e^-_{2p}} - E_{D^0_{2p}}, \quad (11)$$

The energies $E_{e^-_{1s}}$ and $E_{e^-_{2p}}$ correspond to the ground and first excited state energies of a one electron system in a QD. Similarly, the energies of the D^0 system for the ground and first excited state are denoted as $E_{(D^0_{1s})}$ and $E_{\left(D^0_{2p}\right)}$, respectively. This study also investigates the transition energy i.e. $\Delta E\left(D^0\right)$ of the D^0 system from $1s \rightarrow 2p$ state which is given by

$$\Delta E_{D^0} = E_{D^0_{2p}} - E_{D^0_{1s}}, \quad (12)$$

In conclusion, all results were obtained through numerical calculations.

3 Numerical Results and Analysis

Expanding on the foundational QD theory, this section focuses on GaAs QDs, leveraging their potential for practical applications. Our numerical simulations augment earlier analytical results, utilizing GaAs material parameters are assuming $\varepsilon = 12.4$ and $m^* = 0.067m_0$. We quantify magnetic field and SOIs in Tesla and Rydberg-Bohr units $\left(R_y^* a_B^*\right)$ not only provides valuable insights into GaAs QD behavior but also has significant implications for future device development. To illustrate our findings, Fig. 1 shows the ground and first excited state energies of a D^0 system in a QD as a function of the cyclotron frequency w_c, with fixed parameters: $V_0 = 60R_y^*$, $R = 6a_B^*$, $x_R = 0.9$, $x_D = 1$ and for several values of the confinement potential shape parameter γ. Additionally, Fig. 2 displays 3D plots of the GS and FES energies as a function of the SOIs for a fixed shape parameter ($\gamma = 2$) and $w_c = 8R_y^*$, while all other parameters held constant (as in Fig. 1). As predicted the GS energy exhibits a lower value than the FES energy. Figure 1, results show that as the cyclotron frequency increases, the energy also increases linearly. This is due to the fact that the cyclotron frequency corresponds to the magnetic field, and as the magnetic field increases, additional potential is introduced into the system. Additionally, we observe that as the shape parameter γ increases from 1 to 4, the energy decreases. However, once γ value reaches 4 and continues to increase to 8 or infinity, the energy remains constant. Moreover, in Fig. 2, we observe that the energy increases with increasing Dresselhaus spin-orbit interaction, which is associated with bulk inversion asymmetry, while it decreases with increasing Rashba spin-orbit interaction, linked to structural inversion asymmetry. This indicates that bulk inversion asymmetry enhances the energy, whereas structural inversion asymmetry reduces it.

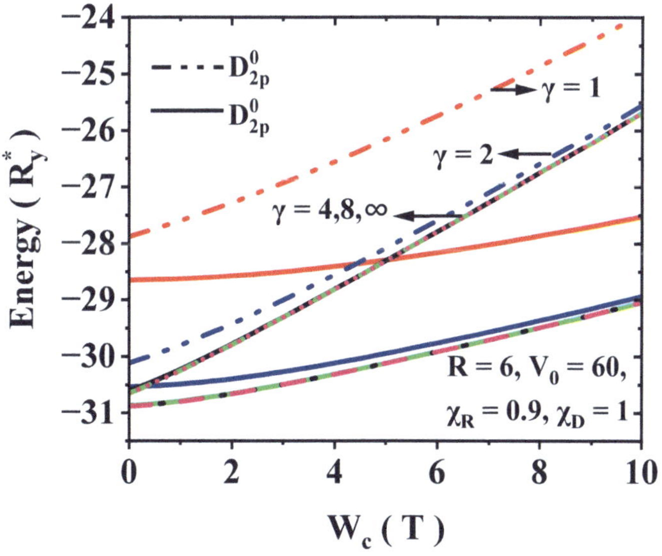

Fig. 1. GS energy (D_{1s}^0) and FES energy $\left(D_{2p}^0\right)$ in a QD as a function of w_c for $V_0 = 60R_y^*$, $R = 6a_B^*$ $x_R = 0.9$, $x_D = 1$ and for a few values of γ.

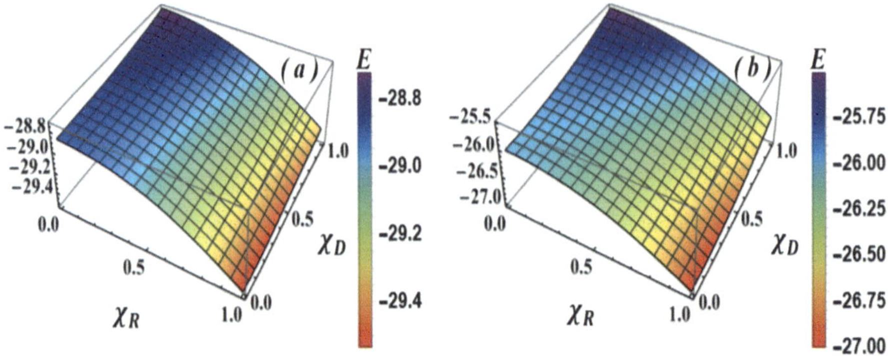

Fig. 2. 3D plots of GS energy (D^0_{1s}) and FES energy $\left(D^0_{2p}\right)$ in a QD as a function of SOI coupling strengths (x_R, x_D) and for two impurity positions for $V_0 = 60R^*_y$, $R = 6a^*_B$ $w_c = 8R^*_y$, and for various γ values.

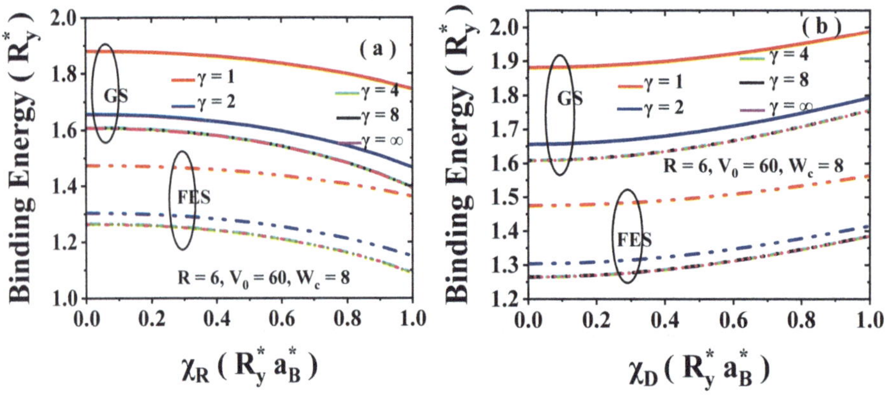

Fig. 3. (a) GS and FES binding energies of a D^0 system vs RSOI, (b) GS and FES binding energies of a D^0 system vs DSOI in a QD for $R = 6a^*_B$, $w_c = 8R^*_y$, $V_0 = 60R^*_y$ and for few values of γ.

In Fig. 3 (a) and (b), we plot the variation of the ground state and first exited state energies as a function of SOIs with all other variables held constant $R = 6a^*_B$, $w_c = 8R^*_y$, $V_0 = 60R^*_y$ and, for several values of the shape parameter γ. From the plot, we conclude that the binding energy is positive for both the GS and FES, indicating that the on-center D^0 system remains stable across different confinement potential shapes. Similar to the energy behaviour, we notice that the binding energy decreases with the RSOI and rises with the DSOI. Additionally, as the shape parameter γ increases from 1 to infinity, the binding energy decreases, following the same trend as the energy. In Fig. 4, we plotted the variation of the transition energy as a function of SOIs with variables $V_0 = 60R^*_y$, $R = 6a^*_B$ $w_c = 8R^*_y$, and $\gamma = 2$. Similar to the behavior of energy and binding energy, the transition energy is lowering as RSOI increases and increasing with DSOI. The transition energy decreases as the shape parameter gamma increases from 1 to infinity.

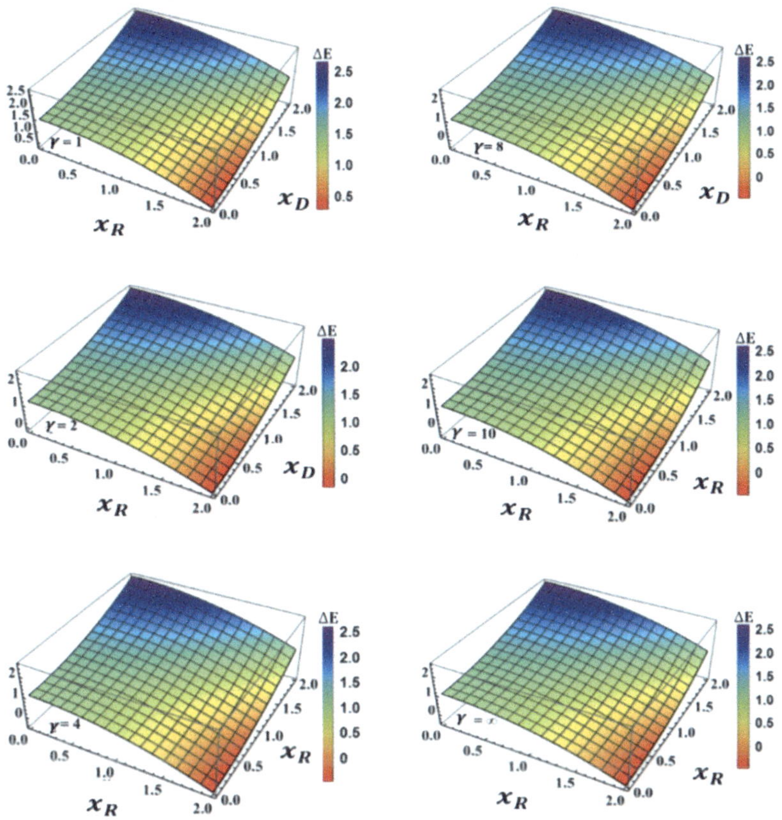

Fig. 4. 3D plots of transition energy in a QD as a function of SOI coupling strengths (x_R, x_D) and for $V_0 = 60R_y^*$, $R = 6a_B^*$ $w_c = 8R_y^*$, $\gamma = 2$.

4 Conclusion

In conclusion, the results of this study have practical implications, demonstrating the stability of both the energy and binding energy of the system in a quantum dot across various confinement potential shapes. The increase in cyclotron frequency leads to a linear rise in energy, a finding that could have significant implications in the presence of a magnetic field. The decrease in binding energy and transition energy with increasing Rashba spin-orbit interaction (RSOI) and shape parameter, and the enhancement with Dresselhaus spin-orbit interaction (DSOI), are key insights that confirm the stability of the on-center system under diverse conditions, opening up new possibilities for quantum dot applications.

Acknowledgments. A. Kachu would like to acknowledge and thank the National Institute of Technology Warangal, India, for providing financial support through the Institute Fellowship, which enabled the completion of this work.

References

1. Winkler, R.: Spin-Orbit Coupling Effects in Two-Dimensional Electron and Hole Systems. Springer-Verlag, Berlin (2003)
2. Gambardella, P., Miron, I.M.: Philos. Trans. r. Soc. a. **369**, 3175–3197 (2011). https://doi.org/10.1098/rsta.2010.0336
3. Feng, R., Akimoto, S.I., Gozu, T., Mozume, T., Hasama, H.: Ishikawa. Opt. Express **21**, 15840–15846 (2013). https://doi.org/10.1364/OE.21.015840
4. De Jesus, J., Chen, G., Hernandez-Mainet, L.C., Shen, A., Tamargo, M.C.: J. Cryst. Growth **425**, 207–211 (2015). https://doi.org/10.1016/j.jcrysgro.2014.12.021
5. Sayrac, M., Turkoglu, A., Mora-Ramos, M.E., Ungan, F.: Opt. Quant. Electron. **53**, 485–495 (2021). https://doi.org/10.1007/s11082-021-03136-y
6. Brown, J.W., Spector, H.N.: J. Appl. Phys. **59**, 1179–1186 (1986). https://doi.org/10.1063/1.336555
7. Hsieh, C.Y.: J. Appl. Phys. **92**, 484–490 (2002). https://doi.org/10.1063/1.1480121
8. Sidor, Y., Partoens, B., Peeters, F.M.: Phys. Rev. B **71**, 165323–165331 (2005). https://doi.org/10.1103/PhysRevB.71.165323
9. Wu, J., Chen, S., Seeds, A., Liu, H.: J. Phys. D Appl. Phys. **48**, 363001–363030 (2015). https://doi.org/10.1088/0022-3727/48/36/363001
10. Kim, J., Song, S., Kim, Y.-H., Park, S.K.: Small Struct. (2020). https://doi.org/10.1002/sstr.202000024
11. Ye, M., et al.: Sustain. Energy Fuels **1**, 1217 (2017)
12. Zahid, M.U., Ma, L., Lim, S.J., Smith, A.M.: Nat. Commun. **9**, 1830–1840 (2018). https://doi.org/10.1038/s41467-018-04185-w
13. Wang, B., Cai, H., Waterhouse, G.I.N., Qu, X., Yang, B., Lu, S.: Small Sci. **2**, 2200012–220037 (2022). https://doi.org/10.1002/smsc.202200012
14. Gu, J., Liang, J.Q.: Phys. Lett. Sect. A Gen. At. Solid State Phys. **335**, 451 (2005)
15. Xie, W.F.: Chin. Phys. **15**, 203–208 (2006). https://doi.org/10.1088/1009-1963/15/1/033
16. Boyacioglu, B., Chatterjee, A.: J. Appl. Phys. **112**, 083514–083519 (2012). https://doi.org/10.1063/1.4759350
17. Boyacioglu, B., Chatterjee, A.: Phys. B Condens. Matter **407**, 3535–3538 (2012). https://doi.org/10.1016/j.physb.2012.05.018
18. Boda, A., Chatterjee, A.: Phys. B Condens. Matter **448**, 244–246 (2014). https://doi.org/10.1016/j.physb.2014.02.059
19. Jahan, K.L., Boda, A., Shankar, I.V., Raju, C.N., Chatterjee, A.: Sci. Rep. **8**, 5073–5085 (2018). https://doi.org/10.1038/s41598-018-23348-9
20. Sharma, H.K., Boda, A., Boyacioglu, B., Chatterjee, A.: J. Magn. Magn. Mater. **469**, 171–177 (2019). https://doi.org/10.1016/j.jmmm.2018.07.070
21. Kachu, A., Boda, A.: Phys. E Low-Dimensional Syst. Nanostructures. **155**, 115826–115835 (2024). https://doi.org/10.1016/j.physe.2023.115826
22. Adamowski, J., Sobkowicz, M., Szafran, B., Bednarek, S.: Phys. Rev. B **62**, 13233–13237 (2000). https://doi.org/10.1103/PhysRevB.62.4234
23. Ciurla, M., Adamowski, J., Szafran, B., Bednarek, S.: Physica E **15**, 261–268 (2002). https://doi.org/10.1016/S1386-9477(02)00572-6
24. Kwaśniowski, A., Adamowski, J.: J. Phys.: Condens. Matter **20**(21), 215208 (2008). https://doi.org/10.1088/0953-8984/20/21/215208
25. Xie, W.: Superlattices Microstruct. **65**, 271–277 (2014). https://doi.org/10.1016/j.spmi.2013.11.012
26. Rashba, E. I.: Sov. Phys. Solid State **2**, 1109 (1960)
27. Dresselhaus, G.: Phys. Rev. **100**, 580–586 (1955). https://doi.org/10.1103/PhysRev.100.580

28. Koga, T., Nitta, J., Takayanagi, H.: Phy. Rev. Lett. **88**, 126601–126604 (2002). https://doi.org/10.1103/PhysRevLett.88.126601
29. Das, B., Datta, S.: Appl. Phys. Lett. 665, 12 (2012)
30. Gvozdic, D.M.: Appl. Phys. Lett. **90**, 053105–053108 (2007). https://doi.org/10.1063/1.2437132
31. Bastard, G.: Phys. Rev. B **24**, 1714 (1981)
32. Sundqvist, P.A., Narayan, V., Stafstrom, S., Willander, M.: Phy. Rev. B **67**, 165330–165337 (2003). https://doi.org/10.1103/PhysRevB.67.165330
33. Kumar, D.S., Boda, A., Mukhopadhyay, S., Chatterjee, A.: Superlattices Microstruct. **88**, 174–179 (2015). https://doi.org/10.1016/j.spmi.2015.09.003
34. Saini, P., Chatterjee, A.: Superlattice. Microst. **146**, 106641–106652 (2020). https://doi.org/10.1016/j.spmi.2020.106641
35. Kachu, A., Vemula, S., Chebrolu, N.R., Boda, A.: Phys. B Condens. Matter. **693**, 416376–416383 (2024). https://doi.org/10.1016/j.physb.2024.416376

A Brief Study on Nonlinear Optical (NLO) Property of BOX-Type Tetracationic Cyclophane Structure

Bathula Rajasekhar[1](✉), G. Premkumar[2], and Toka Swu[2]

[1] Department of Chemistry, Dharma AppaRao College, Krishna District, Nuzvid 521201, Andhra Pradesh, India
rajashekhar415@gmail.com
[2] Department of Chemistry, Pondicherry University, Puducherry 605014, India
tokaswu@gmail.com

Abstract. The work includes the structure optimization and first static hyperpolarizability calculations of the tetracationic cyclophane BOX-type molecule with two different symmetries such as C_1, and C_{2h}. The first static hyperpolarizability calculations studies were also carried out on crystal structures with C_{2h} symmetry. In this computational work, the Gaussian09 programme was used by applying the DFT-based B3LYP method and 6 – 31G* basis set used for C, H and N atoms. The study finds that among all, BOX–3 molecule exhibits higher β value (β_{total} = 0.09743 × 10 − 30 e. s. u). Also, molecules with higher symmetry showed a higher β value compared with lower symmetry-contained molecules.

1 Introduction

The demand for molecular-based nonlinear optical (NLO) materials is growing, and the development of such materials with higher NLO properties is much sought [1–3]. The relation between the molecular structure and the NLO property of the molecules is well documented in the literature for various molecular structures [4–7]. Some of the key points for designing the molecular structure with higher NLO properties are also reported by some researchers. Supramolecular systems are one category that provides the tools to remove the centrosymmetry and thereby enhance the NLO property [8, 9]. Cyclophanes (organic cyclic structures) are from the supramolecular category (host type) known to be involved in the host-guest interactions through the non-covalent bonding mode [10]. Few theoretical studies for various cyclophanes are also reported in the literature [11]. They are in the fields of catalysis, pharmaceuticals, optics, and electronics [12]. Tetracationic cyclophanes derived from the 4,4′-bipyridinium dications and bisparaphenylmethylenes were extensively studied for host-guest interactions with various types of molecules as the cavity with 6.8 Å is available in the cyclophane structure [13, 14]. In these host-guest structures, the guest molecule will occupy the cavity that is accessible in the host, whereby these host and guest molecules behave structurally and electronically differently. In this context, we were encouraged to study the NLO property of BOX-type and also the BOX in BOX-type organic cyclophane systems (Host type) that were recently

P. Joseph D. et al. (Eds.): IEMDST 2024, SPPHY 567, pp. 242–249, 2025.
https://doi.org/10.1007/978-981-96-5863-3_31

published by J. Fraser Stoddart's research group [14] for their interesting electronic features. However, due to the limited resources, we could study the NLO property of only one cyclophane BOX-type structure in this work.

2 Materials and Methods

In this study, tetracationic structure (Scheme 1) with two different symmetries such as C_1 and C_{2h} were used for structure optimization and hyperpolarizability calculations. To compare the β values between theoretical and crystallographic structures, CIF (crystallographic information file) of the crystal structure containing XYZ coordinates was used for hyperpolarizability calculations. All the computational calculations were performed using the GAUSSIAN09 (programme suite) [15]. For structural optimization and hyperpolarizability calculations, DFT DFT-based B3LYP method and basis set, 6-31G* [16] was used. Lower basis set (6-31G*) was used in this work because of the huge computational cost. Hyperpolarizability calculations were performed by giving the following keyword # B3LYP 6-31G* Polar = (DCSHG, Cubic) CPHF = RdFreq" in route section and 0.1 Hartrees (455.6300 nm, [17]) frequency at the bottom of XYZ coordinates. UV-visible calculations were carried out using the keyword # td = (nstates = 20, root = 3) b3lyp/6-31g(d) geom = connectivity'. TD-DFT analysis was also carried out for all molecules, and electronic transitions with oscillatory strength ≥ 0.002 were considered, as these transitions contribute more to the β value of the studied molecules. The spectra and the orbital shapes were generated using GaussSum [18] and GaussView 5.0 [19], respectively.

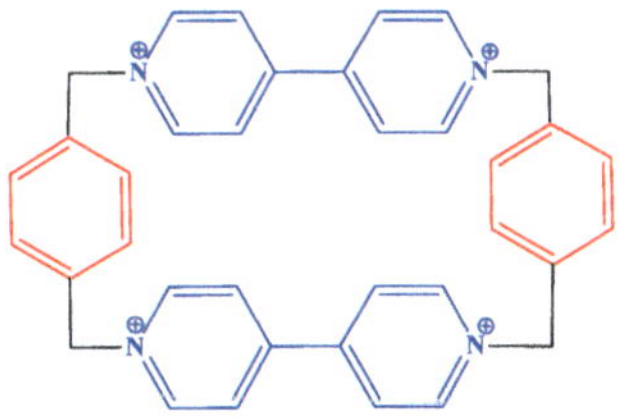

Scheme 1. Tetracationic cyclophane (BOX type molecule) [14].

3 Results and Discussion

The studied molecules and their single-crystal X-ray diffraction structures were reported J. F. Stoddart's research group [14]. The hyperpolarizability calculations of the reported molecules (only the cation part) were performed using GAUSSIAN09 [16]. For structural optimization and Hyperpolarizability calculations, DFT DFT-based B3LYP method was used.

Hyperpolarizability Analysis: For BOX – 1 molecule, XYZ coordinates were taken from the CIF (Crystallographic Information File), whereas in the case of the other two molecules (BOX – 2 molecule and BOX – 3 molecule), XYZ coordinates were imported from the optimized structures. These coordinates are given in supplementary information, and structures are given in Fig. 7. Further, for the calculation of the dipole moment (μ), polarizability (α), and first static hyperpolarizability (β), the XYZ coordinates (Table 1) were obtained from the DFT calculations. The following standard equations [7] were employed and the results are presented in the Tables 2 and 3.

$$\mu = \left(\mu_x^2 + \mu_y^2 + \mu_z^2\right)^{1/2} \tag{1}$$

$$\alpha = 1/3(\alpha_{XX} + \alpha_{YY} + \alpha_{ZZ}) \tag{2}$$

$$\beta_{total} = [(\beta_{XXX} + \beta_{XYY} + \beta_{XZZ})^2 + (\beta_{YYY} + \beta_{YZZ} + \beta_{YXX})^2 + (\beta_{ZZZ} + \beta_{ZYY} + \beta_{ZXX})^2]^{1/2} \tag{3}$$

The polarizability and hyperpolarizability tensors (α_{xx}, α_{yy}, α_{zx}, β_{xxx}, β_{xyy}, β_{xzz}, β_{yyy}, β_{yzz}, β_{yxx}, β_{zzz}, β_{zyy}, β_{zxx}) (Table 1) were obtained from frequency dependant DFT calculations, and the units were converted from atomic units (a.u.) into electrostatic units (e.s.u.), (α: 1 a. u = 0.1482 × 10 – 24 e.s.u; β: 1 a. u = 8.6393 × 10 – 33 e. s.u) and the results are presented in tabular format (Table 2). All the studied molecules exhibited lower β values (except BOX–3 molecule) compared to the standard organic compound, Urea [7]. BOX–2 and BOX-3 molecules exhibited the highest HOMO - LUMO energy difference (ΔE), and thus, as expected, BOX-2 showed the lowest β value (Table 3). However, the BOX–3 molecule has the highest β value despite having the highest ΔE. BOX–1 molecule has a lower β value even though it has the lowest ΔE. The observed values are contrary to the fact that lower ΔE produces higher β value [20]. This anomaly can be understood from the TD–DFT study.

TD – DFT Study: TD–DFT calculations were carried out by using the keyword *[# td = (nstates = 20, root = 3) b3lyp/6-31g(d) geom = connectivity]*, and the field-dependent electronic transitions and their oscillatory strengths were used to understand the observed anomaly [21]. A summary view of the TD-DFT study is given in Table 4.

This study revealed that the BOX–3 molecule exhibited a higher number of electronic transitions (Table 4, highlighted in bold), which has enhanced the β value though the molecule (BOX–3) has a higher ΔE. Similar results were found in the literature [21], where more electronic transitions and higher oscillatory strength (either one or both) have enhanced β values. For the other two molecules (BOX–1 and BOX–2), the TD–DFT data revealed that the BOX–1 molecule has the least electronic transitions and highest oscillatory strength values, and the BOX–2 molecule has the highest number of electronic transitions and lowest oscillatory strength values. In the case of BOX–1 molecule, fewer electronic transitions have decreased the β value whereas, in the case of BOX–2 molecule, less oscillatory strength has reduced the β value. From the TD–DFT study, we obtained the UV–visible spectra for all the molecules (Fig. 1), and it was observed that all the molecules were able to absorb the visible region light (400–700 nm).

Molecular Orbitals Analysis: The box structures exhibited low β values even though they were structurally large and possessed the ability of π - conjugation, which is known

Table 1. Tensors of BOX-type molecules

BOX–1 Molecule (C_{2h})	BOX–2 Molecule (C_1)	BOX–3 Molecule (C_{2h})
Dipole moment (μ) tensors	**Dipole moment (μ) tensors**	**Dipole moment (μ) tensors**
X = 0.0000	X = −0.0110	X = 0.0000
Y = 0.0000	Y = 0.0000	Y = 0.0000
Z = −0.2663	Z = 0.2756	Z = −0.2981
Polarizability (α) tensors	**Polarizability (α) tensors**	**Polarizability (α) tensors**
XX = −121.3839	XX = 82.6377	XX = −132.7082
YY = 89.0653	YY = 50.0708	YY = 82.6388
ZZ = 32.3186	ZZ = −132.7085	ZZ = 50.0694
Hyperpolarizbility (β) tensors	**Hyperpolarizbility (β) tensors**	**Hyperpolarizbility (β) tensors**
XXX = 0.0000	XXX = −1.2910	XXX = 0.0000
YYY = 0.0000	YYY = 0.0000	YYY = 0.0000
ZZZ = −4.6039	ZZZ = 1.3619	ZZZ = −6.1427
XYY = 0.0000	XYY = 0.6063	XYY = 0.0000
XXY = 0.0000	XXY = −0.0003	XXY = 0.0000
XXZ = −0.5353	XXZ = 3.9733	XXZ = −0.6655
XZZ = 0.0000	XZZ = −0.0230	XZZ = 0.0000
YZZ = 0.0000	YZZ = 0.0001	YZZ = 0.0000
YYZ = −4.1053	YYZ = 2.8422	YYZ = −4.4698
XYZ = 0.0274	XYZ = 0.0010	XYZ = 0.3673

Table 2. μ, α and β values of BOX-type molecules

Molecules	Dipole moment (μ, Debye)	Polarizability ($\alpha \times 10^{-24}$ e. s. u)	Hyperpolarizbility ($\beta_{total} \times 10^{-30}$ e. s. u)	HOMO-LUMO Gap (Kcal/mol)
BOX–1 (C_{2h})	0.0000	0	0.07987	60.6481
BOX–2 (C_1)	0.0000	0	0.07091	74.9364
BOX–3 (C_{2h})	0.0000	0	0.09743	74.9364

Table 3. HOMO – LUMO energy difference comparison of the studied molecules

Molecule	Energy (RB3LYP) (a. u)	RMS Gradient (a. u)	Dipole Moment (Debye)	HOMO Energy (a.u.)	LUMO Energy (a.u.)	ΔE = HOMO-LUMO (a.u.)	HOMO-LUMO Gap (Kcal/mol)
BOX 1 (C_{2h})	−1608.26944	–	0	−0.6228	−0.5261	0.0966	60.6481
BOX – 2 (C_1)	−1608.78325	0.00000023	0	−0.6135	−0.4941	0.1194	74.9364
BOX – 3 (C_{2h})	−1608.78325	0.00000029	0	−0.6135	−0.4941	0.1194	74.9364

for enhancing the β values [8]. Thus, the electron density distributions in both the structures (Optimized and Post excited structure) were examined by comparing their HOMOs and LUMOs (Fig. 2, Fig. 3, Fig. 4, Fig. 5 and Fig. 6). However, no considerable differences in electron density were observed for BOX–2 and BOX–3 molecules.

Table 4. Summary of the TD-DFT study.

Molecules	Number of electron transitions	state	ΔE (eV)	Λ (nm)	*f*
Box – 1	6	Excited State 12	4.1512	298.67	1.2491
Box – 2	8	Excited State 18	4.4429	279.06	0.0537
Box – 3	8	Excited State 18	4.4429	279.06	0.0537

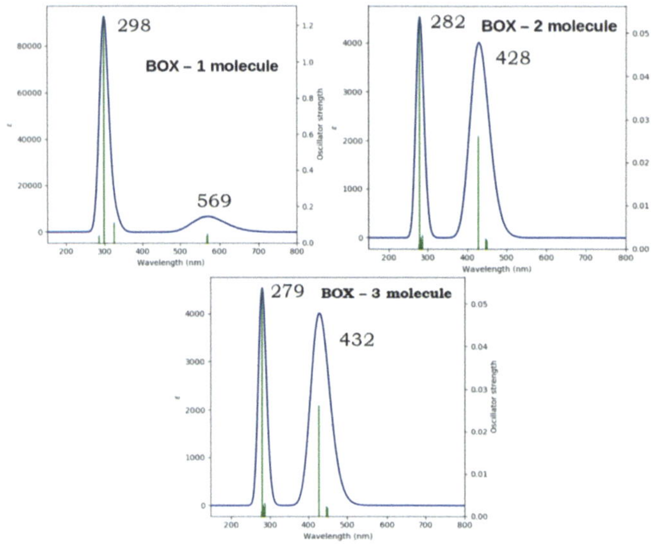

Fig. 1. UV–Visible spectra of studied molecules.

Comparison of HOMO – LUMO Orbitals

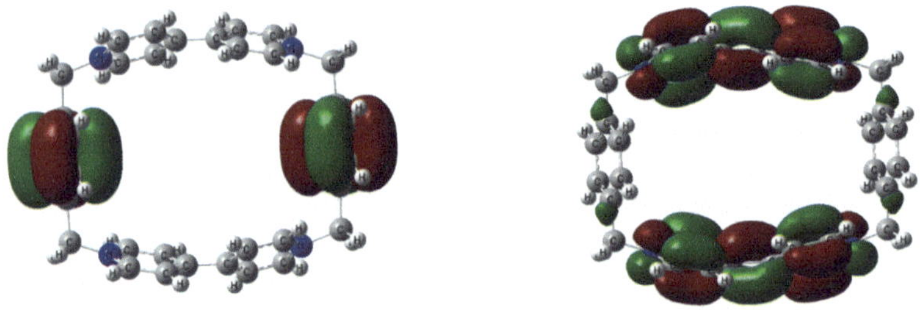

(a) **(b)**

Fig. 2. (a) HOMO of BOX–1 molecule (b) LUMO of BOX–1 molecule. *These HOMO and LUMO are obtained from the Hyperpolarizability calculations check point file (.chk)*

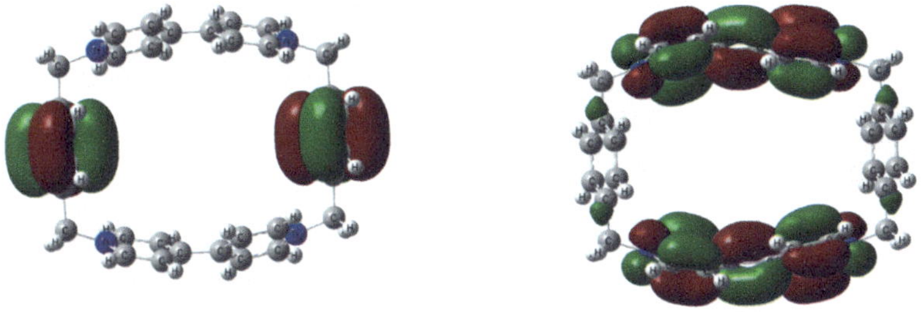

(a) (b)

Fig. 3. (a) HOMO of BOX–2 molecule (from the optimization calculations) (b) LUMO of BOX–2 molecule

(a) (b)

Fig. 4. (a) HOMO of BOX–2 molecule (from the hyperpolarizability calculation). (b) LUMO of BOX–2 molecule (from the hyperpolarizability calculations)

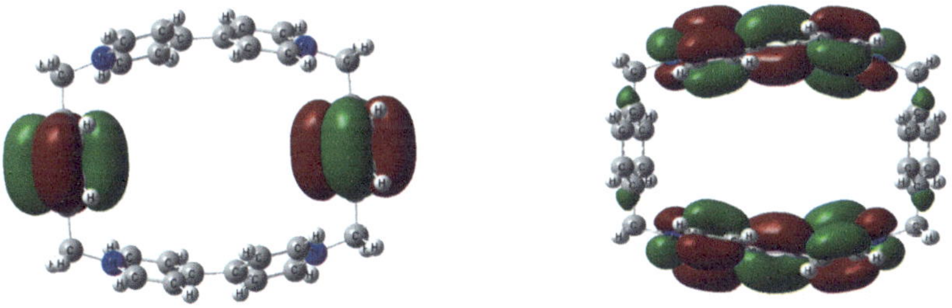

(a) (b)

Fig. 5. (a) HOMO of BOX–3 molecule (from the hyperpolarizability calculation). (b) LUMO of BOX–3 molecule (from the hyperpolarizability calculations)

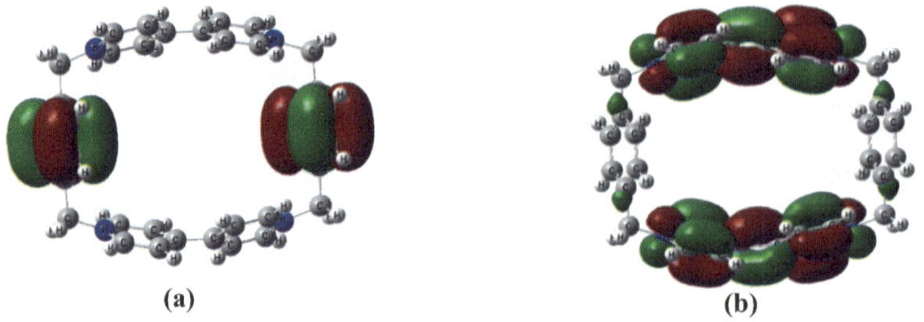

Fig. 6. (a) HOMO of BOX–3 molecule (from the hyperpolarizability calculations). (b) LUMO of BOX–3 molecule (from the hyperpolarizability calculations)

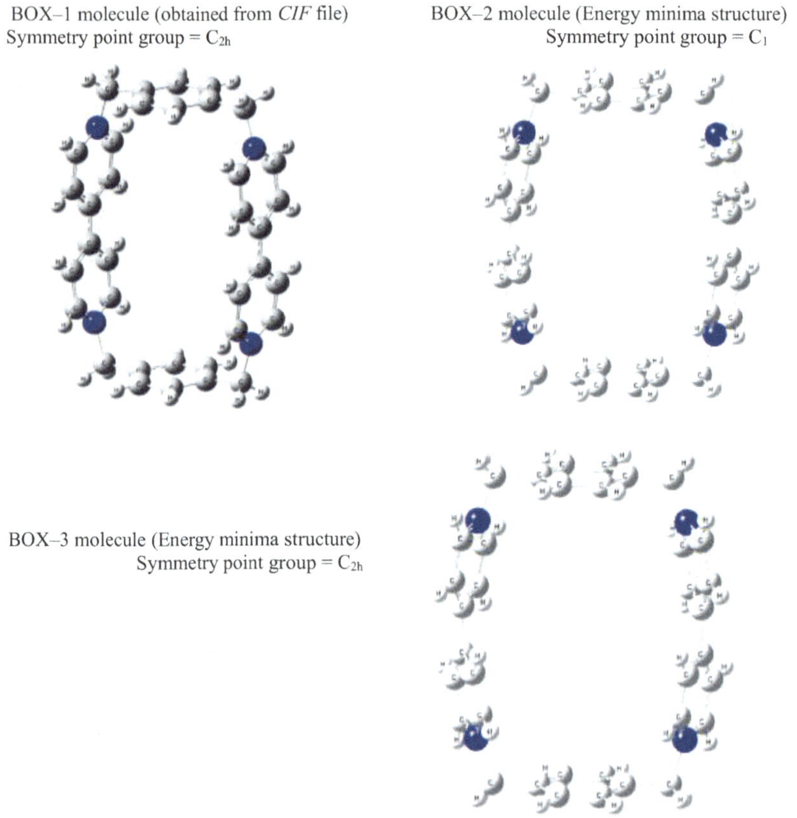

Fig. 7. Structures of the studied BOX-type molecules.

4 Conclusion

The BOX-type molecules (cations) were optimized, and their hyperpolarizability calculations were carried out using the DFT-based B3LYP method. BOX–3 molecule has a higher HOMO-LUMO energy gap, but contrary to the normal trend, it exhibited a higher β value than the other BOX-type structures. The TD-DFT study revealed that this anomaly is due to the difference in the number of electronic transitions: The higher the number of electronic transitions, the higher the β value. The study also suggests that BOX-type structures (cationic-type cyclophanes) are not promising materials for NLO study and point toward the molecules with electron-donating groups as the substituents, resulting in push-pull effects. Finally, this study has enlightened us in getting insights into understanding the anomalies in the NLO properties, and it will be a good pointer for designing and developing a material with high β value.

Acknowledgments. Bathula Rajasekhar (BR) is grateful to the entire Department of Chemistry, Pondicherry University, for allowing me to access the Department computer lab for the study. BR thanks Mr. Samuel Kiran Bathula, Mr. L. Nagaraju, Mr. Lone Kranthiraj and Mr. Ch. SriKumar for their hospitality in the hard times and also BR acknowledges Dharma AppaRao College Leadership for their timely support during the progress of the work.

References

1. Conference Paper: Synthesis of New Alkynyl Ru (II) Organometallic Compounds with Extended Dipolar Chains for Second Order Nonlinear Optical (2014)
2. Peris, E.: Chem. Rev. **248**, 279–297 (2004). https://doi.org/10.1016/j.ccr.2004.01.006
3. Barlow, S., Marder, S.R.: Chem. Commun. 1555–1562 (2000). https://doi.org/10.1039/b004907g
4. Rajasekhar, B., Bodavarapu, N., Sridevi, M., et al.: J. Mol. Struct. **1156**, 690–699 (2018)
5. Rajasekhar, B., Swu, T.: International Conference on 'Frontier Areas in Chemical Technologies', vol. 2, pp. 155–157 (2017). ISBN 978-81-928690-7-0
6. Rajasekhar, B., Patowary, N., et al.: Mol. Phys. (2018). https://doi.org/10.1080/00268976.2018.1460497
7. Rajasekhar, B., Hijaz, P.K.M., et al.: J. Mol. Struct. **1168**, 212–222 (2018)
8. Xiong, R.-G., Zuo, J.-L., et al.: New J. Chem. **23**, 1051–1052 (1999). https://doi.org/10.1039/A905955E
9. Xiong, R.G., Liu, C.M., Zuo, J.L., You, X.Z.: Inorg. Chem. Commun. **2**, 292–297 (1999)
10. Sun, J., et al.: J. Am. Chem. Soc. **139**, 12704–12709 (2017). https://doi.org/10.1021/jacs.7b06857
11. Li, X., Staykov, A., Yoshizawa, K.: Bull. Chem. Soc. Jpn **85**, 181–188 (2012). https://doi.org/10.1246/bcsj.20110256
12. Kotha, S., Shirbhate, M.E., Waghule, G.T.: Beilstein J. Org. Chem. **11**, 1274–1331 (2015)
13. Lipke, M.C., et al.: ACS Cent. Sci. **4**, 362–371 (2018). https://doi.org/10.1021/acscentsci.7b00535
14. Cai, K., Lipke, M.C., Liu, Z., et al.: Nat. Commun. **9**, 1–8 (2018). https://doi.org/10.1038/s41467-018-07673-1
15. Frisch, M.J., et al.: Gaussian 09, Revision A.02, Gaussian, Inc., Wallingford (2009)
16. Rajasekhar, B., Swu, T.: International Conference on 'Frontier Areas in Chemical Technologies', vol. 2, pp. 155–157. ISBN 978-81-928690-7-0 (2017)
17. https://www2.chemistry.msu.edu/faculty/reusch/virttxtjml/cnvcalc.htm
18. O'Boyle, N.M., Tenderholt, A.L., Langner, K.M.: J. Comput. Chem. **29**, 839–845 (2008)
19. Frisch, A., et al.: GaussView 5 Reference. Gaussian Inc., Wallingford (2009)
20. Nkungli, N.K., Ghogomu, J.N.: J. Theor. Chem. **2016**, 1–19 (2016). https://doi.org/10.1155/2016/7909576
21. Nkungli, N.K., Ghogomu, J.N.: J. Theor. Chem. 1–19. Article ID 7909576 (2016).https://doi.org/10.1155/2016/7909576

Investigation of 2D Hexagonal I_2 - VI Group Compound via First Principles Calculation

G. Sneha and R. D. Eithiraj(✉)

Department of Physics, School of Advanced Sciences, Vellore Institute of Technology, Chennai 600127, India
eithiraj.rd@vit.ac.in

Abstract. The wide range of applications scaling from electronics, energy storage, sensing, catalysis, and thermal management to medical applications, the 2D materials with key aspects such as atomic thickness, high surface area, electronic, optical, thermal properties and mechanical strength have potential to revolutionize industries. The researchers are trying to unveil new areas of such materials as they have vibrant properties and versatility. This study focuses on understanding the monolayer formed by Rb and Te in the hexagonal phase. WIEN2k software package was deployed to analyse the physical properties of the monolayer. The monolayer was optimized with Monkhorst-Scheme and the electronic band structure displays a direct band gap and is a semiconductor with the energy band gap value of 1.03 eV. On computing other properties such as, dynamical, optical and the monolayer exhibiting low direct band gap, it can have promising applications in the fields of nano-optoelectronics.

1 Introduction

The discovery of Graphene and its derivatives persuaded many scientists to further investigate more about the low-dimensional materials in its atomically-thin layered forms. The transition metals combined with chalcogens form TMCs (Transition-metal chalcogenides) and TMDCs (Transition-metal dichalcogenides) exhibiting stupendous properties. These monolayers exist in many phases such as 1T, 1H, 2T, 3R etc., that shows varied range of applications [1]. Progressive researches in alkali-metals with chalcogens has also been carried out by the researchers [2–4]. Interestingly, these monolayer structures have defined structural and electronic features that paves the way to further understand their uses. The lithium and sodium chalcogenide single layers (A_2X, A = Li, Na and X = S, Se, Te) are direct band gap materials with applications in opto-mechanical resonators, optoelectronic and electronic devices. The Na_2S monolayer has both 1T and 1H forms the latter shows semiconducting properties and has same stoichiometry alike TMDs [5, 6]. Potassium chalcogenide single layers are indirect band gap materials, when K_2S is interfaced with Cs_2S the K_2S/Cs_2S interfaced monolayer displays its potentiality in solar cell applications. In recent study, the Rb_2S and Rb_2Se single-standing layers have been predicted and they possess intriguing characteristics. All these monolayers formed by alkali-metals, and chalcogens are dynamically, thermally stable with positive phonon frequencies and no structural distortion at 300 K validated via molecular dynamics study [7].

P. Joseph D. et al. (Eds.): IEMDST 2024, SPPHY 567, pp. 250–255, 2025.
https://doi.org/10.1007/978-981-96-5863-3_32

To examine the 1H-Rb_2Te monolayer through the theoretical analysis this study has been performed. The structural, electronic, dynamical and optical properties have been keenly computed. The prediction, existence and stability of the hexagonal Rb_2Te monolayer aids in confirming its applications in optoelectronic devices.

2 Materials and Methods

Amidst the firm literature of two-dimensional materials, the single-layers formed by I_2 -VI elements are promptly being predicted and are well-reported. As of our knowledge there is no previous data for hexagonal Rb_2Te monolayer.

To compute the ground-state parametric features of this compound the WIEN2k computational software based on density functional theory (DFT) was utilized [8]. The PBE-GGA exchange correlation was employed. Single layer formed by Rb and Te doesn't require the vdWH corrections, as they are only used for bi- and tri-layers. The initial parameters such as energy cutoff of −9 Ry, $R_{MT}K_{max}$ of 7 and l_{max} of 10 were set to obtain the minimum stable ground state energy or point of convergence of self-consistent field (SCF). The Monkhorst-scheme was deployed for the k-point convergence of 12 × 12 × 1. Further calculations of electronic band structure, optical and thermal attributes were computed via spaghetti and OPTIC codes interfaced with the software itself. The PHONOPY code in WIEN2k aided in analysing the stability of the monolayer in its optimized state.

3 Results and Discussion

1H-phase of Rb_2Te forms the hexagonal lattice in the space group 187. The atomic positions were optimized via mini positions scheme in WIEN2k. The optimized atomic lattice constants were obtained from the Fig. 1 in which the geometry optimization of total energy variation with respect to the lattice constants is shown. Figure 2(a) and (b) illustrates the structural phases of 1H-Rb_2Te monolayer in which a unit cell comprises of two Rb atoms adjoined with one Te atom forming a hexagon ring that is similar to

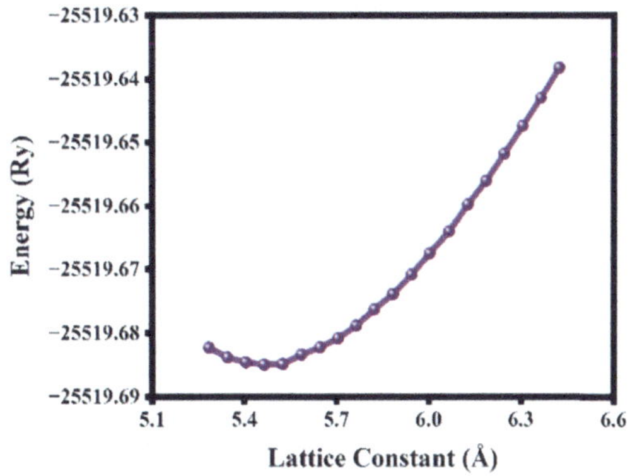

Fig. 1. Depiction of the geometric optimization plot of hexagonal Rb_2Te monolayer.

1H-Na_2S. The optimized lattice constants are a = b = 5.46 Å and the vacuum was set as 15 Å.

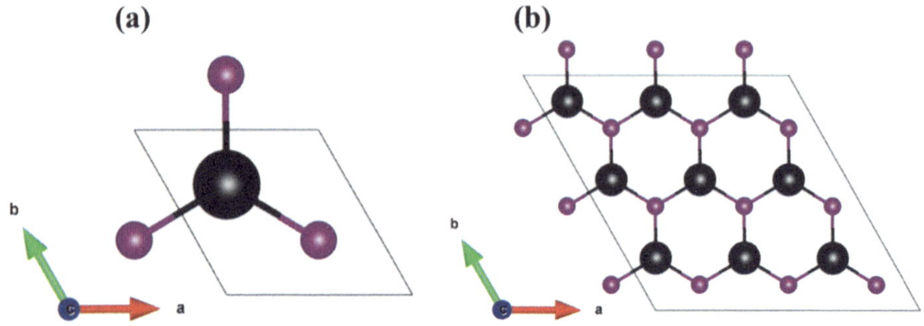

Fig. 2. The 1H structure of Rb_2Te monolayer (a) unit cell, (b) 3 × 3 × 1 supercell.

The electronic band structure plot is represented in Fig. 3. The high symmetry wave vectors are chosen in the direction of $\Gamma - M - K - \Gamma$ in which Γ is highly symmetrical as many of the energy waves in the conduction and valence bands form curves or dips in the Γ point. The negative energies correspond to the valence levels and the positive energies are the conduction band levels plotted from −2 eV to 6.5 eV. There is a deep energy band curve in the Γ point of conduction band (CB) near the fermi level that is the minima of the CB known as the conduction band minimum (CBM) point. The valence band maximum (VBM) is also situated in the Γ point making the 1H-Rb_2Te monolayer a direct band semiconductor. The electronic band gap of the single layer is 1.03 eV computed via PBE-GGA. There are some flat bands present in the VB near the fermi level that implies there is zero energy dissipation.

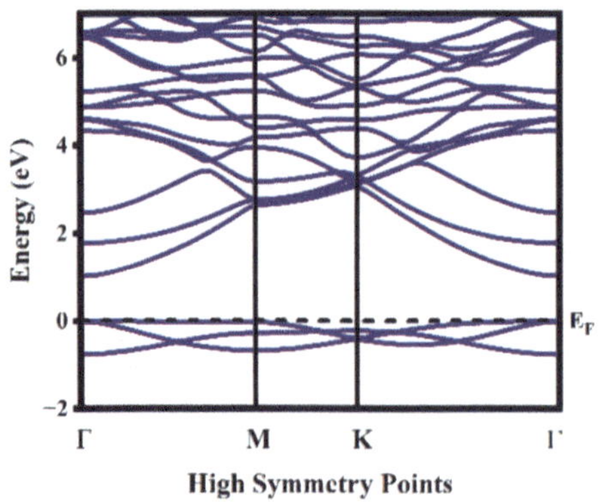

Fig. 3. High symmetry points versus energy (eV) plot of electronic band structure of 1H-Rb_2Te monolayer computed via PBE-GGA.

The phonon dispersion confirms the dynamics of the monolayer through the vibrations in the medium. Figure 4 depicts the phonon dispersion spectra of the 1H-Rb_2Te

monolayer. The wave vector path was chosen as $\Gamma - M - K - \Gamma$ and the vibrations are represented in terms of terahertz. The negative frequency modes denotes that the monolayer in dynamically unstable at this particular lattice constant point. The dynamical stability can be achieved through phase transitions.

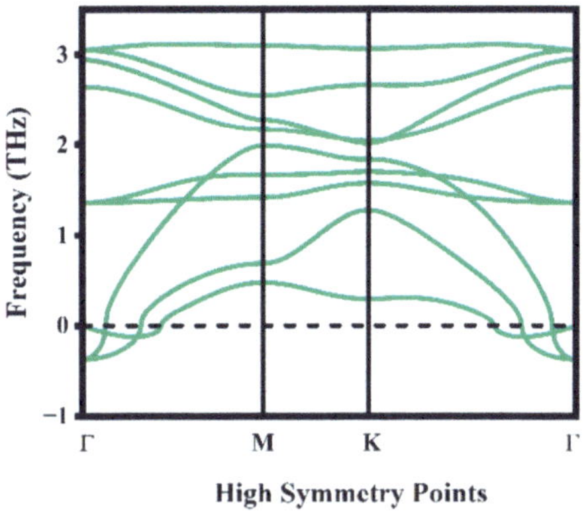

Fig. 4. The dynamical stability plot computed via PHONOPY.

The further investigation of the Rb was carried out by computing the optical responsiveness of the monolayer to the electromagnetic spectrum effects. The dielectric constants defined by the equation,

$$\varepsilon(\omega) = \varepsilon_1(\omega) + i\varepsilon_2(\omega) \tag{1}$$

These dielectric functions are interlinked to the absorptiveness and reflectivity of the material. The $\varepsilon_1(\omega)$ is the real dielectric function that takes the positive values and has highest peaks in infrared, visible and ultraviolet regions shown in Fig. 5(a). The energy stored in a medium can be inferred from this real part. The plasma frequency states whether a material displays semiconducting or metallic characteristics over the spectrum. There is no shift in $\varepsilon_1(\omega)$ from positive to negative hence the plasma frequency for the monolayer is zero. The static real dielectric function $\varepsilon_1(0)$ value is 2.40. The imaginary dielectric function $\varepsilon_2(\omega)$ is linked to the energy loss function shown in Fig. 5(b). The excitation peaks are ranging infrared to ultraviolet. The maximum excitation is seen in the ultraviolet zone with the value of 2.55 at 4.06 eV and the $\varepsilon_2(0)$ is 0.02. Figure 5(c) depicts the absorption spectra versus the energy of the electromagnetic radiation in eV. The incident energy losses and the absorption coefficients of a crystalline material are linked to the $\varepsilon_2(0)$. The maximum absorption (α) is seen in the ultraviolet range at 5.34 eV with the absorption coefficient of $44.10 \times 10^4 \text{cm}^{-1}$. The initial peaks also span in visible range that makes the monolayer a considerate material for solar cell applications. With good light absorbing characteristics in UV regime, the monolayer can also be considered for UV applications. Reflectance (R) of the monolayer in Fig. 5(d) has the static value R(0) of 0.04 and increases to 0.13 at 5.34 eV which implies that the monolayer absorbs 4%–13% of the incident light and the rest of the amount is transmitted through the medium of the material. There is very less reflectivity in the infrared and visible regions

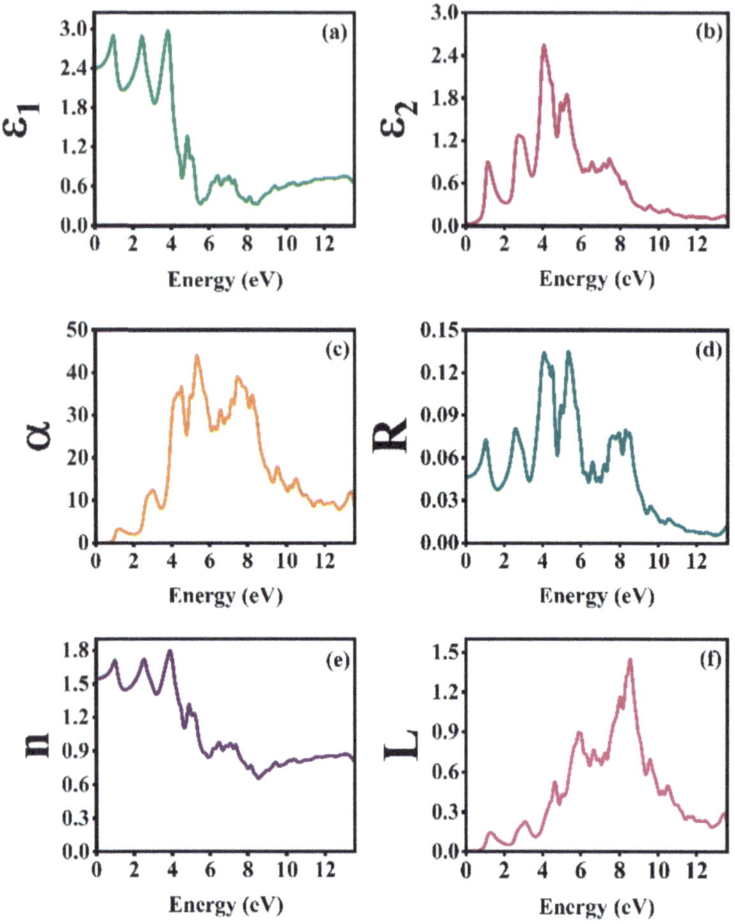

Fig. 5. The optical properties which include (a) real dielectric function (b) imaginary dielectric function (c) absorption coefficients, (d) reflectance, (e) refractive index and (f) optical energy loss.

of about 4%–7% that makes the monolayer a suitable material for anti-wave reflectors. When the density of states is high, the refractive index (n) at those points is high. The n(0) is 1.55 and gradually reduces in the UV region which is majorly due to the ions in the monolayer that can be seen in Fig. 5(e). The energy loss plot in Fig. 5(f) implies that the energy losses are minimum when the absorption is maximum. This is validated as it can be seen that the loss is in peak at the UV spectral range above 8 eV and is minimal in the near-UV region. On referring to these optical features, the hexagonal rubidium telluride monolayer can be a useful material for optoelectronic applications.

4 Conclusion

In conclusion, the theoretical analysis of the hexagonal Rb_2Te monolayer has been carried out via the DFT approach. It is well observed that the 1H (single layer, hexagonal) Rb_2Te has optimized lattice constants and relative bond lengths. The electronic band gap between VBM and CBM is 1.03 eV with direct band. Additionally, the vibrational stability was validated on computing the phonon dispersion plot. The negative phonon frequencies aided in confirming the semiconducting monolayer to be dynamically unstable and might undergo phase transitions. Strong absorption is seen in the visible and UV EM spectral range, weak reflectivity in UV range and high refractive index peaks traversing over all the ranges confirms that the direct band monolayer can be a promising compound for optoelectronic applications.

References

1. Chia, X., Eng, A.Y.S., Ambrosi, A., Tan, S.M., Pumera, M.: Chem. Rev. **115**(2015), 11941–11966 (1966). https://doi.org/10.1021/acs.chemrev.5b00287
2. Miró, P., Audiffred, M., Heine, T.: Chem. Soc. Rev. **43**, 6537–6554 (2014). https://doi.org/10.1039/C4CS00102H
3. Naseri, M., Lin, S.: Chem. Phys. Lett. **722**, 58–63 (2019). https://doi.org/10.1016/j.cplett.2019.02.047
4. Tan, C., Lai, Z., Zhang, H.: Adv. Mater. **29**, 1701392 (2017). https://doi.org/10.1002/adma.201701392
5. Bui, P.T., Nguyen, D.K., Guerrero-Sanchez, J., Hoat, D.M.: Appl. Surf. Sci. **606**, 154809 (2022). https://doi.org/10.1016/j.apsusc.2022.154809
6. Nguyen, D.K., Guerrero-Sanchez, J., Hoat, D.M.: Phys. Chem. Chem. Phys. **24**, 27505–27514 (2022). https://doi.org/10.1039/D2CP04613J
7. Rawat, A., Arora, A., De Sarkar, A.: Appl. Surf. Sci. **563**, 150304 (2021). https://doi.org/10.1016/j.apsusc.2021.150304
8. Blaha, P., Schwarz, K., Madsen, G.K., Kvasnicka, D., Luitz, J.: WIEN2k. An augmented plane wave+ local orbitals program for calculating crystal properties **60** (2001)

Dynamics of Passive and Self-propelled Tracer Particles in Ordered Crowded Environments

Baburao Simma, Radha Ramana, Vrinda Garg, and Surya K. Ghosh(✉)

Department of Physics, National Institute of Technology, Warangal, Warangal, Telangana 506004, India
skghosh@nitw.ac.in

Abstract. Crowded cellular environment plays a vital role in governing the dynamics of tracer particles. In this study, we use extensive Langevin dynamics simulations to explore the anomalous diffusion behaviour of passive and self-propelled particles in ordered crowded environments. The crowded environment is created with an array of immobile atoms arranged periodically in space, designed to mimic the crowded conditions of cellular environments. Our findings reveal that, as the packing fraction and binding strength increase, the motion of passive probes becomes more restricted, with dynamics becoming increasingly non-Gaussian and sub-diffusive. From the analysis of mean squared displacement and diffusion constant, we uncover a fascinating non-monotonic behaviour as we increase the packing fraction by changing the crowder size for both passive and self-propelled particles. Finally, we examine the intermittent anomalous diffusion by analysing the time-averaged mean squared displacement. This study has broad implications for understanding probe dynamics in complex biological environments.

1 Introduction

Diffusion of macromolecules plays a crucial role in various biological processes, such as drug delivery, protein-protein interactions, and the self-assembly of supramolecular structures [1, 2]. Recent research has explored the impact of macromolecular crowding (MMC) on the diffusive behaviour of nanoscale tracers. A more in-depth investigation is required, given that intracellular spaces are densely populated with crowding agents like proteins and ribosomes, which occupy up to 30–40% of the total space [3]. The dynamics in such crowded environments often exhibit anomalous behaviour, which can be characterized by the mean squared displacement (MSD) $r^2(t) = Dt^\beta$, where D is the diffusion coefficient and β is the scaling exponent which determine the nature of diffusion. For example, with $\beta < 1$, sub-diffusive behaviour includes tracer particle movement in cytoplasm, insulin granule transport in intracellular spaces, and motion in mucus membranes. In systems influenced by external or internal energy, where directed motion is observed, the fluctuation-dissipation theorem is violated. In these cases, $\beta > 1$ indicates super-diffusion, as seen in systems such as molecular motors, passive particles in active baths, and energy-driven catalytic enzymes [4, 5].

Recent studies focused on the enhanced diffusion of self-propelled rod-like particles in obstructed environments [6, 7]. Specifically, they investigated how random crowders

P. Joseph D. et al. (Eds.): IEMDST 2024, SPPHY 567, pp. 256–264, 2025.
https://doi.org/10.1007/978-981-96-5863-3_33

affect the dynamics of active rods. However, how the competition between activity and attractive centres alter the dynamics of self-propelled particles in ordered, homogeneous crowders is yet to be understood. In this study, we model a system that accounts for the transient binding affinity of tracer particles to physical obstacles. Using Langevin dynamics simulations, we analyse the diffusion of both passive and self-propelled tracers in a 3D cubic lattice formation of non-inert spherical particles with adjustable sizes (σ_c). This obstruction-and-binding diffusion model allows us to investigate the ensemble-averaged mean squared displacement, effective tracer diffusivity (D), and ergodic behaviour under varying conditions. Key model parameters, such as crowder size (σ_c), tracer-crowder binding affinity (ε_a), and degree of self-propulsion (Pe), are systematically varied to uncover the underlying physical mechanisms. Figure 1 provides a schematic representation of the system, along with the trajectories of both passive and self-propelled tracers. From a broader perspective, this model offers insight into the behaviour of tracer particles in a homogeneous complex media.

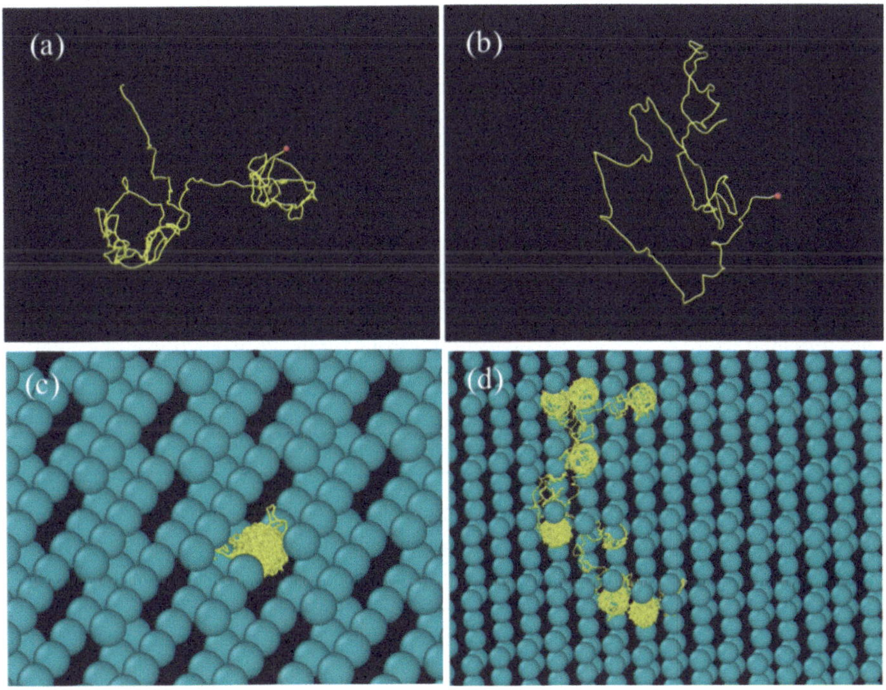

Fig. 1. Schematic representation of passive (a) and self-propelled (b) particle trajectories in free space and the corresponding trajectories of passive and self-propelled particles in an environment with stickiness of strength ($\varepsilon_a = 10$) at fixed packing fraction (Φ) are shown in (c) and (d) respectively.

2 Materials and Methods

We model the crowded environment by arranging the immobile spherical obstacles in a cubic lattice formation to study the dynamical behaviour of a point-size tracer particle. The distance between the centre of two neighbouring obstacles is the same as the lattice constant L. Naturally, the maximum size of a crowder can be $\sigma_{max} = L/2$ and the corresponding packing fraction is $\varphi_{max} = 0.52$. By varying the size of each obstacle, we can vary the packing fraction of the medium. For the sake of simplicity, we keep the size of all the obstacles the same to maintain the spatial homogeneity of the system. In addition, these obstacles are non-inter in nature; they can attract the tracer if they are in proximity. The crowder particles are fixed in our simulations and impose periodic boundary conditions in all directions. In our simulation, we release the point tracer particle at the centre of the lattice to ensure that it has the same distance from the surrounding neighbours. LAMMPS software has been used to carry out the simulations, the Lennard-Jones parameters σ, ϵ, and m are the fundamental units of length, energy, and mass, respectively. All other physical quantities are defined based on these fundamental units of σ, ε, and m and are presented in a non-dimensional form.

The attractive interaction between the mobile tracer particle and obstacles is implemented by full Lennard-Jones (LJ) potential as

$$V(r) = \begin{cases} 4\varepsilon_{ij}\left[\left(\frac{\sigma_{ij}}{r_{ij}}\right)^{12} - \left(\frac{\sigma_{ij}}{r_{ij}}\right)^{6}\right], & if\ r_{ij} \le r_{ij}^{cut} \\ 0, & otherwise. \end{cases} \quad (1)$$

Here r_{ij} is the distance between the ith and j^{th} particle, and ε_{ij} is the interaction strength between two particles of interaction diameter $\sigma_{ij} = (\sigma_i + \sigma_j)/2$. The attraction between beads of the crowded atoms and tracer particle is implemented by attractive Lennard-Jones potential. Here $r_{ij}^{cut} = 2.5\sigma_{ij}$ is the cutoff radius for the crowder-tracer pair interaction. The system is equilibrated for 2×10^5 steps; the simulations are carried out for 10^7 time steps, and the timestep $\Delta t = 0.001$ is used. All the simulations are implemented using the Langevin thermostat and equation of motion integrated using the velocity verlet algorithm in each time step.

We implement the following underdamped Langevin equation to simulate the dynamics of the ith particle with mass m of position $r_i(t)$ at time t [8]:

$$m\frac{d^2 r_i(t)}{dt^2} = -\gamma\frac{dr_i}{dt} - \sum\nolimits_j \nabla V\left(r_i - r_j\right) + f_i(t) + F_a\hat{n} \quad (2)$$

Here, V(r) is the interaction potential between a pair of atoms i and j. We have implemented two cases: passive and self-propelled particles. The former case was implemented by $Pe = 0$ and later $Pe \neq 0$ respectively, here $Pe = F_a\sigma^2/K_BT$ is the dimensional quantity to measure the propulsion where F_a is the self-propulsion force. We consider a low friction coefficient; as a result, the dynamics are underdamped. The Thermal fluctuation is considered by random force fi(t) which follows the fluctuation-dissipation theorem, i.e., $\langle f_i(t)\rangle = 0$ and., $\left\langle f_i(t).f_j(t)\right\rangle = 6\gamma K_BT\delta_{\alpha\beta}\delta(t_1 - t_2)$. Where K_B is the Boltzmann constant, T is the temperature, δ represents the Dirac delta function, with α and β as the cartesian components. Temperature is constant in our simulations and hydrodynamic interactions have been neglected for simplicity.

3 Results and Discussion

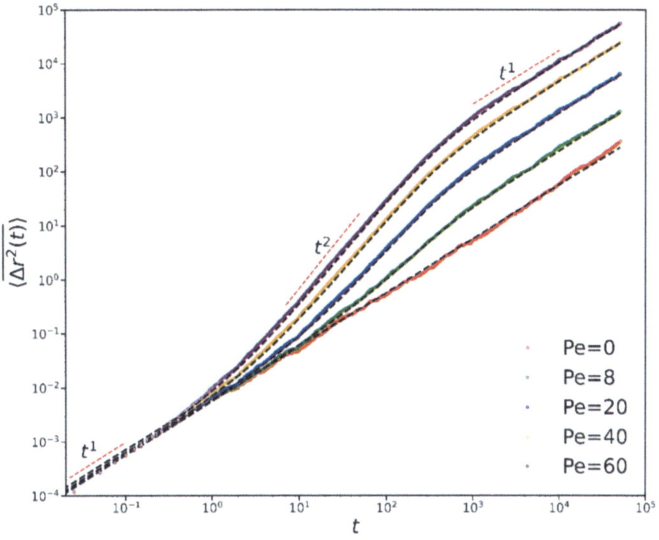

Fig. 2. Ornstein-Uhlenbeck solution for self-propelled particles (dotted lines) shows an excellent agreement with the MSD obtained from our simulation (solid lines) for a self-propelled particle in free space across various Péclet number (Pe) values.

For validation of our model of a self-propelled particle (SPP), we first simulate an SPP in free space. Then, we calculate the mean squared displacement (MSD) of the particle for various propulsion strengths (Pe). In free space, the analytical expression for the active Ornstein-Uhlenbeck (OU) process is [9]

$$\overline{\langle \Delta r^2(t) \rangle} = 6D_t t + \frac{2F_a^2 t_R}{\gamma^2}\left[t + t_R\left(e^{-\frac{t}{t_R}} - 1\right)\right]. \tag{3}$$

For a passive particle in free space (i.e., with Pe $= 0$), the MSD is always diffusive $\Delta r^2(\tau) \sim t^1$. For the self-propelled tracer particle, the MSD shows a three-step variation. At the shorter time scale, the MSD shows diffusive behaviour ($\sim t^1$); at the intermediate time scale, it is superdiffusive ($\sim t^2$) and finally, at the longer time scale, it becomes diffusive again ($\sim t^1$). To verify our simulation, in Fig. 2, we reproduced the solution of the Ornstein-Uhlenbeck (OU) process. Our simulation data perfectly match the analytical expression of OU solution for self-propelled particles for different Pe (Fig. 2).

4 Mean Squared Displacement

To get insights into the underlying dynamics of passive and active Brownian particles, at first, we calculate the Mean Squared Displacement (MSD) $\overline{\Delta r^2(\mathrm{t})} = \langle |\mathbf{r}_\mathrm{i}(\mathrm{t}) - \mathbf{r}_\mathrm{i}(0)|^2 \rangle$, of

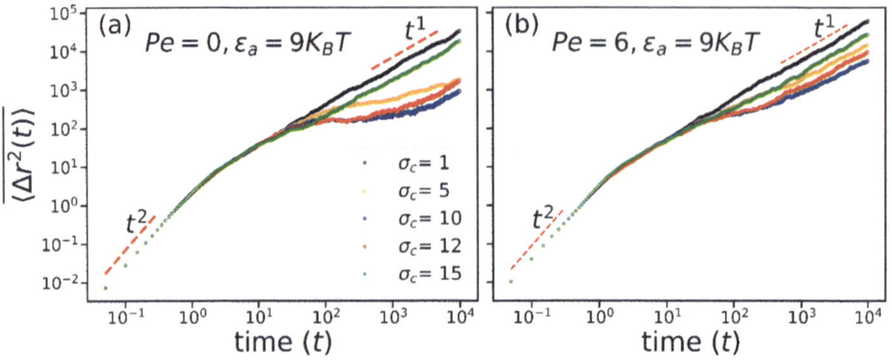

Fig. 3. The Mean squared displacement of tracer particle of (a) passive (b) self-propelled particle at constant stickiness ($\varepsilon_a = 9\mathrm{K_BT}$) at different packing fractions (Φ).

the tracer particle [10]. For the ensemble average, we have taken averages of individual trajectories $\langle\overline{\Delta r^2(\mathrm{t})}\rangle = \frac{1}{\mathrm{N}}\sum_{\mathrm{i}=1}^{\mathrm{N}} \overline{\Delta r^2(\mathrm{t})}$. In normal diffusive case, MSD varies linearly with time, but in the case of anomalous diffusion, MSD varies $\langle\overline{\Delta r^2(\mathrm{t})}\rangle = \mathrm{t}^{\beta}$ with $\beta < 1$ for Sub diffusive and or $\beta > 1$ for superdiffusive.

To examine the effect of crowding on the dynamic of a passive tracer particle, we study the MSD for different packing fractions (Φ) at a fixed strength of stickiness ($\varepsilon_a = 9K_BT$), as shown in Fig. 3(a). As expected, we noticed that as the crowder size increased, tracer particles trapped at the surface of the sticky crowders exhibited sub diffusive behaviour at the lower and intermediate packing fractions. However, the fixed crowder's attractive potential surface starts to overlap at the higher packing fraction, creating a valley of semi-unipotential surfaces. As a result, the passive particle can again move from one attractive surface to another due to their overlap, contributing to a higher value of MSD with respect to the MSD of relatively lower value of packing fractions. This result implies that there is an optimal crowding density just before the sticky surface starts to overlap, contributing most effectively to restricting the particle motion. Increasing the packing fraction further, counterintuitively, contributes to higher MSD for a long time.

The analysis so far highlights the interplay between the thermal agitation of a passive particle and the attractive forces of sticky crowders. Figure 3(b) shows the variation of MSD at various crowder sizes at fixed attraction strength ($\varepsilon_a = 9K_BT$) and fixed propulsion ($\mathrm{Pe} = 6$). Introducing self-propulsion adds an intriguing layer to this dynamic, particularly in how it affects the particle's ability to navigate through an environment with varying packing fractions of fixed, sticky crowders. In the case of self-propulsion with a fixed propulsion strength, we observe that the tracer particle gains enhanced mobility, allowing it to escape from one lattice site to another more effectively. This increased mobility is reflected in the mean squared displacement (MSD) for different packing fractions, demonstrating rapid growth at a constant attraction strength (Fig. 3(b)). Self-propulsion enables the particle to overcome the attractive interactions and obstacles presented by the crowders.

However, as the packing fraction increases, we also note higher MSDs emerge in the particle's dynamics, likely due to the increased overlap among crowders. At higher packing fractions, the crowding effects become pronounced, which complicates the particle's movement despite its self-propulsive capabilities. Interestingly, as self-propulsion is further intensified, the trapping behaviour diminishes, leading the tracer particle to behave more like a free particle. This transition indicates that at sufficiently high self-propulsion strengths, the influence of attractive interactions becomes negligible, allowing the particle to navigate through the crowded environment with less hindrance. Overall, the balance between self-propulsion, attraction strength, and packing fraction plays a crucial role in dictating the dynamics of the tracer particle in this complex system.

5 Diffusion Constant

The Mean Squared Displacement (MSD) profile is a fundamental tool for analysing diffusion dynamics. With the aid of profiles of MSDs, we can estimate the normalized diffusion coefficient at the longer time scales, according to $\mathrm{D} = \frac{1}{6}\lim\limits_{t\to\infty}\frac{\mathrm{MSD(t)}}{\mathrm{t}}$ [11]. By estimating the normalized diffusion coefficient $\mathrm{D/D_0}$ at longer time scales, you can examine how different model parameters, such as σ_c, ε_a and the Péclet number Pe, influence the system's behaviour.

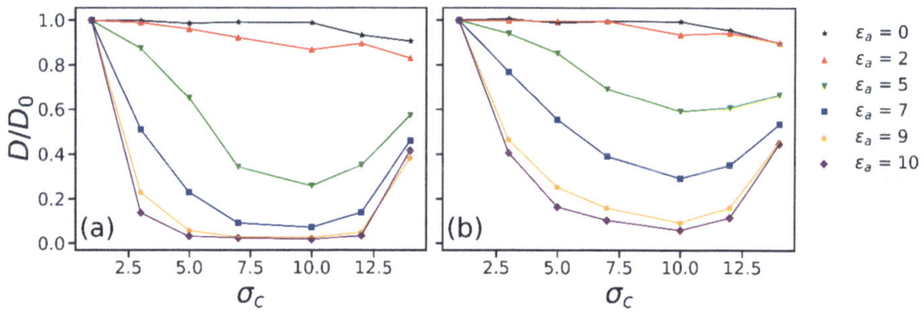

Fig. 4. Relative diffusivity of passive tracer particle at (a) passive (b) self-propelled particle as a function of crowder size (σ_c) at different stickiness (ε_a).

We plot $\mathrm{D/D_0}$ of passive tracer particles as a function of crowder size at different binding strengths in Fig. 4(a), where $\mathrm{D_0}$ is the diffusivity of the tracer particle in free space. It is clear from the figure that the diffusivity shows counterintuitive non-monotonic behaviour, which can be attributed to the anomalous behaviour of tracer particles due to the crowdedness. This behaviour is due to the creation of the overlapping potential valley where the particle can move effectively, even in the presence of large attractive crowders.

The effect of self-propulsion on the long-time diffusivity is shown in Fig. 4(b). For a fixed propulsion strength, (Pe = 6) we studied the variation of $\mathrm{D/D_0}$ as a function of crowder size and attraction strengths (ε_a). It is observed that, as in the case of passive particles, the diffusivity of the self-propelled particle also shows non-monotonic behaviour because the moderate self-propulsion is not enough to make the particle escape from the lattice sites. However, at higher propulsion strengths, the nonmonotonic nature vanishes (the figure is not shown).

6 Time Averaged MSD

We compute the time-averaged mean squared displacement (TAMSD) to unravel any non-ergodic behaviour of the system due to the presence of attractive centres.

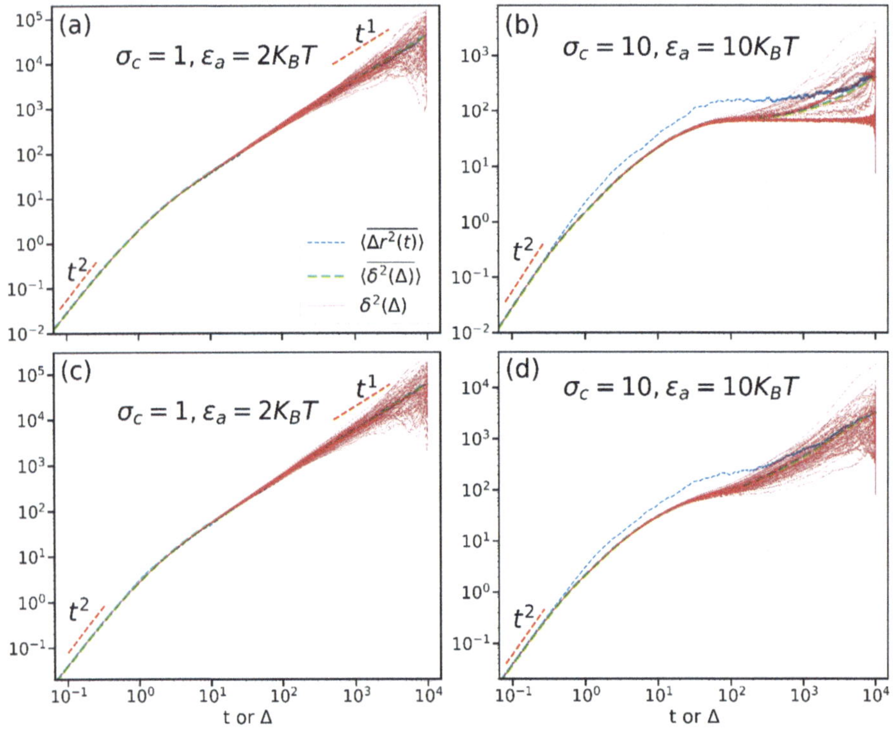

Fig. 5. The TAMSD of Passive probe depicted (top row) with, $\varepsilon_a = 2\mathrm{K_BT}$, $\sigma_c = 1$ (left) and $\varepsilon_a = 10\mathrm{K_BT}$, $\sigma_c = 10$ (right). Self-propelled particle of Pe = 6 depicted (bottom row) with, $\varepsilon_a = 2\mathrm{K_BT}$, $\sigma_c = 1$ (left) and, $\varepsilon_a = 10\mathrm{K_BT}$, $\sigma_c = 10$ (right).

The TAMSD can be defined as $\overline{\delta r^2(\Delta)} = \frac{1}{\mathrm{T_{max}}-\Delta}\int_0^{\mathrm{T_{max}}-\Delta}[\mathbf{r}(\mathrm{t}+\Delta)-\mathbf{r}(\mathrm{t})]^2\mathrm{dt}$, from the individual trajectory $\mathbf{r}(\mathrm{t})$ [12, 13]. Here, the length of the trajectory T_{max}, which is also referred to as the total simulation time, Δ refers to the lag time, which defines the size of the time-window slides along the trajectory $\mathbf{r}(\mathrm{t})$. Further, we average the TAMSDs over the available ensembles N, defined as $\left\langle\overline{\delta r^2(\Delta)}\right\rangle = \frac{1}{N}\sum_{i=1}^{N}\overline{\delta r^2(\Delta)}$.

We have plotted the individual TASMD ($\overline{\delta r^2(\Delta)}$), MSD ($\left\langle\overline{\Delta r^2(\tau)}\right\rangle$), as well as the ensemble average of the individual TAMSD ($\left\langle\overline{\delta r^2(\Delta)}\right\rangle$) in Fig. 5. We initially compare the above three statistical properties for passive tracer, with Pe = 0, $\varepsilon_a = 2\mathrm{K_BT}$ a lower crowder size of $\sigma_c = 1$ in Fig. 5(a). We can observe ergodic behaviour at lower stickiness and crowder size as TAMSD and ensemble-averaged MSD match perfectly. As we further increase the attraction strength ($\varepsilon_a = 10\mathrm{K_BT}$) and crowder size ($\sigma_c = 10$), in Fig. 5(b), the dynamics remain ergodic at the shorter time scale, and MSD and TAMSD deviate from each other at the intermediate time scale, leading to non-ergodic phenomena. Finally, in the diffusive limit, at the longer time scale, it restores to ergodicity.

Subsequently, we show the comparative study of the above three properties for self-propelled tracer particles at propulsion strength (Pe = 6) in Fig. 5(c)–(d). As observed in the passive tracer particles, the self-propelled tracer particle also shows ergodic behaviour at the lower attraction strength $2K_BT$ and lower crowder size $\sigma_c = 1$ (Fig. 5(c)). For higher attraction strength ($\varepsilon_a = 10K_BT$) and the crowder size ($\sigma_c = 10$), unlike passive, the deviation of MSD from MSD decreased at the intermediate times. With an additional increase in propulsion strengths, the trapping due to sticky crowders vanishes; as a result, the dynamics are ergodic for the complete time scale (the figure not shown here). Hence, we can infer that self-propulsion revives ergodicity even in the complex media.

7 Conclusion

In the current work, we employ a three-dimensional Langevin dynamics simulation to explore the behaviour of both passive and self-propelled tracer particles in a crowded, immobile lattice environment. The simulation is performed on a simple cubic lattice with varying crowder sizes, ensuring both homogeneity and a sufficiently complex medium to capture the essential features of crowding effects. We focus on the impact of three critical factors—crowdedness, stickiness, and self-propulsion—on the tracer particle dynamics. As expected, we observe that the effective diffusion coefficient of the tracer particle decreases as the degree of crowding and stickiness increases. Interestingly, this relationship is non-monotonic when crowd sizes and stickiness reach higher values, indicating complex interactions between the tracer particle and the environment for both passive and self-propelled particles. At intermediate time scales, the dynamics of the tracer particles become non-ergodic due to trapping by immobile crowders. This non-ergodicity persists for passive particles and self-propelled particles with moderate propulsion strength. However, when propulsion strength is sufficiently high, the tracer particles overcome trapping effects, leading to ergodic motion. Understanding the transport properties of tracer particles, particularly in crowded and sticky environments, is crucial for a wide range of applications, from material science to biological systems. The microstructure and interaction forces within the medium highly influence both passive and active particle dynamics, and our work highlights the need for further investigation into these complex transport processes.

Acknowledgements. SKG would like to thank MATRICS, SERB, DST (Project No. MTR/2022/000833) for financial support. We want to acknowledge NIT Warangal for the research support.

References

1. Kim, Y., Joo, S., Kim, W.K., Jeon, J.H.: Macromolecules **55**(16), 7136–7147 (2022). https://doi.org/10.1021/acs.macromol.2c00610
2. Maitra Roy, S., et al.: ACS Appl. Nano Mater. **6**(20), 18988–18998 (2023). https://doi.org/10.1021/acsanm.3c03481

3. Ghosh, S.K., Cherstvy, A.G., Metzler, R.: Phys. Chem. Chem. Phys. **17**(3), 1847–1858 (2015). https://doi.org/10.1039/C4CP03599B
4. Yadav, R.S., Das, C., Chakrabarti, R.: Soft Matter **19**(4), 689–700 (2023). https://doi.org/10.1039/d2sm01626e
5. Castañeda-Priego, R., Pica Ciamarra, M., Corezzi, S., Ghosh, S.K., Liu, Y.: Front. Phys. **12**, 1388370 (2024). https://doi.org/10.3389/fphy.2024.1388370
6. Khalilian, H., Fazli, H.: J. Chem. Phys. **145**(16), 164909 (2016). https://doi.org/10.1063/1.4966188
7. Ghosh, S.K., Cherstvy, A.G., Petrov, E.P., Metzler, R.: Soft Matter **12**(38), 7908–7919 (2016). https://doi.org/10.1039/C6SM01522K
8. Garg, V., Mathew, R., Ibrahim, R., Singh, K., Ghosh, S.K.: iScience **27**(4), 109348 (2024). https://doi.org/10.1016/j.isci.2024.109348
9. Bechinger, C., Di Leonardo, R., Löwen, H., Reichhardt, C., Volpe, G., Volpe, G.: Rev. Mod. Phys. **88**(4), 045006 (2016). https://doi.org/10.1103/RevModPhys.88.045006
10. Cherstvy, A.G., Thapa, S., Wagner, C.E., Metzler, R.: Soft Matter **15**(12), 2526–2551 (2019). https://doi.org/10.1039/C8SM02096E
11. Chen, Y., Yan, R., Zhao, N.: Phys. Chem. Chem. Phys. **24**(12), 7415–7429 (2022). https://doi.org/10.1039/d2cp00323f
12. Ghosh, S.K., Cherstvy, A.G., Grebenkov, D.S., Metzler, R.: New J. Phys. **18**(1), 013027 (2016). https://doi.org/10.1088/1367-2630/18/1/013027
13. Joo, S., Durang, X., Lee, O.C., Jeon, J.H.: Soft Matter **16**(40), 9188–9201 (2020). https://doi.org/10.1039/D0SM01200A

Exploration of New Zero Bandgap 2D Material: Theoretical Insight

S. Chellaiya Thomas Rueshwin and R. D. Eithiraj(✉)

Department of Physics, School of Advanced Sciences, Vellore Institute of Technology (VIT), Chennai, Tamil Nadu 600127, India
eithiraj.rd@vit.ac.in

Abstract. The 2D materials manifest prominent physical properties compared to their bulk counterparts due to dimensional confinement effect along one direction, which in turn allures researchers towards these materials. Many researchers have been working extensively, both theoretically and experimentally, on 2D materials. In this investigation, the 1H phase of the Na_2O monolayer is computationally studied using Density Functional Theory (DFT). WIEN2K software is employed to examine the structural and electronic properties of the 1H phase of the Na_2O monolayer. The 1H-Na_2O monolayer crystallizes in the space group 187 with a lattice parameter, $a = b = 3.54$ Å and $c = 15$ Å. The exchange-correlation functional used for this analysis is PBE-GGA. The computational study reveals that the 1H phase of the Na_2O monolayer is zero bandgap in nature. The optical properties of 1H-Na_2O were investigated using the OPTIC code in WIEN2k, the peak value of the absorption coefficient was found to be 21.22×10^4 (1/cm) at 9.46 eV.

1 Introduction

Since the advent of graphene [1], 2D materials have attracted researchers due to their unique properties compared to their bulk counterparts. The unique properties of 2D materials are due to the confinement effect along a certain direction. These 2D materials show tunable electrical, optical, thermal and mechanical properties. 2D materials are utilized for many applications such as energy storage, electronics, optoelectronics, spintronics, and other applications. Extensive research is being conducted on 2D materials both experimentally and theoretically. The synthesis and investigation of the low-dimension materials are facilitated by the developments in the sophisticated experimental facilities. Theoretical studies help in the enhanced understanding of the synthesized or predicted lower dimension materials. Many 2D materials, including graphene, CdX (X = Se, Te), transition metal dichalcogenides (TMDs), and MXenes, have exhibited superior thermoelectric performance [1–6]. In the field of developing photocathodes, alkali metal oxides play a crucial role, as they facilitate catalytic reactions and enhance the oxidation of semiconductor surfaces. As a consequence, they show significant promise for applications in batteries, solid-state gas sensors, and fuel cells. Numerous studies have been conducted on alkali-based metal oxides. Recently, due to their distinctive properties, alkali-based 2D metal oxides have enticed researchers. 2D monolayers of alkali oxides like Li_2O,

P. Joseph D. et al. (Eds.): IEMDST 2024, SPPHY 567, pp. 265–270, 2025.
https://doi.org/10.1007/978-981-96-5863-3_34

Na_2O, K_2O, Rb_2O and Cs_2O have been extensively studied theoretically [7–12]. Li_2O has been theoretically predicted in two structural phases, T-phase and H-phase, with both showing potential for optoelectronic applications [8]. Additionally, H-phase Na_2O and bilayer K_2O have been investigated as a 2D pseudospin-1 metal, respectively, suggesting their potential for electronic devices and sensitive mechanical sensors [13]. The electronic, optical, and thermoelectric properties of 1T-Rb_2O have been studied theoretically [11]. Further theoretical studies have investigated the synthesis of M_2X (M = Na, K, Cs; X = O, S, Se, Te) monolayers and their heterojunction bilayer configurations for excitonic and tandem solar cells [10]. These hetero-bilayers showcase beneficial characteristics, such as enhanced power conversion efficiency, enhanced carrier mobility, and substantial visible light absorption, making them promising candidate for photovoltaic applications [14]. In the progress of studying diverse 2D alkali-based oxides, 1H-Na_2O has been extensively studied in this research. In this research, a theoretical investigation of the 1H-Na_2O monolayer using GGA is carried out, assessing its electronic, and optical properties to determine its potential for optical applications.

2 Computational Details

To explore the structural, electronic, and optical properties of 1H-Na_2O, WIEN2k, a simulation software based on Density Functional Theory (DFT) was utilized. In WIEN2k, Full potential linearized plane wave (FP-LAPW) method was utilized to solve the self-consistent Kohn-Sham equations [15, 16]. The optimization of atomic positions and lattice parameters was performed using Generalized Gradient Approximation (GGA) functionals [17]. The maximum angular momentum was set to $l_{max} = 10$ and the product $R_{MT}K_{MAX} = 7$ and $G_{max} = 12$, were used. To distinguish core from valence states, a cutoff energy of −7 Ryd was applied, and a vacuum of 15 Å along the non-periodic axis was introduced to mitigate interactions among the layers [8]. For integrating special points across the Brillouin zone, the Monkhorst-pack method with E_k convergence was employed [18]. To optimize the structure, a k-point grid of 10 × 10 × 1 was employed. Subsequently, benchmarks for energy, force, and charge were defined as 0.0001 Ry, 1 mRy/au, and 0.001 eV, respectively. The computational assessment of the optical and thermal properties of 1T-Na_2O was performed using GGA exchange-correlation. The code OPTIC was used in the computational calibration of 1H-Na_2O with k-points settings of 20 × 20 × 1.

3 Structural Details

The structural properties of the 1H-Na_2O were computationally calibrated using DFT. The structural properties of the 1H-Na_2O were investigated employing GGA exchange-correlation [15, 17]. The 1H-Na_2O crystallizes in the space group P-6m2 (187) which is similar to that of materials like 1H-MoS_2, 1H-WS_2, with a central hexagonal structure [19] was shown in Fig. 1. For the ground state theoretical calculations, optimization of the positions of the atoms and the lattice parameters was essential. The optimized structure is illustrated in Fig. 1. The optimized lattice parameters were determined to be a = b = 3.54 Å and c = 15 Å. To mitigate the interaction between the layers, a vacuum was incorporated in the nonperiodic direction.

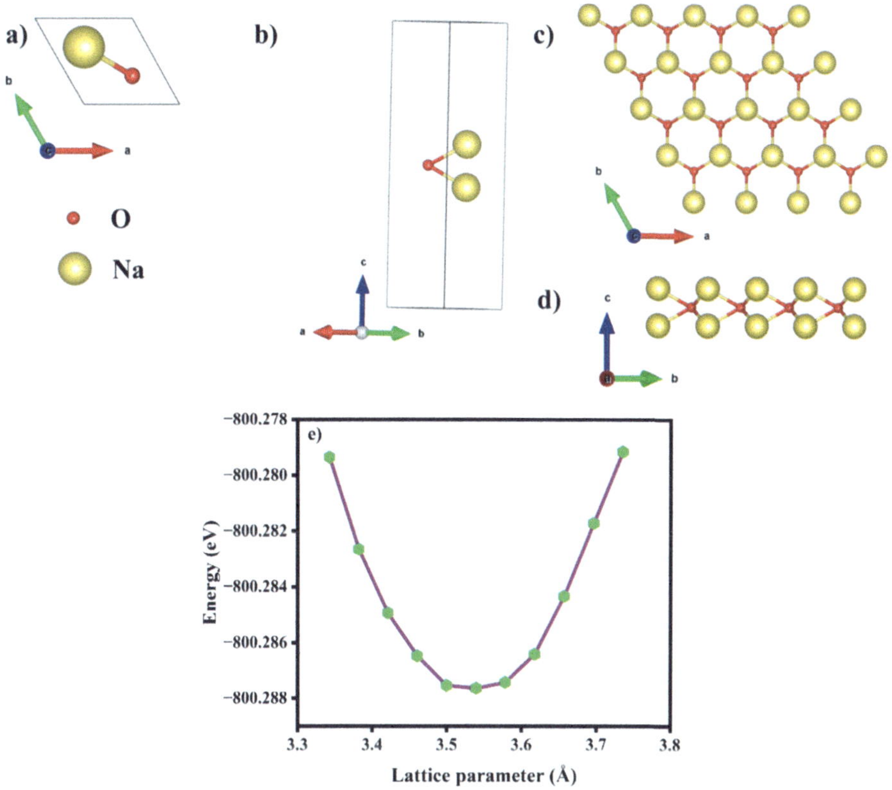

Fig. 1. Illustrates (a) top view of unit cell, (b) side view of unit cell, (c) top view of 4 × 4 × 1 supercell, (d) side view of 4 × 4 × 1 supercell and (e) Lattice parameter optimization curve, of 1H-Na_2O.

4 Electronic Properties

WIEN2k software was employed for the thorough computational investigation of the electronic properties. To assess the electronic properties, the GGA functional was utilized. Figure 2 illustrates key electronic properties, such as electronic band structure, density of states (DOS) and projected density of states (PDOS). The band gap of 1H-Na_2O, revealed through computational calibration using GGA functional, is 0 eV. The previously reported 1H-phase of Na_2O has a direct band gap of 0.25 eV, and the bulk counterpart was reported with a direct band gap of 2.42 eV [7, 9]. During the band structure assessment, the fermi energy (E_F) is adjusted to zero, presenting the conduction bands in the positive range and the valence bands in the negative range. The DOS and PDOS were calibrated for a more efficient evaluation of the electronic properties. Insights into the electron energy state distribution are provided by DOS and PDOS. GGA exchange-correlation was employed to calibrate DOS and PDOS. From the DOS calculation, it can be inferred that 1H-Na_2O has a band gap of 0 eV. It can be inferred from the PDOS calculation that the contribution of the O to the valence band is more than the contribution to the conduction band. While the contribution of Na to the conduction band is more than the valence band.

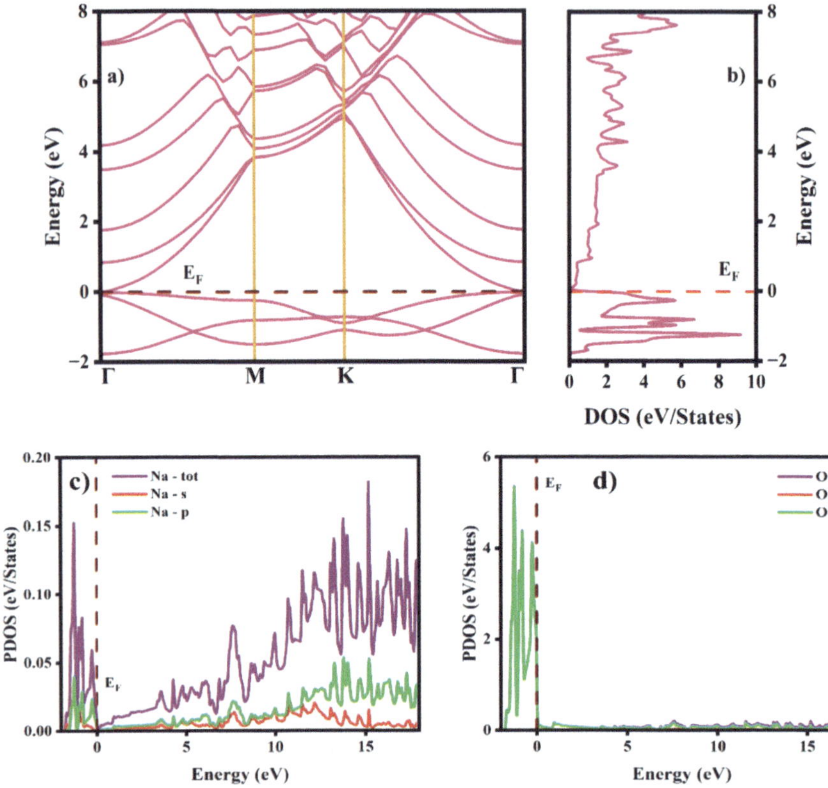

Fig. 2. Illustrates (a) Band structure of 1H-Na_2O (b) DOS of 1H-Na_2O (c) PDOS of Na (d) PDOS of O.

5 Optical Properties

Insights into the optical characteristics of a material is essential for developing optical devices. The computational analysis of the optical properties of 1H-Na_2O was calibrated using the OPTIC code in WIEN2k [16]. The optical properties such as complex dielectric function $\varepsilon(\omega)$, and absorption coefficient $\alpha(\omega)$ were calibrated. GGA functional was utilized to compute the optical properties in WIEN2k [17, 20, 21]. The complex dielectric function is the optical response to incident electromagnetic radiation of a material. The mathematical representation of the complex dielectric function is

$$\varepsilon(\omega) = \varepsilon_1(\omega) + i\varepsilon_2(\omega) \tag{1}$$

Insight into polarization effects is provided by the real part of the dielectric function (ε_1), while insights into energy dissipation or absorption is provided by the imaginary part (ε_2). At zero frequency, the calculated static real part of the dielectric function is 13.31 which the peak value $\varepsilon_1(\omega)$, which is in the visible region, and then fluctuates throughout the energy range. The $\varepsilon_2(\omega)$ show the transition between the occupied state and unoccupied state throughout the Brillouin zone with a set of k-vectors. The peak value of $\varepsilon_2(\omega)$ was observed at 0.09 eV, with a value of 9.65, which lies in the visible region. This indicates that the transition between the electron-rich states of the valence band and the electron-deficient state of the conduction band. Consequently, the value of $\varepsilon_2(\omega)$ fluctuates throughout the energy range. Absorption coefficient $\alpha(\omega)$ can be expressed algebraically as (Fig. 3)

$$\alpha(\omega) = \sqrt{2}\omega\left(\left[\varepsilon_1^2 - \varepsilon_2^2\right]^2 - \varepsilon_1\right)^{1/2} \tag{2}$$

The inter-band and intra-band transitions result from the absorption of a material, which instigates the interaction between electrons and photons. The quantity of light intensity absorbed by a material per fixed length is determined by the absorption coefficient. From the figure it can noted that, absorption fluctuates with the increase in the energy. The peak value of absorption coefficient was found to be 21.22 × 10^4 (1/cm) at 9.46 eV.

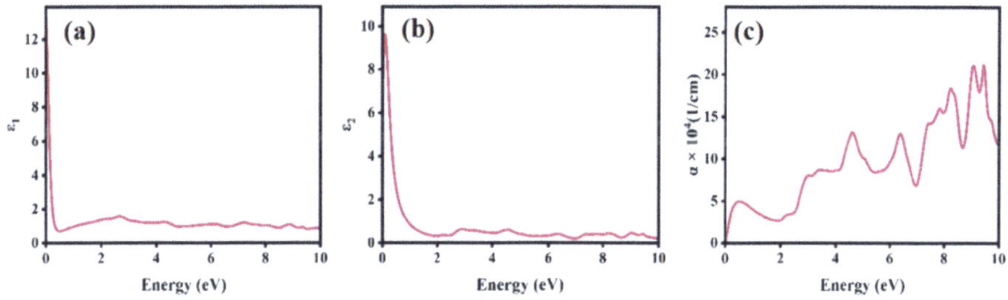

Fig. 3. Illustrates the (a) real part of dielectric function (b) imaginary part of dielectric function and (c) absorption coefficient.

6 Conclusion

The computational analysis of the 1H phase of the Na_2O monolayer has been conducted using DFT. The structural, electronic and optical properties were investigated with the WIEN2K software. The 1H-Na_2O crystallizes in space group P-6m2 (187), with lattice constants a = b = 3.54 Å and c = 15 Å. In this study, the GGA exchange-correlation functional was employed. The results unveiled that the 1H phase of Na_2O monolayer exhibits a zero bandgap. The optical properties of 1H-Na_2O investigated utilizing GGA functional. The peak absorption coefficient was found to be 21.22 × 10^4 (1/cm) at 9.46 eV.

References

1. Novoselov, K.S., et al.: Science **306**(2004), 666–669 (1979). https://doi.org/10.1126/science.1102896
2. Ahuja, U., Dashora, A., Tiwari, H., Kothari, D.C., Venugopalan, K.: Comput. Mater. Sci. **92**, 451–456 (2014). https://doi.org/10.1016/j.commatsci.2014.06.005
3. Wang, W., et al.: Microelectron. Eng. **223** (2020). https://doi.org/10.1016/j.mee.2020.111215
4. Ashwin Kishore, M.R., Larsson, K., Ravindran, P.: ACS Omega **5**, 23762–23768 (2020). https://doi.org/10.1021/acsomega.0c02804
5. Zhu, H., et al.: Ceram. Int. **48**, 27217–27239 (2022). https://doi.org/10.1016/j.ceramint.2022.06.070
6. Fu, B., et al.: Small **17** (2021). https://doi.org/10.1002/smll.202006054
7. Eithiraj, R.D., Jaiganesh, G., Kalpana, G.: Phys. B Condens. Matter **396**, 124–131 (2007). https://doi.org/10.1016/j.physb.2007.03.024
8. Roondhe, B., Sanyal, S.P., Jha, P.K., Ahuja, R., Shukla, S., Saxena, S.: Appl. Surf. Sci. **556** (2021). https://doi.org/10.1016/j.apsusc.2021.149737
9. Yan, X., Cao, W., Li, H.: Coatings **12** (2022). https://doi.org/10.3390/coatings12091294
10. Rawat, A., Arora, A., De Sarkar, A.: Appl. Surf. Sci. **563** (2021). https://doi.org/10.1016/j.apsusc.2021.150304
11. Chellaiya Thomas Rueshwin, S., Eithiraj, R.D.: J. Phys. Chem. Solids **187** (2024). https://doi.org/10.1016/j.jpcs.2023.111855
12. Zintl, E., Harder, A., Dauth, B.: Zeitschrift Für Elektrochemie Und Angewandte Physikalische Chemie **40**, 588–593 (1934). https://doi.org/10.1002/bbpc.19340400811
13. Hua, C., Li, S., Xu, Z.A., Zheng, Y., Yang, S.A., Lu, Y.: Adv. Sci. **7** (2020). https://doi.org/10.1002/advs.201901939
14. Singh, A.K., Mathew, K., Zhuang, H.L., Hennig, R.G.: J. Phys. Chem. Lett. **6**, 1087–1098 (2015). https://doi.org/10.1021/jz502646d
15. Kohn, W., Sham, L.J.: Phys. Rev. **140** (1965). https://doi.org/10.1103/PhysRev.140.A1133
16. Blaha, P., Schwarz, K., Madsen, G.K., Kvasnicka, D., Luitz, J.: **60** (2001). http://www.wien2k.at/reg_user/textbooks/usersguide.pdf
17. Perdew, J.P., Burke, K., Ernzerhof, M.: Generalized gradient approximation made simple (1996). https://doi.org/10.1103/PhysRevLett.77.3865
18. Monkhorst, H.J., Pack, J.D.: Special points for Brillouin-zone integrations (1976). https://doi.org/10.1103/PhysRevB.13.5188
19. Kumar, A., Ahluwalia, P.K.: Eur. Phys. J. B **85** (2012). https://doi.org/10.1140/epjb/e2012-30070-x
20. Ambrosch-Draxl, C., Sofo, J.O.: Linear optical properties of solids within the full-potential linearized augmented planewave method (2004). https://doi.org/10.1016/j.cpc.2006.03.005
21. Husain, M., et al.: RSC Adv. **12**, 32338–32349 (2022). https://doi.org/10.1039/D2RA05602J

Effect of Electromagnetic Fields on the Polaron in an Asymmetrical Gaussian Confinement Potential Quantum Well

Swapna Vemula, Anusha Kachu, and Aalu Boda(✉)

Department of Physics, National Institute of Technology, Warangal, Telangana 506004, India
abdop@nitw.ac.in

Abstract. In this work, we have embarked on a novel investigation of the electromagnetic field effects on the ground state energy (GSE) and ground state binding energy (GSBE), magnetic moment, and magnetic susceptibility of the weak coupling polaron in an asymmetrical Gaussian confinement potential quantum well. Our results, which demonstrate the increase of GSE and susceptibility with the rise of the magnetic field, and the intriguing behavior of GSBE and magnetic moment with quantum well parameters and electromagnetic fields, are sure to pique the interest of our professional colleagues in the field of condensed matter physics and quantum mechanics.

1 Introduction

The study of polarons in quantum wells has garnered significant attention in recent years due to their potential applications in optoelectronic devices and quantum computing. A *polaron* is a quasiparticle consisting of an electron surrounded by a cloud of phonons, quanta of lattice vibrations. The confinement potential and external fields [1, 2] significantly influence the properties of polarons. This research investigates the impact of electromagnetic fields [3–5] on the properties of polarons in asymmetrical Gaussian confinement potential quantum wells. The asymmetry of the confinement potential and the presence of electromagnetic fields introduce additional complexity, making this a rich and fascinating study area. Our research aims to provide a comprehensive understanding of the interplay between electromagnetic fields, confinement potential, and polaron properties, which is crucial for developing novel quantum technological implications with the variational properties like magnetic moment and magnetic susceptibility of polaron with the applied field.

Investigating the behavior of polarons in semiconductor quantum wells with asymmetric Gaussian confinement potentials poses significant theoretical challenges due to their complexity. Researchers employ approximate methods to overcome these challenges, including perturbation theory, the Lee-Low-Pines approach, (LLP) variational techniques, and Landau-Pekar calculations [6–8]. Recently, the linear combination operator and unitary transformation methods have been introduced to this field of study. This research explores how external fields, such as electric, magnetic, and noise, influence polaron dynamics in these quantum wells. Specifically, the impact of electric fields on the

P. Joseph D. et al. (Eds.): IEMDST 2024, SPPHY 567, pp. 271–278, 2025.
https://doi.org/10.1007/978-981-96-5863-3_35

ground state energy and binding energy of weakly coupled polarons is being examined using innovative methods. Researchers aim to drive innovation in quantum technology and semiconductor device development [9–11] by advancing the understanding of polaron behavior in these systems. Particularly in this paper, we approached LLP to solve the considered system's Hamiltonian [12].

Confinement potential in a quantum well refers to the potential energy that confines particles, such as electrons or excitons, within a narrow region of space, typically in a semiconductor material. This confinement is achieved by creating a potential barrier, often through heterojunctions or quantum dots, which restricts the movement of particles in one or more dimensions [13, 14]. In a quantum well, the confinement potential is typically modeled as a one-dimensional potential well, where the particles are free to move in the plane of the well but are confined in the perpendicular direction. The confinement potential can be categorized into four types: square well potential [15, 16], parabolic potential [17], Gaussian potential [18], cylindrical [19], hyperbolic [20] and Asymmetrical potential. Each type has its unique characteristics and applications.

The confinement potential is crucial in determining quantum wells' electronic and optical properties, such as energy level quantization, wave function confinement, bandgap engineering, optical transitions, and emission spectra [21, 22]. By controlling the confinement potential, researchers can tailor the properties of quantum wells for various applications, including quantum computing and information processing, optoelectronics and photonics, nanotechnology, and materials science [23–25]. In polaronic behavior, the magnetic moment and magnetic susceptibility are essential due to the changes in electron-phonon interaction. The magnetic moment and Magnetic susceptibility characterize the strength enhancement of material behavior.

2 Model

We examine a quantum system where an electron is trapped in a quantum well with an Asymmetrical Gaussian Confinement (AGC) potential. The electron, moving within a piezoelectric crystal, interacts with optical phonons while being influenced by a magnetic field (MF) oriented along the X-axis, described by vector potential A, and an electric field (EF) applied along the Y-axis [26–28]. Notably, using the effective-mass approximation, the Hamiltonian describing the electron-phonon interaction in this AGC quantum well can be expressed as:

$$H = \frac{1}{2m}\left(P + \frac{eA}{c}\right)^2 + \sum_q \hbar\omega_q a_q^+ a_q + \left(\frac{4\pi\alpha}{V}\right)^{\frac{1}{2}} \hbar s \sum_q \frac{1}{\sqrt{q}}\left(a_q + a_{-q}^+\right) e^{iq.r} + V(x) + exF, \tag{1}$$

The first term in equation describes the electron momentum energy in the MF, the second term represents the acoustic lattice vibration energy of the crystal, the third term denotes electron-phonon interaction energy, and the fourth is the confinement potential, $V(x)$ is the x-directional Gaussian confinement potential that represents the growth direction of the QWs the fifth term is the contribution of external EF to the Hamiltonian. Here 'm' is the electron mass in the band, $\omega_q = s.q$ is the linear dispersion used for the frequencies of piezo-acoustic phonons with a DCOW q_0 and s the sound velocity. a_q

and a_q^+ are the annihilation and creation operators for the acoustic phonon of the wave vector q. $r = (x, y, z)$ and P are momentum and position of an electron. The piezoelectric coupling constant is given by $\alpha = \frac{1}{2}\frac{e^2\langle e_{ijk}^2\rangle}{\varepsilon^2 Cs}$, where ε is the dielectric constant, $\langle e_{ijk}^2 \rangle$is the tensor average of the piezoelectric, C is an average elastic constant. The confinement potential in the X-direction is giving by

$$V(X) = \begin{cases} -V_0\, exp\left(-\frac{x^2}{2R}\right) & x \geq 0 \\ \infty & x < 0 \end{cases}, \tag{2}$$

where V_0 is the depth of the QW with AGC potential, and R is the range of the AGC potential [29–34]. Applying symmetric gauge transform [35–38] to Hamiltonian, it can be expressed as

$$\begin{aligned} H &= \frac{p^2}{2m} + \frac{m\omega_c^2}{8}\left(y^2 + z^2\right) + \frac{\omega_c}{2}L_x + \sum\nolimits_q \hbar\omega_q a_q^+ a_q \\ &+ \left(\frac{4\pi\alpha}{V}\right)^{\frac{1}{2}} \hbar s \sum\nolimits_q \frac{1}{\sqrt{q}}(a_q + a_{-q}^+)e^{iq.r} + V(x) - exF, \end{aligned} \tag{3}$$

where $\omega_c = \frac{eB}{m}$ is the cyclotron frequency. Performing the modified LLP transformation on the weak coupling polaron system, we introduce the two unitary transformations to eliminate electron-phonon coordinates as

$$U_1 = exp\left(-i\sum_q q.rb_q b_q^+\right), \quad U_2 = exp\sum\left(f_q b_q^+ - f_q^* b_q\right), \tag{4}$$

Here U_1 is the first unitary transformation, U_2 is the second unitary transformation, and $f_q(f_q^*)$ is the vibrational function. Using Eq. (4) of unitary transformations. We can rewrite the Hamiltonian as

$$\check{H} = U_2^{-1}U_1^{-1}HU_1U_2, \tag{5}$$

To evaluate ground state energy we can choose a suitable wave function after performing trials suggested by variational method, by minimising the expectation value of Hamiltonian is

$$\phi_0 = \pi^{-\frac{3}{4}}\lambda^{\frac{3}{2}} exp\left(-\frac{\lambda^2}{2}(y^2 + z^2\right) exp\left(-\frac{\lambda^2}{2}x^2\right) \tag{6}$$

The chosen ground state wave function (ϕ_0) and zero phonon state ($|0\rangle$) satisfy these conditions. $a_q|0\rangle = 0$, $\langle\phi_0 \mid \phi_0\rangle = \langle 0 \mid 0\rangle = 1$ and $\sum_q q|f_q|^2 = 0$. We also use the total momentum conservation relation as

$$E = \langle\phi_0|\langle 0|\check{H}|0\rangle|\phi_0\rangle \tag{7}$$

After performing simple algebra we can get an expression for ground state energy as

$$E = \frac{3}{4}\eta^2 + \frac{\omega_c^2}{8\eta^2} - V_0\left(1 + \frac{1}{2\eta^2 R^2}\right)^{-\frac{1}{2}} + \frac{\sqrt{\pi}}{2\eta}F - \frac{2\alpha}{\pi}\ln\left(1 + \frac{q_0}{2}\right), \quad (8)$$

The variation of Eq. (8) concerning λ gives

$$\lambda^4 - \frac{V_0}{3R^2}\left(1 + \frac{1}{2\lambda^2 R^2}\right)^{-\frac{3}{2}} - \frac{\sqrt{\pi}}{3}\lambda F - \frac{\omega_c^2}{6} = 0, \quad (9)$$

After solving Eq. (9) and assuming E_e and E_{ph} refer respectively to the energies of the electron and phonon. The Ground State Binding Energy E_B can be obtained following the formula

$$E_B = \frac{4\alpha}{\pi}\ln\left(1 + \frac{q_0}{2}\right) - \frac{\omega_c^2}{8\lambda^2} + V_0\left(1 + \frac{1}{2\lambda^2 R^2}\right)^{-\frac{1}{2}} - \frac{\sqrt{\pi}}{2\lambda}F, \quad (10)$$

Once the ground state energy is calculated, the weak coupling polaron's magnetic moment and magnetic susceptibility can be calculated using the following formula [34–38].

$$M = -\frac{\partial E}{\partial B};\ \chi = -\frac{\partial^2 E}{\partial B^2}. \quad (11)$$

3 Results and Discussion

A comprehensive numerical analysis has been performed to explore the sensitivity of polaron properties to critical parameters, including the external magnetic field (B), confinement characteristics such as potential depth (V_0), range (R), degree of confinement, Debye cut-off wave number (q_0), and electron-phonon interacting coupling strength (α). The study reveals how these parameters affect the polaron's ground state energy, binding energy, magnetic moment, and magnetic susceptibility. Importantly, this research offers valuable insights into the physical behavior of polarons in asymmetrical Gaussian quantum wells, enhancing our understanding of these complex systems. The graphical analysis with numerical results has been discussed in Fig. 1 and 2 consisting (a) and (b).

Figure 1(a) shows the variation of ground state energy with an applied magnetic field as a function of potential depth, and this behavior was observed within an electric field and without supplementing an electric field. Here, the variable Magnetic field is in tesla, observed ground state energy in $\left(R_y^*\right)$ units, and the parameters range $R = 4$, degree of confinement, Debye cut-off wave number ($q_0 = 50$), and within weak coupling strength is about $\alpha = 0.1$. We have observed that the increment in the magnetic field lowers the ground state energy due to magnetic confinement, i.e., the magnetic field introduces an additional confinement potential, compressing the electron wave function and increasing the kinetic energy. Also, the magnetic field modifies the electron-phonon interaction, leading to increased phonon-assisted scattering and higher energy.

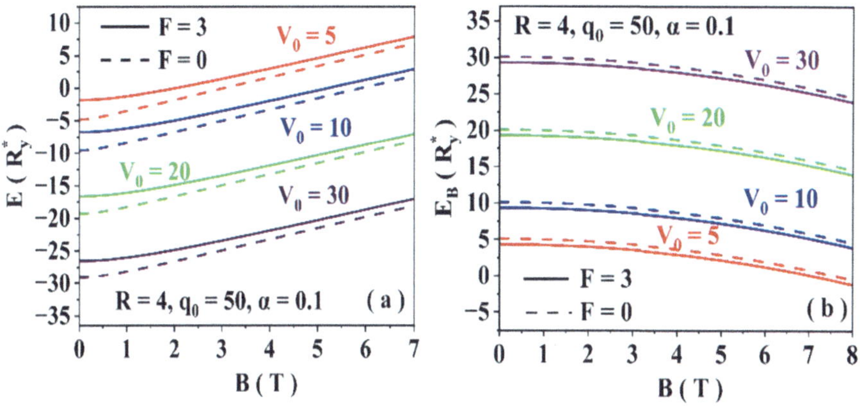

Fig. 1. (a) Ground state energy of polaron as a function of magnetic field for varying potential depth within electric filed and without electric field, (b) Ground state binding energy of a polaron as a function of magnetic field for varying potential depth within electric filed and without electric field.

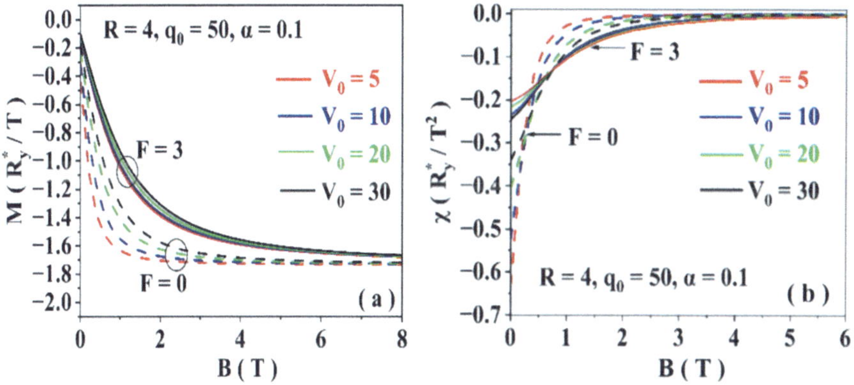

Fig. 2. (a) Magnetic moment of a polaron as a function of magnetic field for varying potential depth within electric filed and without electric field, (b) Magnetic susceptibility as a function of magnetic field for different values of confinement potential.

Figure 1(b) shows the variation of ground state energy with the applied magnetic field as a function of potential depth, and this behavior was observed within the electric field and without supplement of electric field and the parameters range $R = 4$, degree of confinement, Debye cut off wave number ($q_0 = 50$), , and within weak coupling strength is about $\alpha = 0.1$. We have observed that the ground state binding energy is decreasing as we increase the magnetic field as we expected. We can also observe that potential depth increases as the ground state binding energy (GSBE) increases. The electron is more tightly bound, leading to stronger confinement. The wave function becomes more localized, reducing the electron's spatial extent. Theoretically, we can predict that electron-phonon interaction strengthens here, leading to enhancing polaronic effects. In theoretical view, we can understand that to obtain the effect of electron-phonon interaction, we define the polaron self-energy as, which is also known as phonon binding energy because this is the energy required to make the electron accessible from the polaron bound state decreases (more negative) due to stronger confinement.- increases due to enhanced electron-phonon interaction, and E decreases, leading to the increment in GSBE [39].

Figure 2(a) indicating that the magnetic moment of a polaron is a function of a magnetic field for varying potential depth within and without an electric field. For range ($R = 4$), Debye cut-off wave number ($q_0 = 50$), and within weak coupling strength is about $\alpha = 0.1$. As the magnetic field increases, the magnetic moment decreases; this behaviour is enhanced with varying potential depth and applied electric field. Theoretically, the Gaussian potential weakens the magnetic field's influence on the electron's orbital motion, and more vital magnetic fields enhance confinement, suppressing orbital currents and reducing magnetic moment.

Figure 2(b) depicts the Magnetic susceptibility as a function of magnetic field for different values of confinement potential (V_0), and as the magnetic field increases the magnetic susceptibility increasing, we can observe that the external electric field is aggrandizing this behaviour and also, in both the situations the curves are showing saturation level at 0.1, Here even for high magnetic field material is not accepting beyond negative range so, we can confirm that the considered material is diamagnetic as it is within negative susceptibility, and theoretically strong magnetic fields enhance spin alignment and orbital magnetization [40].

4 Conclusion

A comprehensive study of the magnetic properties of a weak-coupling piezoelectric polaron in an acoustic Gaussian confinement potential quantum well has been conducted. The ground state energy was calculated using the Lee-Low-Pines transformation, enabling the determination of magnetic moment and susceptibility. The key findings are that as the magnetic moment decreases with increasing magnetic field strength under specific conditions, this behaviour persists across varying electric fields and potential depths. Magnetic susceptibility exhibits an inverse relationship with magnetic field strength, increasing as the field intensifies. This susceptibility trend holds consistent across diverse electric fields and potential depths. These results provide valuable insights into the polaron's magnetic response, demonstrating a clear and predictable pattern. The research offers a foundational understanding of the interplay between magnetic fields, electric fields, and potential depths on the magnetic properties, such as magnetic moment and susceptibility of polarons in Gaussian confinement potential quantum wells.

Acknowledgments. I would like to extend my sincere gratitude to Anusha Kachu for her valuable assistance with the analytical calculations. I am also deeply thankful to Dr. Aalu Boda for his invaluable guidance and insightful discussions throughout this work. Additionally, I appreciate the support of the National Institute of Technology for providing access to the necessary research facilities.

References

1. Li, Q., et al.: Phys. Rev. B **101**, 125301–125312 (2020). https://doi.org/10.1103/PhysRevB.101.125301
2. Hameau, S., et al.: Phys. Rev. Lett. **83**, 4152–4155 (1999). https://doi.org/10.1103/PhysRevLett.83.4152
3. Atoyan, M.S., Sarkisyan, H.A., Hameau, S., Harmand, J.C.: Phys. E **22**, 860 (2004). https://doi.org/10.1016/j.physe.2004.03.004
4. Atoyan, M.S., Kazaryan, E.M., Khodzhanov, A.F., Petrosyan, S.G.: Phys. E **31**, 83–85 (2006). https://doi.org/10.1016/j.physe.2005.10.026
5. Gambardella, P., Micron, I.M.: Philos. Trans. R. Soc. A **369**, 3175 (2011). https://doi.org/10.1098/rsta.2010.0336
6. Landau, L.D., Lifshitz, E.M.: Quantum Mechanics. Butterworth-Heinemann (1977)
7. Kittel, C.: Introduction to Solid State Physics. Wiley (2005)
8. Devreese, J.T.: Encyclopedia of Materials Science, pp. 1–6. Elsevier (2013)
9. Feng, R., Wang, J., Zhu, Y., Shen, F., Zhu, Z.: Opt. Express **21**, 15840–15846 (2013). https://doi.org/10.1364/OE.21.015840
10. Ma, X., Li, Y., Yang, Y., Zhao, Y.: J. Semicond. **36**, 102004 (2015). https://doi.org/10.1088/1674-4926/36/10/102004
11. Sze, S.M., Ng, K.K.: Physics of Semiconductor Devices. Wiley (2007)
12. Streetman, B.G., Banerjee, S.: Solid State Electronic Devices. Prentice Hall (2015)
13. Bastard, G.: Phys. Rev. B **24**, 1714 (1981). https://doi.org/10.1103/PhysRevB.24.4714
14. Weisbuch, C., Vinter, B.: Quantum Semiconductor Structures. Academic Press (1991)
15. Khordad, R.: J. Phys. Conf. Ser. **435**, 012008 (2013). https://doi.org/10.1088/1742-6596/435/1/012008
16. Zhang, Y.: J. Math. Phys. **61**, 102101 (2020). https://doi.org/10.1063/1.5144552
17. Nizametdinova, M.A., et al.: J. Phys. Conf. Ser. **245**, 012011–012015 (2010). https://doi.org/10.1088/1742-6596/245/1/012011
18. Ping, E.X., Dalal, V.: J. Appl. Phys. **73**, 5289–5291 (1993). https://doi.org/10.1063/1.353520
19. Xie, H.-J., Wang, C.-Y., Li, S.-S.: Superlattices Microstruct. **39** (2006)
20. Chen, T., Chen, W., Chen, S., Wang, Z.: Phys. Rev. B Condens. Matter **407**, 263 (2012). https://doi.org/10.1016/j.physb.2012.06.032
21. Bastard, G.: Phys. Rev. B **24**, 5693–5697 (1981). https://doi.org/10.1103/PhysRevB.24.5693
22. Singh, J.: Electronic and Optoelectronic Properties of Semiconductor Structures (1983)
23. Chuang, S.L.: Physics of Photonic Devices. Wiley (2009)
24. Coldren, L.A., Corzine, S.W., Mashanovitch, M.L.: Diode Lasers and Photonic Integrated Circuits. Wiley (1995)
25. Haug, H., Koch, S.W.: Quantum Theory of Optical and Electronic Properties. World Scientific (2009)
26. Guo, A., Du, J.: Superlattices Microstruct. **64**, 158–166 (2013). https://doi.org/10.1016/j.spmi.2013.08.030
27. Zhai, W.: Phys. B Condens. Matter **454**, 50–55 (2014). https://doi.org/10.1016/j.physb.2014.07.030
28. Jian, W., Li, S., Xiong, Q., et al.: Phys. B **446**, 59 (2014). https://doi.org/10.1016/j.physb.2014.04.041
29. Kachu, A., Boda, A.: Phys. E **155**, 115826–115836 (2024). https://doi.org/10.1016/j.physe.2024.115826
30. Kachu, A., Chebrolu, N.R., Boda, A.: J. Magn. Mater. **589**, 171565–171573 (2024). https://doi.org/10.1016/j.jmmm.2024.171565

31. Kachu, A., Chebrolu, N.R., Boda, A.: Micro Nanostruct. **186**, 207733–207746 (2024). https://doi.org/10.1016/j.micronano.2024.207733
32. Saini, P., Boda, A., Chatterjee, A.: Micro Nanostruct. **174**, 207487–207495 (2023). https://doi.org/10.1016/j.micronano.2023.207487
33. Boda, A., Chatterjee, A.: J. Magn. Mater. **485**, 407–412 (2019). https://doi.org/10.1016/j.jmmm.2019.03.026
34. Boda, A., Chatterjee, A.: J. Nanosci. Nanotechnol. **15**, 6472–6477 (2015). https://doi.org/10.1166/jnn.2015.10900
35. Farias, G.A., Degani, M.H., Hipolito, O.: Heliyon **8**, e10505 (2022). https://doi.org/10.1016/j.heliyon.2022.e10505
36. Fock, V.: Z. Physik **49**, 323 (1928). https://doi.org/10.1007/BF01390713
37. Darwin, C.G.: Proc. R. Soc. A **118**, 654–629 (1928). https://doi.org/10.1098/rspa.1928.0024
38. Lee, T.D., Rosenbluth, M., Yang, C.N.: Phys. Rev. **90**, 297 (1953). https://doi.org/10.1103/PhysRev.90.297
39. Manfouo, O.M., Razafindralambo, H.A.F., Diallo, A.M., et al.: Phys. Rev. B **95**, 125303–125309 (2017). https://doi.org/10.1103/PhysRevB.95.125303
40. Wu, J.H., Guo, K.X., Liu, G.H.: Phys. Rev. B **446**, 59 (2014). https://doi.org/10.1016/j.physb.2014.04.041

Magnetic and Transport Behaviour Study on Rare-Earth Based RE_2TMIrO_6 (RE = Gd, Dy and Ho & TM = Co and Ni) Double Perovskite Systems: DFT Calculations

Debasmita Bala[1](✉), Bheema Lingam Chittari[2], and Krishnamurthy Jyothinagaram[1]

[1] Department of Physics, School of Sciences, National Institute of Technology, Andhra Pradesh, Tadepalligudem 534101, India
baladebasmita@gmail.com

[2] Department of Physical Sciences, Indian Institute of Science Education and Research Kolkata, Mohanpur, West Bengal 741246, India

Abstract. Using first-principles calculations, the structural, electronic and magnetic properties of hetero-tri-spin of $3d$-$4f$-$5d$ based RE_2TMIrO_6 (RE = Gd, Dy and Ho & TM = Co and Ni) double perovskite systems with generalised gradient approximation (GGA) along with Hubbard parameter (U) i.e., GGA + U approach. The electronic transport study claims that half-metallic behaviour is observed along with ferrimagnetic ordering of sublattices in the ground states. The effects of U on the ground state structural parameters as well as the electronic band gap are also investigated. Rare-earth element of $4f$ orbitals show a major contribution near to the fermi level whereas, the Ir ($5d$)-O ($2p$) strong hybridization yields to the narrow band gap. The interaction of $3d$-$4f$ spin states gives rise to the splitting on HOMO (Highest Occupied Molecular Orbital) and LUMO (Lowest Unoccupied Molecular Orbital) bands located at (1/2,0,1/2) and (1/2,0,1/2) increases the superimpose of bands. The occurrence of Jahn-Teller distortion in non-linear octahedral and tetrahedral complex systems plays a pivotal role in reducing symmetry and influencing the transport behaviour.

1 Introduction

The enriched magnetic and electronic properties of the perovskite oxides have been studied for last few decades in double perovskites (DPs); $A_2BB'O_6$ by varying the combination between two transition metals (TMs) at B and B' sites and rare-earth (RE) at A site [1]. The stoichiometric ratio of cations in rock-salt arrangement of DPs is controlled by the cation size or charge i.e., if the charge difference among TMs is less than two, the normal ordering is dominant or else for greater than two, the rock-salt arrangement is dominant. The majority of the magnetic ordering originate from the direct or indirect super exchange interaction among TM elements through oxygen ions as the nearest neighbour interaction (bond angle ~ 90°) is much smaller than the next nearest neighbour interaction (bond angle ~ 180°). The super exchange interaction can be antiferromagnetic (AFM) or ferromagnetic (FM) based on the same or different TM orbitals present

P. Joseph D. et al. (Eds.): IEMDST 2024, SPPHY 567, pp. 279–287, 2025.
https://doi.org/10.1007/978-981-96-5863-3_36

in the system. Initially, the strong correlation effect of 3*d* electrons with pronounced 5*d* electron showing spin orbit coupling (SOC) effect in the hybrid 3*d*-3*d*, 3*d*-5*d*, 3*d*-4*d*, 4*d*-5*d* TM DP leads to the wide varieties of exotic properties possessing intrinsic interplay among charge, spin, orbital and lattice degree of freedom. Remarkably, the Sr_2CrBO_6 (B = Re, Os, W) series of Cr based DPs shows half metallic behaviour for B = Re and Os [2–4] and ferrimagnetic (FiM) insulator for B = W [5] with transition temperature T_C ~ 635 K, 725 K and 458 K, respectively. With the substitution of Ca^{2+} in place of Sr^{2+}, Ca_2CrBO_6 (B = Re, Os) [6] behaves as a FM insulating behaviour with the decrease in T_C near to 360 K and 490 K, respectively. The half-metallic FM nature of Sr_2FeBO_6 (B = Re, Mo, W)[7–9] results from the generalised double-exchange interactions involving a hybridization driven delocalization among Fe-t_{2g} sublattices. The 3*d*-5*d* interaction is quite interesting as 3*d* element exhibits more localized electron correlations than 5*d* elements. For example, low-dimensional AFM behaviour is observed in 3*d*-5*d* based Ba_2CuXO_6 (X = W, Te) [10] and Sr_2CuXO_6 (X = W, Te) compounds [11]. The Sr_2NiIrO_6 material undergoes structural phase transition from room temperature to high temperature but shows AFM transition near T_N ~ 58 K. The contradictory statement on the long-range AFM interaction among the face-centred cubic (FCC) Ir^{6+} sublattices with the FM (Ni-Ir) coupling energy concludes the distorted low-temperature AFM behaviour of the system. The presence of 5*d* element enhances the long-range magnetic coupling in the DP systems as the orbitals are highly delocalized and moderately or weakly correlated. The involvement of magnetic or non-magnetic rare-earth *f* orbitals in DPs contributes in structural changes from monoclinic to orthorhombic or cubic structure and the systems show either half-metallic FM ordering or FiM ordering among the RE-TM or TM-TM sublattices. Some exotic complex magnetic behaviour with lattice distortion and spin frustration are observed due to antisite disorder in FM/AFM exchange interaction among TMs in FM insulators like La_2BMnO_6 (B = Co, Ni) systems [12, 13]. The selection of the Iridium-based DP system is driven by the presence of Ir^{4+} instead of Ir^{5+}, which results in a narrow electron bandwidth in $J_{1/2}$, promoting a Mott insulator state. The splitting of 5*d* levels of Ir into t_{2g} triplet and e_g doublet are facilitated by the presence of crystal field [14]. The occurrence of Jahn-Teller distortion in non-linear octahedral and tetrahedral complex systems plays a pivotal role in reducing symmetry and influencing the transport behaviour. Some exceptional DPs systems with $Ir^{5+}$with $3d^5$ electronic configuration, like Sr_2YIrO_6, Ba_2YIrO_6 exhibit long-range magnetic ordering at low temperature due to the presence of distorted octahedra IrO_6 [15, 16]. Other distorted iridium-based DPs such as RE_2NiIrO_6 (RE = Y, La, Pr) behaves as AFM mott insulators but the transition temperatures T_N significantly decreases by increasing the ionic radius of the rare-earth element from Pr to Y [17, 18].

Numerous double perovskites of alkaline metals and rare-earth have been widely studied and reported with magnetoelectric, multiferroics and magnetic frustration behaviours [19–21]. The presence of two or three magnetic elements enhance the super-exchange interaction with John-Teller (JT) distortion [22]. Here, we are approaching the Ir-based double perovskites to investigate another two magnetic contributions from 4*f*-RE and 3*d*-TM elements. Density functional theory (DFT) is used to calculate the ground state properties with GGA + U formalism where U provides the strength of coulombic repulsion among the localized *d* and *f*-orbitals. We discuss about the total and orbital density of state and partial density of state calculation and contribution of different magnetic RE and TM sublattices near to the fermi level. We have analysed the variation of net magnetic moment, volume and lattice parameters with respect to the change in ionic radii of the elements in the systems.

2 Materials and Methods

By employing projected-augmented wave formalism of first principle calculations [23], the ground-state electronic structure and magnetic properties of RE_2TMIrO_6 (RE = Gd, Dy, Ho & TM = Ni/Co) are analysed by using VASP (Vienna ab initio simulation package) code [24]. Here, *4f 5d 6s*, *3d 4s* and *5d 6s* were used as valence electrons for RE, TM and Ir ions, respectively. The exchange correlation interactions between localized electrons present in the partially filled d-orbitals were treated using GGA + U method [25] and Perdew-Burke-Ernzerhof (PBE) parameters where GGA and U stand for generalized gradient approximation and Hubbard-onsite Coulombic repulsive potential, respectively.

3 Theory and Calculations

The expansion of the Kohn-Sham wave functions for valence electrons involves utilizing a basis set comprising plane waves, with the cutoff energy 500 eV. A 6 × 6 × 4 Monkhorst-pack k-point mesh centred at Γ were chosen for sampling over Brillouin zone for structural optimization until the remaining inter-atomic forces left after computation were diminished to below 10^{-3} eV/Å. The Gaussian broadening was 0.05 eV. The self-consistent electronic minimization employed a convergence criterion of 10^{-6} eV. To investigate the effect of Hubbard parameter (U), we calculated using U_{RE}, U_{TM} and U_{Ir} of values 6, 4 and 1.5 eV, respectively [26, 27].

4 Results and Discussion

With GGA + U method, the spin-polarized calculations predicted that RE_2TMIrO_6 exhibits semiconducting nature with AFM interaction between Ni/Co and Ir ions and FiM interaction between RE (Gd, Dy, Ho) and Ni/Co/Ir ions. The band gap is often underestimated by the local spin-density approximation (LSDA) in systems exhibiting strongly localized *d* orbitals [28] and Local density approximation (LDA) or generalized gradient approximation (GGA) mislocate the positions of both the conduction band minimum (CBM) and the valence band maximum (VBM) even though they predict the band gap accurately. So, it is preferable to use GGA + U approach and thereby to obtain the ground state free energy, band gap, volume and individual sublattices' magnetic moments for all the five systems are tabulated in Table 1.

To determine the symmetry of the system, the initial magnetic moment of each atom is set as 5 μ_B/f.u. The specified total spin (S) is considered as 1 for Co^{2+} and Ni^{2+} and ½ for Ir^{4+}. The optimized lattice parameters are very closed to the experimentally reported lattice parameter on the Gd_2NiIrO_6 system; $a = 5.3263$ Å, $b = 5.7189$ Å and $c = 7.6514$ Å [29]. The claimed monoclinic symmetry of the system involves with several exchange interactions with the nearest neighbours or next-nearest neighbours i.e. TM-TM, RE-TM, TM-Ir, RE-Ir etc. [30].

Table 1. The calculated ground state energy, band gap, volume and the individual magnetic moments of all RE (rare-earth), TM (Ni/Co) and Ir sublattices in the five double perovskites are presented below.

System	Energy (meV)	Total Band gap (eV)	Volume (Å^3)	Individual magnetic moment (μ_B)		
				RE	TM	Ir
Gd_2CoIrO_6 (GCIO)	−171.41	0.008	241.81	7.04	−1.81	−0.69
Dy_2NiIrO_6 (DNIO)	−164.729	0.004	231.07	5.09	−1.77	−0.59
Dy_2CoIrO_6 (DCIO)	−167.236	0.046	235.65	5.07	−1.86	−0.66
Ho_2NiIrO_6 (HNIO)	−160.584	0.107	226.71	4.07	−1.77	−0.58
Ho_2CoIrO_6 (HCIO)	−164.452	0.149	230.03	4.07	−1.79	−0.59

At first, to determine the ground state energy, different configurations have been considered with fixed Coulombic interaction parameter (U) of 6 eV for RE and 4 eV for TM and varying the U parameter for Ir from 0.5 to 6 eV [21, 31] in GGA + U formalism. Based on the minimal ground state energy, we are approaching FiM ordering i.e., AFM arrangement of RE with neighbouring Ni/Co and Ir-FM sublattices. The VESTA generated ground state monoclinic structure of the unitcell is represented in Fig. 1(a), where the blue balls represent RE = Gd, Dy and Ho sublattices occupied the body centred position and aligned FM with each other with tetrahedra REO_6. The green and yellow balls represent the edge-shared octahedra of Ni/Co and Ir sublattices aligned antiferromagnetically with each other. The calculated magnetic moments for the whole five systems are shown in Fig. 1(b). There is a systematic decrease in magnetic moment with decreasing ionic radii of the rare-earth elements. The spin only calculated magnetic moments for Gd_2CoIrO_6 (GCIO), Dy_2NiIrO_6 (DNIO), Dy_2CoIrO_6 (DCIO), Ho_2NiIrO_6 (HNIO) and Ho_2CoIrO_6 (HCIO) are 10.62, 8.82, 7.21, 6.33 and 5.73 μ_B which are approximately similar to the obtained net magnetic moment along x-axis from DFT calculations (shown in Fig. 1(b)). The lattice parameter a, b and c thus obtained for five different systems are deployed in the Fig. 1(c)-(e). The lattice parameters are indicating the centrosymmetric monoclinic structure of space group $P2_1/n$. There is a systematic decrease in the values of lattice parameter b whereas fluctuations in a and c parameters are observed.

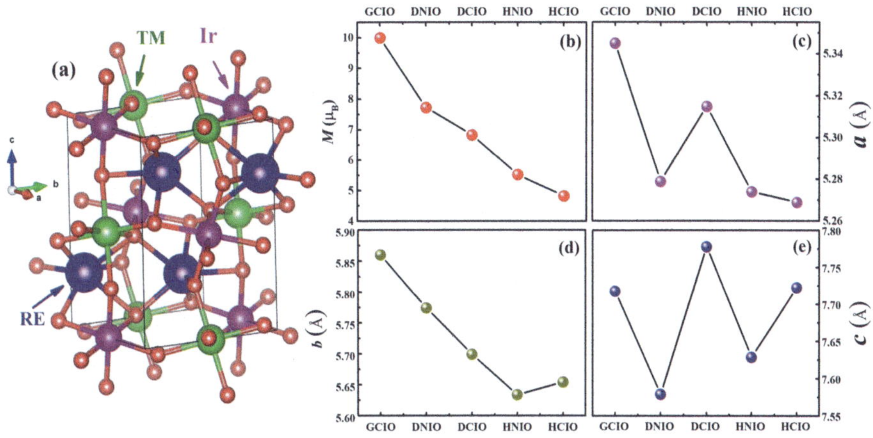

Fig. 1. (a) Body centred monoclinic (BCM) crystal structure representing all the five double perovskite structures where the blue, green and pink balls represent Gd/Dy/Ho, Ni and Ir, respectively. The panel (b-e) presenting the variation of magnetic moments and all the lattice parameters for the systems.

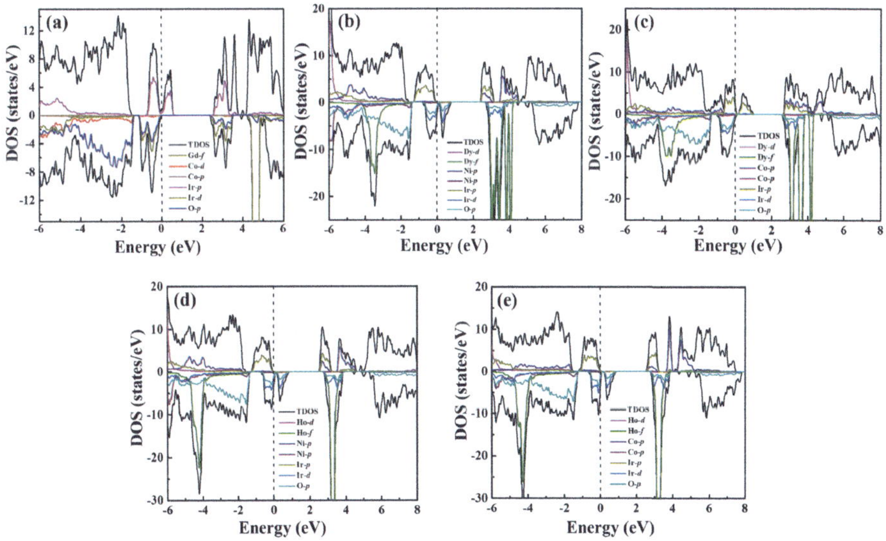

Fig. 2. Total calculated DOS (TDOS) and orbital DOS of RE (*f*-orbital) and TM (*d*-orbital)contribution for both spin-up and spin-down channels represented with $U_{\text{eff}} = 6$, 4 and 1.5 eV for (a) Gd_2CoIrO_6 (b) Dy_2NiIrO_6 (c) Dy_2CoIrO_6 (d) Ho_2NiIrO_6 and (e) Ho_2CoIrO_6, respectively.

A slight deviation in (Ni/Co)-O-Ir bond angles ranging from 144.7°–159.1° can be comparable with the bond angles in Y_2NiIrO_6 (139.9°–142.6°) and La_2NiIrO_6 (151.6°–152.8°) [18]. As, Ir is in 4 + state, t_{2g} level of 5*d* orbital splits and gives rise to the narrow band formation as the electron is hopping from Ir-t_{2g} to Ni-e_g level as a result

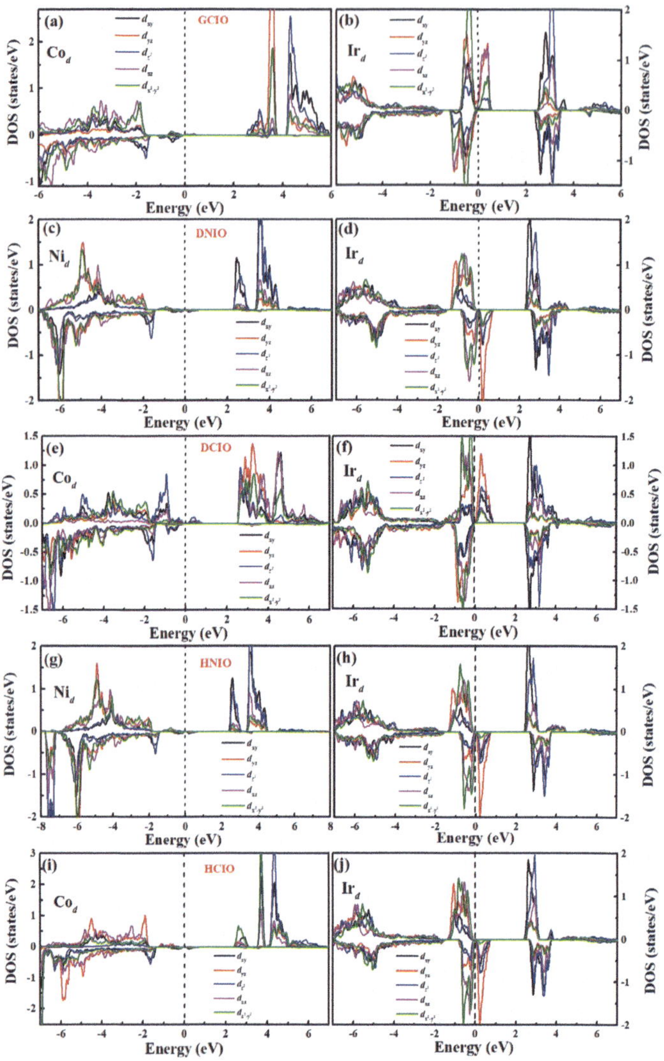

Fig. 3. The spin-polarized partial density of states (PDOS) calculated with GGA + U approximation represented as a function of energy ranging from −6 to 6 eV, across the fermi level set at zero, the panel (a, c, e, g, i) and (b, d, f, h, j) show the calculated d-orbitals resolved PDOS of TM^{2+}-3d and Ir^{4+}-5d orbitals for GCIO, DNIO, DCIO, HNIO, and HCIO DPs, respectively.

of lattice distortion generated from mixing of t_{2g}-e_g states. Instead of $J_{eff} = ½$, $S = ½$ state is being considered due to Hund exchange. The atomic radii are chosen here for RE, Ni/Co, Ir and O as 1.79–1.67, 1.25–1.21, 1.36 and 0.74 Bohrs, respectively. The spin-polarized total density of states (TDOS) and the orbital DOS contribution from RE-f orbital, TM-d orbital in FiM ordering in both spin-up and spin-down channels with energy ranging from −6 to 6 eV shown in Fig. 2(a-e). The obtained band gap for

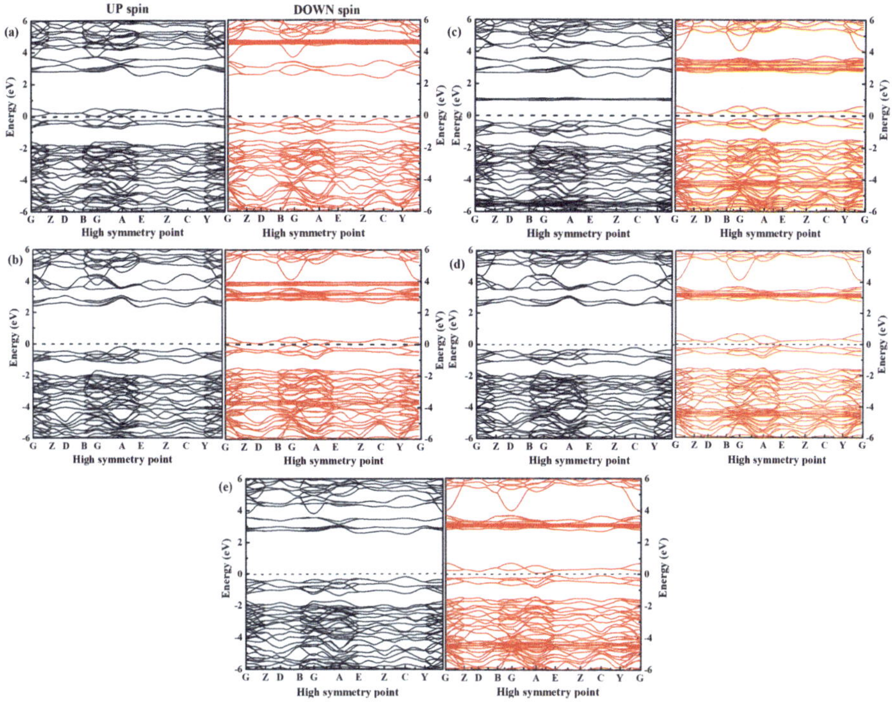

Fig. 4. The spin-up conduction band and spin-down valence band structures for ground states of (a) Gd_2CoIrO_6 (b) Dy_2NiIrO_6 (c) Dy_2CoIrO_6 (d) Ho_2NiIrO_6 and (e) Ho_2CoIrO_6, respectively are plotted with energy ranging from -6 to 6 eV and fermi energy is set at zero.

all the systems is tabulated in Table 1 which indicates the semiconducting behaviour of all DPs.

The spin-up channel conduction band shows band gap of 0.008 eV for GCIO and 0.046 eV for DCIO sample but for DNIO, HNIO and HCIO it varies from 2.517–2.710 eV. Similarly, the spin-down channel valence band shows band gap of 2.454 eV for GCIO and 1.114 eV for DCIO sample but for DNIO, HNIO and HCIO it varies from 0.004–0.149 eV. The band gap increases with decreasing ionic radii of the RE sublattices. Near to the fermi level at $E_g = 0$ eV, the Ir-5*d* orbital contribution is more as the influence of strong SOC. According to Hund's rule, due to the octahedral arrangement of NiO_6/CoO_6, the *d*-orbital splits into triply degenerate t_{2g} (upper d_{xy} and lower d_{yz} and d_{zx} bands) and doubly degenerate e_g (upper $d_{x^2-y^2}$ and lower d_{z^2} bands) i.e., $t_{2g}^5 e_g^2$ for Co^{2+} ($3d^7$), $t_{2g}^6 e_g^2$ for Ni^{2+} ($3d^8$) and $t_{2g}^3 e_g^2$ for Ir^{4+} ($5d^5$) [14, 21]. Thus, the calculated spin only moment for Co^{2+}, Ni^{2+} and Ir^{4+} are 3.89 μ_B, 2.82 μ_B and 1.02 μ_B, respectively. We have represented partial DOSs (PDOS) for Co-3*d*, Ni-3*d* and Ir-5*d* states as shown in Fig. 3(a-j), both Ni/Co- t_{2g} and e_g states located from -0.6 to -6.6 eV in the spin-up channel and -0.2 to -6.7 eV discontinuously present in spin-down channel for all the DPs systems, whereas Ni/Co- t_{2g} and e_g states occupies only the spin-up channel of the conduction band energy ranging from 2.2 eV to 7.8 eV. But both Ir- t_{2g} and e_g states occupy both spin-up and spin-down channel. The presence of Ni/Co-e_g energy levels signifies the hybridization with both O-2p and Ir-5d states.

The exchange splitting between t_{2g} up and t_{2g} down states for Ni_d/Co_d and Ir_d orbitals are 1.6–1.8 eV and 1.2–1.4 eV, respectively while the exchange splitting between e_g up and e_g down states for Ni_d/Co_d and Ir_d orbitals are 0.3–0.7 eV and 1.5–1.9 eV, respectively, resulting from the octahedral crystal-field effect. This negligible contribution of Ni_d/Co_d character near to the fermi-energy level as compared to Ir_d character confirms the mott-Hubbard insulating behaviour of the systems. The bandwidth of Ni_d/Co_d states are much wider than the Ir_d levels, thus, the t_{2g} - e_g crystal splitting gap is more in TM_d than Ir_d states. The orbital and spin contribution of all the elements are represented as the dispersion of band structure in both valence band and conduction band energies with respect to different high symmetry points for different U_{eff} are shown in Fig. 4(a-e). It is observed that all the band structures are showing indirect band gap with high symmetric points ranging from Γ to Y. All the DPs are stable in FiM structure with half-metallic as the gap in spin-up for GCIO and DCIO and spin-down for DNIO, HNIO and HCIO are metallic but enlarge strikingly in the respectively opposite spin channels.

5 Conclusion

We have performed theoretical density functional calculations and predicted the electronic and magnetic ground state properties of the five different double perovskites (DPs) of RE_2TMIrO_6 (RE = Gd, Dy, Ho & TM = Ni/Co) using GGA + U formalism. It is confirmed from the ab-initio first-principles calculation that all the systems are stable in FiM ground state with indirect band gap. The influence of Hubbard parameter U, the half-metallic nature of the systems with insulating behaviour enhance due to the adjustment of band structure and lattice parameters without affecting the structural, electronic and magnetic characteristics.

Acknowledgments. Debasmita Bala thanks to NIT Andhra Pradesh for financial support. Dr Krishnamurthy Jyothinagaram acknowledges DST-SERB sponsored core research grant, India and also thanks to Department of Physics, NIT Warangal for providing the opportunity for publishing research work through international conference on Emerging Multifunctional Materials and Devices for Sustainable Technologies (IEMDST-2024). We acknowledge the support provided by the Kepler Computing Facility, maintained by the Department of Physical Sciences, IISER Kolkata, for the various computational requirements.

References

1. Vogl, M., Morrow, R., Aczel, A.A., Rodriguez, R.B., Wolter, A.U.B., Wurmehl, S., Aswartham, S., Büchner, B.: Phys. Rev. Mater. **4**, 1–10 (2020). https://doi.org/10.1103/PhysRevMaterials.4.054413
2. Serrate, D., De Teresa, J.M., Ibarra, M.R.: J. Phys.: Condens. Matter **19**, 023201 (2007). https://doi.org/10.1088/0953-8984/19/2/023201
3. Guo, S.D.: Chin. Phys. Lett. **31**, 017101 (2014). https://doi.org/10.1088/0256-307X/31/1/017101

4. Vaitheeswaran, G., Kanchana, V., Delin, A.: J. Phys. Conf. Ser. **29**, 50–53 (2006).https://doi.org/10.1088/1742-6596/29/1/008
5. Saad M.H.E.: Adv. Tissue Eng. Regen. Med. Open Access. 4, 27–32 (2018). https://doi.org/10.15406/atroa.2018.04.00073
6. Morrow, R., Soliz, J.R., Hauser, A.J., Gallagher, J.C., Susner, M.A., et al., J. Solid State Chem. 238, 46–52 (2016).https://doi.org/10.1016/j.jssc.2016.02.025
7. Azad, A.K., Eriksson, S.G., Mellergård, A., Ivanov, S.A., Eriksen, J., Rundlöf, H.: Mater. Res. Bull. **37**, 1797–1813 (2002).https://doi.org/10.1016/S0025-5408(02)00872-3
8. Balcells, L.I., Navarro, J., Bibes, M., Roig, A., Martínez, B., Fontcuberta, J.: Appl. Phys. Lett. **78**, 781–783 (2001).https://doi.org/10.1063/1.1346624
9. Ding, L., Khalyavin, D.D., Manuel, P., Blake, J., Orlandi, F., Yi, W., Belik, A. A.: Acta. Mater. **173**, 20–26 (2019). https://doi.org/10.1016/j.actamat.2019.04.044
10. Mustonen, O.: Royal Soci. Chem. 1–3 (2013). https://doi.org/10.1039/c8cc09479a
11. Xu, Y., et al.: J. Phys. Condens. Matter **29**, 105801 (2017). https://doi.org/10.1088/1361-648X/aa58cb
12. Ullah, M., Ayaz Khan, S., Murtaza, G., Khenata, R., Ullah, N., Bin Omran, S.: J. Magn. Magn. Mater. **377**, 197–203 (2015). https://doi.org/10.1016/j.jmmm.2014.10.069
13. de Azevedo Filho, J.B., Souza, R.F., Queiroz, J.C.A., Costa, T.H.C., et al.: J. Magn. Magn. Mater. **527**, 167770 (2021). https://doi.org/10.1016/j.jmmm.2021.167770
14. Ding, X., et al.: Phys. Rev. B **99**, 014438 (2019). https://doi.org/10.1103/PhysRevB.99.014438
15. Cao, G., et al.: Phys. Rev. Lett. **112**, 056402 (2013). https://doi.org/10.1103/PhysRevLett.112.056402
16. Dey, T., et al.: Phys. Rev. B **93**, 014434 (2016). https://doi.org/10.1103/PhysRevB.93.014434
17. Ferreira, T., Calder, S., Parker, D.S., Upton, M.H., Sefat, A.S., Loye, H.C.Z.: Phys. Rev. Mater. **5**, 064408 (2021). https://doi.org/10.1103/PhysRevMaterials.5.064408
18. Liu L., Yang, K., Lu, D., et al.: Phys. Rev. B **108**, 174428 (2023). https://doi.org/10.1103/PhysRevB.108.174428
19. su, J., Yang, Z.Z., Lu, X.M., Zhang, J.T., Gu, L., et al.: ACS Appl. Mater. Interfaces **7**, 13260–13265 (2015). https://doi.org/10.1021/acsami.5b00911
20. Kleemann, W.: Binek, C.: Springer Tracts Mod. Phys. **246**, 163–187 (2013).https://doi.org/10.1007/978-3-642-32042-2_5
21. Ou, X., Li, Z., Fan, F., Wang, H., Wu, H.: Sci. Rep. **4**, 7542 (2014). https://doi.org/10.1038/srep07542
22. Kanungo, S., Yan, B., Jansen, M., Felser, C.: Phys. Rev. B. **89**, 214414 (2014).https://doi.org/10.1103/PhysRevB.89.214414
23. Blöchl, P.E.: Phys. Rev. B. **50**, 17953–17979 (1994).https://doi.org/10.1103/PhysRevB.50.17953
24. Kresse, G., Furthmüller, J.: Phys. Rev. B. **54**, 11169–11186 (1996).https://doi.org/10.1103/PhysRevB.54.11169
25. Noura, H.: Superlattices Microstruct. **76**, 425–435 (2014).https://doi.org/10.1016/j.spmi.2014.10.004
26. Ghimire, M.P., Thapa, R.K.: Adv. Sci. Lett. **21**, 2727–2729 (2015).https://doi.org/10.1166/asl.2015.6337
27. Calderon, C.E., Plata, J.J., Toher, C., Oses, C., Levy, O., Fornari, M., et al.: Comput. Mater. Sci. **108**, 233–238 (2015). https://doi.org/10.1016/j.commatsci.2015.07.019
28. Zhang, J.T., et al.: Phys. Rev. B **82**, 224413 (2010). https://doi.org/10.1103/PhysRevB.82.224413
29. Ferreira, T., Morrison, G., Yeon, J., Zur Loye, H.C.: Cryst. Growth Des. **16**, 2795–2803 (2016). https://doi.org/10.1021/acs.cgd.6b00121
30. Ou, X., Wu, H.: Sci. Rep. **4**, 4609 (2014).https://doi.org/10.1038/srep04609
31. Mugavero, S.J., Fox, A.H., Smith, M.D., zur Loye, H.C.: J. Solid State Chem. **183**, 465–470 (2010). https://doi.org/10.1016/j.jssc.2009.11.025

Investigating the Crystal Bonding, Structure and Electronic Properties of Sr_2FeMoO_6: A Computational Study

S. Vinothini[1,2], V. Jayalakshmi[3], V. Anbarasu[4], A. Rajeshkanna[5], and K. Saravana Kumar[1](✉)

[1] Department of Physics, Sri S. Ramasamy Naidu Memorial College, Sattur, Tamil Nadu 626203, India
kskumar23@hotmail.com
[2] Department of Science and Humanities, P.S.R. Engineering College, Sivakasi, Tamil Nadu 626140, India
[3] Department of Physics, SRM Institute of Science and Technology, Ramapuram, Chennai 600089, India
[4] Department of Physics, PSG College of Arts and Science, Coimbatore, Tamilnadu 641014, India
[5] Department of Computer Science, Sri S. Ramasamy Naidu Memorial College, Sattur, Tamil Nadu 626203, India

Abstract. This study investigates the structural and electronic properties of double perovskite Sr_2FeMoO_6 (SFMO) using the full-potential linearized augmented plane wave method (FP-LAPW), which is grounded in Density Functional Theory (DFT) and implemented in the WIEN2K computational code. Study analysis reveals that the compound belongs to the space-group I4/mmm (space-group No. 139) and demonstrates stability through structural examination. Furthermore, Sr_2FeMoO_6 is characterized as a hard material due to its high-magnitude bulk modulus in its ground state structure. Importantly, this material is an attractive candidate for spintronic device development requiring materials like SFMO that exhibit significant resistivity reduction and magnetic ordering well below room temperature.

1 Introduction

Perovskites with the general formula ABO_3 have appeared as a paradigmatic material class of the 21st century, showing unparalleled promise in various applications. This surge of interest over the past few decades has been fuelled by their exceptional performance in energy storage, environmental remediation, and optoelectronic devices, using their outstanding photovoltaic and catalytic properties. The $A_2(BB')O_6$ double perovskites, where A stands for an alkali-earth metal and B/B' include transition metals from different rows, have garnered significant attention in recent years. The structural versatility and intriguing physical properties of these materials make them attractive candidates for various applications, including microwave dielectric ceramics, photoactive devices, fuel cell electrodes, and luminescent systems [1–7]. The double perovskite

P. Joseph D. et al. (Eds.): IEMDST 2024, SPPHY 567, pp. 288–293, 2025.
https://doi.org/10.1007/978-981-96-5863-3_37

compound Sr_2FeMoO_6 exhibits a unique crystal structure characterized by an ordered arrangement of Fe and Mo ions on the B-site lattice, as found in rock-salt lattices. This configuration arises from the substantial disparity in ionic radii and valency between Fe and Mo cations, prompting the system to adopt an energetically favourable state through Fe-Mo cation ordering.

2 Method of Calculation

The crystallographic stability and electronic structure of the perovskite compound Sr_2FeMoO_6 were investigated using first-principles calculations [8]. The full-potential Linearized Augmented Plane Wave (LAPW) method, implemented in the WIEN2k code within the generalized gradient approximation (GGA)[9], was employed to determine the total energy and electronic density of states (DOS). Specific parameters used in these calculations included Muffin-tin radii of 2.46 a.u for Sr, 1.87 a.u for Fe, 1.83 a.u for Mo, and 1.61 a.u for O. For the total energy calculation, a high-resolution k-space mesh of 159 special points in the irreducible Brillouin Zone (BZ) was used. In contrast, a denser grid of 405 k-points was used to generate the electronic DOS, enabling detailed analysis of the material's electronic properties. Convergence of self-consistent calculations was determined to be achieved when the absolute charge-density difference per formula unit between successive integration loops decreased below 0.0001 |e|, where e denotes the fundamental electronic charge. This criterion served as a critical indicator of numerical stability and ensured the reliability of subsequent results.

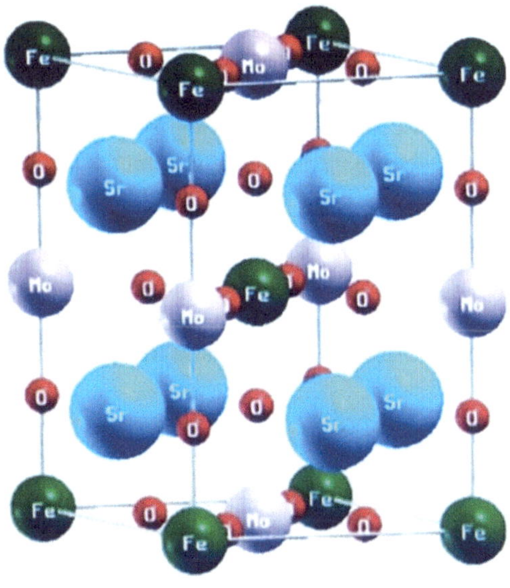

Fig. 1. Crystal Structure of Sr_2FeMoO_6.

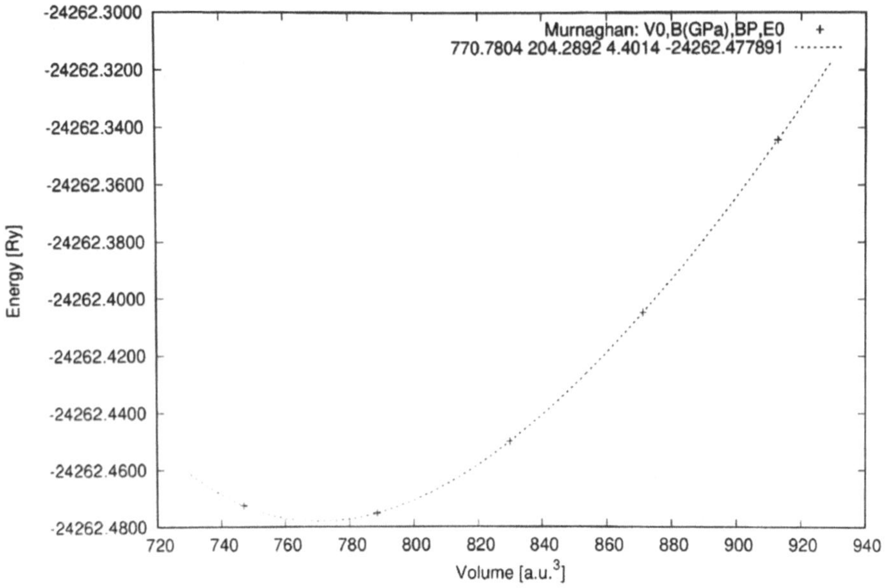

Fig. 2. Total energy optimization of Sr_2FeMoO_6

3 Total Energy and Lattice Parameter Evolution

A computational study of Sr_2FeMoO_6 double perovskite in its tetragonal phase with I4/mmm space group was conducted by [10–15] as illustrated in Fig. 1. The structural and energetic properties of this compound were characterised through thermodynamic modelling using Murnaghan's equation (ME) to minimize the total energy (TE). Figure 2 presents a plot of TE vs. unit cell volume, proving the fitting behaviour of the system with respect to the tetragonal structure under the Wannier-Campos-GGA (WC-GGA) potential. The lattice parameters are a = 5.58 Å, c = 7.90 Å and the coordinates values are shown in Table 1. Notably, the findings show that Sr_2FeMoO_6 exhibits a high bulk modulus of 204 GPa, signifying a rigid and mechanically robust crystal structure.

Table 1. Atomic Coordinate Refinement for Structural Analysis

Perovskite Sr2FeMoO6	Site	x	y	z
Sr	4d	0	0.5	0.25
Fe	2a	0	0	0
Mo	2b	0	0	0.5
O1	4e	0	0	0.25
O2	8h	0.25	0.25	0

4 Electronic Properties of Sr2FeMoO6

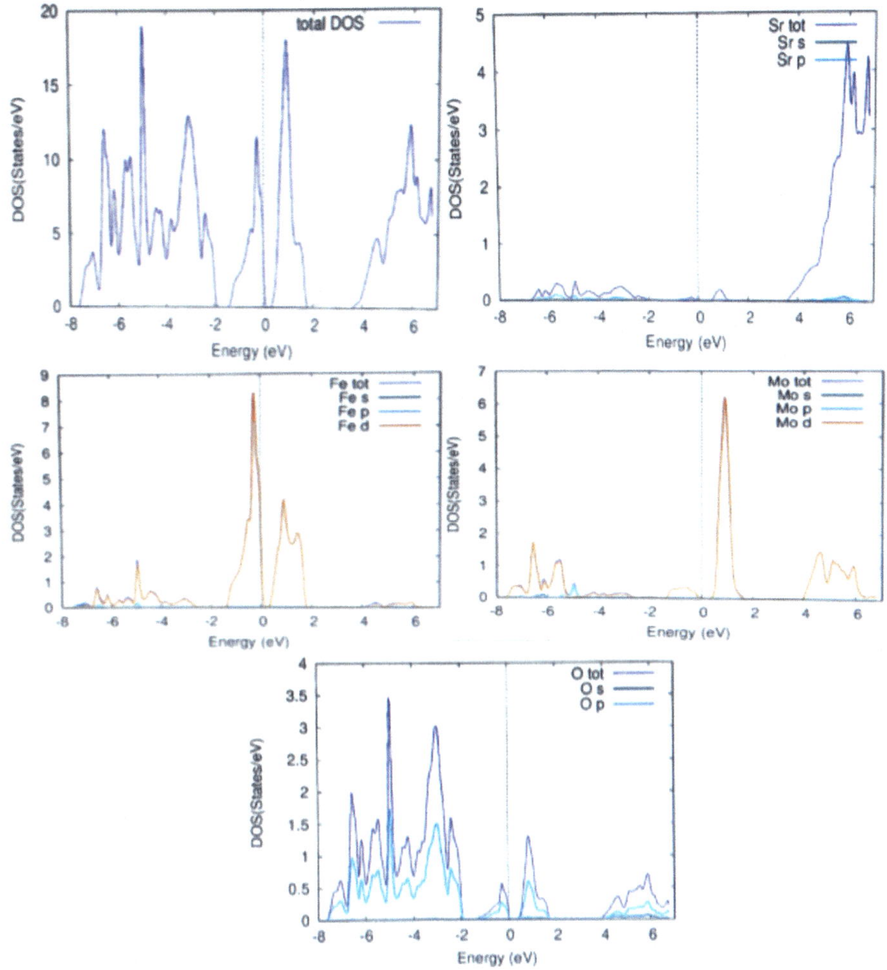

Fig. 3. Total and Partial Density of States in Sr_2FeMoO_6.

The electronic properties of double perovskite compounds can be determined through their band structure and density of states (DOS). A detailed analysis of the total DOS for Sr_2FeMoO_6, calculated using the generalized gradient approximation (GGA), is presented in Fig. 3. The energy range considered spans from -8 eV to 8 eV, with all values referenced to the respective Fermi level. Notably, our results show that Sr_2FeMoO_6 exhibits half-metallic behaviour. Furthermore, a bandgap of approximately 0.7 eV is observed in this compound. A closer examination of Fig. 3 reveals that the lowest valence band lies within the energy range -7 eV to -0 eV, forming most of the valence band states. The dominant contribution to these states arises from the Fe-3d orbitals and O-2p

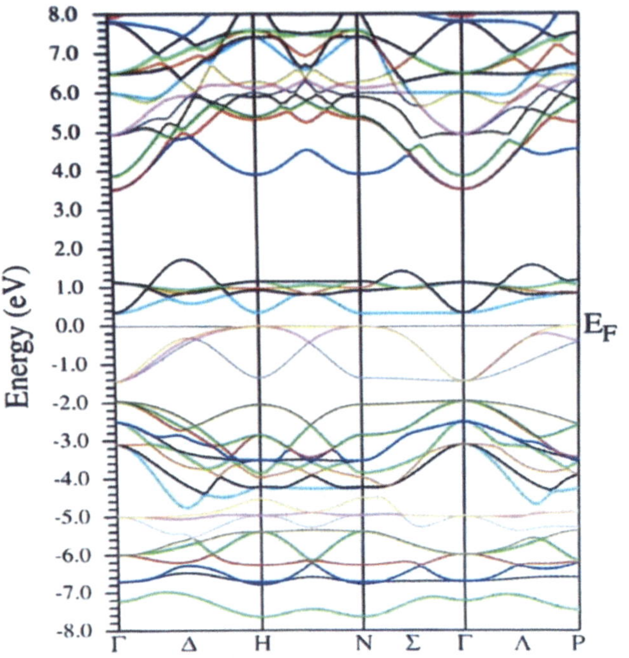

Fig. 4. Band Structure for Sr_2FeMoO_6.

orbitals, while the Mo-4d states contribute marginally. Conversely, the conduction band is primarily comprised of electrons in the 4d orbitals of Mo.

The electronic band gap is a crucial parameter for elucidating various material properties, including electrical resistivity, magnetic behaviour, and optical absorption characteristics [16]. Figure 4 represents the calculated band structure, which offers valuable insights into the electron distribution within the material. By setting the Fermi energy (E_F) at the topmost valence band, with an energy value of 0 eV, we can gain a deeper understanding of the electronic configuration within the system. The atomic configurations of Sr, Fe, Mo, and O are identified as [Kr] $5s^2$, [Ar] $3d^6$ $4s^2$, [Kr] $4d^5$ $5s^1$, and $1s^2$ $2s^2$ $2p^4$, respectively. Consequently, the analysis reveals that the upper valence band, spanning an energy range from -7 eV to 0 eV, is attributed to hybridisation states involving Fe-3d, Mo-4d, and O-2p electrons.

5 Conclusion

First-principles calculations were conducted to investigate the electronic properties of Sr_2FeMoO_6 in its tetragonal phase using the Full-Potential Linearized-Augmented-Plane-Wave (FP-LAPW) method. Theoretical predictions for lattice parameters, bulk modulus, and bandgap energy were obtained, showing that this material shows both ductility and hardness. Analysis of the total density of states revealed a half-metallic behaviour in Sr_2FeMoO_6, accompanied by a small band gap of approximately 0.8 eV.

This study is expected to motivate further experimental research into double perovskite materials, with potential applications arising from their unique electronic properties.

References

1. Khalyavin, D.D., Han, J.P., Senos, A.M.R., Mantas, P.Q.: MSF 455–456, 30–34 (2004). https://doi.org/10.4028/www.scientific.net/MSF.455-456.30
2. Iwakura, H., Einaga, H., Teraoka, Y.: J. Novel Carbon Res. Sci. **3,** 1–5 (2011)
3. Ma, X., Wang, C.M., et al.: J. Mater. Sci. **29**, 17818 (2018). https://doi.org/10.1007/s10854-018-9893-3
4. Markandeya, Y., Reddy, Y.S., Bale, S., Bhikshamaiah, G.: Synthesis, characterization, conductivity, magnetic and magnetoresistance studies of Sr2Fe1 − xInxMoO6 (x = 0.0 and 0.2) double perovskite. J. Mater. Sci.: Mater. Electron. **29**(8), 6711–6719 (2018). https://doi.org/10.1007/s10854-018-8657-4
5. Mahato, D.K., Sinha, T.P.: J. Mater. Sci. **24**, 4399 (2013). https://doi.org/10.1007/s10854-013-1416-7
6. Bugaris, D.E., Hodges, J.P., Huq, A., zur Loye, H.-C.: J. Solid State Chem. **184**, 2293 (2011). https://doi.org/10.1016/j.jssc.2011.06.015
7. Khalyavin, D.D., Han, J., Senos, A.M.R., Mantas, P.Q.: J. Mater. Res. 18, 2600 (2011). https://doi.org/10.1557/JMR.2003.0364
8. Blaha, P., Schwarz, K., Madsen, G.K.H., Kvasnicka, D., Luitz, J.: WIEN2K, An Augmented Plane Wave + Local orbitals Program for calculating crystal properties (Wien, Austria: Karlheinz Schwarz, Tech. Universitat Wien) (2001)
9. Perdew, J.P., Burke, K., Ernzerhof, M.: Phy. Rev. Lett. **77**, 3865–3868 (1996)
10. Galasso, F., Douglas, F.C., Kasper, J.: J. Chem. Phys. **44**, 1672 (1966)
11. Yanchun, H., Wang, S., Wang, X., Tang, Y.: Physica B **583**, 4122023 (2020)
12. Erchidi Elyacoubi, A.S., Masrour, R., Jabar, A., Ellouze, M., Hlil, E.K.: Mater. Res. Bull. **99**, 132–135 (2018)
13. Archer, T., et al.: Phys. Rev. B. **84**, 115114 (2011)
14. Stoeffler, D., Etz, C.: J.Phys: Condens. Matter **18**, 11291–11300 (2006)
15. Martins, H.P., et al.: J. Alloys Compd. **640**, 511–516 (2015)
16. Yaakob, M.K., Taib, M.F., Lu, L., Hassan, O.H., Yahya, M.Z.: Mater. Res. Express **2**(11), 116101 (2015)

Optimization of Optoelectronic Properties in $BaHfS_3$ with Substitutional Alloying: A First Principles Study

A. K. Sneha[1], Saji Joseph[2], Thomas Mathew[3], and Vincent Mathew[1](✉)

[1] Department of Physics, Central University of Kerala, Periye, Kasargod, Kerala 671316, India
vincent@cukerala.ac.in

[2] Department of Physics, Pavanatma College, Idukki, Kerala 685604, India

[3] Department of Physics, St Stephens College, Uzhavoor, Kerala 686634, India

Abstract. $BaHfS_3$ is a Transition Metal Chalcogenide Perovskite (TMCP) with an experimentally measured bandgap of 2.06 eV. Lowering its bandgap could make $BaHfS_3$ a promising candidate for use as a solar cell absorber material. In this study, we perform a DFT-based first principles investigation to explore the bandgap tuning in $BaHfS_3$ through substitutional alloying. The d-block element Hf is switched with different percentages of Ti atoms and the resulting variations in bandgap and optoelectronic properties are thoroughly examined. Our calculations reveal that the bandgap of $BaHf_{1\text{-}x}Ti_xS_3$ decreases from 2.0 eV to 1.2 eV as the Ti proportion (x) increases from 0 to 0.5. Corresponding changes in the material's optical properties are also observed. This work highlights the possibility of tailoring optoelectronic properties of $BaHfS_3$ with substitutional alloying, enabling us to optimize the material's properties for specific applications.

1 Introduction

Chalcogenide perovskites are an advanced family of semiconductors with the general composition ABX_3. Due to their promising optoelectronic properties [1], lead-free nature, non-toxicity, stability and availability they hold significant potential for various optoelectronic applications [2]. In Transition Metal Chalcogenide Perovskites (TMCPs), the B site is occupied by a transition metal. TMCPs with optimal bandgaps are good candidates for solarcell materials [3]. $BaHfS_3$ is a member of TMCP family with an experimentally measured bandgap of 2.06 eV [4]. While this bandgap is slightly higher than the optimal range of solarcell applications tuning it to lower values could enhance its potential as a solar cell absorber with improved light absorption [5]. There are various methods to tailor the bandgap of TMCPs, such as structural and chemical tuning [6], substitutional alloying [7, 8], doping [9] etc. In this work, we investigate bandgap tuning in $BaHfS_3$ through substitutional alloying. The d-block element Hf is replaced with the same group element Ti in different proportions. Density Functional Theory is used to analyse the bandgap tuning and corresponding variation in optoelectronic properties. Recently in 2024, Kong et al. successfully synthesized $BaHf_{1\text{-}x}Ti_xS_3$ powders, x ranging from 0 to 0.09 demonstrating thermal stability in air up to 550^0C [10]. This provides

P. Joseph D. et al. (Eds.): IEMDST 2024, SPPHY 567, pp. 294–299, 2025.
https://doi.org/10.1007/978-981-96-5863-3_38

experimental support for our study. In our study we explore a wider range of alloying proportions and place a stronger focus on the variations in optical properties.

2 Computational Methods

The first-principles calculations were performed using Density Functional Theory (DFT) with the projected augmented wave (PAW) method [11–13]. These calculations were carried out via the Vienna ab initio simulation package (VASP) with the MedeA interface. To address exchange-correlation interactions, the Perdew–Burke–Ernzerhof (PBE) functional within the generalized gradient approximation (GGA) framework was employed [14]. All the structural, electronic and optical calculations are done in a $2 \times 1 \times 1$ supercell including spin-orbit coupling effect. The supercells are created using MedeA interface and the substitution is done with VESTA software [15]. SUMO package is employed for graphical representation of electrical and optical properties [16]. Structural relaxation was performed with the conjugate gradient method. Structures are relaxed in self-consistent loop to reach the Hellmann–Feynman forces below 0.01 eV/Å. To address the van der Waals interactions, the Becke-Johnson damping (DFT-D3(BJ)) approach is employed [17]. The wave functions were expanded using a plane-wave basis set, with a cut-off energy of 400 eV. The Brillouin zone was sampled with a Γ-centered $5 \times 7 \times 10$ grid, and the energy accuracy was established to 10^{-8} eV.

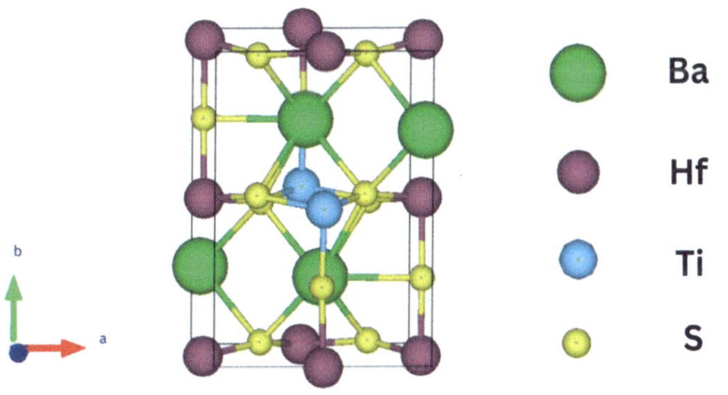

Fig. 1. The crystal structure of $BaHf_{0.75}Ti_{0.25}S_3$ unit cell with two Hf atoms substituted with Ti atoms.

3 Results and Discussion

$BaHfS_3$ crystallizes in the Pnma space group and features a distorted perovskite structure. The lattice parameters obtained are a = 7.02 Å, b = 9.90 Å and c = 6.97 Å which are in good agreement with experimental values. The substitutional alloying is done in a $2 \times 1 \times 1$ supercell of $BaHfS_3$. This supercell contains 8 Ba, 8 Hf and 24 S atoms. By substituting different number of Hf atoms with Ti atoms we obtain samples of $BaHf_{1-x}Ti_xS_3$ with substitution levels of 12.5%, 25%, 37.5% and 50%. The crystal structure of $BaHf_{0.75}Ti_{0.25}S_3$ is given in Fig. 1. The Hf atom occupies four different Wyckoff positions, resulting in four distinct samples for single-atom substitution. However, all of these structures exhibit identical optoelectronic properties. This indicates that the optoelectronic properties are not influenced by the specific position of substitution, but rather by the number of substituted atoms. Therefore, we have adopted a random substitution approach. All the substituted structures were relaxed to get the most stable arrangement. Formation energy is calculated to confirm the stability of the substituted samples. The negative values in Table 1 indicate that these samples are energetically favorable.

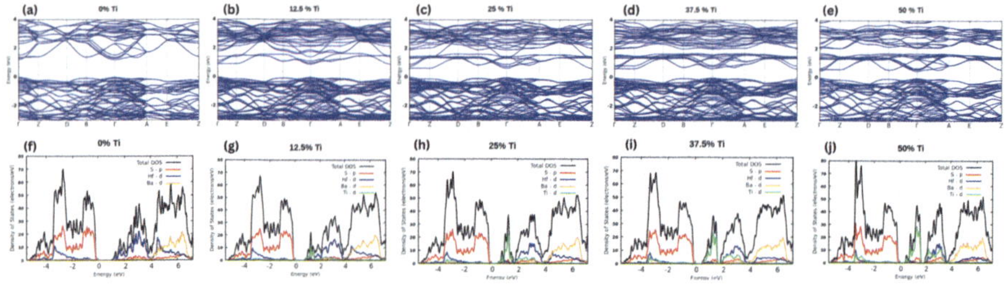

Fig. 2. Band structure of $BaHf_{1-x}Ti_xS_3$ for different alloying proportions (a) x = 0 (b) x = 0.125 (c) x = 0.25 (d) x = 0.375 and (e) x = 0.50. Density of states of $BaHf_{1-x}Ti_xS_3$ for different alloying propotions (f) x = 0 (g) x = 0.125 (h) x = 0.25 (i) x = 0.375 and (j) x = 0.50.

Table 1. Formation energy, volume and bandgap of $BaHf_{1-x}Ti_xS_3$ with varying substitution proportion.

x in $BaHf_{1-x}Ti_xS_3$	Formation Energy (kJ/mol)	Volume (Å^3)	Bandgap (eV)
0	−849.39	967.82	2.0
0.125	−6675.98	956.95	1.7
0.250	−3278.06	945.48	1.4
0.375	*−6447.10*	*933.95*	1.3
0.500	*−1548.64*	*921.80*	1.2

From the electronic structure calculations, it is obvious that the bandgap of $BaHf_{1-x}Ti_xS_3$ decreases with increasing substitution proportion x. The bandgap has reduced from 2.0 eV to 1.2 eV as x varied from 0 to 0.5 (as shown in Table 1). From the band structures in Fig. 2 the bandgap remains a direct bandgap for all the samples. The conduction band minimum shifts to lower energies as the proportion of Ti substitution increases. The density of states plots provides insights into the orbital contributions. The valence band is primarily composed of contributions from S-p and Hf-d orbitals, while the conduction band is mainly influenced by Ba-d and Hf-d orbitals. The Ba-d orbital contribution becomes significant only above 4 eV. As the substitution percentage increases, the Ti-d orbital begin contributing to the conduction band while the Hf-d orbital contribution diminishes. Initially for 12.5% sample, contribution of Hf-d orbital dominates over that of Ti-d orbital. However, by the time the substitution reaches 50%, the Ti-d orbital contribution exceeds Hf-d orbital.

The three-dimensional structure of $BaHfS_3$ is formed by corner-sharing HfS_6 octahedra. When Hf atoms are swapped with Ti atoms, TiS_6 octahedra are created. The difference in ionic radius and electronegativity of Hf and Ti atoms is directly impacts the bandgap values. Ti atoms have a smaller ionic radius and higher electronegativity than Hf atoms, which leads to shorter bond lengths in TiS_6 octahedra compared to HfS_6 octahedra. Consequently, the structural volume decreases due to substitution. The change in structural volume is shown in Table 1. This reduction in volume is the primary reason for the bandgap decrease observed with substitutional alloying.

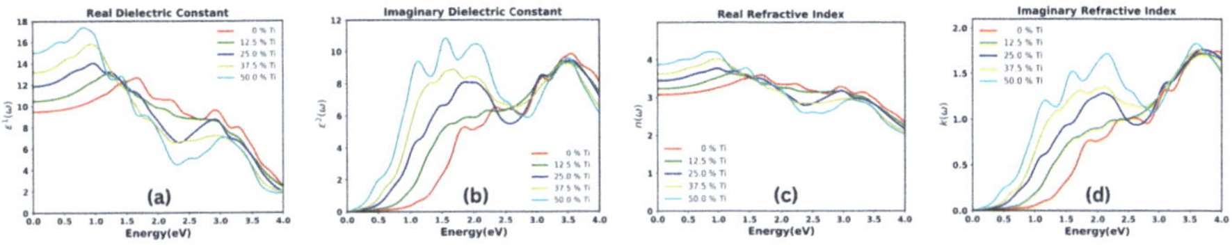

Fig. 3. Variation of different optical properties of $BaHf_{1-x}Ti_xS_3$ for different alloying proportions (a) Real dielectric constant (b) Imaginary dielectric constant (c) Real refractive index (d) Imaginary refractive index.

The variation in optical properties is also studied in detail. The reduction in bandgap is expected to influence the optical properties. The DFT calculated bandstructure data are used to calculate the optical characteristics [18–20]. The variation in dielectric constant and refractive index are shown in Fig. 3. Both real and imaginary part variations are included.

Initial threshold value of imaginary part of dielectric constant has moved to a smaller energy region with increasing substitution percentage. In the Infra-Red and visible region, alloyed samples prevail over the pure $BaHfS_3$. Whereas in the UV region, both unmodified and modified samples exhibit comparable values. In the real part, there is an enhancement from 9.2 to 15.0 in the static value $\varepsilon(0)$ with the substitution percentage. For the dynamic ε value, substituted samples lead over unmodified $BaHfS_3$ in IR region. Whereas in UV and visible regions, the pure sample exhibits a higher dielectric constant.

Regarding the refractive index, the imaginary component transitions to a lower energy range as a result of the reduction in the bandgap. Imaginary component of refractive index of the modified samples dominate in IR and visible energy levels. For the UV region, the unmodified sample leads over modified samples in magnitude. The zero point value of the real component of the refractive index has increased from 3.1 to 3.9. Real component of the refractive index, in IR and some part of the visible region, modified samples prevail. In the concluding region of the visible and UV regions, the unmodified sample exhibits larger values.

In Fig. 4 the variation in absorption coefficient is also shown. Since imaginary part of refractive index is directly related to absorption coefficient, similar trend is observed here. The first three figures focus on a lower energy range, while the last three cover a

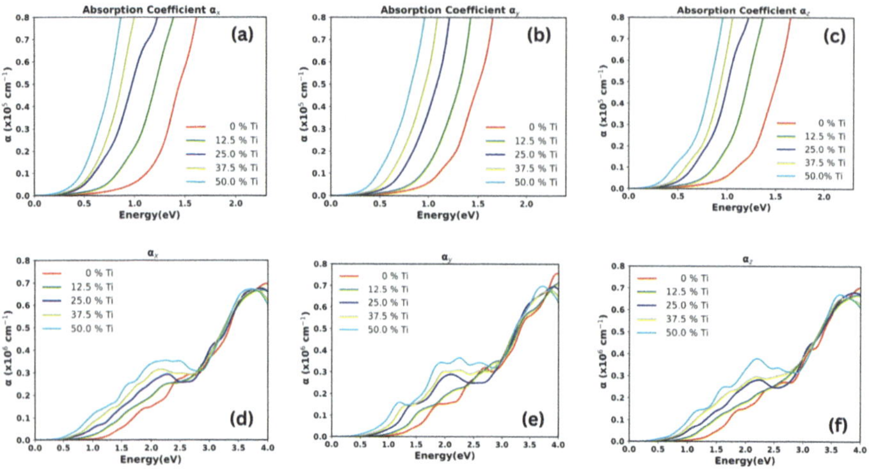

Fig. 4. Variation of absorption coefficients α_x, α_y and α_z of $BaHf_{1-x}Ti_xS_3$ for different alloying proportions (a), (b) and (c) highlights variation in short energy range, (d), (e) and (f) highlights variation in longer energy range.

higher energy range. As the bandgap decreases with increasing substitution, the initial threshold energy for optical absorption shifts to a lower energy range. This means that absorption begins at a lower energy value, covering a broader energy spectrum for absorption. From the last three figures, we could see that the magnitude of absorption coefficient is larger for modified samples in the Infra-Red and visible range of the energy spectra. In UV region, both modified and unmodified samples show almost similar magnitude and variation. Overall, the results demonstrate the potential for tailoring the photovoltaic properties of $BaHfS_3$ through Ti substitution. Depending on the application, the substitution percentage can be adjusted to fine-tune the material's properties.

4 Conclusion

In this study, we successfully tuned the bandgap of $BaHfS_3$ through Ti alloying. As the substitution percentage increased from 0% to 50%, the bandgap decreased from 2.0 eV to 1.2 eV. Thus, it can be used as solarcell absorbers. This reduction in bandgap is primarily attributed to the differences in ionic radius and electronegativity between Ti and Hf atoms. The Ti-S bond length is shorter than the Hf-S bond, leading to a reduction in structural volume with increasing substitution, which drives the bandgap variation. This change is also reflected in other optoelectronic properties. With Ti replacement, both the refractive index and the static dielectric constant increase in magnitude. In the IR and most of the visible regions, the modified samples exhibit improved dielectric constant, refractive index, and, consequently, a higher absorption coefficient. Hence with proper substitutional alloying we can optimize the optoelectronic properties of $BaHfS_3$ according to our desired application.

Acknowledgments. With deep appreciation, author Sneha A. K. thanks KSCSTE for providing the PhD Fellowship.

References

1. Abhishek, S., Mir, W.J., Chakraborty, R., et al.: Chem. Mater. **31**(3), 565–575 (2019).https://doi.org/10.1021/acs.chemmater.8b04178
2. Suhail, M., Abbas, H., Khan, M.B., Khan, Z.H.: Chalcogenide perovskites for photovoltaic applications: a review. J. Nanop. Res.**24**(7), 142 (2022).https://doi.org/10.1007/s11051-022-05525-0
3. Adjogri, S.J., Edson, M.L.: Materials **14**(24), 7857 (2021). https://doi.org/10.3390/ma14247857
4. Huo, Z., Wei, S.-H., Yin, W.-J.: J. Phys. D Appl. Phys. **51**(47), 474003 (2018). https://doi.org/10.1088/1361-6463/aae1ee
5. Sun Y.-Y., Agiorgousis, M.L., Zhang, P., Zhang, S.: Nano Lett. **15**(1), 581–585 (2015).https://doi.org/10.1021/nl504046x
6. Niu, S., Huyan, H., Liu, et al.: arXiv preprint arXiv 1804.09362. https://arxiv.org/abs/1804.09362
7. Wei, X., Haolei, H., Samanthe, P., Aaron, S., David, F., Watson, et al.: ACS Omega **5**(30), 18579–18583 (2020). https://doi.org/10.1021/acsomega.0c00740
8. Sneha, A.K., Saji, J., Sujith, C.P., Thomas, M., Vincent, M.: Comput. Mater. Sci. **245**, 113286 (2024). https://doi.org/10.1016/j.commatsci.2024.113286
9. Yu, Y., et al.: Chin. Chem. Lett. **32**(11), 3558–3561 (2021).https://doi.org/10.1016/j.cclet.2021.02.040
10. Kong, S., Haoxi, D., Zhonghai, Y., et al.: Ceramics Int. **50**(7), 10889–10896 (2024). https://doi.org/10.1016/j.ceramint.2023.12.405
11. Kresse, G., Jürgen, F.: Phys. Rev. B **54**(16), 11169 (1996). https://doi.org/10.1103/PhysRevB.54.11169
12. Kohn, W., Becke, A.D., Parr, R.G.: J. Phys. Chem. **100**(31), 12974–12980 (1996).https://doi.org/10.1021/jp960669l
13. Burke, K.: J. Chem. Phys. **136**, 15 (2012).https://doi.org/10.1063/1.4704546
14. Perdew, J.P., Kieron, B., Matthias, E.: Phys. Rev. Lett. **77**(18), 3865 (1996).https://doi.org/10.1103/PhysRevLett.77.3865
15. Momma, K., Izumi, F.: J. Appl. Crystallogr. 44(6), 1272–1276 (2011)
16. Ganose, A.M., Jackson, A.J., Scanlon, D.O.: J. Open Source Software **3**(28), 717 (2018). https://doi.org/10.21105/joss.00717
17. Grimme, S.S., Ehrlich, L.G.: J. Comput. Chem. **32**(7), 1456–1465 (2011). https://doi.org/10.1002/jcc.21759
18. Gajdoš, M., Hummer, K., Kresse, G., Furthmüller, J., Bechstedt, F.J.P.R.B.: Phys. Rev. B Cond. Matter Mater. Phys. **73**(4), 045112 (2006). https://doi.org/10.1103/PhysRevB.73.045112
19. Mathew T., Joseph, S., Mathew, V.: Superl. Microstr. **153**, 106859 (2021). https://doi.org/10.1016/j.spmi.2021.106859
20. Sujith, C.P., Joseph, S., Sneha, A.K., Mathew, T., Mathew, V.: J. Phys. Chem. Solids **174**, 111161 (2023). https://doi.org/10.1016/j.jpcs.2022.111161

Advancements in Biogas Production: Enhancing Efficiency Through Anaerobic Digestion Technologies

Tavva Chandra Lekha[1], Ravi Gugulothu[1(✉)], Tejaswini Malothu[1], and Achiya Khanam[2]

[1] Department of Mechanical Engineering, JNTUH College of Engineering Hyderabad, Hyderabad, Telangana, India
ravi.gugulothu@gmail.com
[2] Department of Coatings and Polymeric Materials, North Dakota State University, Fargo, USA

Abstract. The convergence of energy conservation, sustainable water management, and multifaceted environmental protection poses critical challenges globally. As populations grow and economies expand, the demand for energy escalates, placing immense strain on traditional fossil fuel resources. Biomass, a renewable source derived from organic matter, offers a compelling solution. It reduces waste disposal burdens, lowers greenhouse gas emissions, and provides an avenue for decentralized energy generation. Anaerobic digestion (AD) is a cornerstone technology for transforming biomass into biogas – a versatile fuel composed primarily of methane and carbon dioxide. This biological process occurs in oxygen-deficient environments, where microorganisms break down complex organic materials into simpler compounds. Biogas presents notable advantages, including utilization for heating, electricity production, or conversion into transportation fuels. This paper investigates the production of biogas through anaerobic digestion, emphasizing its role in fostering sustainable energy practices. It examines the suitability of diverse biomass feedstocks, the impact of pre-treatment techniques, and the integration of cutting-edge technologies aimed at maximizing biogas yields. The goal is to underscore the significance of this process in advancing both environmental sustainability and energy security.

1 Introduction

Water and energy are utmost important things on this world to survive the human and animals [1, 2]. Among these energy is an essential factor for the development of societies, for this renewable energy is a key source [3]. Due to availability of fossil fuels, thermal power generation has vigorous development in the fast decades and more research is conducted on the same. But due to non-renewable energy sources categories, fossil fuel limitations and negative environmental impression, researchers are thinking alternative solutions for the power generation to meet world demand [4]. In recent decades with the aggravate of shortage energy and the continuous augmentation of green house emanation, the utilization of numerous sources of renewable energy is elegant most promising sources to sustain any country [5]. Therefore, great emphasis on protection of

P. Joseph D. et al. (Eds.): IEMDST 2024, SPPHY 567, pp. 300–314, 2025.
https://doi.org/10.1007/978-981-96-5863-3_39

environment and exploitations of renewable energy is gradually transforming towards cleaner energy [6]. They concluded that anaerobic digestion process requires low energy essential, low cost explore of biogas production when differentiate to the conventional processes and anaerobic digestion process will produce biogas which is a adequate alternative inexhaustible energy derivation.

Celik et al. [7] investigated the methane prospective of pistachio hull waste which is produced from the de-hulling process for various concentrations with and without pre-treatment and concluded that the pistachio de-hulling deteriorate is a renewable energy source, no adjustment was required to pistachio de-hulling deteriorate with methane yield ranging from 42.2 to 73.4 ml/g COD which is lesser than theoretical value i.e., 395 ml/g, augments the sCOD/tCOD ratio 26% to 5% subsequently, They detect 213.4 ml/g COD as highest yield of methane. Sakir Yilmaz et al. [8] numerically studied using poultry manure with the help of anaerobic digestion to produce biogas. They observed that the RSM with central composite design (CCD) can predict quite well and models the biogas offering from AD to PM. They observed the maximum progressive biogas offering, CH_4%, and COD% elimination are 8965.87 mL, 71.298%, and 68.4711% subsequently.

Sabeeha N.B.A. Khadaroo et al. [9] explored the ramifications of thermal pre-treatment and solid loading reshaping on a thermophilic continuing set up. They noticed that 40S:60L solid loading which is thermally pre-treated significantly augmented biogas offering, methane freshness, chemical and biological oxygen demand (COD and BOD), and cumulative sling solids (TSS) elimination ability. They achieved 48 L of biogas production, and 77.2% methane purity, the COD, OD, and TSS elimination ability are 97.5, 95.1, and 94.5% respectively. Sameena N. Malik et al. [9] investigated the breakdown of complex lignocellulosic structures to augmentation of biogas production on yard waste with effects of Fenton, ozone, and peroxone pre-treatment processes. They observed that the cellulose, hemicelluloses, and lignin tranquil of yard decay decreases during the pre-treatment activity with augmented TRS attentiveness, and in turn augment the biogas yield, the highest biogas yield is 210 mL/g.

Kajsa Risberg et al. [10] compared the biogas yield from wheat straw, steam–extracted in fusion with ordure as a base for biogas yield at both mesophilic and thermophilic temperatures of 37, 44, or 52 °C at 25 days of retentions. They observed the straw furnish a similar yield as ordure but the steam access didn't augment the gas offering, where very low methane yield is acquire with current innovation and economic conditions. Apurva Mishra et al. [12] reviewed the pioneer's work on biogas yield from bio-waste through complex anaerobic digestion. They noticed that the elimination of degradation unusually hydrogen sulphide (H_2S), and carbon dioxide (CO_2) from the biogas will remarkably augment the ability of biogas, the offering and elevation of biogas anticipate upon the numerous feedstocks, anaerobic digestion, microbial diversity, salvation techniques along with the application of numerous preservatives.

Juan Luis et al. [12] evaluate the production of biogas from scenedesmus biomass due to its low biodegradability using an anaerobic co-digestion process. Scenedesmus biomass exhibits low biodegradable properties. They distinguish that the high-rise methane relent was enhanced by 66.4%, and 63.9% from the Scenedesmus, and Opuntia maxima subsequently. Sachankar Buragohain et al. [13] deliberate the anaerobic mono and co-digestion of various biomass along biochemical methane potential tests performed at various concentrations, and cattle dung: feedstock ratios. They observed that the co-digestion of raw materials at 1:1 ratio yields superior productivity than the other studied cowpats: feedstock ratios. Sachankar et al. [14] investigated the biogas production by integrating 1 kW solar photovoltaic plant with 3.5 kVA biogas fuelled by consideration of constant load of the rated power conditioning. They observed that the photovoltaic systems are sufficient to meet lower loads, but hybrid system is required for higher loads. This hybrid system is 104.59 Tons of CO_2 mitigation on the environment.

Siswo Sumardiono et al. [15] studied the effects on biogas production using corn stalks using physical, biological, and chemical pre-treatment. They noticed that the corn stalks with 20 mesh by adding EM-4, 7% v/v achieved the extreme and cumulative biogas yields are 19.86 L/kg, and 215.77 L/kg respectively. They concluded that the biogas yield enhanced from corn stalks through physical, biological, and chemical pre-treatment processes. Siswo Sumardiono et al. [16] studied the biogas production through pre-treatment process of sugarcane bagasse to laboratory scale. They observed that physical pre-treatment can enhance biogas yield, and it was enhanced with both at C/N ratio conditions. Among these two C/N ratio, 25 C/N ratio is better than the 30 C/N ratio i.e. 53 to 90.5% higher. Cioabla et al. [17] experimentally investigated the production of biogas from residual waters in a combination of various substrates/inoculums in a laboratory. They observed that the digestion accomplishment of residual waters enhance the biogas productivity when combined with other wastes, for further usage of the same material is a test at a pilot scale to identify their potential. They concluded that 40% wheat, 40% corn, and 20% sunflower husk is the best anaerobic co-digestion produced the highest concentration of methane for highest quantity of biogas yield.

Santosh Kumar et al. [18] experimentally investigated the gasoline and raw biogas at a compression ratio of 10 to estimate the engine performance and characteristics by varying the engine speed. They observed that 18% depletion in brake power, a 66% enhancement in brake-specific fuel dissipation, and 12% depletion in brake-thermal capability when tested with raw biogas. The emission characteristics are the unburned hydrocarbon and CO_2 emissions ensure enhanced by 6.8%, and 40% subsequently. Santosh Kumar et al. [19] conducted further investigation on a SI engine using raw biogas at wide open throttle conditions and various crank ratios ranging from 10 to 14 by varying the operating speed and ignition time ranging from 23° to 47° crank angles. They noticed that the 10 to 12 crank angle enhances the power, and capability of the raw biogas is 12.72% and 5.68% subsequently, and depleted fuel utilization is 5.42% at 1400 rpm. Enhancement in NOx and unfired hydrocarbon ejection by 10.17%, and 15.6% subsequently at same speed. Finally, they concluded the optimum crank angle is 12 at 1400 rpm for wide open throttle condition. Seyed Sajad Hashemi et al. [20] investigated the baker's yeast production from wastewater to biogas in a two-step dissipate biorefinery task which involves the elimination of COD, and nutrients. In the first step, they noticed the maximum biomass yield, mycoprotein, and COD elimination are 5.13 g/L, 43.8%, and 36.3% respectively. In the second step, AD, and COD elimination are 22.4, and 44.2%. They observed that the protein-rich biomass and methane are 5.13 kg, and 1.42 m^3 from 1m^3 of BYW.

Guangze Guo et al. [21] investigated biogas production by thermal pre-treatment with a extortionate solid AD membrane bioreactor (AnMBR) with different pre-treatment methods of venture nozzle, and steam injector process (VNT and SIT) at a temperature of 70 °C. They observed that the higher biogas production is 23.5 to 30.5% than the unprocessed WAS, and VS biogas yield is 0.46 to 0.47 L/g-VS in pre-treatment. They concluded that better energy recovery and non chemical matter elimination capability were attain for the WAS process using better-solid AnMBR with in-pipe thermal pre-treatment. Siman Yang [22] investigated trough an anaerobic digestion activity utilizing food waste without accompaniment and co-digestion to evaluate solid appease, temperature, and pH. They observed that the methane up to 699.7 mL.gVs^{-1}.d^{-1} which is 70.4% of the methane yield recovered. Hui Sun et al. [23] explored the synergetic accomplish of ensiling recently ingather maize stover (FHM) and unnecessarily droop maize stover (EWM) on biogas yield, to acquire dry matter delighted as 30 to 40% by mixing FHM, and EWM. They found that the organic moment loss is 1.1 to 2.2%, and 16.5% methane yield enhanced. Hui Sun et al. [24] evaluated the achievement of unnecessary droop maize stover (EWMS) with biogas fym and bio-methane potential where chicken and pig manure biogas is used. They observed that 41%, and 15% of barley with chicken, and pig wastes biogas muck outside solid liquid disaffiliation, and 7.1 to 9.6% enhancement in specific methane yield with 5.0 to 7.3% offset storage loss.

Hui Sun et al. [25] studied anaerobic digestion to recover bio-methane from crop residue to avoid greenhouse gas (GHG). They observed 29.2%, and 2.25% natural gas consumption, and energy consumption per annum respectively. Hui Sun et al. [26] explored the possessions of acid based hydrolysis and biological degradations using pre-treatment of biogas production, synergetic possessions of acid-based hydrolysis and biological degradation is discover in various process with non chemical acid inclusion of 20 g/kg of dry matter (DM). They noticed that 10 to 13%, and 6 to 8% of bio-methane enhanced with addition of non chemical acid of 80 g/kg, and raw material respectively. Hui Sun et al. [27] explored the part of carbon/ nitrogen (C/N) ratio and barrier role throughout the ensiling activity for coupling biorefinery and biogas exhibition, where maize stover was ensiled for 360 days at an inceptive C/N ratio of 25:1 by urea with a barrier ranging from 100 to 160 g lactic acid/kg dry affair by $CaCO_3$. They noticed that 65–78% of biorefinery products enhanced. Ahmed El Nemr et al. [28] investigated the modified biochar which is obtain from saw dust biomass utilize sonication and Ozonation process, with green algae as an additive. They concluded that the ozonation process of biochar enhances the biogas yield process, which is given accuracy R2 = 99.3%. They achieved the highest yields 1059 mL/g VS and 1054 mL/g VS for 50 mg of SSDB, and 150 mg of OSDB.

Avicenna et al. [29] investigated the impact of biogas yield by corn husk (CH) by adding NaOH through a pre-treatment stage. They observed that the optimum biogas yield is 67.6 mL/m with a methane attentiveness of 63.4% at 33 wt% concentration of NaOH. Azam Jeihanipour et al. [30] a novel activity has emerged for detachment of the cellulose from blended fibers waste textiles by pre-treatment process. They found that 85%, and 89% of cellulose yields for 2 day enzymatic hydrolysis and 1 day fermentation respectively. For further Azam Jeihanipour et al. [31] explored the offering of biogas from deteriorate textiles by utilizing the capability of Stirred Batch Reactor (SBR) and Up-flow AD Sludge Blanket Bed (UASB) method by differentiate the single and two-stage AD in reactors. They observed that the maximum methane production is 80%. Deepanraj et al. [32] experimentally explored the effects of operation parameters and pre-treatment on biogas generation, volatile solids, and COD degradation through anaerobic digestion by employing Taguchi-based Grey relational exploration to identify the optimum ambience. They noticed that the parameters of food waste, pH, temperature, C/N ratio are 7.5% TS, 7, 50 °C, and 20.19 respectively. Bin Chen et al. [33] investigated the life cycle evaluation of biogas generation from agro-industry towards fruitful energy execution. They suggested that the in all the aspect utilization of biogas is beneficial. They concluded that the coupled system is optimized by enhancing the fermentation efficiency of biogas digesters.

Biswanath Saha et al. [34] investigated the methods to augment the biogas yield on Lantana camara through anaerobic digestion by various heating pre-treatment techniques. They noticed that the ATC pre-treatment of Lantana camara is most efficient were compared with other methods, ATC pre-treatment increased the solubilization, and volatile fatty acids (VFA) by 45.44%, and 56.75% at 110 °C for 80 min respectively compared with untreated. Finally, they concluded that the methane yield improved by 3656, and 2895 mL for 5, and 7 weeks respectively. Veluchamy et al. [35] investigated the impact of corn silage on biogas generation through anaerobic digestion at mesophilic conditions by varying non chemical loading rates (OLR) ranging from 3.5 to 8.5 $kgm^{-3}d^{-1}$. They concluded that the enhancement in biogas generation is achieved with each upsurge in OLR, CH_4 accumulation declined, and H_2S content augmented under elevated loading.

Ciro Fernando Bustillo-Lecompte et al. [36] optimize the biogas yield and eliminate the cumulative non chemical carbon and nitrogen from the meat processing industry by reducing wastewater towards green practices using Response surface methodology (RSM). They noticed that maximum TOC, TN removals, minimum residual, and maximum biogas yield are 85.03%, 72.10%, 19.54 mg/L, and 116.56 mL/min respectively at optimum conditions. Keerthana Devi et al. [37] investigated the Agrowaste sources to generate biogas by anaerobic digestion. They noticed that biogas production depends on metabolic reactions, utilization of microorganisms, salvation and upgrading activity, and elimination of carbon dioxide from the gas mixture. They focussed on the pre-treatment, utilising of water, production activities, and applications for the advancement of biogas production. They observed that economic, technical, and environmental factors affect the biogas production. They concluded that among all agro- wilt, sugarcane straw generated 94% of biogas by utilizing advanced techniques of biogas production.

Coskun et al. [38] explored the capability of milk-processing wastewater to produce hydrogen and biogas. They noticed that 54.2 million m^3, and 12,670 tons of biogas/year, and H_2/year were generated respectively from milk processing waste when 15.1 million dollars assessment of energy was contribute per year. The overall hydrogen production energy efficiency is 19 and 70% for various reforming and ambient temperatures. Yu et al. [39] examine the biogas yield from bio-wilt and sewage sludge in an AD reactor at 30 and 70 wt% respectively by varying non chemical loading rates. They noticed that more biogas yield with all OLRs by the thermophilic activity. They concluded that the thermophilic activity is more stable when compared with the mesophilic process as a bacterial cell count is 6.25 times higher when compared with the cell count in the mesophilic activity.

Daniel E. Arias et al. [40] explored the anaerobic co-digestion of agricultural wastes which contributes greenhouse gases (GHG) to produce biogas by varying organic loading rate, and total solids content ranging from 0.25 to 4.7 kg, and 1.5 to 9.0% respectively. They observed that 0674 $m^{-3}kg^{-1}$, 62% the high rise cumulative biogas and methane content respectively at 0.25 kg/m^3d. Dieter Wolf et al. [41] explored the yield of biogas from biomass by employing two great technologies through digesters for biogas production from energy crops or organic waste by using two novel technologies. The appraisal of the volumetric flow of raw biogas is performed using ultrasonic biogas meters and found to a high degree of precision which is helpful to develop a model for activity automation. Almomani et al. [42] investigated the impact of advanced oxidation activity (AOPs) involving ozone, fenton, and combined of iron and H_2O_2 on the division of various solid wilts. They noticed that substrates lost 15.2 to 29.5%, 33.6 to 37.5%, and 16.5 to 25% of solids, total chemical oxygen requirement, and total biopolymers (Humic acids and proteins) respectively. They observed 23 to 30%, and 11.2 to 25% enhancement in cumulative methane yield, and digestion efficiency.

Florian Filarsky et al. [43] studied the yield of the continuous biogas by combining it with a pioneering gas hydrate by launching fermentation and optimization to microbiological cultivated with sewage sludge of wastewater. To effectuate higher disintegration factors, tetrahydrofuran, a substance used as a kinetic promoter, and additional catalysts are added. They concluded that for biogas production, the two experimental series should be combined and disintegration capability should be evaluated. Francesca Valenti et al. [44] explore the aptness of utilizing citrus pulp as a viable matrix of blend required to feed AD to enhance biogas production by increasing the biogas plants which does not exhibit a negative impact on the environment by using specific GIS tools and illustrative statistics. They noticed that there is a difference in citrus generation from various citrus fruits because of different species and generation abilities. They concluded that the production of citrus pulp from localized citrus processing areas ensures biogas generation.

Hasan Pasalari et al. [45] examined biogas production by increasing the biodegradation of raw landfill leachate (RLL) and biomethane potentiality (BMP) in AD reactors, optimized the operational parameters by employing central composite design (CCD), and RSM. They noticed that the O_3/H_2O_2 biodegradability indicator increased ranging from 0.41 to 0.68 for optimum conditions of pH, oxygen flow, Reaction time, and hydrogen peroxide concentration are 8, 0.25 L/m, 25 min, and 2.5 g/L respectively. They concluded that the pre-treatment O_3/H_2O_2 and mesophilic AD activity release a realistic and well regulated activity for enhanced BMP of RLL.

Heng Li et al. [46] enlarge a mathematical prototype and conducted simulation to explore the accomplishment, and fates of carbon, nitrogen and phosphorous in moist and high cumulative solid AD activity. They noticed that anaerobic digestion model-1 is developed when inorganic substrates such as pig manure are added, high-TS and liquid-solid activities are integrated and C, N, and P (methane, ammonia, and hosphorous concentrations) exhibits good precision. They concluded that by promoting pre-treatments such as hydrolysis and biodegradability of additive for pig manure, increases the biogas production.

Kiros Hagos et al. [47] scrutinized the advancement and challenges associated with AcoD technology for the biogas production. They noticed that as requirement for AcoD technology increases, the organic substances become volatile, highlights that the Biochemical methane probable (BMP) test is important to evaluate biodegradability and perishing rate of non chemical materials. They concluded that by adding economic and environmental friendly nanoparticles to substrate can optimize and improves the biogas production from different wastes. Lan Mu et al. [48] examined the different co-digestion of derived non-chemical wastes in a semi-continuing by utiling real-time quantifiable polymerase chain reaction (qPCR). They noticed that when the SS (25% VS basis) is introduced to two feedstocks in anaerobic co-digestion, it augmented the activity stability and archaea/ cumulative microbe ratio which is anticipated to the regulating impact created by trace metals in SS and increased the methane capitulate by 2.04 times than AD of YW. They observed that in co-AD of FW and YW, synergistic accomplish increased in the form of methane capitulate.

Larissa Renata Santos Andrade et al. [49] investigated the production of biogas by employing nanoparticles that rapidly convert biomass into bioenergy with the metabolism of microorganisms and the relationship between TiO_2 nanoplates anatase [001] saturate on Luffa Cylindrica and the reprocess of this nanocatalyst in creation of biogas from cassava waste water. They observed that when Luffa Cylindrica ($LufTiO_2$) is added, it enhanced biogas production. They concluded that the usage of nanoplates through a combination of nanotechnology and simple fibers stimulates the biogas generation from cassava waste water. Petros Gikas et al. [50] explored the offering biogas from biodegradable materials which causes dangerous impacts on the environment are segregated through Mechanical Biological Treatment (MBT) and rotary drum reactor (RDR) activity of MSW pre-treatment technology by therapy them in well-managed digesters under thermophilic conditions with varying few parameters the wetness contented, TS, VS, and C/N are 64.7, 44.4%, 55.6 to 35.3%, 27 to 41.3%, and 24.5 to 42.7 subsequently. They noticed that the biogas produced was between 533.0 and 675.6 mL/g VS after 20 days of AD and 90% biogas yield during the first 13 days.

Sagor Kumar Pramanik et al. [51] reviewed the characteristics of food wilt (FW) for the generation of biogas through co-digestion and pre-treatment of anaerobic digestion. They noticed that both co-digestion and pre-treatment process enhances the food wilt hydrolysis rate and methane generation which results in biogas production, converting the food waste into biogas by anaerobic pre-treatment is an achievement that leads to challenges for production. Seyed Saeid Hosseini et al. [52] investigated the generation of biogas for future operations at laboratory techniques. They noticed that achieving raw biogas, there involves the retrieval of solvate methane and liquids at the exhibition stage, as well as the ill treatment of semipermeable and gas-liquid membranes. They concluded that delivering methanation and hydrogenation processes by membranes helps in the upgradation of biogas. Srisowmeya et al. [53] examined the display of biogas from rice wastewater and cow dung slurry using a two-stage AD (TSAD) activity to augment biogas capitulate and reduce the impacts of acidogenesis and methanogenesis. They noticed that when rice wastewater is used to single-stage digestion, accumulation of volatile fatty acid (VFA) and activity unpredictability took place due to high organic load and biodegradability, biogas yield enhanced by 58.69% and cumulative energy healing by 1.58 times.

Siying Zhang et al. [54] explore the impact of anaerobic co-digestion of food with kitchen and garden wilt (FW, KW, and GW) under thermophilic dry conditions for biogas production. They noticed that FW-KW and FW-GW co-digestion increased the biogas exhibited by 24.69% and 44.96% and increased the OLR tolerance by inhibiting ammonia and volatile fatty acids gathering. They observed that co-digestion increased the Defluviitoga bacteria which enhances substrate hydrolysis. They concluded that the FW-GW co-digestion processes augmented the biogas yield by accelerating gene abundance and ABC transporters- associated genes. Steven Nartker et al. [55] explored the energy yield and enhancement of methane content of biogas from primary sewage sludge within a mesophilic anaerobic co-digestion ranging from 0 to 60% of the non chemical loading rate (OLR), the optimum glycerol loading. They noticed the biogas production yield is 82 to 280% augment in specific gas yield. They concluded that mesophilic anaerobic co-digestion of sewage sludge is an environmentally friendly process to yield biogas.

Fig. 1. Deenbandhu type construction [13]

Xing Su et al. [56] investigated biogas production by prediction of a kinematic model associated with anaerobic digestion in cold climate under dynamic characteristics of solar radiation by considering air temperature and other variable parameters. They achieved 55% of the average volumetric biogas rate. They concluded that the optimizing operation strategy of solar-assisted biogas plants. Figure 1 and Fig. 2 are used for production of biogas through digesters by using cattle dung as feedstock, the productivity is utilizing for cooking purposes as one of the house hold application. Figure 3 depicts the maximum cumulative biogas yield at 15% TS for duckweed.

To measure the biomass crystallinity index [13]

$$C_i = \left(\frac{I_{002} - I_{am}}{I_{002}} \right) \times 100 \quad (1)$$

where I_{002} is crystalline peak of maximum intensity and I_{am} is the minimum intensity.

Fig. 2. Biogas digesters of 4 m^3 volume [13]

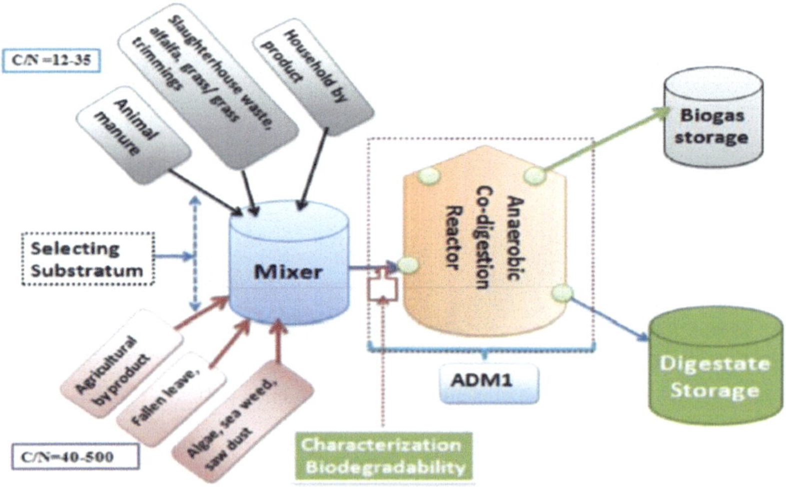

Fig. 3. Various organic materials for anaerobic co-digestion system [47]

Smaller is the better condition [32]

$$\frac{S}{N} = -10\log\left(\frac{1}{n}\sum_{i=1}^{n} y_i^2\right) \tag{2}$$

Larger is the better condition [32]

$$\frac{S}{N} = -10\log\left(\frac{1}{n}\sum_{i=1}^{n} \frac{1}{y_i^2}\right) \tag{3}$$

Normal is the better condition [32]

$$\frac{S}{N} = 10\log\left(\frac{\bar{y}}{s_y^2}\right) \tag{4}$$

The anaerobic digestion efficiency [42]

$$\%\eta_{AD} = \frac{UMY}{TY_{Methane}} \times 100\% \tag{5}$$

where, UMY is the ultimate methane yiled and $TY_{Methane}$ is the theoretical methane yield.

The volume of the biogas production [50]

$$V_{biogas} = \frac{(P_2 - P_1)T_a}{P_a \times T_r} V_r \tag{6}$$

where, V_{biogas} is volume of dialy biogas production in ml, P_1 is after released headspace pressure of the previous day in kPa, P_2 is headspace pressure before biogas release in kPa, P_a is ambient pressure in kPa, T_r is temperature of the reactor in K, V_r is headspace volume in ml.

The average methane content [50]

$$WA = \frac{\sum_{i=1}^{n} DBP_i \times MP_i}{\sum_{i=1}^{n} DBP_i} \tag{7}$$

where, WA is average methane content (%), DBP_i is biogas production rate of the i day in ml/gVS.day, MP_i is daily methane content of the i day (%).

The biohydrogen and biomethane production analyzed using the following equation of kinetic model [53]

$$P(t) = P_{\max} + \exp\left[-\exp\left(\frac{R_{\max} \times e}{P_{\max}}(\lambda - t) + 1\right)\right] \tag{8}$$

where P(t) is the cumulative hydrogen or biomethane yield over time t (mL/g VS), P_{max} is maximum biohydrogen or biomethane yield (mL/g VS), R_{max} is maximum biohydrogen or biomethane production rate (mL/d), λ is lag phase (d), e – kinetic constant i.e. 2.71828

Biogas production

$$V_{STP} = \frac{V_T \times 273 \times (760 - P_w)}{(273 + T) \times 760} \tag{9}$$

where V_{STP} is the volume of biogas (L) measured at standard temperature (^{0}C) and pressure (1 atm), V_T is the volume of biogas measured at temperature T(L), P_w is the water vapour pressure as a function of temperature (mm Hg). T is the temperature of the gas (^{0}C).

Gas production efficiency [56]

$$v_d = \frac{B_0 S_0}{HRT\,\beta_{CH_4}}\left(1 - \frac{K}{HRT \times \mu_m - 1 + K}\right) \tag{10}$$

where,

$$K = 0.6 + 0.0206e^{0.051S_0} \tag{11}$$

For cow manure [56]

$$\mu = 0.0026e^{0.1319T_i} \tag{12}$$

When temperature ranging from 15 to 30 °C.

Figures 4 and 5 indicates the biogas yield from duckweed and rice straw feedstocks for 36 days. In Fig. 4, biogas produced is initially increased till 16th day then gradually decreased till ending day and 15% TS produced maximum cumulative biogas. In Fig. 5, 10% TS produced maximum biogas on 4th day.

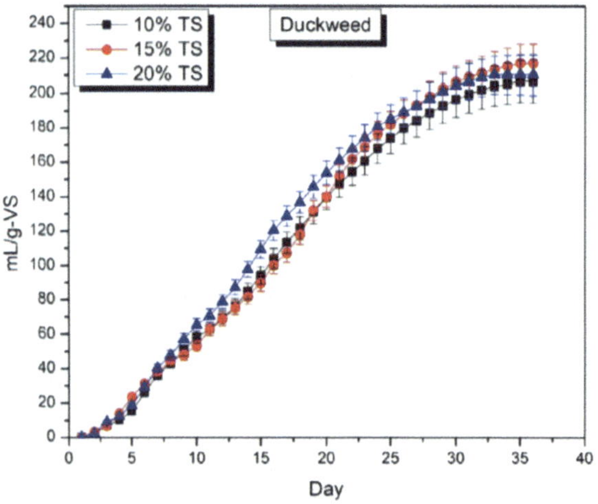

Fig. 4. Daily biogas versus duckweed [13]

From Fig. 6, it is clear that both DLT and gas production efficiency (GPR) are enhanced after optimization of feeding time. The average volumetric GPR is increased by 0.77 m^3/m^3.d, i.e. 15% with respect to 0.67 m^3/m^3.d of non-optimization. Figure 7 shows the volumetric biogas production before and after integrated optimization of both aspects. From the Fig. 7, the average volumetric biogas yield enhanced to 1.04 m^3/m^3.d, i.e. 55% higher were compared to before optimization.

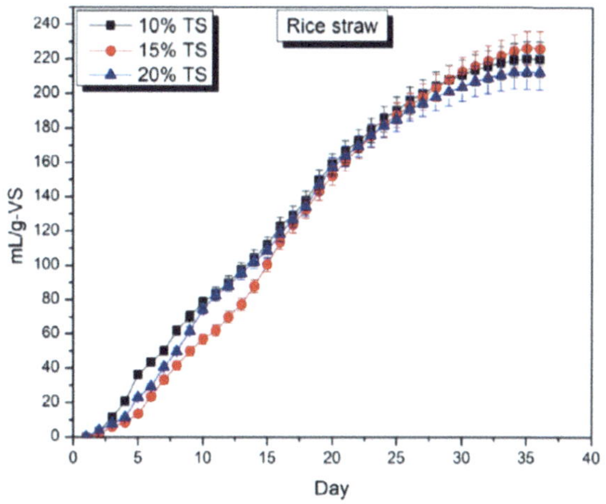

Fig. 5. Cumulative biogas versus rice straw [13]

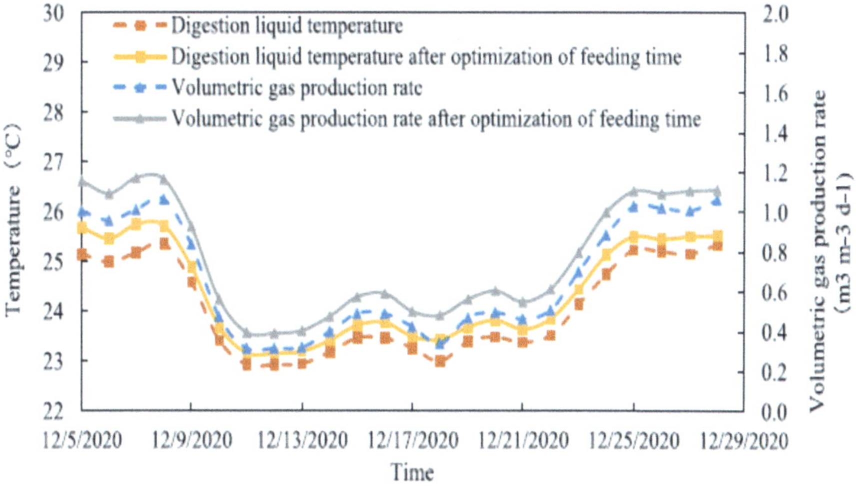

Fig. 6. Biogas yield before and after optimization of feeding time [56]

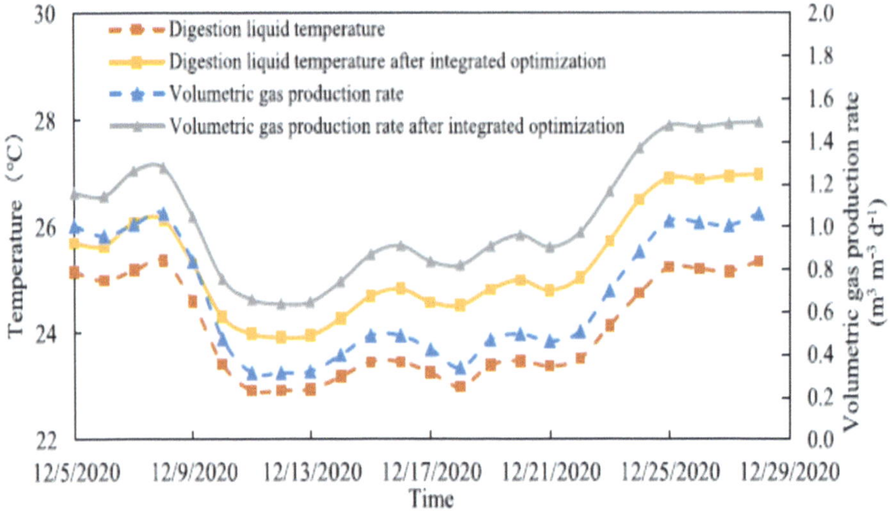

Fig. 7. Biogas yield before and after integrated optimization [56]

2 Conclusion

The exploration of biomass for biogas production through anaerobic digestion highlights the inherent potential of renewable energy sources to address critical global challenges. This review has illuminated how biomass, while often perceived as waste, can be harnessed as a valuable resource. Challenges in maximizing methane yields and optimizing process stability must be addressed, yet ongoing innovation in pre-treatment, reactor design, and process control point to a future where these hurdles are overcome. By embracing anaerobic digestion technology, alongside a commitment to sustainability, communities can reduce environmental footprints, decentralize energy production, and foster greater resource efficiency. The widespread adoption of biogas generation requires a multi-faceted approach driven by technological advancements, policy support, and public awareness. Such a collaborative effort signals a profound shift toward a future where energy production and environmental stewardship are deeply intertwined.

References

1. Somanchi, N.S., Gugulothu, R., Sagi, S.L.S., Kumar, T.A., Vijaya, K.: Proceedings of 4th International Conference on Hydrology and Watershed Management with a Focal Theme on Ecosystem Resilience-Rural and Urban Water Requirements, 29th October to 1st November, pp. 470–474 (2014)
2. Gugulothu, R., Somanchi, Devendar N.S., Kaluri, G.P., Deepika, N.: Mater. Today: Proce. 4, 2 Part-A, 314–321 (2017)
3. Somanchi, N.S., Anjaneya Prasad, B.R., Gugulothu, R.K., Nagula, S., Phanindra Dinesh, K.: Int. J. Energy Power Eng. **4**, 33–37 (2015)
4. Gugulothu, R., Somanchi, N.S., Reddy, V.K., Akkiraju, K.K.: Mater. Today: Proc. **4**(2), Part-A 1045–1050 (2017)

5. Gugulothu, R., Somanchi, N.S., Devi, S.R., Vilasagarapu, R.D.: Journal of Sustainable Manuf. Rener. Energy, Nova Sci. Publishers, Inc **3**(1–2), 121–128 (2014)
6. Peng Z., Chen, X., Yao, L.: Sustain. Fut. **3**, 100042 (2021)
7. Celik I., Demirer, G.N.: Biocatal. Agric. Biotechnol. **4**(4), 767–772 (2015)
8. Yilmaz, S., Sahan, T.: Biomass Bioenerg. **138**, 105601 (2020)
9. Khadaroo, S.N.B.A., Grassia, P., Gouwanda, D., Jing, H., Poh, P.E.: Biomass Bioenergy **151**, 106167 (2021)
10. Malik, S.N., Madhu, K., Mhaisalkar, V.A., Vaidya, A.N., Mudliar, S.N.: Biomass Bioenerg. **142**, 105780 (2020)
11. Risberg, K., Sun, L., Levén, L., Horn, S.J., Schnürer, A.: Technol. **149**, 232–237 (2013)
12. Mishra, A., Kumar, M., Bolan, N.S., Kapley, A., Kumar, R., Singh, L.: Biores. Technol. **338**, 125514 (2021)
13. Ramos-Suárez, J.L., Martínez, A., Carreras, N.: Energy Convers. Manage. **88**, 1263–1270 (2014)
14. Buragohain, S., Mahanta, P., Mohanty, K.: Environ. Technol. Innov. 24, 101981 (2021)
15. Buragohain, S., Mohanty, K., Mahanta, P.: Hybridization of solar photovoltaic and biogas system: experimental, economic and environmental analysis. Sustain. Energy Technol. Assess. **45**, 101050 (2021)
16. Sumardiono, S., Matin, H.H.A., Hartono, I., Choiruly, I.I., Budiyono.: Mater. Today: Proc. **63**, S477–S483 (2022)
17. Sumardiono, S., Matin, H.H.A., Sulistianingtias, I., Nugroho, T.Y., Budiyono B.: Mater. Today: Proc. (2023)
18. Cioabla, A.E., Dumitrel, G.A., Popescu, F., Lelea, D., Trif-Tordai, G.: Mater. Today: Proc. **4**(Part 1), 6757–6763 (2017)
19. Hotta, S.K., Sahoo, N., Mohanty, K.: Renew. Energy **134**, 1307–1319 (2019)
20. Hotta, S.K., Sahoo, N., Mohanty, K., Kulkarni, V.: Renew. Energy **150**, 854–867 (2020)
21. Hashemi, S.S., Karimi, K., Taherzadeh, M.J.: Bioresource Technol. **337**, 125356 (2021)
22. Guo, G., Li, Y., Zhou, S., Chen, Y., Qin, Y., Li, Y.-Y.: Biores. Technol. **346**, 126583 (2022)
23. Yang, S., et al.: Biores. Technol. **373**, 128712 (2023)
24. Sun, H., et al.: Biores. Technol. **285**, 121338 (2019)
25. Sun, H., et al.: Biores. Technol. **305**, 123042 (2020)
26. Sun, H., Wang, E., Li, X., Cui, X., Guo, J., Dong, R.: Renew. Sustain. Energy Rev. **148**, 111360 (2011)
27. Sun H., Li, J., Cui, X., Stinner, W., Guo, J., Dong, R.: J. Clean. Prod. **319**, 128826 (2021)
28. Sun, H., Li, R., Wang, E., Guo, J., Zhou, Y., Dong, R.: J. Clean. Prod. **339**, 130770 (2022)
29. Nemr, A., et al.: Ultrason. Sonochem. **90**, 106197 (2022)
30. Avicenna, M.M., Ihsan, S.I., Setyobudi, R.H.: Procedia, Chem. **14**, 91–100 (2015)
31. Jeihanipour, A., Karimi, K., Niklasson, C., Taherzadeh, M.J.: Waste Manag. **30**, 2504–2509 (2010)
32. Jeihanipour, A., Aslanzadeh, S., Rajendran, K., Balasubramanian, G., Taherzadeh, M.J.: Renew. Energy **52**, 128–135 (2013)
33. Deepanraj, B., Sivasubramanian, V., Jayaraj, S.: Energy Conver. Manag. **141**, 429–438 (2017)
34. Chen, B., Chen, S.: Energy Policy **62**, 707–716 (2013)
35. B. Saha, B., Barua, V.B., Khwairakpam, M., Haq, I., Kalamdhad, A.S., Varjani, S.: Environ. Res. **216**(Part: 3), 114661 (2023)
36. Veluchamy, C., Gilroyed, B.H., Kalamdhad, S.: Fuel **253**, 1097–1103 (2019)
37. Bustillo-Lecompte, C.F., Mehrvar, M.: J. Clean. Prod. **141**, 278–289 (2017)
38. Devi, K., Manikandan, M., Oviyapriya, S., Selvaraj, M.M., et al.: Bioresource Technol. **363**, 127871 (2022)
39. Coskun, C., Bayraktar, M., Oktay, Z., Dincer, I.: Int. J. Hydrogen Energy **37**(21), 16498–16504 (2012)

40. Yu, D., Kurola, J.M., Lahde, K., Kymalainen, M., Sinkkonen, A., Romantschuk, M.: J. Environ. Manag. **143**, 54–60 (2014)
41. Arias, D.E., Veluchamy, C., Habash, M.B., Gilroyed, B.H.: J. Environ. Manage. **284**, 112027 (2021)
42. Wolf D., von Canstein, H., Schroder, C.: Journal of Natural Gas Sci. Eng. **3**(5), 625–632 (2011)
43. Almomani, F., Bhosale, R.R., Khraisheh, M.A.M., Shawaqfah, M.: Fuel **253**, 964–974 (2019)
44. Filarsky, F., Schmuck, C., Schultz, H.J.: Chem. Eng. Res. Des. **134**, 257–267 (2018)
45. Valenti, F., et al.: Land Use Policy **66**, 151–161 (2017)
46. Pasalari, H., Ghasemian, M., Esrafili, A., Gholami, M., Farzadkia, M.: Ecotoxicol. Environ. Saf. **247**, 114222 (2022)
47. Li, H., Chen, Z., Dun, F., Wang, Y., Zheng, Y., Li, Q.: Renew. Energy **146**, 2330–2336 (2020)
48. Hagos K., Zong, J., Li, D., Liu, C., Xiaohua L.: Renew. Sustain. Energy Rev. **76**, 1485–1496 (2017)
49. Lan, M., Zhang, L., Zhu, K., Ma, J., Ifran, M., Li, A.: Sci. Total Environ. **704**, 135429 (2020)
50. Andrade, L.R.S., Cruz, I.A., Fuess, L.T., Iqbal, H.M.N., Bharagava, R.N., et al.: Biocatal. Agric. Biotechnol. **48**, 102653 (2023)
51. Gikas, P., Zhu, B., Batistatos, N.I., Zhang, R.: J. Environ. Manage. **216**, 96–104 (2018)
52. Pramanik, S.K., Suja, F.B., Zain, S.M., Pramanik, B.K.: Bioresource Technol. Rep. **8**, 1003 (2019)
53. Hosseini, S.S., Tabar, M.A., Vankelecom, I.F., Denayer, J.F.: Separ. Purific. Technol. **310**, 123139 (2023)
54. Srisowmeya, G., Chakravarthy, M., Bakshi, A., Nandhini Devi, G.: Biomass Bioenergy **152**, 106184 (2021)
55. Zhang, S., et al.: Biores. Technol. **361**, 127684 (2022)
56. Nartker, S., Ammerman, M., Aurandt, J., Stogsdil, M., Hayden, O., Antle, C.: Waste Manage. **34**(12), 2567–2571 (2014)
57. Xing, S., Shao, X., Geng, Y., Tian, S., Huang, Y.: 2022 Renew. Energy **197**, 59–68 (2022)

Fabrication and Experimental Study on Al6082 + B_4C Metal Matrix Composite

K. Anusha[1], S. Hemalatha[2(✉)], and K. Vahini[1]

[1] Department of Mechanical Engineering, Lendi Institute of Engineering and Technology, Vizianagaram 535005, India

[2] Department of Mechanical Engineering, Raghu Engineering College, Visakhapatnam, AP 531162, India

hemalatha.somireddi@gmail.com

Abstract. Composites represent a new class of materials that perform better than traditional alloys and pure materials. Although there are many matrix materials used in composites, the most widely used of these is Aluminium (Al), Iron, Magnesium, Titanium, and Copper, due to the nature of having a lightweight material with reliable mechanical properties as well as high thermal and electrical conductivity. Its ductility, malleability, low coefficient of thermal expansion, and superior corrosion resistance make it more versatile and suitable for applications in various industries. These desirable properties make Aluminium a preferred matrix material for various applications. In the family of Aluminium Alloys (AA), AA 6082, part of the 6000 series alloys, is particularly popular in automotive and other engineering applications due to its excellent mechanical properties and usability. Some particulate reinforcements added in aluminum matrices are SiC, Al_2O_3, Boron, B_4C, and TiC, which enhance the stiffness, specific strength, fatigue resistance, conductivity, creep resistance, and wear property of aluminum matrices. Among them, Boron Carbide has a very high hardness, excellent chemical resistance, low density, and strong improvement of wear and corrosion resistance of composites. In this work, The B_4C reinforcing effect on AA6082 matrix composites at various weights is examined. The AA6082 matrix composites were reinforced using a stir-casting technique by incorporating 1 wt%, 2 wt%, 3 wt%, and 4 wt% B_4C into the matrix. The compositions were characterized to examine the mechanical properties, such as yield strength, tensile strength, percentage elongation, and hardness, in order to analyze how the mechanical properties are affected by the B_4C reinforcement.

1 Introduction

Aluminium matrix composites (AMCs) have attracted significant attention from material scientists and researchers in recent years due to their lightweight properties and enhanced mechanical and physical characteristics. AMCs combine the low density of aluminium with reinforcement materials such as ceramics or carbides, resulting in significantly improved strength, hardness, and wear resistance. The B_4C particles can be considered as one of the best reinforcing materials in the market because it possesses maximum hardness, excellent chemical resistance, and lower density. This paper focuses on the

P. Joseph D. et al. (Eds.): IEMDST 2024, SPPHY 567, pp. 315–323, 2025.
https://doi.org/10.1007/978-981-96-5863-3_40

effects of B_4C reinforcements in the AA6082 matrix composites by varying the weight percentages. B_4C is added at 1%, 2%, 3%, and 4% weight fractions into the AA6082 matrix, using a stir-casting setup, and then these prepared composites are weighed for measuring density, tensile strength, and hardness.

Raj Kumar et al. [1] studied the stir-casting process of aluminum matrix composites reinforced with 5–15 wt% B_4C. In their observations, they reported that higher content of B_4C showed an increase in hardness but a decrease in impact strength due to higher brittleness. Further, their investigation reported that higher cutting speeds reduced the chip-tool contact length and cutting forces; however, higher feed rates and depth of cut increased cutting forces due to friction and tool wear. Saikeerthi S P [2] has explored the composites for Al6061 alloy with the varied size of B_4C and SiC particles, 37 μm to 250 μm. They concluded that for both varied particle sizes and percent weight enhancement of micro hardness, maximum was 129 VHN for 250 μm in this exploratory study.

S. Rama Rao et al. [3] have prepared aluminum-B_4C composites with 2.5, 5, and 7.5 wt% reinforcement and reported uniform dispersion of particles, improvement in hardness, and reduction in the density of the composites. Niranjan et al. [4] have found that when ceramic particles are added to a ductile matrix then the strength improves due to residual stress and dislocation density generated by the difference in thermal expansion of the matrix and reinforcement. Doel and Bowen [5] have observed that aging has increased tensile strength and proof stress of Al7075-SiC composites but reduced in the case of ductility.

Zhang Fang et al. [6] studied friction behavior of Al6061 composites with SiC and Al_2O_3 under high loads. It was found that friction and wear change with the appearance of ploughing, adhesion, and particle fracture. Lin et al. [7] showed that the presence of reactive films by EP additives in oil-lubricated Al6061-graphite composites has a significant effect on friction, but no reactive film is formed under boundary lubrication. Chen et al. [8] reported that the friction on a steel surface reduces due to carbon nanofiber reinforcement-based coatings on AA6061 aluminium through the formation of lubrication layers.

Candan et al. [9] reported that adding SiC to Al-4 wt% Mg produced composite corrosion resistance in NaCl solutions through the formation of Mg_2Si phases, which withhold electrons. Kiourtsidis et al. [10] determined that pitting corrosion in the composite AA2024 was due not because of the addition of SiC but the intermetallic phase near the particle location. They attributed the pits closer to the interdendritic zones galvanically where Al_2Cu acted as cathodes and α-phases were anodic. Aylor et al. [11] concluded that the SiC in AA6061 composites did not increase the susceptibility of pitting in marine environments, although pits were formed at the SiC/Al interface.

2 Material Selection and Experimentation

2.1 Aluminium as a Metal Matrix

The American Aluminium Association proposed the nomenclature for Aluminium Matrix Composites, to be designated based on their constituents: the accepted designation of the matrix, the abbreviation of the reinforcement, the arrangement and volume fraction (%) of the reinforcement, and the symbol for the shape of the reinforcement. More typical examples, based on their chemical compositions and application categories, would include Aluminum Alloy AA6082. Aluminum AA6082 is a wrought aluminum-magnesium-silicon alloy from the 6000 or 6xxx series. Within that series, it is perhaps one of the most common after 6005, 6061, and 6063. Though less popular in the ASTM standards of North America, AA6082 is produced generally by the means of extrusions and rolling. It is a wrought alloy, and while it cannot usually be cast, under very few conditions it can sometimes be forged or clad. AA6082 cannot be work-hardened, but is normally heat-treated to obtain tempers that give increased strength at the expense of reduced ductility. Tables 1 and 2 shows the chemical composition and the properties of AA6082.

Table 1. Aluminium 6082 chemical composition

Elements	Al	Cr	Cu	Fe	Mg	Mn	Si	Ti	Zn	Residuals
Weight %	95.2 to 98.3	0.25% max	0.1% max	0.5% max	0.6 to 1.2%	0.4 to 1.0%	0.7 to 1.3%	0.1% max	0.2%	0.15 max

Table 2. Properties of 6082 Aluminium

Properties	Density	Melting point	Electrical Resistivity	Young's modulus	Ultimate Tensile strength	Yield strength
Value	2.72 gm/cc	555 °C	0.038x10–6 Ω-m	71 GPa	140 to 330 MPa	280 MPa

3 Boron Carbide (B_4C) as Reinforcement

Boron carbide or B_4C is one of the hardest known ceramic compounds and consists of a crystalline composition made from boron and carbon. The compound is mostly utilized as a manufactured product for various applications in abrasion- and wear-resistant applications, with its most recognized usages being in ballistic vests and tank armor. Owing to high hardness and relative light weight, B_4C-reinforced aluminium matrix composites have attracted significant attention, especially through low-cost casting routes. The Table 3 following is a summary of different properties of the B_4C compound.

4 Method for Fabrication

In order to explore the mechanical characterization of the developed Al6082 metal matrix composite, various mechanical tests have been conducted and are discussed below.

Table 3. Properties of B_4 C

Properties	Density (g/cm^3)	Melting Point (°C)	Hardness (kg/mm)	Fracture Toughness (MPa)	Young's Modulus (GPa)	Electrical Conductivity (at 25 °C) (S)	Thermal Conductivity (at 25 °C) (W/m-K)
Value	2.52	2763	2900 - 3580	2.9 – 3.7	450 - 470	140	30 - 42

Stir Casting: In this fabrication technique is also called vortex technique, in which reinforcement (ceramic particulates and/or short fibers etc.) is mixed with a molten metal matrix with the help of mechanical stirring. This technique required a heating furnace, a stirrer attached electric motor with speed controller and temperature sensors. In this method, ceramic particulates (which are generally preheated for good wettability) are added into the liquid molten metal and mixed then by stirrer then the composite slurry is allowed to solidify. Generally, up to 30% ceramic particulates is possible to add in a variety of molten aluminium alloys. To reduce porosity, refinement of micro-structure, and uniform distribution of the reinforcement in Al matrix the AMMCs are further extruded after fabrication. Segregation of the reinforced particulates due to surfacing or settling during the melting and casting are the major concerns associated with this process. The particle distribution in the molten matrix depends upon the geometry, position of impeller, speed and characteristics of the added particles. The setup for stir casting is shown in Fig. 1.

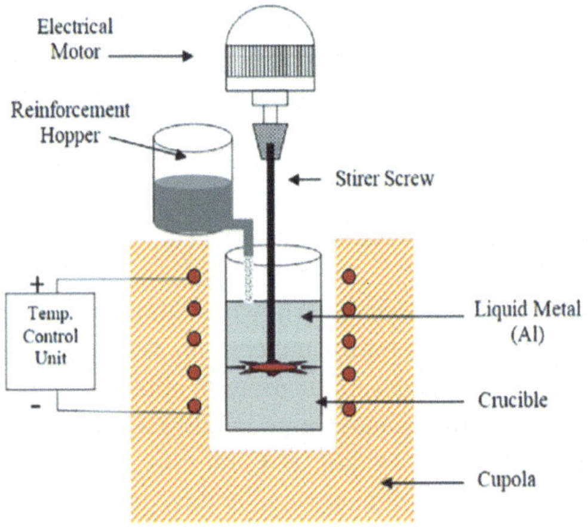

Fig. 1. Stir casting setup

5 Mechanical Characterization

Material characterization encompasses the measurement and determination of physical, chemical, mechanical, and microstructural properties of materials. Therefore, it brings about a better understanding that leads to the solution of some of the main issues, such as the causes of material failure and the process-related problems, thus allowing material manufacturers to make informed decisions on material choices and processing.

Tensile Testing: Tensile Testing: involves the use of a Universal Testing Machine by subjecting a material to a controlled uniaxial tensile force until material failure. The test yields crucial data regarding its mechanical properties, including yield strength, ultimate tensile strength, elongation, and reduction of area.

Hardness Test: The surface properties of a material that are important include its hardness, defined as the resistance to indentation, penetration, or scratching. The most common method for measuring hardness is the Brinell Hardness Test. In this test, a hard indenter is pressed into the surface of the material under a specified load, and then the size of the permanent indentation produced is measured. Materials that are harder than each other produce shallower indentations while the softer materials yield deeper indentations.

$$\text{BHN} = \frac{2P}{\pi D(D - \sqrt{D^2 - d^2})} \tag{1}$$

where, d = Average value of the vertical and horizontal diameter of the indentation, F = Test force applied (in Newtons), D = Diameter of the test ball (indenter).

6 Results and Discussion

The experimental results for the densities and mechanical properties of the Al6082/B_4C metal matrix composite (MMC) are summarized in Tables 4 and 5 and graphs below.

7 Tensile Testing

Ultimate Tensile Strength: The ultimate tensile strength of a metal matrix composite material is significantly influenced by the amount of reinforcement used. The results for the ultimate tensile strength of the Al6082 + B_4C specimens, as shown in Table 4 below, indicate a remarkable increase in UTS compared to the base material (Al6082).

Table 4. Variation of ultimate strength with respect to composition

Composition	Al6082	Al6082 + 1% B_4C	Al6082 + 2% B_4C	Al6082 + 3% B_4C	Al6082 + 4% B_4C
Ultimate strength (MPa)	148	156	162	168	172

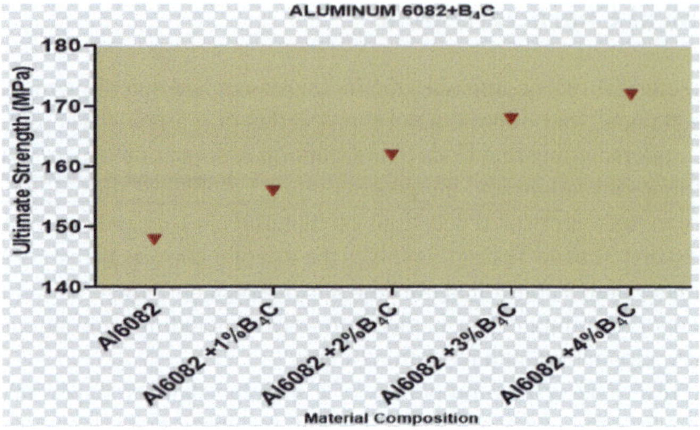

Fig. 2. Graphical representation of increased ultimate strength.

Figure 2 is showing how ultimate tensile strength of specimens AL6082 + B_4C is increasing when compared to the base material.

Yield Strength: The yield strength of a metal matrix composite material is significantly influenced by the amount of reinforcement used. The values shown in the Table 5 indicate a remarkable increase in the yield strength of Al6082 + B_4C specimens when compared to the base material.

Table 5. Change in the yield strength with respect to composition

Composition	Al6082	Al6082 + 1% B4C	Al6082 + 2% B4C	Al6082 + 3% B_4C	Al6082 + 4% B_4C
Yield Strength (MPa)	123	132	140.22	143.91	148

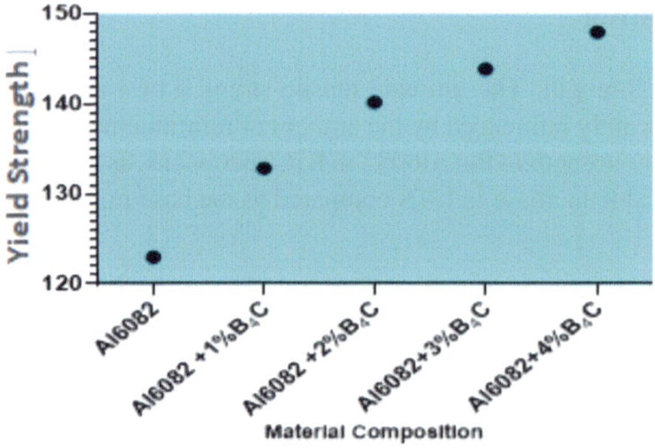

Fig. 3. Graphical representation of increased yield strength.

Figure 3 is showing how yield strength of specimens AL6082 + B_4C is increasing when compared to the base material.

% Elongation: The Elongation of a metal matrix composite material depends on the amount of reinforcement materials, which is shown in Table 6 illustrates indicate that there is a remarkable decrease in the elongation when compared to the base material.

Table 6. Variation of the % of elongation with respect to composition

Composition	Al6082	Al6082 + 1% B_4C	Al6082 + 2% B_4C	Al6082 + 3% B_4C	Al6082 + 4% B_4C
% Elongation	14.6	12.41	11.68	11.0	10.1

The elongation of Al6082 + B_4C specimens is decreased when compared to the base material. The elongation values for the composites are shown in Fig. 4, highlighting this reduction.

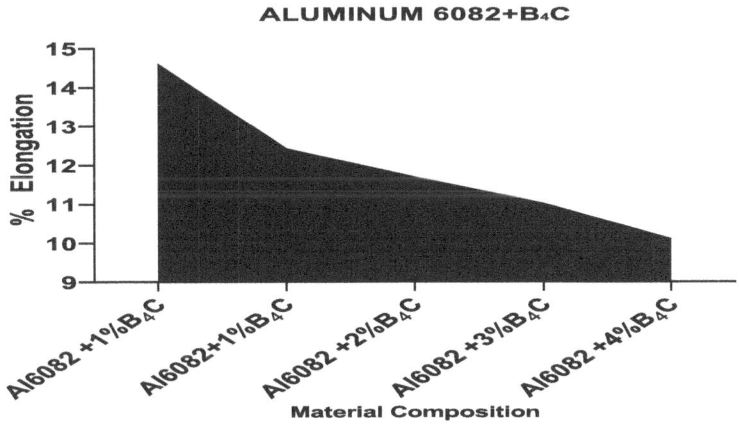

Fig. 4. Graphical representation of decreased % of elongation

Hardness Test: The hardness of the specimens with AL6082 and B_4C increased significantly compared to the non-reinforced metal matrix (base metal) as shown in Table 7.

Table 7. Variation of hardness with the addition of B_4C

Composition	Al6082	Al6082 + 1% B_4C	Al6082 + 2% B_4C	Al6082 + 3% B_4C	Al6082 + 4% B_4C
Hardness BHN	70	75	83	89	93

Figure 5 illustrates that the Hardness (BHN) value for the hybrid composite material is higher when the reinforcement material is incorporated.

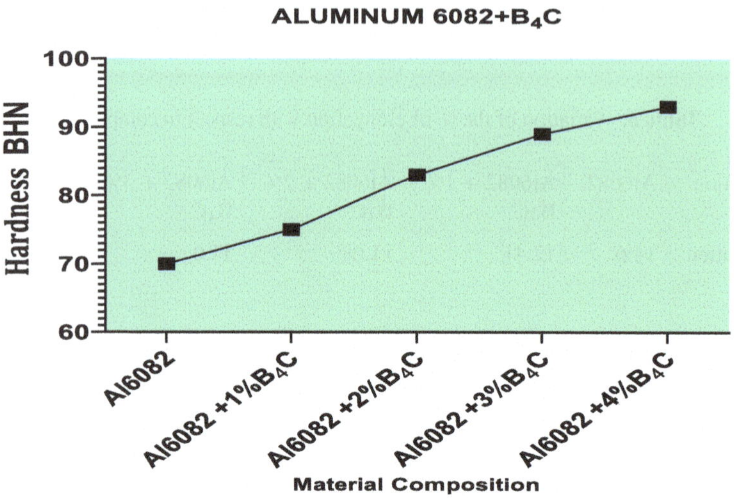

Fig. 5. Graphical representation of increase in hardness of the material.

8 Conclusion

This work prepared an Al6082 metal matrix composite reinforced with the non-metallic ceramic material B_4C through the stir-casting technique and investigated its mechanical characterization. The following pertinent observations for this study are as follows:

- With an increase in B_4C, tensile strength and yield strength for the composite increased; in contrast, elongation decreased, whereas hardness increased.
- The composite added showed tremendous improvements in tensile strength, yield strength, and hardness when compared to the base metal Al 6082.
- The characteristics of B_4C or other matrix materials found in AMMCs were much better than that in their unreinforced metals and alloys.
- Reinforcing increased mechanical resistance in aluminum/aluminum alloy matrices.
- The micro-hardness of composites increases with an increase in the B_4C particle size, but the corresponding ultimate tensile strength drops.

B_4C reinforced Al6082 produces some optimistic mechanical properties. A detailed report of the investigation on the proposed AMMC mechanical properties varying in weight percentage of reinforcement materials is presented in this report. Future studies can evaluate more significant variations of mechanical properties and employ additional reinforcement techniques to improve the performance of such composites.

References

1. Rajkumar, K., Maria Antony Charles, J., Vinoth Kumar, K., John Charles Praveen, J., Padmanathan, A.: Proceedings of the National Conference on Emerging Trends in Mechanical Engineering, pp. 306–312 (2015)
2. Saikeerthi, S.P., Vijayaramnath, B,, Elanchezhian, C.: Int. J. Eng. Res. **3**(1), 70–73 (2014)
3. Rama Rao, S., Padmanabhan, G.: Int. J. Mater. Biomater. Appl. **2**(3), 15–18 (2012)
4. Niranjan, H.B., et al.: Tensile and corrosive properties of Al Hematite composite. In: Proceedings of the First Australasian conference on Composite Materials (ACCM-1), Osaka, Japan, 7–9 October, vol. 546 (1998)
5. Doel, T.J.A., Bowen, P.: 1996 Compos. Part-A: Appl. Sci. Manuf. **27**(8), 655–665 (1996).https://doi.org/10.1016/1359-835X(96)00040-1
6. Zhang, F., Liangehi, Z., Yiu, W.M.: Wear **176**(2), 231–237 (1994). https://doi.org/10.1016/0043-1648(94)90151-1
7. Lin, J.F., Shih, M.G., Chen, Y.W.: Wear **198**(1–2), 58–70 (1996).https://doi.org/10.1016/0043-1648(96)06932-3
8. Chen, M.Y., Bai, Z., Tan, S.C., Unroe, M.R.: Wear **252**(7–8), 624–634 (2002).https://doi.org/10.1016/S0043-1648(02)00021-2
9. Candan S., Bilgic, E.: Mater. Lett. **58**, 2787–2790 (2004).https://doi.org/10.1016/j.matlet.2004.04.009
10. Kiourtsidis G., Skolianos, S.M.: Mater. Sci. Eng. A **248**, 165–172 (1998).https://doi.org/10.1016/S0921-5093(98)00494-8
11. Aylor, D.M., Moran, P.J.: J. Electrochem. Soc. **321**(6), 1277–1281 (1985).https://doi.org/10.1149/1.2114101

Arduino Built Automatic Tool Change CNC Plotting Machine

R. Tamilselvan(✉), B. Varun, and S. Veerakumar

Department of Mechanical Engineering, Sri Ramakrishna Institute of Technology, Coimbatore, Tamil Nadu 641010, India
tamilselvanrathinasamy@gmail.com

Abstract. The evolution of CNC technology has transformed manufacturing. This study explores integrating automatic tool change into a CNC-based plotting machine for enhanced efficiency and versatility. The system automates transitions between drawing tools, facilitating diverse plotting without manual intervention. This feature is valuable for drafters, producing attractive and clear drawings. The study examines the impact on productivity, precision, and execution of intricate designs. The findings contribute insights to CNC machining, particularly in plotting and design. The proposed machine marks a significant advancement for adaptable and efficient manufacturing in drafting and design.

1 Introduction

The introduction of an automatic tool change system holds the promise of transforming the landscape of CNC machining. By enabling the machine to autonomously switch between different tools, we anticipate a substantial reduction in downtime associated with manual interventions. This enhancement not only accelerates operational workflows but also opens new avenues for the application of CNC-based plotting machines, making them more adaptable to the diverse needs of modern manufacturing and design. This paper aims to explore the intricacies of implementing an automatic tool change mechanism in CNC-based plotting machines. From design considerations to mechanical and electronic components, and the intricacies of programming, we delve into the critical aspects that define the successful integration of this feature. Moreover, we seek to evaluate the impact of automatic tool change on overall productivity, precision, and the machine's ability to execute intricate designs.

2 Literature Survey

This section, a succinct overview of the studies conducted by different authors is discussed. Badamasi, et al. [1] an insightful analysis of the fundamental working principles of Arduino, a popular choice for simplifying electronics projects. It briefly explains the hardware and software along with its definition makes it easy to understand. Ismailov, et al. [2] an overview of the Arduino microcontroller, emphasizing its working principles and diverse applications. It highlights the important reasons to use Arduino board. The

P. Joseph D. et al. (Eds.): IEMDST 2024, SPPHY 567, pp. 324–333, 2025.
https://doi.org/10.1007/978-981-96-5863-3_41

paper underscores the Programming Arduino using C/C++ Languages. It generally uses two main functions. Though it holds limitations, it is a good tool for learning and experimenting. Kanchan D Mahajan, et al. [3] The X-Y Plotter has powerful ability to draw larger diagrams with fast pace to help artists and designers. This two-axis controller is efficient and accurate since it uses two stepper motor for x and y direction and a servo motor for pen hold. Makelangelo software is used to create G codes out of given input. Precise system can be embedded to get accurate and precise coordinates in future. Jahnavi, et al. [4] the goal is to design a three-axis computer-controlled potable machine, serving as a valuable resource for learning fundamental robot functions. While similar to a CNC machine, this plottable machine, essentially a CNC router with a writing tool, is open-source and replicable. Idris. O.Y. and Kotoye A.A., [5] Introduces a CNC plotter designed for versatile applications in drawing, writing, and sketching. It comprises an input port, a microcontroller [Arduino], a motor, and a movable pen. The specific focus of this work is on using the machine to produce text as an output, demonstrating its practical utility in industrial settings. Ibrahim, M. A. T. B. et al. [6]. The design and implementation of a CNC writing machine using an Arduino microcontroller. It builds upon the versatility of Arduino in CNC applications, aiming to contribute to the practical implementation of precise control in writing machines. This project builds upon existing knowledge to contribute a practical implementation, showcasing the potential of Arduino-based CNC systems in various applications. Anjanah. D, et al. [7] Designing a cost-effective three-axis CNC plotter machine, utilizing a stepper motor, Arduino UNO, and microcontroller software. Its computer numerical control principle allows input from a computer program. The CNC plotter has a plotting area of 8×9 cm.

3 Materials and Methods

The methodology section outlines the systematic approach employed in the development of the CNC pen plotter, covering the design, construction, and testing phases. The project was executed in a series of well-defined steps to ensure precision, efficiency, and the successful realization of the pen plotter. The project commenced with a comprehensive analysis of the requirements. This involved identifying the primary objectives, functional specifications, and user expectations. Clear project goals were established, including the ability to produce accurate and intricate pen-drawn designs, user-friendly control, and compatibility with open-source CNC tools. The primary objective of the CNC pen plotter project was to design and develop a versatile and accurate plotting system capable of creating intricate drawings and designs on various surfaces. The specific objectives included automatic tool change, precision and user friendly. The functional specifications were derived from the project objectives and include linear motion system, automatic tool change, electronics integration, GRBL plotter software compatibility. Understanding user expectations was paramount to designing a system that would meet the needs of the intended audience. Key considerations include ease of use, versatility, reliability.

Utilizing SolidWorks, a detailed 3D model of the CNC pen plotter was developed. This stage involved creating virtual representations of each component, ensuring precise dimensions, and validating the overall design concept. The design implementation of the CNC pen plotter involved a comprehensive approach to component design and assembly, seamlessly facilitated by SolidWorks. To provide a tangible representation of our virtual design, STL files were generated for each 3D-modeled part, ensuring a smooth transition from the digital realm to the physical prototype. GRBL-Servo Firmware Implementation: The Arduino microcontroller was programmed with the specialized "GRBL-servo" firmware to accommodate the automatic tool change mechanism. This modification enabled control of the servo motor, ensuring seamless transitions between different drawing tools. GRBL-Plotter Software Integration: GRBL-Plotter, a dedicated control software, was integrated into the system. This involved sending G-codes to instruct the CNC machine's actions and using the software's graphics converter for efficient G-code generation from images or graphics.

4 3D Printing and Assembly

STL File Generation: The first step is exporting the SolidWorks-designed components into STL [Standard Triangle Language] files. These files encapsulate the geometry and details of each component in a format compatible with 3D printing technologies. The STL files serve as the blueprint for constructing the physical parts. Material Selection: Choosing the appropriate 3D printing material is crucial for achieving the desired structural integrity and functionality. Common materials include PLA [Polylactic Acid] for its ease of use and versatility, or ABS [Acrylonitrile Butadiene Styrene] for enhanced durability. Printing Parameters: It is crucial to calibrate the 3D printer with ideal values for infill density, layer height, and printing speed. The final strength and quality of the printed components are affected by these factors. To get the best results, it could be necessary to test and make adjustments repeatedly. The CNC pen plotter underwent rigorous testing to validate its performance, precision, and overall functionality. Real-time monitoring and motion simulation tools were employed to assess and refine the design. In the realization of the "Arduino built Automatic Tool Change CNC Pen Plotter" project, the design implementation phase played a pivotal role in shaping the physical components and ensuring their seamless integration. "SolidWorks", a robust Computer-Aided Design [CAD] software, served as the cornerstone for the creation and assembly of various elements within the system.

5 Integration of 3D Parts

The design implementation of the CNC pen plotter involved a comprehensive approach to component design and assembly, seamlessly facilitated by SolidWorks. To provide a tangible representation of our virtual design, STL files were generated for each 3D-modeled part, ensuring a smooth transition from the digital realm to the physical prototype. SolidWorks allowed for the export of each designed component in STL format, capturing the intricate details and specifications of the 3D models. This file format, recognized across various CAD and 3D printing platforms, served as a universal bridge between the virtual and physical realms. The exported STL files were instrumental in reconstructing the complete assembly in physical form. Each file represented a specific component, from the chassis and linear motion system to the tool holder and electronics enclosure. The design implementation phase, powered by SolidWorks, played a crucial part in transforming conceptual ideas into tangible, functional components. The software's versatility, precision, and simulation capabilities facilitated an efficient and iterative design process, ultimately contributing to the achievement of the CNC pen plotter project. The seamless integration of SolidWorks in component design and assembly underscored its significance in the overall development lifecycle.

6 Hardware Implementation

The NEMA 17 stepper motor is a widely used and standard stepper motor size in the CNC [Computer Numerical Control] and robotics community. "NEMA" represents the National Electrical Manufacturers Association, and "17" refers to the faceplate size of the motor. The MGN15H linear rail is a miniature, high-profile linear motion component commonly used in machinery and automation systems. It consists of a rail, along which a carriage with ball bearings moves smoothly, providing precise and stable linear motion. This type of linear rail is often employed in applications such as CNC machines and 3D printers.

The A4988 is a stepper motor driver integrated circuit that enables precise control of stepper motors, supporting micro stepping for smoother motion. Widely used in applications like 3D printers and CNC machines, it allows for adjustable current control, interprets step and direction signals, and includes protection features such as thermal shutdown and overcurrent protection. Function: Translates digital signals for stepper motor control. An electromechanical tool that precisely controls acceleration, velocity, and angular or linear position is called a servo motor. It continuously modifies its position in response to the input signal using feedback from an encoder or resolver. Servo motors are commonly employed in robotics, automation, and other systems requiring accurate and controlled motion.

The Arduino Uno is an open-source microcontroller board featuring a versatile ATmega328P microcontroller. Equipped with digital and analog I/O pins, USB interface for programming, and multiple power options, it provides an accessible platform for electronics projects. An Arduino CNC shield is a specialized expansion board that interfaces with an Arduino microcontroller to control CNC [Computer Numerical Control] machines. It typically incorporates stepper motor drivers, power connectors, limit switch inputs, and features for spindle control.

7 Software Implementation

To bring the CNC pen plotter to life, the installation of firmware onto the Arduino microcontroller is a crucial step. While the widely adopted open-source GRBL firmware is a common choice for DIY CNC machines, a modified version called "GRBL-servo" is essential for pen plotters like ours. This tailored version of GRBL is specifically designed to control a servo motor, a critical component for precise pen manipulation. The standard GRBL firmware primarily focuses on spindle control for traditional CNC machines. However, the "GRBL-servo" modification becomes imperative for our pen plotter due to its unique requirements. The key alteration involves adjusting the default PWM frequency of the spindle control pin [typically digital pin 11] from the standard

1 kHz to 50 Hz. This modification ensures compatibility with and effective control of the servo motor used in the pen plotter setup.

In conjunction with the "GRBL-servo" firmware, the GRBL-Plotter control software serves as the interface for sending G-codes to instruct the CNC pen plotter's actions. This dedicated software, designed specifically for plotter control, includes a graphics converter that simplifies the generation of G-code from images or graphics. Being an open-source tool available on Github, GRBL-Plotter provides flexibility for further customization based on project requirements. The GRBL-Plotter software serves as the indispensable interface that bridges the user's commands with the CNC pen plotter's actions. Specifically designed for plotter control, GRBL-Plotter is a powerful open-source tool that streamlines the process of sending G-codes to instruct the CNC machine during plotting operations.

8 Operating Theory

The CNC-based automated tool change plotting machine operates through a sophisticated system that seamlessly integrates computer control and automated tool change mechanisms. At its core is a CNC controller, often powered by a microcontroller like Arduino, interpreting G-code instructions generated from CAD files. Stepper motors drive linear motion systems, which consist of linear rails and lead screws, allowing precise movement in the X, Y, and Z axes. The plotting tool, whether a pen, pencil, or specialized instrument, is intricately controlled by the CNC controller to follow a predefined toolpath. This toolpath, calculated based on the G-code instructions and design requirements, is executed with high precision. What distinguishes this machine is the automated tool change mechanism. A tool holder or magazine stores various tools, and a robotic or mechanical arm moves to select and replace tools as needed, enhancing efficiency and versatility. Sensors, including limit switches and encoders, provide feedback to controller to ensure accurate positioning and prevent collisions. The entire process, from CAD integration and G-code interpretation to toolpath planning and execution, allows for the creation of intricate designs on the plotting surface, making the CNC-based automated tool change plotting machine a powerful and adaptable tool for applications ranging from drafting to artistic endeavors.

The working of an CNC based Automatic tool change plotting machine Fig. 1 (a) is break down into the following process that includes: Control System: The heart of the system is a CNC controller, often powered by a microcontroller [e.g., Arduino or specialized CNC controller] running a pre-programmed code generated from CAD [Computer-Aided Design] files. Stepper Motors: Stepper motors control the movement of various components in the machine. These may include motors for X, Y, and Z axes to control the position of the tool and, potentially, additional motors for tool changes. Linear Motion Systems: Linear motion systems, such as linear rails and lead screws, facilitate precise and controlled movement of the plotting tool in the X, Y, and Z directions. This ensures accurate positioning of the tool over the plotting surface. Plotting Tool: The plotting tool could be a pen, pencil, or other writing instrument. In some cases, it might also be a cutting tool or other specialized tool depending on the machine's application. Tool Change Mechanism: The automated tool change mechanism allows the machine to switch between different tools seamlessly. This mechanism often involves a tool holder or magazine where various tools are stored. A robotic or mechanical arm moves to pick up and replace tools based on the programmed instructions.

Sensors and Feedback Systems: Sensors, like limit switches and encoders, provide feedback to the controller, ensuring accurate positioning and preventing collisions. In order to confirm that a tool change was successful, these sensors may also be utilized in the procedure. Computer-Aided Design [CAD] Integration: The CNC machine receives instructions from CAD files, specifying the toolpath, designs, and any tool change requirements. These commands are interpreted by the CNC controller, which then converts them into exact movements and tool changes. G-code Interpretation: The CNC controller reads G-code instructions, a set of commands that dictate tool movements and operations. The G-code is generated by CAM [Computer-Aided Manufacturing] software based on the original CAD design. Toolpath Planning: The CNC controller calculates the toolpath based on the G-code instructions, taking into account the specific tool being used, the material properties, and any tool changes required during the process. Execution of Toolpath: The CNC controller executes the toolpath, commanding the stepper motors to move the plotting tool precisely along the programmed path. This results in the creation of the desired design or pattern on the plotting surface.

Automated Tool Change Sequence: When a tool change is needed, the CNC controller initiates the automated tool change sequence. This involves moving the plotting tool to a designated tool change position, activating the tool change mechanism to replace the tool, and then resuming the plotting process. By combining these elements, a CNC-based automated tool change plotting machine can efficiently execute complex designs with multiple tools, providing versatility and precision in various applications such as drafting, prototyping, and artistic creations.

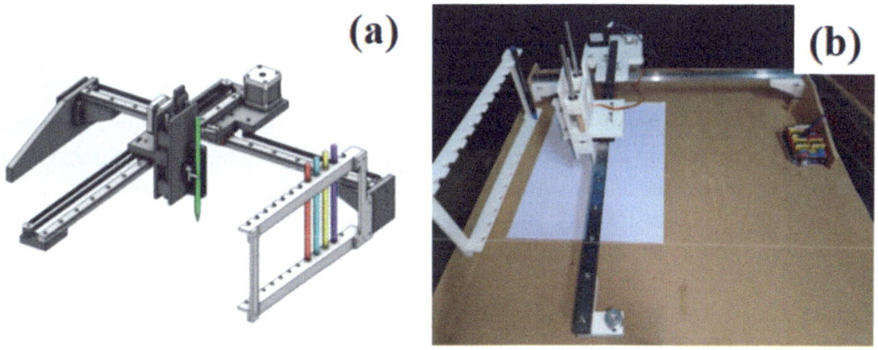

Fig. 1. (a) Design of Automatic Tool Change Pen Plotter and (b) Automatic Tool Change Plotting Machine.

9 Testing and Validation

The testing and validation phase plays a pivotal role in ensuring the robustness, precision, and reliability of the CNC pen plotter. This chapter details the comprehensive evaluation conducted to verify the functionality of various components and the overall performance of the system.

Automatic Tool Change Testing: Executing meticulously scripted scenarios ensured rigorous validation of the CNC pen plotter's automatic tool change mechanism. The systematic tests verified the machine's capability to execute accurate and efficient transitions between different drawing tools, a critical feature enhancing adaptability and efficiency in precision-based tasks. Linear Motion System Testing: Subjecting the pen plotter to predefined patterns and intricate designs allowed for a comprehensive evaluation of the precision and smoothness of its linear motion system. This step played a pivotal role in ensuring the machine's ability to execute intricate designs with various drawing tools while maintaining accuracy. Homing and Limit Switches Verification: Thorough testing of homing and limit switches validated the establishment of a precise reference point and the prevention of potential collisions within the predefined work area. These crucial safety features contribute to the overall reliability and safe operation of the CNC pen plotter.

Tool Holder Mechanism Testing: The assessment focused on the tool holder mechanism's ability to securely hold various drawing tools and release them accurately during the automatic tool change process. Stability and reliability during tool changes were key aspects ensuring consistent performance. Vibration and Stability Check: Conducting tests to identify vibrations or instabilities during tool changes ensured the stability of the tool holder mechanism. This step aimed to eliminate potential disruptions that could impact the overall accuracy and reliability of the CNC pen plotter. GRBL-Servo Firmware Evaluation: Thorough testing confirmed the compatibility and effectiveness of the GRBL-Servo firmware, specifically designed for controlling the servo motor during automatic tool changes. The seamless control of the servo motor represents a critical aspect of the CNC pen plotter's functionality.

GRBL-Plotter Software Assessment: Evaluation of the GRBL-Plotter software included rigorous testing for accurate G-code generation from images or graphics. Scrutinizing the user interface and real-time monitoring features ensured the software's capability to provide a user-friendly and efficient control platform. Real-time Monitoring and Motion Simulation: Utilizing monitoring tools for dynamic visualization of tool movements during plotting allowed for real-time assessment. Additionally, simulating intricate designs facilitated the identification of any irregularities, contributing to the refinement of the pen plotter's precision. User Experience Testing: This step involved assessing the overall ease of operation, covering design uploading and tool changes. User feedback was actively gathered to pinpoint areas for usability improvements, ensuring a user-friendly and accessible interface.

Versatility and Reliability Tests: Comprehensive testing was conducted to evaluate the CNC pen plotter's versatility with various drawing tools and materials. Ensuring consistent performance across diverse scenarios validated the machine's reliability in meeting different plotting requirements. Iterative Refinement: Incorporating feedback from the testing phases, the iterative refinement process aimed to enhance design and functionality. Iterative adjustments were systematically made to address identified issues, optimizing overall performance and user experience. This iterative approach was crucial in achieving a finely-tuned CNC pen plotter ready for deployment in creative applications. This systematic approach to testing and validation ensured that the CNC pen plotter met design specifications, delivering accurate and reliable plotting capabilities with a user-friendly interface.

10 Applications

The CNC pen plotter across various domains. The versatility and precision of the machine open avenues for creative and practical applications, showcasing its relevance in different industries. Architectural Drafting: The CNC pen plotter finds application in architectural drafting, producing detailed and accurate drawings. It enables architects to visualize and communicate design concepts effectively, contributing to the efficient development of architectural plans. Engineering Prototyping: In engineering, the CNC pen plotter serves as a valuable tool for prototyping components and designs. Its ability to create precise and scaled drawings aids in the rapid iteration of prototypes, streamlining the product development process. Educational Tool: As an educational tool, the CNC pen plotter offers hands-on learning opportunities in CNC technology, automation, and design. It provides students with a practical understanding of manufacturing processes and computer-aided design.

Artistic Creations: The machine's adaptability to various drawing tools makes it a powerful tool for artists. It can create intricate and detailed artworks, allowing artists to explore new possibilities in digital and traditional art forms. Cartography and Mapping: In cartography, the CNC pen plotter is employed to produce precise maps and charts. Its accuracy ensures that maps convey geographical information with high precision, making it a valuable tool for mapmaking. Textile Design: The CNC pen plotter contributes to textile design by creating detailed patterns and designs. It offers a unique approach to designing textiles, allowing for intricate and customized fabric patterns. Scientific Illustrations: In scientific research, the CNC pen plotter assists in creating accurate scientific illustrations. Whether it's illustrating biological specimens or complex scientific concepts, the machine ensures precision in representation. Custom Signage and Lettering: For businesses and signage purposes, the CNC pen plotter excels in creating custom signage and lettering. Its ability to work with various materials makes it a versatile solution for producing visually appealing and customized signs. Decorative Woodworking: Woodworking enthusiasts can utilize the CNC pen plotter for decorative woodworking projects. It enables the creation of intricate patterns and designs on wooden surfaces, enhancing the aesthetic appeal of woodworking projects. Graphic Design: In graphic design, the CNC plotter can bring digital designs to physical form. Designers can experiment with different drawing tools to create unique and visually appealing graphics with a tangible output. The diverse applications outlined demonstrate the CNC pen plotter's adaptability and significance across various industries. From architectural drafting to artistic creations, the machine's precision and versatility make it a valuable asset in enhancing efficiency and creativity in different fields. As technology continues to evolve, the CNC pen plotter is poised to play a pivotal role in shaping the future of design and manufacturing applications.

11 Results and Discussion

The successful execution of automatic tool changes demonstrates the machine's enhanced adaptability and efficiency. Precise and smooth linear motion, confirmed through predefined patterns and intricate designs, underscores the CNC pen plotter's ability to handle complex tasks with accuracy. The establishment of a precise reference point and collision prevention by homing and limit switches contributes significantly to the systems overall reliability and safety. The tool holder mechanism's secure holding, accurate release, and stability during changes affirm its crucial role in the CNC pen plotter's reliability. The compatibility and effectiveness of the GRBL-Servo firmware, coupled with the accuracy of the GRBL-Plotter software, showcase a successful integration meeting project requirement. Real-time monitoring tools and simulation provide insights for refining precision and addressing irregularities. Positive user feedback in ease of operation and identified areas for improvement guide iterative refinements. The demonstrated versatility and consistent performance across diverse scenarios underscore the CNC pen plotter's reliability in meeting varied plotting requirements. These findings not only validate project objectives but also lay the foundation for future advancements in precision-based plotting systems.

12 Conclusion

The CNC pen plotter project has culminated in a noteworthy achievement, bringing forth innovations in automatic tool change, precise linear motion, and a user-friendly interface through GRBL-Plotter software. The successful integration of these features not only meets the project's primary objectives but also positions the CNC pen plotter as a promising solution for precision-based tasks in diverse industries, including architecture, engineering, and design. Valuable lessons learned throughout the development process will guide future endeavours, and the project's success underscores the significance of interdisciplinary collaboration and iterative refinement. The CNC pen plotter stands as a testament to the boundless possibilities at the intersection of automation, design, and manufacturing. As technology evolves, this project represents a step forward in enhancing efficiency and adaptability in creative applications.

References

1. Badamasi, Y.A.: The working principle of an Arduino. In: 2014 11th International Conference on Electronics, Computer and Computation ICECCO, pp. 1–4. IEEE, September 2014
2. Ismailov, A.S., Jo'Rayev, Z.B.: Study of Arduino microcontroller board. Sci. Educ. **3**(3), 172–179 (2022)
3. Mahajan, K.D., Vishwakarma, P., Wadkar, B., Rawat, N.: X-Y plotter robot using mXY board **23**(5), 1678–1683 (2023)
4. Jahnavi, A., Nagulmeera, S.K., Abdulla, S.D., Naveen Kumar, T.: Arduino based 2D CNC plotter. IJCRT **11**(4), 185–189 (2023)
5. Idris, O.Y., Kotoye, A.A.: Designing a CNC plotter machine. Int. J. Innov. Eng. Res. Technol. **8**(08), 66–70 (2021)

6. Ibrahim, M.A.T.B., Omer, M.O.S., Mohamed, M.M.A.: Design and implementation of CNC writing machine based on Arduino micro-controller. J. Karary Univ. Eng. Sci. (2021). https://doi.org/10.54388/jkues.v1i2.53
7. Anjanah, D., Kaushal, A., Gautam, A., Arun Kumar, G.: CNC plotter machine. Int. J. Eng. Res. Technol. **10**(11), 357–360 (2022)

Numerical Investigation of Performance Characteristics of TI-6AL-4V and IN - 738 Gas Turbine Blade

G. Ramya Sree(✉), Ch. Sai Krishna, P. Dilleshwar, and B. D. Y. Sunil

Department of Aerospace Engineering, Institute of Aeronautical Engineering, Hyderabad 500043, Telangana, India
ramyasree.swaeroe@gmail.com

Abstract. Material advancements in the field of aviation particularly in gas turbine engine play a critical factor in overall development of an aircraft. The performance of a critical component namely turbine blade in gas turbine engine which works under high pressure and temperature environment undergoes various mechanical failures like fatigue, creep, erosion, oxidation etc. Advanced alloys like Al – Ti (Ti-6Al-4V) and Nickel based alloy (IN – 738) promises certain in range parameters that enhance the turbine blade against mechanical wear and tear. So, a numerical investigation is performed to find out the performance of the materials under working conditions of the actual gas turbine blade. A turbine blade similar to GE F404 has been considered for testing out the material role in thermal management in the blade during operation of the gas turbine engine by performing CFD analysis in ANSYS Fluent. After analysis, by comparing the results of materials Aluminium – Titanium alloy gives better performance than the Nickel based alloy for a solid blade.

1 Introduction

In advanced technology materials are playing vital role in every field for making better lifestyle. Materials like alloys, composites, nano materials etc., under research for future advancements and also some types of materials are used in some fields. In aircraft different parts were manufactured with different materials according to conditions which the part undergoes, like combustion chamber having high temperatures so it needs material which has high melting temperature, good flexural strength, etc.,

Similarly, turbine blades will perform their operation under high temperatures and pressures, the material used for turbine blades should be able to perform the operations for longer period without any blade losses [1]. There are many advanced materials under research, some of the suitable materials were known for the gas turbine blades from the literature survey (by studying various research papers, articles etc.,). Considering two different alloys for gas turbine blade based on the literature survey, to know how materials are responding for the conditions on the blade with the help of modelling and simulation software's (CATIA V5 and Ansys) [2–5].

Gas turbine engine is one of the types of internal combustion engine which converts fuel into mechanical energy by burning fuel in a continuous combustion process [6].

P. Joseph D. et al. (Eds.): IEMDST 2024, SPPHY 567, pp. 334–341, 2025.
https://doi.org/10.1007/978-981-96-5863-3_42

Gas turbine engine has different components like air inlet, compression, combustion, expansion, exhaust. Every component has its own specific functionality. Gas turbine blades are of two types compressor blades and turbine blades. These blades have vast differences in their operations, shape, size and also material used for both blades. The compressor blades are designed to intake and compress ambient air, increasing its pressure and temperature to optimize the combustion process. Turbin blades extract energy from the high pressure, high-velocity exhaust gases produced during combustion [7].

Gas Turbine Blade: Gas turbine blade is a specialized component designed to extract energy from the hot flow i.e., high pressure gases which comes from the combustion process where the blade was shown in the Fig. 1.

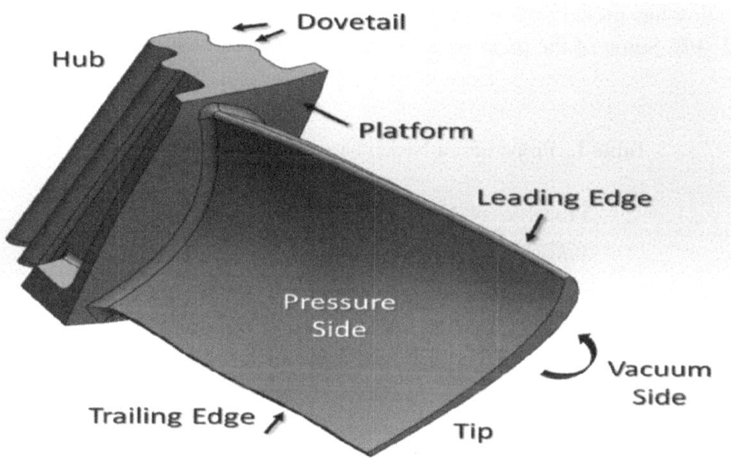

Fig. 1. Gas Turbine Blade Features [10].

Figure 2 Block diagram will give an overview of the paper in a short note. Required data was collected from the study of literature survey (by analysing the research papers), two materials (Nickel based alloy and Aluminium – titanium alloy) were considered in this paper. From the collected data 3D model was generated with appropriate dimensions based on GE-F404 turbofan engine using Catia v5 software, converting Catia file to Ansys file then generating fluid volume and mesh in Ansys, by giving the appropriate set up for the both materials to the according to the expected outcomes, results were analysed from the solution and grid generation method was used to verify the results obtained in analysis and solution. Objective of the paper is to study the distribution of flow over the gas turbine blade and to know the effect of flow distribution. To study the turbine blade performance by selecting the aluminium-titanium alloy and nickel-based alloy.

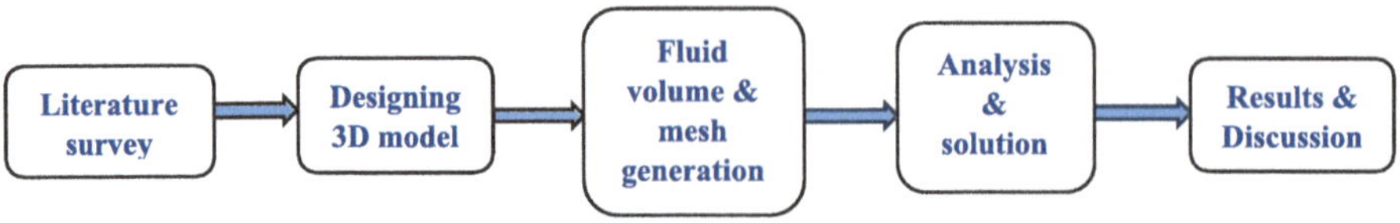

Fig. 2. Block diagram of the paper.

2 Materials

For gas turbine blades two different alloy materials i.e., nickel - based alloy (INCONEL – 738) and Aluminium – titanium alloy (6Al-4V-Ti) materials are used for this paper. Al-Ti alloy has 90% titanium, 6% aluminium and 4% vanadium and other minute materials, Inconel alloy has nickel 58%, chromium 20–23%¸ cobalt 9–11%, iron- 7–9%, molybdenum 2–3%. Some of the properties of these alloys are given in the below Table 1 and Table 2.

Table 1. Properties of Nickel based alloy (Inconel – 738) [8]

S. No	Properties of Alloy (A_1)	Al – Ti
1	Poisson's Ratio	0.34
2	Specific Heat (C_p)	526 J/Kg K
3	Density	4430 kg/m^3
4	Thermal conductivity	7 W/m-K
5	Melting Point	1604–1660 °C

Table 2. Properties of Aluminium – Titanium alloy (6Al-4V-Ti) [9]

S. No	Properties of Alloy (A_2)	IN – 738
1	Poisson's Ratio	0.30
2	Specific Heat (C_p)	510 J/Kg K
3	Density	8850 kg/m^3
4	Thermal conductivity	14.3 W/m-K
5	Melting Point	1230–1315 °C

3 Design and Analysis

Designing of gas turbine blade was done in the Catia v5 R21 software, before starting the model initial dimensions should be ready before starting the blade model (Fig. 3). By using different tools and techniques, from 2D design to 3D design. Dimensions were taken as reference only (Table 3).

Table 3. Dimensions of gas turbine blade design.

S. No	Parameters	Value (mm)
1	Chord length	39
2	Blade length	80
3	Thickness	5
4	Total length (Blade + Root)	112

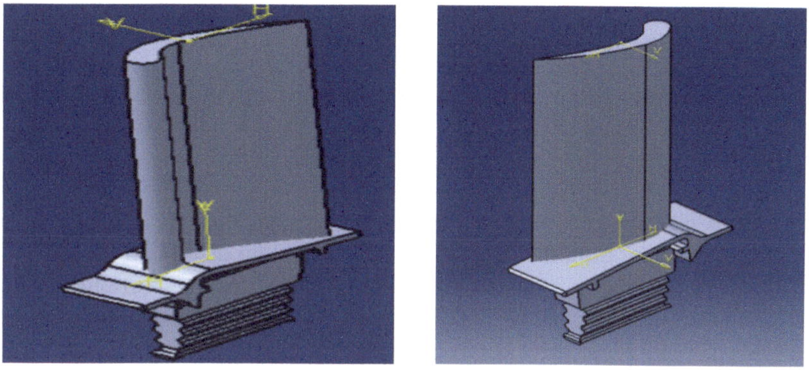

Fig. 3. Turbine blade design in Catia v5 (3D model).

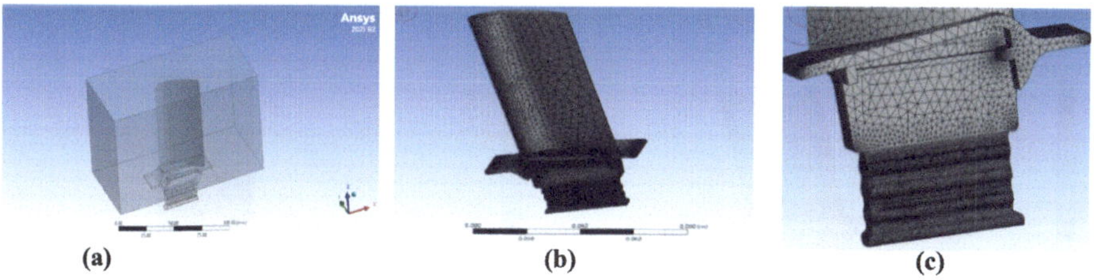

Fig. 4. (a) Design modeler done in Ansys (b) meshing for gas turbine blade (blade + root) (c) mesh for the root of the blade.

After completion of 3D model in Catia v5, design modeler in Ansys was used to generate fluid volume and mesh. Fluid volume was created around the 3D blade to better fluid flow during analysis as shown in Fig. 4(a). Further, mesh will be generated using Ansys R2 software i.e., by using fluent with fluent mesh or fluent mesh. Element size can be changed or can continue with the default element size (9.9557e–003 m) and type of mesh can be changed according to the flexibility of design (tetrahedral, polyhedral, etc.). In Fig. 4(b) and (c) generation of mesh can be seen for the turbine blade. To get refined mesh element size should be reduced for complex designs. After generating mesh, in the setup module giving the properties of material, select the material, giving boundary conditions, mesh interfaces. These are general steps for the turbine blade setup.

After setup, initialize the setup and give the required contours, surfaces, vectors, path lines, animations, particle tracks etc., which are used for capturing the results during software calculation. For this paper temperature contour and pressure contours are selected for the analysing the performance of the turbine blade. To get better results 1000 iterations are considered for 100time steps. After calculation, using contours the results are estimated and comparing the results of Aluminium – titanium (6Al-4V-Ti) and Nickel – based alloy (In -738).

4 Results and Discussion

4.1 Results of Material Ti - 6Al – 4V

In the results pressure contour was considered as shown in Fig. 5(a) and 5(b) pressure distribution on gas turbine blade for the material Ti-6Al-4V (upper surface, lower surface). Initial input for the pressure was given as 30 bars after analysis the output was maximum pressure 3.429135e +08 Pa. By observing the contour of pressure, the upper surface has low pressure and lower surface has high pressure this is because of the design i.e., separation of flow from the leading edge. Temperature distribution of gas turbine blade for material aluminum – titanium alloy can be seen in the Fig. 6(a) and 6(b) (upper surface and lower surface). Initial input was given for the temperature 1773 K and the output was 1.77e+03 K. As there is no change from the input to output. Though there is a minor distribution as the initial temperature was 300 K after attaining the temperature 1773 K it remain constant for the entire analysis.

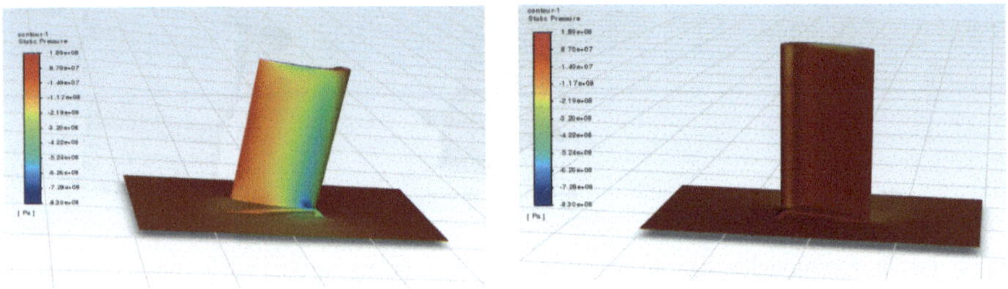

(a) (b)

Fig. 5. Pressure distribution on a Ti-6AL-4V gas turbine blade (a) upper surface (b) lower surface

4.2 Results of Material (IN- 738)

Figure 7(a) and 7(b) are the results of pressure distribution on gas turbine blade for material nickel -based alloy, the initial input was 30 bars and after analysis the output was 7.23e + 07 Pa and also there is a pressure difference from the upper and lower surfaces. Upper surface has low pressure and lower surface has high pressure because of the flow separation.

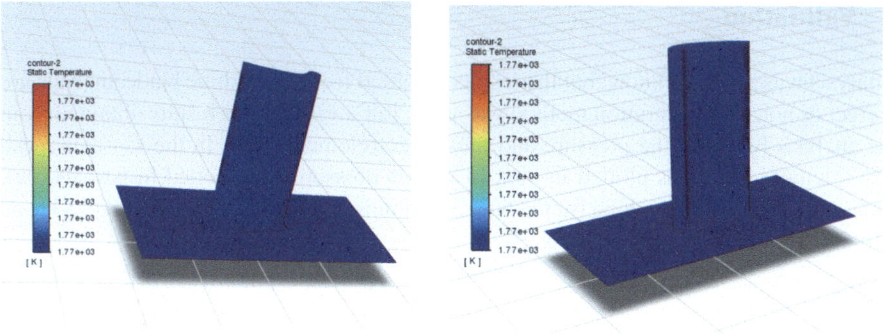

(a) **(b)**

Fig. 6. Temperature distribution on a Ti-6AL-4V turbine blade (a) upper surface (b) lower surface

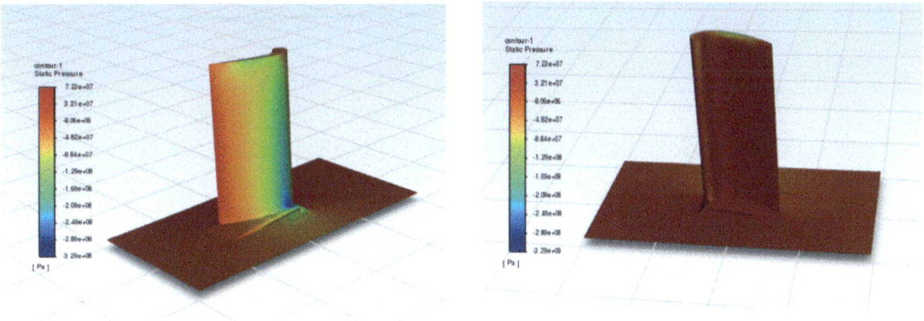

(a) **(b)**

Fig. 7. Pressure distribution on a IN-738 gas turbine blade (a) upper surface (b) lower surface

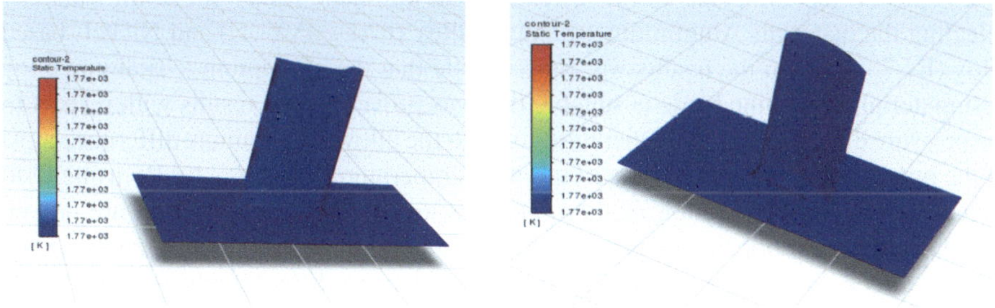

(a) **(b)**

Fig. 8. Temperature distribution on a IN-738 turbine blade (a) upper surface (b) lower surface.

5 Validation

Analysis was done. To know that the results obtained from the turbine blade analysis are correct or not, grid generation method was used. Let's validate the results obtained from the analysis by using the pressure and mesh size (element size). In the x – axis mesh size (element size) was plotted in mm, in y -axis pressure was plotted in pascals (pa). This grid generation was done for the both aluminum – titanium alloy and nickel – based alloy for the results validation (Fig. 9).

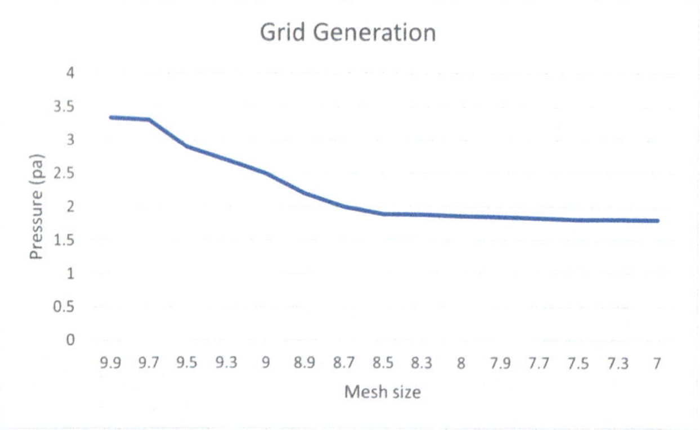

Fig. 9. Grid generation graph for both alloys.

6 Conclusion

In this project turbine blade analysis was done to know the behaviour of the blade by selecting the materials Aluminium – titanium alloy (6Al – 4V – Ti) and Nickel- based alloy (IN-738). From the results we can conclude that the aluminium – Titanium alloy best material for turbine blade is suitable this was proven by the results with respect to the pressure distribution. As temperatures of both the alloys have minute difference most of the turbine blade as constant distribution of temperature (Figs. 8 a, b). By using Grid generation method (validation method) analysis and results of this project are proven as valid or correct. For the further scope this project can be extended by applying the cooling effects to the turbine blade and also by selecting different advanced materials and also for the comparison with other recent advanced materials.

References

1. Rybnikov, A.I., Getsov, L.B., Leontiev, S.A.: Microsc. Microanal. **11**(Suppl 2). Copyright 2005 Microscopy Society of America (2005). https://doi.org/10.1017/S1431927605509292
2. Aabid, A., Jyothi, J., Zayan, J.M., Khan, S.A.: Adv. Mater. Res. **8**(4), 275–293 (2019). https://doi.org/10.12989/amr.2019.8.4.275

3. Teja, P.S., Babu, P.K., Rao, K.K., Babu, M.M., Sreenivasan, M.: Mater. Today: Proc. **33**, 1038–1043 (2020). https://doi.org/10.1016/j.matpr.2020.07.054
4. Yadav, M., Misra, A., Malhotra, A., Kumar, N.: Mater. Today: Proc. **25**, 639–645 (2020). https://doi.org/10.1016/j.matpr.2019.07.530
5. Singh, H.P., Rawat, A., Manral, A.R., Kumar, P.: Mater. Today: Proc. **44**, 63–69 (2021). https://doi.org/10.1016/j.matpr.2020.06.486
6. Narendranath, G., Suresh, S.: Int. J. Eng. Res. Appl. **2**(5), 2021–2027 (2012)
7. Naeem, M.T., Jazayeri, S.A., Rezamahdi, N.: Proceedings of the 2008 IAJC-IJME International Conference. ISBN 978-1-60643-379-9
8. https://nickelinstitute.org/media/4690/ni_inco_497_alloy738.pdf
9. https://asm.matweb.com/search/specificmaterial.asp?bassnum=mtp641
10. https://www.researchgate.net/figure/Gas-turbine-blade-nomenclature_fig4_346594509

Comparison of Linear Regression, Power Regression and Logarithmic Model Developed Using MATLAB to Determine Relationship Between Compressive Strength and Flexural Strength of Blended Concrete

P. Raj Kumar(✉)

Department of Civil Engineering, National Institute of Technology, Warangal 506004, Telangana, India
manukumarpuppala@gmail.com

Abstract. The association between compressive strength (C) and flexural strength (F) is vital for the investigation of materials, particularly in civil engineering and construction. In general, there is an association between C and F, as both are affected by comparable elements such as the composition, porosity, and structural integrity of the material. However, the association between these two factors can differ depending on the material type and characteristics. Notably, the association between C and F is not always linear. Understanding the association between C and F is crucial for the design and analysis of the structural components. Attaining the relationship among these for blended concrete with hefty data requires extremely tedious and complex work. Hence, MATLAB code was developed to resolve this problem using numerous models, such as linear regression (LR), power regression (PR), and logarithmic (L). The input data in MATLAB were supplied from the research work conducted on blended concrete mixes, and the models were run. The results from these models were then compared based on existing standards, and the R^2 value was used to reveal the best fit.

1 Introduction

The building sector's global and unavoidable increase in cement usage has resulted in substantial concerns regarding greenhouse-gas emissions. This necessitates the implementation of rigorous measures, particularly by researchers, to resolve the issue. To comply with the principles of sustainable construction, it is necessary to reduce the use of cement in concrete production by substituting it with alternative materials [1]. One promising solution is the development of blended concretes. For blended concrete, it is necessary to understand its behavior with the incorporation of various materials.

MATLAB (derivative from **Mat**rix **Lab**oratory) is a computerized programming software that was crafted by 'Cleve Moler' and is exercised to accomplish numerous numerical calculations for engineering complexities. It is extremely accurate, well-optimized, and designed to accomplish scientific and engineering computations. It is

P. Joseph D. et al. (Eds.): IEMDST 2024, SPPHY 567, pp. 342–352, 2025.
https://doi.org/10.1007/978-981-96-5863-3_43

widely used in various fields including engineering, mathematics, physics, and finance [2]. MATLAB offers a wide range of mathematical functions and operatives for execution of complex numerical computations. We can accomplish operations on matrices, solve equations, determine integrals and differentials, and execute statistical analyses that are helpful in the field of engineering [3]. MATLAB can fit in with further programming languages like C, C++, Java, and Python. It also offers interfaces and libraries for swapping data. It is extensively utilized in numerous fields, such as engineering, science, finance, and economics. Its graphical user-friendly interface, wide-ranging documentation, and robust community provision make it accessible to both beginners and experienced programmers for a wide range of applications [4].

Computing technologies and programs have become increasingly prevalent. This offers engineers the benefit of holding access to a great variety of knowledge, permitting them to crack difficult issues, and managing complex engineering systems. Construction materials and designs are two arenas that exercise computers and software programs, as engineers and other specialists explore methods to simplify engineering operations [5]. The present work focuses on developing a MATLAB code to determine the relationship between C and F. For these three models, linear regression (LR), power regression (PR), and logarithmic (L) were developed. The experimental test results were utilized as input data to determine the relationship using three models.

2 Materials

In this investigation, grade 53 cement (OPC) was utilized. Fine aggregates (FA) are drawn from nearby and locally available rivers. Coarse aggregates (CA) were gathered from nearby quarries. Silica fume (SF) is procured from a local vendor. Table 1 lists the various properties of the materials established from laboratory testing.

Table 1. Physical properties of materials

S. No	Material	Property	Result	IS Code
1	OPC	Specific gravity	3.12	IS 4031 (Part 11) - 2019
		Fineness	4%	IS 4031 (Part 1) - 2021
		Standard consistency	33%	IS 4031 (Part 4) - 2019
		Compressive strength	58 N/mm^2	IS 4031 (Part 6) -2019
2	FA	Specific gravity	2.52	IS 2386 (Part III) - 2021
		Fineness modulus	2.3	IS 2386 (Part I) - 2021
		Grading	Zone II	IS 383 - 2016
3	CA	Specific gravity	2.79	IS 2386 (Part III) - 2021
		Fineness modulus	7.41	IS 2386 (Part I) - 2021
4	SF	Specific gravity	2.64	-

3 Mix Grade and Quantities

In this investigation, the M30 concrete grade proportion was ascertained based on trials that incorporated recommendations from IS 10262:2019 and IS 456:2000. SF was then substituted in increments of 5% up to 20%. Table 2 shows the quantity of materials deployed for various designated mixes.

Table 2. Quantity of materials

Mix	Quantities (Kg/m^3)			
	OPC	SF	FA	CA
CS0	385.50	0	675.33	1145.64
CS5	366.23	19.27	675.33	1145.64
CS10	346.95	38.55	675.33	1145.64
CS15	327.68	57.82	675.33	1145.64
CS20	308.40	77.10	673.13	1145.64

4 Test Methods – Compressive Strength and Flexural Strength

The compressive strength of a material may be characterized as its ability to withstand failure when subjected to compressive forces. It refers to the quantified maximum resistance shown by a concrete specimen. Cubes of size 150 mm × 150 mm × 150 mm were used in this study. The cubes were tested utilizing compression testing equipment with a maximum capacity of 2000 k N.

The flexural strength test is performed to assess the ability of an unreinforced concrete slab or beam to withstand failure when subjected to bending circumstances. It is also known as transverse rupture strength, bend strength, or modulus of rupture, which is an essential property of a material. The flexural test involves subjecting a prism with dimensions of 100 mm × 100 mm × 500 mm to a load applied at the midpoint of its span.

5 Test Values

Figure 1 displays the compressive and flexural strengths of the concrete blend incorporating SF. The mechanical properties were measured after 28 and 56 days of curing, respectively. The substitution of 10% of OPC with SF has resulted in the highest strengths at different curing durations. This substitution ratio is considered to be the best value. In general, increasing the amount of SF replacements led to a decrease in strength. Furthermore, the strengths have increased with the curing period.

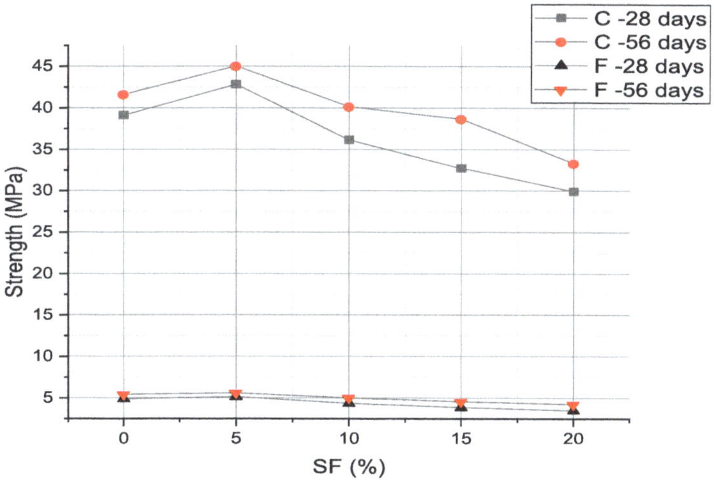

Fig. 1. Strength variation with SF.

6 Methodology Adopted

In this research effort, the programming code in MATLAB was initially developed to determine the relation in equation form and R^2 values for the LR, PR, and L models. The code was rigorously excluded to eliminate both logical and syntactic errors. In the next stage, three sets of input data are collected, presumably one dataset based on the empirical formula and the other two sets based on the research values and its established relations. Then, all sets of input data were run on MATLAB, and the relation along with the R^2 value was extracted. Based on the extracted values, the developed models are compared with the established models and with each other to determine the best model along with efficacy. Figure 2 graphically shows the methodology of this study.

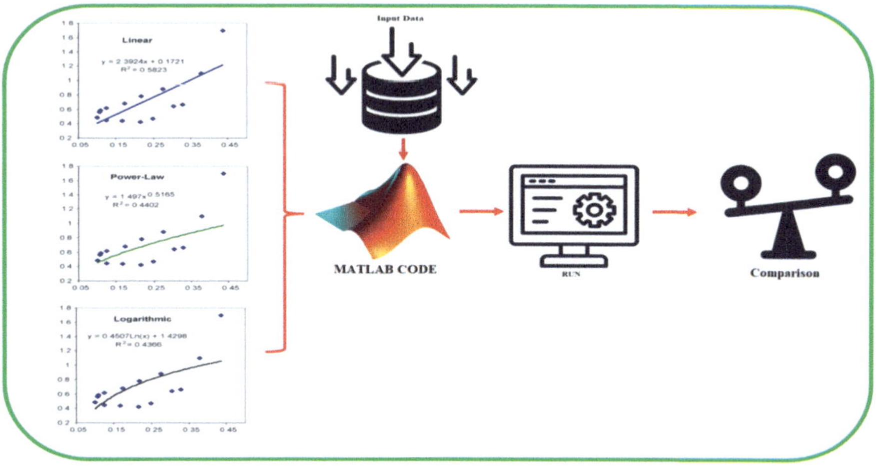

Fig. 2. Graphical methodology.

7 Input Data

To assess the efficacy of the polished code, a minimum of three sets of data values were investigated manually and compared with the developed analytical models. One set of data is based on the empirical formula provided by IS 456:2000 as evidence to prove that the code deployed is satisfactory in terms of already existing relations. The other two sets of data were gathered from the research findings and the established relationship. The second set of data was obtained from a laboratory-controlled situation at 28 days of curing, whereas the third set of data was obtained at 56 days of curing. The second and third sets of data comprise five observations arranged in ascending order of strength with varying quantities of silica fume (SF). Table 3 presents the research data that were fed as input in MATLAB.

Table 3. Research input data

Set No	C values (MPa)	F values (MPa)	Relation	Inference
I	16	2.8	$F = 0.7 \times C^{0.5}$	Indian standard equation
	25	3.5		
	36	4.2		
	49	4.9		
	64	5.6		
II	29.92	3.51	$F = 0.07 \times C^{1.11}$	SF Blended 28 days curing
	32.73	3.87		
	36.14	4.37		
	39.11	4.9		
	42.81	5.15		
III	33.33	4.26	$F = 0.1334 \times C^{0.9826}$	SF Blended 56 days curing
	38.66	4.58		
	40.14	5.01		
	41.62	5.43		
	45.03	5.61		

Programming code – LR Model

```
clc;
clear all;
% Input data
C = [16, 25, 36, 49, 64]; % Enter compressive strength values
F = [2.8, 3.5, 4.2, 4.9, 5.6]; % Enter flexural strength values
% Perform linear regression
A = C;
B = F;
coefficients = polyfit(A, B, 1);
slope = coefficients(1);
intercept = coefficients(2);
```

```
% Calculate R-squared value
Y_predicted = slope * A + intercept;
residuals = B - Y_predicted;
SSR = sum(residuals.^2); % Sum of squared residuals
SST = sum((B - mean(B)).^2); % Total sum of squares
R_squared = 1 - SSR / SST;
% Plot the data
scatter(C, F, 'filled');
xlabel('Compressive Strength (MPa)');
ylabel('Flexural Strength (MPa)');
title('Compressive Strength vs Flexural Strength');
% Plot the regression line
hold on;
plot(A, Y_predicted, 'r');
legend('Data', 'Regression Line');
% Display the equation and R-squared value
equation = sprintf('F = %.2f*C + %.2f', slope, intercept);
r_squared_text = sprintf('R-squared = %.2f', R_squared);
text(16, 5.8, equation);
text(16, 5.5, r_squared_text);
disp(equation);
disp(r_squared_text);
```

Programming code – PR Model

```
clc;
clear all;
% Input data
C = [16, 25, 36, 49, 64]; % Enter compressive strength values
F = [2.8, 3.5, 4.2, 4.9, 5.6]; % Enter flexural strength values
% Perform power regression analysis
[power_coefficients, ~, ~, ~, power_stats] = regress(log(F'), [ones(length(C), 1) log(C')]);
% Extract coefficients and R-squared value
a = power coefficients(1); % Coefficient 'a' of power regression
b = power_coefficients(2); % Coefficient 'b' of power regression
r_squared = power_stats(1); % R-squared value of the regression
% Calculate fitted values using the power regression equation
fitted_values = exp(a) .* C'.^b;
% Plot the data points and fitted curve
scatter(C, F, 'filled'); % Plot data points
hold on;
plot(C, fitted_values, 'r');
legend('Data', 'Regression Line');
hold off;
% Add labels and title
xlabel('Compressive Strength (MPa)');
```

```
ylabel('Flexural Strength (MPa)');
title('Compressive Strength vs Flexural Strength');
% Display the equation of the power regression curve and R-squared value
equation = sprintf('F = %.2f C^{%.2f}', exp(a), b);
r_squared_text = sprintf('R-squared = %.2f',r_squared);
text(16, 5.8, equation);
text(16, 5.5, num2str(r_squared_text));
disp(['Equation: ' equation]);
disp(['R-squared value: ' num2str(r_squared)]);
```

Programming code – L Model

```
clc;
clear all;
% Input data
C = [16, 25, 36, 49, 64]; % Enter compressive strength values
F = [2.8, 3.5, 4.2, 4.9, 5.6]; % Enter flexural strength values
% Perform logarithmic regression analysis
log_C = log(C); % Log-transform the compressive strength
coefficients = polyfit(log_C, F, 1); % Fit a linear regression model
slope = coefficients(1); % Slope of the regression line
intercept = coefficients(2); % Intercept of the regression line
% Compute R-squared value
predicted_F = slope * log_C + intercept;
SSR = sum((predicted_F - mean(F)).^2); % Sum of squared residuals
SST = sum((F - mean(F)).^2); % Total sum of squares
R_squared = 1 - SSR / SST;
% Plot the data points and regression line
scatter(C, F, 'filled'); % Plot data points
hold on;
x = min(C):max(C);
y = slope * log(x) + intercept; % Evaluate the regression line
plot(x, y, 'r');
legend('Data', 'Regression Line');
hold off;
% Add labels and title
xlabel('Compressive Strength (MPa)');
ylabel('Flexural Strength (MPa)');
title('Compressive Strength vs Flexural Strength');
% Display the equation of the regression line and R-squared value
equation = sprintf('F = %.2f * log(C) + %.2f', slope, intercept);
r_squared_text = sprintf('R-squared = %.2f',R_squared);
text(16, 5.8, equation);
text(16, 5.5, num2str(r_squared_text));
disp(['Regression Equation: ' equation]);
disp(['R-squared value: ' num2str(R_squared)]);
```

8 Results and Discussion

The input data are fed into the developed code, and the run operation is carried out in MATLAB R2017(b). The three input datasets are presented in Table 4.

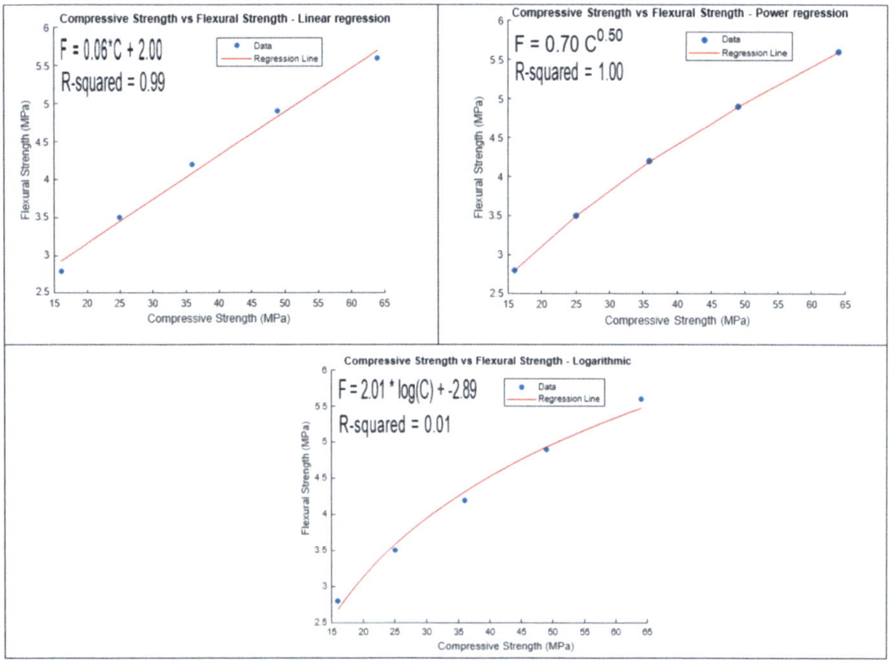

Fig. 3. Linear regression, power regression and logarithmic model for data set I.

Data set I (based on the theoretical formula) for the various models developed is shown in Fig. 3. From this trial, the power regression model has developed an accurate relation with greater precision, followed by the linear regression model, and the least by the logarithmic model. However, this is because the empirical relationship is based on a power regression model. Hence, further analysis with practical data is required. Therefore, datasets II and III were processed to obtain a better view. In the case of dataset II, both the power regression and linear regression models performed much better than the logarithmic model as shown in Fig. 4, but the relationship developed by the power regression model was much closer to the established relationship. For dataset III, the power regression and linear regression models performed equally but outperformed the logarithmic model as shown in Fig. 5. The relationship developed by the power regression model was much closer to the established relationship, with more precision.

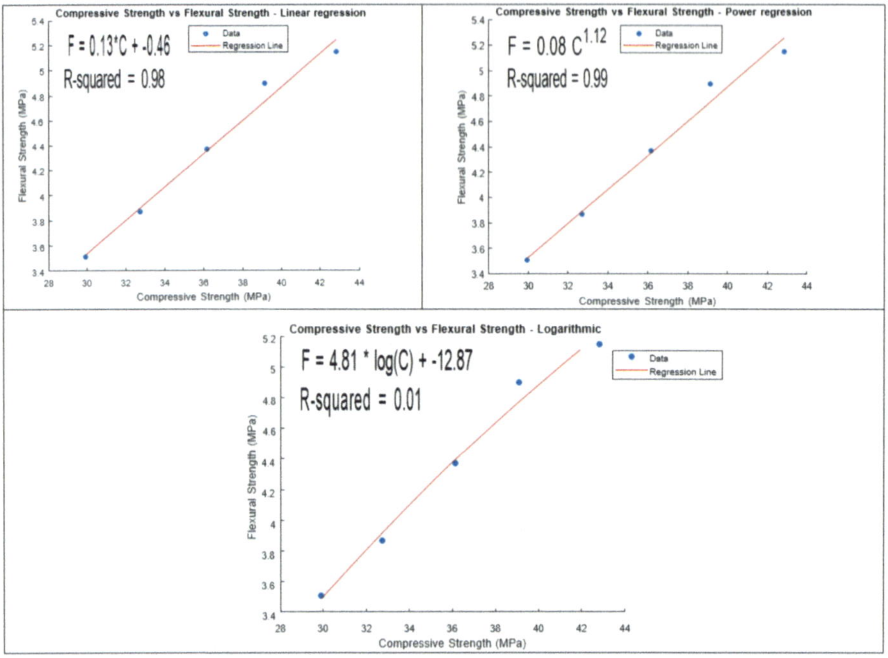

Fig. 4. Linear regression, power regression and logarithmic model for data set II.

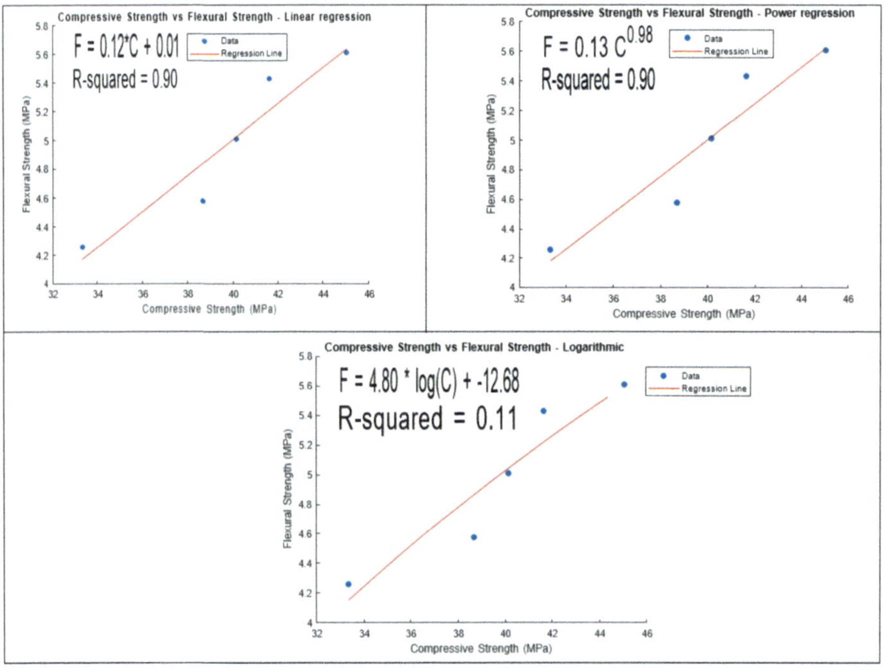

Fig. 5. Linear regression, power regression and logarithmic model for data set III.

Table 4. Research models

Set No	Existing equation	Model	MATLAB equation	R^2 value
I	$F = 0.7 \times C^{0.5}$	Linear regression	$F = 0.06 \times C + 2.00$	0.99
		Power regression	$F = 0.70 \times C^{0.5}$	1
		Logarithmic	$F = 2.01 \times \ln(C) + (-2)$	0.010382
II	$F = 0.07 \times C^{1.11}$	Linear regression	$F = 0.13 \times C + (-0.46)$	0.98
		Power regression	$F = 0.08 \times C^{1.12}$	0.98598
		Logarithmic	$F = 4.81 \times \ln(C) + (-12)$	0.012071
III	$F = 0.1334 \times C^{0.9826}$	Linear regression	$F = 0.12 \times C + 0.01$	0.90
		Power regression	$F = 0.13 \times C^{0.98}$	0.90185
		Logarithmic	$F = 4.80 \times \ln(C) + (-12)$	0.10835

9 Conclusion

The main objective of this study is to provide information on the usage of programming models to simplify the process of developing various relations among strength parameters with large amounts of data. Based on the findings of the current study, the following conclusions were drawn. The use of the developed MATLAB code was beneficial as it facilitated the solving of the problem of identifying the relationship between C and F within a relatively short time and in a cost-effective manner. In a generic sense, it can be stated that for analyzing the relationship between compressive and flexural strengths, the application and stepping up of the LR model was fairly good. However, this performance differed drastically based on the implemented dataset. Regarding dataset I, where the variable quantification was made in proportion to the empirical formulas applied, while LR equaled an evident and relatively high R^2, it was less precise in comparison to the PR model. Comparing all models on all datasets, the experimental results indicated that the PR model performed better than the other models. Therefore, it was found that since it is able to plot the interaction between C and F on a non-linear basis, it resulted in the highest R^2 at all times. Owing to the good accuracy of this model, particularly in practical real-life problems, especially in constriction engineering, it is highly applicable. However, of course in all these categories, the L model stood out as the worst. They contained lower R^2 values, which were indicative of a poor model/data fit. This implies that the logarithmic function is not suitable for estimating the compressive and flexural strengths of blended concrete.

References

1. Raj Kumar, P., Harikrishna, D.: Mater. Today Proc. (2024). https://doi.org/10.1016/j.matpr.2024.05.102
2. Raj Kumar, P.: Int. J. Sci. Res. **11**, 998–1001 (2022). https://doi.org/10.21275/SR22715102352
3. Deepak, S.: Modeling and analysis of different shaped cantilever beams in MATLAB **8**, 139–140 (2020)

4. Tamjid Hossain, S.M., Al-Faruk, A.: Analysis of simple and cantilever beams using matlab graphical user interfaces **2017**, 18–20 (2017)
5. Makenya, A.R., John Paul, M.: J. Civ. Eng. Res. **2017**, 81–98 (2017). https://doi.org/10.5923/j.jce.20170703.01

An Experimental Performance of EDM Process Parameters for Aluminum Composites

Kamalkishor Maniyar[1](✉), Aarti Puri[1], Smita Suhane[1], Swapnil Deshmukh[1], Girish Modak[2], and Ramdas Biradar[3]

[1] Dr. D. Y. Patil Institute of Technology, Pimpri, Pune 411018, Maharashtra, India
kkmaniyar2020@gmail.com

[2] P.E.S.'s Modern College of Engineering, Pune 411018, Maharashtra, India

[3] Pimpri Chinchwad University, Pune 411018, Maharashtra, India

Abstract. The electrical discharge machining process parameters for aluminum composites can be optimized in several responses using the Taguchi-based Grey Relational Analysis that is described in this study. Finding the optimal parametric combination of process parameters that results in the maximum material removal rate, lowest surface roughness rate, and least amount of tool wear is the aim of this study. The experimental design used for the trials is the Taguchi orthogonal array. Aluminum composite prototypes are made by a process called stir casting. To find out how each process parameter specifically affects the replies that are produced, a variance analysis is performed. The confirmation test verifies the experiment's conclusions.

1 Introduction

The purpose of the work should be made clear from the outset. You should provide enough background information about your work, but avoid providing a detailed critique of the literature or an overview of your conclusions. In order to meet the demands of continuously growing industries, composites are updated versions of metal matrix composites that combine two or more reinforced elements to provide better mechanical and wear qualities. When dealing with this kind of advanced material, electrical discharge machining is the most typically utilized non-traditional machining technology [1]. The author assessed the possibility of milling Al_2O_3/6061 Al composites using rotational EDM with a dislike electrode [2]. According to the study, silicon particles and machining parameters have a major influence on the Al-Si alloy's machining properties [3]. It has been shown that powder mixed EDM, as opposed to conventional techniques, increases machining efficiency and surface smoothness [4]. Compared to an untreated electrode, tool wear was found to be reduced when using a coated electrode [5]. As silicon carbide particle fraction increases, electrode wear lowers and material removal rate increases [6]. The author claims that by employing gray relational analysis, the output reactions in the machining process can be enhanced [7]. The study looked at the important impact of silicon carbide on MRR [8]. The primary variables were determined to be the powder concentration, peak current, and pulse duration; the supply voltage was shown to be a

P. Joseph D. et al. (Eds.): IEMDST 2024, SPPHY 567, pp. 353–357, 2025.
https://doi.org/10.1007/978-981-96-5863-3_44

very little factor [9]. By reducing the SiO_2 layer on the surface of the SiC, magnesium was shown to improve the wettability between silicon carbide and aluminum particles [10]. According to the author, hardness levels rise as silicon carbide composition increases [11]. The composite's MRR rises in tandem with increases in the dielectric fluid flushing pressure, pulse on time, and current [12]. They discovered the ideal circumstance that results in the lowest surface roughness measurement [13]. According to the author, the output responses were precisely proportionate to the current and inversely related to the timely pulse [14]. According to the study, the hybrid composites' density and hardness dropped as the amount of graphite increased. In [15] It was found that the process's properties are greatly enhanced by the ideal combination of input parameters. When comparing the grey fuzzy logic technique to the grey relational strategy, they found that the former results in lower output uncertainty and a higher gray fuzzy reasoning grade [16]. The ideal circumstance with the least amount of surface roughness was discovered [17]. The tensile strength of the composites was enhanced by the presence of silicon carbide and graphite in the AL6061 alloy [18]. The best cutting conditions were highlighted as those that result in the least amount of cutting force and surface roughness [19]. Silicon carbide reinforcements were added to an aluminum matrix to increase its hardness and tensile strength [20]. Aluminum composites can be used to create gears and other components of the power transmission system. The ideal circumstance with the lowest surface roughness value was identified [21]. The output responses were inversely correlated with the pulse on time and directly proportional to the current. [22] They examined the hybrid composites' decreased density and toughness. Grey relational analysis can be used to improve the machining features [23]. The high-speed abrasive electrical discharge machining procedure produced better results than the traditional EDM method [24]. The use of graphite greatly improved the mechanical properties of reinforced aluminum when silicon carbide particles were introduced. As the load and grit size grew, accordingly, so did the wear coefficients [25]. Stir casting is a widely used method for producing aluminum hybrid composites. The goal of this research is to determine the ideal process parameter values that produce the best material removal rate and the lowest surface roughness.

2 Materials and Methods

Stir casting is used in the construction of the specimens. By incorporating graphite and silicon carbide in equal weight percentages into the aluminum alloy metal matrix material, the specimen's composition can be changed. A machine agitates the mixtures. In the end, the specimens undergo heat treatment to improve their mechanical characteristics. Commercial aluminum alloy is used as the metal matrix material. Particles of graphite and silicon carbide are utilized as reinforcing elements to improve the specimens' tribological and mechanical properties, respectively. Copper is the best material to machine aluminum composites because it meets the specifications for the electrode material used in the electrical discharge machining process. When applying electrical discharge machining or other non-conventional machining procedures to mill aluminum composites, copper's good electrical and thermal properties are essential. Factors like the total equal weight percentage of SiC-Gr, the pulse length, and the current are taken into account while machining settings. Three distinct layers of implementation make use of them. Material removal rate, tool wear rate, and surface roughness are three of the machining parameters that are measured. Surface tester equipment gathers surface roughness information, while material removal and tool wear rates can be computed by weighting the materials both before and after machining. The trials are carried out and assessed using the Taguchi-based Grey Relational Method. The Taguchi method is used to set up and carry out the experiments. Aluminum composites are machined using an electrical discharge machining technique. By changing the set of machining parameters, the machining characteristics for various experiment sessions are recorded.

3 Results and Discussion

Figure 1 displays the experimental results of the Grey Relational Analysis conducted using the principal technique. Establishing a gray connection begins with normalizing the trial outcomes to fall between 0 and 1. In a second phase, the correlation between the expected and actual experimental data is visualized using the gray relational coefficient derived from the normalized data. The different performance aspects are assessed by calculating the gray relational grade using the grey relational coefficient. The variance analysis and gray relationship grading are conducted using the Mini Tab 14.

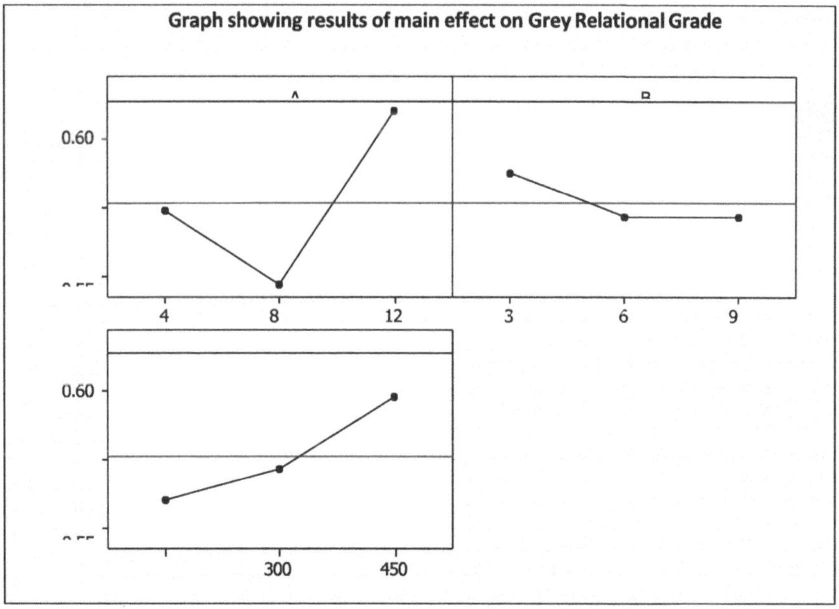

Fig. 1. Graph showing results of main effect on Grey Relational Grade.

4 Conclusion

Stir casting is a successful method for creating aluminum composites including equal weight percentages of silicon carbide and graphite. The Taguchi orthogonal array design facilitates well-thought-out and carried out experiments. Looking at the grey Reducing tool wear, increasing material removal rate, and minimizing surface roughness are the results of multi-response optimization and relationship analysis of the EDM process parameters for aluminum composites. According to a variance study, the equal weight percentage of SiC and Gr together is the most important factor. When compared to other machining parameters, the optimal parametric combination of machining parameters is represented by the same weight percentage of SiC-Gr and current.

References

1. Yan, B.H., Wang, C.C., Liu, W.D., Huang, F.Y.: Int. J. Adv. Manuf. Technol. **16**, 322–333 (2000)
2. Tsai, D., Lui, T., Chen, L.: J. Mater. Trans. **43**, 199–205 (2002)
3. Zhao, W.S., Meng, Q.G., Wang, Z.L.: J. Mater. Process. Technol. **129**, 30–33 (2002)
4. Kilickap, E., Cakir, O., Aksoy, M., Inan, A.: J. Mater. Process. Technol. **164–165**, 862–867 (2005)
5. Seo, Y.W., Kim, D., Ramulu, M.: J. Mater. Manuf. processes **21**, 479–487 (2006)
6. Haq, A.N., Marimuthu, P., Jeyapaul, R.: Int. J. Adv. Manuf. Technol. **37**, 250–255 (2008)
7. Senthil Kumar, K.L., Sivasubramanian, R., Kaliselvan, K.: J. Prot. Electron. Acta **27**, 477–486 (2009)
8. Kumar, H., Davim, J.P.: J. Compos. Mater. **45**, 133–151 (2011)
9. Reddy, A.C., Zitoun, E.: Ind. J. Sci. Technol. **3**, 1184–1187 (2010)
10. Sahib, D.A.: ARPN J. Eng. Appl. Sci. **6**, 41–46 (2011)
11. Maniyar, K.G., Ingole, D.S.: Mater. Today: Proc. **5**, 19836–19843 (2018). https://doi.org/10.1016/j.matpr.2018.06.347
12. Maniyar, K.G., Ingole, D.S.: IOP Conf. Ser.: Mater. Sci. Eng. **377**, 012204 (2018). https://doi.org/10.1088/1757-899X/377/1/012204
13. Maniyar, K.G., Ingole, D.S.: Mater. Today: Proc. **5**, 27700–27709 (2018). https://doi.org/10.1016/j.matpr.2018.10.004
14. Maniyar, K.G., Ingole, D.S.: Advanced Manufacturing and Materials Science: Selected Extended Papers of ICAMMS, pp. 275–284 (2018). https://doi.org/10.1007/978-3-319-76276-0_27
15. Maniyar, K.G.: Mater. Today: Proc. (2024). https://doi.org/10.1016/j.matpr.2024.01.021. ISSN 2214-7853
16. Maniyar, K.G., Marode, R.V., Chikalthankar, S.B.: Int. J. Eng. Adv. Technol. **5**, 112–116 (2016)
17. Maniyar, K.G.: Int. J. Recent Innov. Trends Comput. Commun. **5**, 9 (2017)
18. Maniyar, K.G., Agrawal, S.K., Ingole, D.S.: Int. J. Innov. Technol. Explor. Eng. **5**, 10 (2016)
19. Maniyar, K.G., Deore, H.V.: Int. Res. J. Eng. Technol. **7**(08) (2020)
20. Maniyar, K.G., Ingole, D.S.: Int. J. Recent Innov. Trends Comput. Commun. **4**(07) (2016)
21. Maniyar, K., Jadhav, P., Biradar, R., Gawande, J.: NanoWorld J. **10**(S1), S209–S211 (2024). https://doi.org/10.17756/nwj.2024-s1-037
22. Maniyar, K., Jadhav, P., Biradar, R., Gawande, J.: NanoWorld J. **10**(S1), S206–S208 (2024). https://doi.org/10.17756/nwj.2024-s1-036

23. Maniyar, K., Suhane, S.: NanoWorld J. **10**(S1), S63–S65 (2024). https://doi.org/10.17756/nwj.2024-s1-012
24. Kaushik, N.C., Rao, R.N.: J. Tribol. Int. **103**, 298–308 (2016)
25. Baburaja, K., Venkatasubbaiah, D.K., Kalluri, D.R.: J. Mater. Today Proc. **3**, 4140–4145 (2016)

Investing Experimental Effect of Welding Parameters in Friction Stir Welding on Aluminium Alloy

Kamalkishor Maniyar[1](✉), Namrata Saravade[1], Smita Suhane[1], Swapnil Deshmukh[1], Girish Modak[2], and Ramdas Biradar[3]

[1] Dr. D. Y. Patil Institute of Technology, Pimpri, Pune, Maharashtra 411018, India
kkmaniyar2020@gmail.com

[2] P.E.S.'s Modern College of Engineering, Pune, Maharashtra 411018, India

[3] Pimpri Chinchwad University, Pune, Maharashtra 411018, India

Abstract. The friction stir welding procedure is a solid state joining technique that involves connecting two plates together using an infinite tool. In FSW, the material is welded but does not completely melt and recast. The quality of the welded connection is determined by the lap shear fracture force, which can only be optimized by carefully controlling all relevant process parameters to achieve the maximum strength. In the current work, friction stir welding settings were optimized through the application of Response surface approach. Tool pin profile, Tool rotation speed, and Dwell duration were the variables employed in this experiment. Based on the lap shear fracture force, spot welding response requirements are determined. The influence of several parameters on the strength of the joint was studied.

1 Introduction

The most significant invention of 1991 is thought to have been the solid state welding method known as friction stir welding [1]. It is increasingly being used to join High Density Polyethylene sheets, which can be difficult to fuse together conventional fusion welding [2]. FSW uses a non-consumable tool to generate frictional heat at the welding point, which causes the work piece to severely deform plastically [3]. The plates that require joining are positioned on a rigid backing plate and fastened in order to prevent the neighboring joint faces from separating [4]. A rotating instrument with a handle that is cylindrical and a pin (probe) that protrudes specifically is gradually placed into the joint line [5]. The required weld depth and the pin length are similar [6]. The tool's shoulder is forced up against the plates. The mechanical strength of the plates is diminished by the friction heating generated by the rotating tool [7]. The spinning tool is inserted into the work piece at a high angle of attack until the tool shoulder touches the upper surface of the work piece, creating a weld area [8]. The material stirring process starts when the tool plunges to a predetermined depth [9]. The tool is still rotating inside the work parts at this moment. [10].

P. Joseph D. et al. (Eds.): IEMDST 2024, SPPHY 567, pp. 358–362, 2025.
https://doi.org/10.1007/978-981-96-5863-3_45

The tool's rapid spinning creates heat by softening, heating, and agitating the materials around as a solid-state connection forms [11]. Due to the direct correlation between the weld input parameters and the weld joint quality throughout the welding process, welding is often a multi-input, multi-output procedure. Welding input parameters must be controlled in order to ascertain the mechanical characteristics of welded joints [12, 13]. So, choosing the right process parameters is necessary to produce a welded junction that is of a high caliber [14]. In order to overcome this challenge and define the desired output variables, different optimization techniques can be used with the construction of mathematical models to characterize the relationship between the input parameter and output variables [15]. The study optimizes the process parameter with the use of Response Surface Methods. In [16] High-density polyethylene is a commonly utilized polymer owing to its easily accessible and reasonably priced nature [17]. Heat-resistant polyethylene has unique mechanical properties [18]. With the continued advancement of engineering plastics, there is an increasing demand for dependable, fast, highly productive, and cost-effective joining processes, similar to those used with metals [19].

2 Materials and Methods

Tool rotation speed, dwell time, and tool pin profile are the three level process characteristics were examined in this study [20]. The influence of the primary parameters was ascertained by conducting trial tests in accordance with the design principle of the studies [21]. According to the matrix's design, nine experiments were carried out. To reduce machining error and the mean reaction, or tensile strength, as shown in the accompanying Table, each experiment is run twice under the same conditions. To investigate the effects of the FSW process parameters on friction stir welding joints values is analyzed.

3 Results and Discussion

Contour plots' recognizable round shape raises the possibility that causes could exist apart from responses. A contour map effectively illustrates the region of optimal factor settings. First-order models usually consist of a simple collection of parallel lines; for second-order response surfaces, however, such a plot might be more complicated. The next step usually involves identifying whether the stationary point is a saddle point, maximum response, or minimum response in order to characterize the response surface that surrounds the point once it has been discovered. Making use of a contour plot is the simplest method for classifying this. Researching the response surface requires the use of contour plotting.

It is often suggested that factor effects are independent when elliptical outlines show the possibility of factor interactions. Two parameters are taken from the center level, two from the X and Y axes, and the response from the Z axis to create the response surfaces for both models. The best reaction point is made evident by the response surfaces. Optimal process parameter sets that yield a maximum or lowest response value are found using response surface methodology (RSM). The process parameters that best match the maximum Lap shear fracture force are taken into consideration in this inquiry.

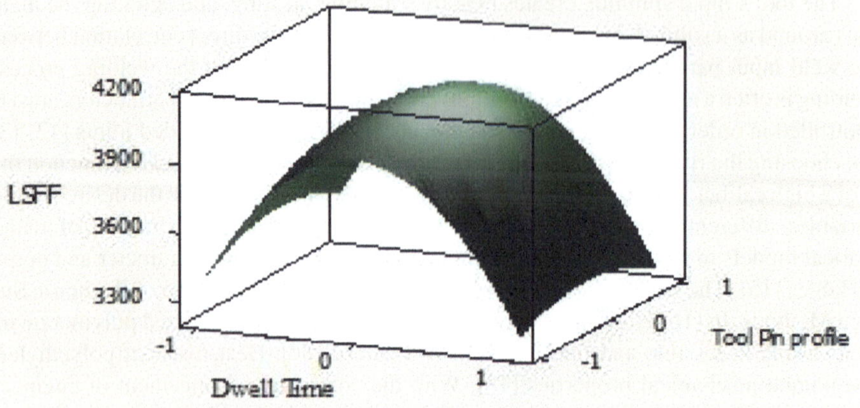

Fig. 1. The relationship between LSFF, Tool Pin profile, and Dwell Time

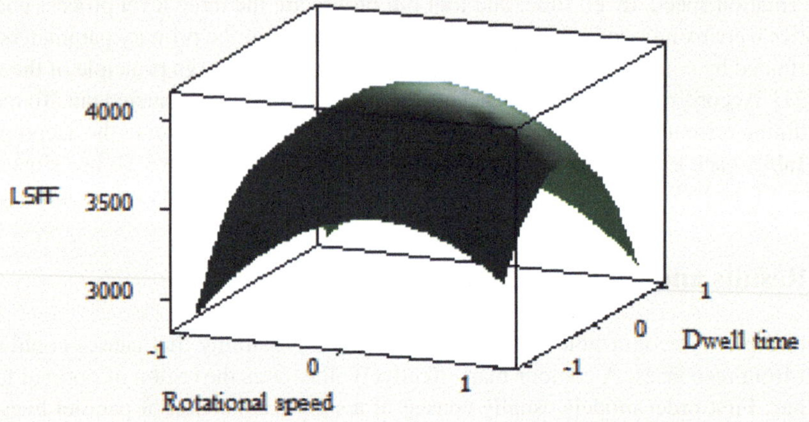

Fig. 2. The relationship between LSFF, Dwell Time and, Rotational speed

In order to get the greatest Lap shear fracture force, certain optimum process parameters must be applied.

The relationship between LSFF, Tool Pin profile, and Dwell Time is depicted in the response surface graph in Fig. 1. The study examines how dwell duration and tool pin profile vary in relation to fracture force. It shows that variations in lap shear fracture force are less affected by changes in tool pin profile. The dwell effect is comparable to the reaction surface graph from earlier. A longer dwell period results in more force; beyond that, it decreases. In both scenarios tool pin profile and dwell time lap shear fracture force increases to a certain point before decreasing. The relationship between

LSFF, Dwell Time and Rotational Speed is depicted in the response surface graph in Fig. 2. The tensile strength seen in Fig. 2 is increased by the combination of the tool pin profile and rotating speed.

4 Conclusion

The Design of Experiment methodology is used to conduct the experiments in this case. The response surface technique offers a mathematical model for response prediction. The results of the Response Surface Methodology are summarized below. Response surface analysis is used in FSSW of HDPE sheets to examine the impact of three process parameters: rotating speed, dwell duration, and tool pin profile, on LSFF of HDPE welded connection. ANOVA, regression analysis, and design of experiment techniques have been used to construct a mathematical relationship that predicts the LSFF of friction stir welded HDPE sheets using FSW process factors. Regression models were used to present and evaluate the impacts of FSW parameters on the lap shear fracture force of friction stir welded joints in detail.

References

1. Elatharasan, G., Senthil Kumar, V.S.: Proc. Eng. **38**, 3477–3481 (2012)
2. Rezgui, M.-A., Trabelsi, A.-C., Ayadi, M., Hamrouni, K.: Int. J. Prod. Qual. Eng. **2**, 55–61 (2011)
3. Mostafapour, A., Azarsa, E.: Int. J. Phys. Sci. **7**, 647–654 (2012)
4. Al-Jarrah, J.A., Swalha, S., Mansour, T.A., Ibrahim, M., Al-Rashdan, M.: Adv. Theor. Appl. Mech. **6**, 13–26 (2013)
5. Balasubramanian, V., Elangovan, K., Babu, S.: J. Mater. Eng. Perform. **17**, 820–830 (2008)
6. Palanivel, R., Mathews, P.K., Murugan, N.: J. Eng. Sci. Technol. **4**(1), 25–31 (2011)
7. Lakshminarayanan, K., Balasubramanian, V.: Trans. Nonferr. Met. Soc. China **19**, 9–18 (2009)
8. Palanivel, R., Mathews, K.P., Murugan, N.: MECHANIKA **18**, 517–523 (2012)
9. Maniyar, K., Jadhav, P., Biradar, R., Gawande, J.: NanoWorld J. **10**, S206–S208 (2024). https://doi.org/10.17756/nwj.2024-s1-036
10. Maniyar, K., Suhane, S.: NanoWorld J. **10**, S63–S65 (2024). https://doi.org/10.17756/nwj.2024-s1-012
11. Maniyar, K.G., Ingole, D.S.: Mater. Today: Proc. **5**, 19836–19843 (2018). https://doi.org/10.1016/j.matpr.2018.06.347
12. Maniyar, K.G., Ingole, D.S.: IOP Conf. Ser.: Mater. Sci. Eng. **377**, 012204 (2018). https://doi.org/10.1088/1757-899X/377/1/012204
13. Maniyar, K.G., Ingole, D.S.: Mater. Today: Proc. **5**, 27700–27709 (2018). https://doi.org/10.1016/j.matpr.2018.10.004
14. Maniyar, K.G., Ingole, D.S.: Advanced Manufacturing and Materials Science: Selected Extended Papers of ICAMMS, pp. 275–284 (2018). https://doi.org/10.1007/978-3-319-76276-0_27
15. Maniyar, K.G.: Mater. Today: Proc. (2024). ISSN 2214-7853. https://doi.org/10.1016/j.matpr.2024.01.021
16. Maniyar, K.G., Marode, R.V., Chikalthankar, S.B.: Int. J. Eng. Adv. Technol. **5**, 112–116 (2016)
17. Maniyar, K.G.: Int. J. Recent Innov. Trends Comput. Commun. **5**(9) (2017)

18. Maniyar, K.G., Agrawal, S.K., Ingole, D.S.: Int. J. Innov. Technol. Explor. Eng. **5**(10) (2016)
19. Maniyar, K.G., Deore, H.V.: Int. Res. J. Eng. Technol. **7**(08) (2020)
20. Maniyar, K.G., Ingole, D.S.: Int. J. Recent Innov. Trends Comput. Commun. **4**(07) (2016)
21. Maniyar, K., Jadhav, P., Biradar, R., Gawande, J.: NanoWorld J. **10**, S209–S211 (2024). https://doi.org/10.17756/nwj.2024-s1-037

Electromagnetic Transient Eddy Current Oscillations Differential Probe Technique to Detect Superficial Cracks in Steel Configurations

Manu Sainadh Pilla, Yesudas Bammidi, Sujitha Nimushakavi, Gopala Krishna Podagatlapalli, Kiran Kumar Yelanati, Kanaka Raju Pappala, and Chandra Sekhar Angani(✉)

Department of Physics, GSS, GITAM Deemed to be University, Visakhapatnam, Visakhapatnam, India
cangani@gitam.edu

Abstract. Stainless Steel structures such as pipelines are utilised in several industries to transport steam, crude oil and gas across long distances. They usually built over many miles onshore and offshore through different weather conditions. However, continued exposure to harsh environmental conditions such as high temperatures and pressure may cause corrosion and local thinning of pipe walls, which may lead initiation of cracks. This issue can be addressed effectively by regular inspection of structures using non-destructive testing (NDT) techniques. This manuscript focuses on application of an electromagnetic NDT method known as Transient Eddy Current Oscillations (TECO) technique for detecting surface cracks in stainless steel. Unlike the conventional Eddy Current Testing (ECT) approach, TECO identifies cracks using the decay of oscillations which are generated by a L-C tank circuit. A differential probe is fabricated with two hall sensors to identify the cracks in metal. The test specimen is a SS304 steel plate, featuring 1 mm to 4 mm crack depths starting from the surface, was fabricated using the electric discharge machine (EDM) technique. The obtained results were analysed in both frequency and time domain, continuous wavelet transform (CWT) technique is applied and several features were identified to interpret the crack depths. The identified features are able to clearly distinguish the crack depth in the test material.

1 Introduction

Pipelines play a significant role in transporting essential resources like oil and gas from one place to another, which is vital for economic and industrial development [1]. Safe and effective conveyance of these resources is necessary for various industries to promote the nation's development. Steel pipelines are one of the prime components to carry steam and hazardous gases over long distances in several industries like oil-gas and nuclear power plants [2]. Metal failure is most commonly seen in these industries due to wall thinning and the formation of cracks due to corrosion and erosion [3]. Regular inspection

P. Joseph D. et al. (Eds.): IEMDST 2024, SPPHY 567, pp. 363–369, 2025.
https://doi.org/10.1007/978-981-96-5863-3_46

helps in the early detection of these failures using NDT methods to prevent financial losses and environmental damage [4]. Several NDT techniques exist for pipeline testing, like Radiography Testing (RT), Ultrasonic Testing (UT), Eddy Current Testing (ECT), and Magnetic Particle Testing (MPT) [5–8] etc. Every technique has their own pros and cons, for example RT uses a radiation source, thus it has the safety issues and is time-consuming, whereas UT utilises couplant to launch ultrasonic wave in to the test material. While the MPT applied only to ferromagnetic materials and is not suitable for bulk materials.

ECT is one of the prominent techniques in electromagnetic NDT (ENDT) which is widely used in materials inspection. ECT runs on the principle of faraday's laws of electromagnetic induction. When a time-varying current (sinusoidal) applied to a coil, a primary magnetic flux is developed around it, this magnetic field induces an eddy current when another conductor is brought nearer to the coil (eddy currents flows in closed circular paths). These induced currents in return produce a counter magnetic field which opposes the primary magnetic field generated by the coil [9]. The circulating currents will be disturbed/perturbed if there is any crack or inhomogeneity occurs in the test material. This perturbation affects the resultant magnetic field, by measuring this magnetic field in the test material, to know the physical condition of the material (existence of cracks). This magnetic field measurement can be done by using a pick-up coil or magnetic field sensors like Hall and magneto-resistive sensors like Giant Magneto-Resistance (GMR), Anisotropic Magneto-Resistance (AMR), Tunnelling Magneto-Resistance (TMR) [10–13]. These sensors attain the interest of researchers due to their small size, measuring bandwidth, and ease of handling. The efficiency of ECT is substantially influenced by parameters such as penetration depth and driving current frequency. The penetration depth or depth at which eddy currents may reduce by a factor of 1/e (37% of surface current density) is called skin depth, which is given by a well-known relation $\delta = \sqrt{1/\pi f \mu \sigma}$, where '$f$' is the coil driving current frequency, 'μ' magnetic permeability and electrical conductivity 'σ' of the test specimen [14]. Lower frequencies allow for deeper penetration of field, making them ideal for detecting deep flaws, whereas higher frequencies are more sensitive to surface and near-surface abnormalities.

In the present work, a novel electromagnetic NDT known as Transient Eddy Current Oscillations (TECO) [15] is used. TECO uses a decay of oscillations to induce the eddy current in to the test the sample. The decay of free running oscillations was created by parallel connection of a capacitor (C) with the driving coil (L). A differential probe fabricated with two hall sensors [16] test a stainless steel plate with cracks varying from 1 to 4 mm from the surface. Experimental results were analysed in time domain and frequency domains to understand the relation between identified feature and the crack depth in the test specimen.

2 Experimental Technique and Setup

The basic equipment of the TECO system has a DC power supply (GW GPE-4323), arbitrary wave generator (Agilent 3320A), digital storage oscilloscope (Keysight DSO X2014A), a differential Probe, Instrumentation amplifier (INA129) and Hall sensor (SS49E). The block diagram of the TECO system and photograph of the laboratory setup are shown in Fig. 1 (a) and (b) respectively. Two differential amplifiers extract signals from two hall sensors HS_1 and HS_2, another differential amplifier used to subtracts both the sensor outputs (HS_1-HS_2) as a resultant signal. The parallel connection of C and L is controlled through a semiconducting switch. A signal generator regulates the semiconductor switch responsible for ON/OFF. DC power supply 3 V corresponding to 1 A current is connected to the coil through the switch during ON state. Storage Oscilloscope is used to store data from the probe.

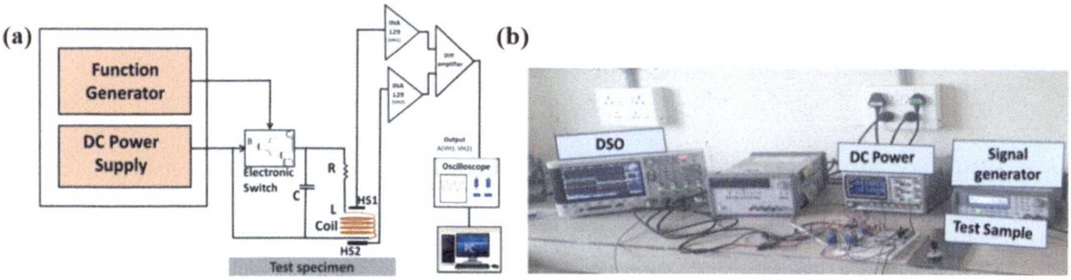

Fig. 1. (a) Block Diagram of TECO (b) Photograph of Experimental System.

The probe is a combination of a capacitor (C) of capacitance 0.154 μ F and copper coil (L) of an inductance of 1.57 mH. The coil is a copper wire of 300 turns encircled on a bobbin and two hall sensors placed in the middle of the coil shown in Fig. 2. A square wave with a 50% duty cycle at 1 Hz frequency is produced using a signal generator to control the ON/OFF states of semiconductor switch (which is a transistor TIP122). When the switch is ON, the transistor conducts the current hence the DC power supply charges the capacitor. When the switch is OFF, the capacitor starts discharging through the coil. This phenomenon creates free-running decay of oscillations with a frequency which is equal to $1/2\pi\sqrt{LC}$. The frequency of oscillations was obtained around 10.2 kHz. The data is obtained by triggering at the falling edge of the square wave to capture the oscillations. By placing the test material below the probe, eddy currents are induced in accordance with the oscillations generated by the probe. As the probe configuration is differential, having two hall sensors where the bottom sensor is directly on the specimen while the upper sensor is facing to the air. The hall sensors measure both the primary field (from the coil) and the secondary field (due to eddy currents in the specimen). Each sensor output is obtained by two different amplifier circuits and both these sensor outputs are connected to another instrumentation amplifier to get the difference voltage of two hall sensors ($V_{H1} - V_{H2}$). The test specimen is a SS304 steel of 230 mm length, 80 mm width with a thickness of 6 mm and the surface simulated cracks having width (w) of 1 mm and length (l) of 15 mm with different depths (d) 1 mm, 2 mm, 3 mm and 4 mm from surface.

3 Experimental Results and Discussion

When the switch is OFF, during the negative edge of the square wave signal the decay of oscillations were captured in the DSO. The results were obtained by placing the probe on cracks with different depth. The results are as shown in Fig. 3 (a), (b). It is observed that

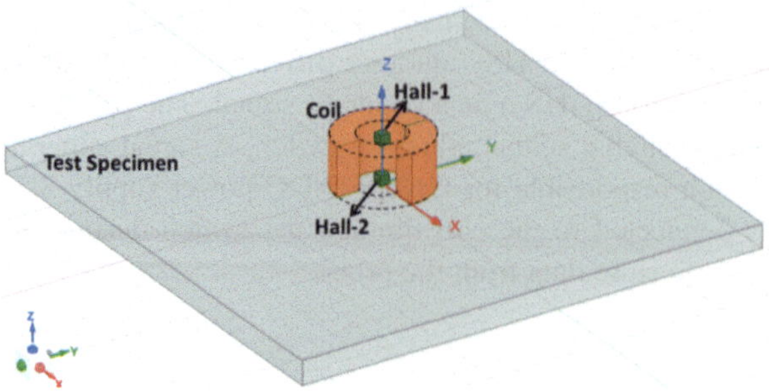

Fig. 2. Schematic view of Differential Probe

the oscillations amplitude is minimum when the probe is in air (without test specimen) due to the almost same magnetic field is detected by both the hall sensors Hall-1 and Hall-2, thus the difference is nullified. While by placing the probe on the specimen where there is no crack, the differential signal amplitude is larger due to the Hall-2 is directly facing the specimen surface, hence it may see more reflected magnetic field than Hall-1. Now considering both these conditions as two extremes (NO Crack area and Air signals), all the other oscillation amplitudes from the different cracks will fall in between these two extremes. The induced currents are more for if more volume of conducting material available beneath the coil (NO crack condition), if there is a crack (material loss) introduced in the material the density of induced currents will be decreased with increasing the crack volume ($l \times w \times d$), thus it is observed that the amplitudes of oscillations decreased with increasing the crack depth. To get clear indication of the decay of oscillations, the upper envelopes are plotted as a function of crack depth as shown in Fig. 4(b).

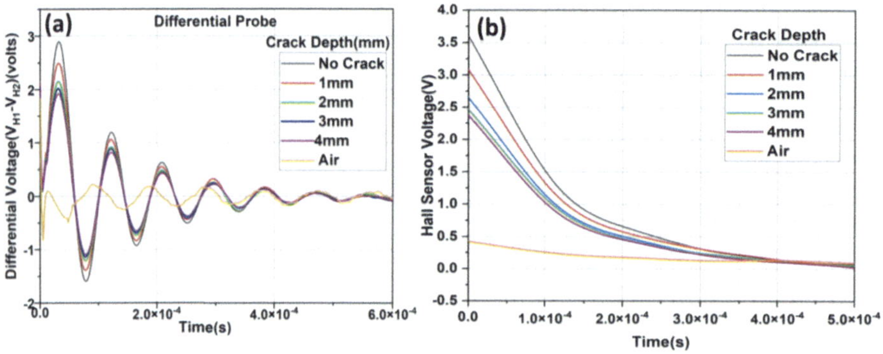

Fig. 3. (a) Amplitudes of different crack depths using TECO and (b) upper envelope of oscillations.

As an emerging technique the TECO method require more intense studies, in search of new features, Time-frequency studies were performed using continuous wavelet transform (CWT) [17] to extract more insight on the results. In the current study, we use morlet

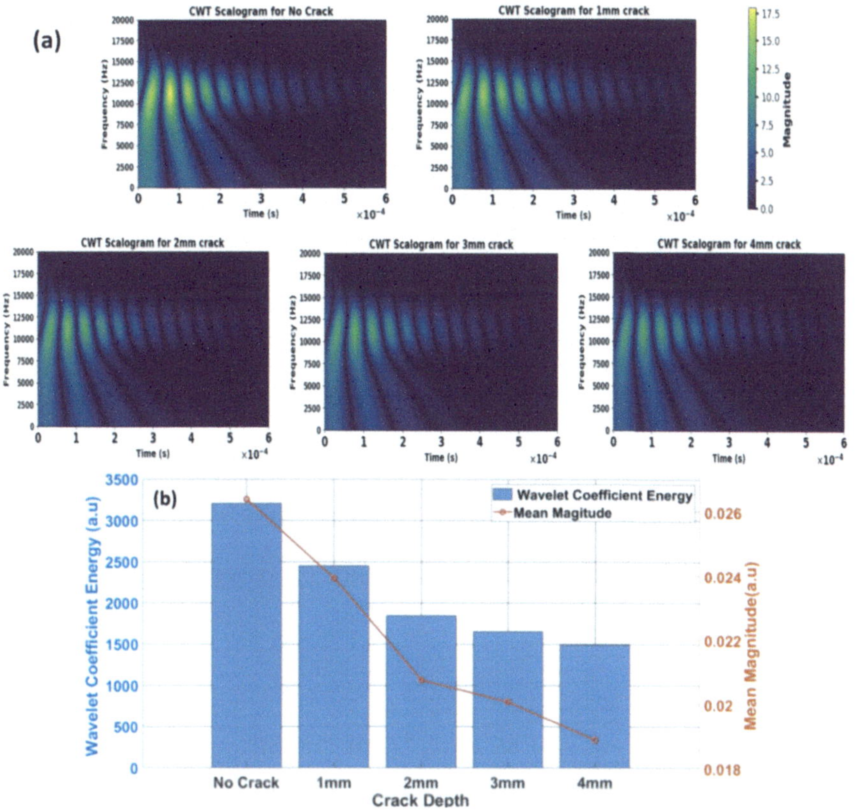

Fig. 4. (a) Scalograms of different crack depths (b) mean magnitude and wavelet coefficient energy as a function of crack depth.

wavelet to generate scalogram images; this technique is widely used to transform time-domain signals to time-frequency information with high resolution images. Figure 4 (a) shows the different time-frequency plots for various crack depths. It is observed that the intensity of magnitude is high for no crack image where it is decreased with increasing the crack depth; the magnitude varies due to the CWT coefficients. This justifies the signal amplitude of oscillations declines continuously as crack depth increases. The scalograms clearly showed that larger cracks result in less energy in the signal due to less induced currents, as seen by the smaller magnitudes. To gain insight into the results, a plot is taken for the mean magnitude against and Wavelet coefficient energy as a function of crack depth as shown in Fig. 3 (b), it clearly indicates that the energy decreases along with the mean magnitude with increasing the crack depth. The energy of the signal is decreased due to increase in crack volume causes the increase the resistance of material because less conductive material available for the induced currents.

4 Conclusion

A recently developed electromagnetic NDT method called Transient Eddy Current Oscillations (TECO) technique with a differential probe has been used for the detection of superficial cracks in a plate made by steel. The probe is made with two hall sensors which are placed at the bottom and upper sides along the axial center of the coil. The output of these two were connected to an instrumentation amplifier IC to get the difference signal, which is utilized for the testing of SS304 material having artificial cracks with depths varying from 1 to 4 mm from the surface. Amplitude changes in decay of oscillations and time-frequency analysis using wavelet transform have been performed to interpret the results with crack depths, and promising results were obtained in identifying cracks, this encourages the TECO method can be a good NDT tool for detecting cracks in stainless steel configurations.

Acknowledgements. The authors acknowledge the financial support from UGC-DAE CSR through a Collaborative.

Research Scheme (CRS) project number CRS/2022-23/03/851 and also, Anusandhan National.

Research Foundation (ANRF)-SERB-CRG grant CRG/2023/005447 for this research work.

References

1. Xie, M., Tian, Z.: Eng. Fail. Anal. **85**, 212–227 (2018). https://doi.org/10.1016/j.engfailanal.2018.09.003
2. Olivares, G.Z., Gayosso, M.J.H.: Mater. Sci. Appl. **6**, 760–772 (2015). https://doi.org/10.4236/msa.2015.68076
3. El-Gebeily, M., Khulief, Y.A.: Appl. Math. Model. **40**, 5335–5348 (2016). https://doi.org/10.1016/j.apm.2015.11.003
4. Ma, Q., Tian, G., Zeng, Y., Li, R., et al.: Sensors **21**, 3862 (2021). https://doi.org/10.3390/s21113862
5. Hou, W., Zhang, D., Wei, Y., Guo, J., Zhang, X.: Appl. Sci. **10**, 1878 (2020). https://doi.org/10.3390/app10051878
6. Abbas, M., Shafiee, M.: Sensors **18**, 3958 (2018). https://doi.org/10.3390/s18123958
7. García-Martín, J., Gómez-Gil, J., Vázquez-Sánchez, E.: Sensors **11**, 2525–2565 (2011). https://doi.org/10.3390/s110302525
8. Yang, Y., Yang, Y., Li, L., Chen, C., Min, Z.: Appl. Sci. **12**, 1293 (2022). https://doi.org/10.3390/app12031293
9. Sophian, A., Tian, G.Y., Taylor, D., Rudlin, J.: Insight **43**, 189–195 (2001). https://doi.org/10.1784/insi.43.5.189
10. Park, D.G., Angani, C.S., Rao, B.P.C., Vértesy, G., Lee, D.-H., Kim, K.-H.: J. Nondestruct. Eval. **32**, 350–353 (2013). https://doi.org/10.1007/s10921-013-0203-7
11. Rifai, D., Abdalla, A.N., Ali, K., Razali, R.: Sensors **16**, 298 (2016). https://doi.org/10.3390/s16030298
12. He, D.F., Tachiki, M., Itozak, H.: Rev. Sci. Instrum. **80**, 036102 (2009). https://doi.org/10.1063/1.3073450
13. Wu, B., Wang, Y.J., Liu, X.C., He, C.F.: Smart Mater. Struct. **24**, 075007 (2015). https://doi.org/10.1088/0964-1726/24/7/075007

14. Rifai, D., Abdalla, A.N., Khamsah, N., Aizat, M., Fadzli, M.: Indian J. Sci. Technol. **9**, 1–6 (2016). https://doi.org/10.17485/ijst/2016/v9i9/87436
15. Angani, C.S., Ramos, H.G., Ribeiro, A.L., et al.: Sens. Actuators A **233**, 217–223 (2015). https://doi.org/10.1016/j.sna.2015.03.020
16. Angani, C.S., Park, D.G., Kim, C.G., et al.: J. Magn. **15**, 1–6 (2010). https://doi.org/10.4283/JMAG.2010.15.4.01
17. Feng, B., Ribeiro, A.L., Ramos, H.G.: NDT E Int. **100**, 64–73 (2018)

Analysis of Unmanned Aquatic Aerial Vehicle

Bommalapally Bhavana(✉), Ch. Sai Krishna, and B. D. Y. Sunil

Department of Aerospace Engineering, Institute of Aeronautical Engineering, Hyderabad, Telangana 500043, India
bbyadav06@gmail.com

Abstract. A drone that possesses both the qualities of flying in the air like all the Unmanned Aerial Vehicles (UAV) and moves underwater like all the Unmanned Underwater Vehicles (UUV) is what this drone project is about. Developing a UAAV with fixed wings that can operate well in both water and the air is the primary goal of this research. When moving from air to water flight, the drone may sink into the water thanks to its floodable chambers in the fuselage and wings. When it emerges from the water, the water swiftly drains out of the open wingtips. Instead of using two vehicles—an unmanned aerial vehicle and an underwater remotely controlled vehicle to observe anything that needs both aerial and underwater monitoring, it would be far more efficient to use only one vehicle. The Buoyancy force needs to be balanced in order to keep the drone moving through the water, which challenges the structural integrity. When in the air, the Center of Gravity is very important. At all times, the plane has to be nose-heavy or neutrally balanced.

1 Introduction

This study examines the development and testing of a miniature amphibious demonstration aircraft without a pilot. An Unmanned Aqua Aerial Vehicle (UAAV) is a kind of unmanned aircraft designed to transport electronic equipment and freight while operating from water. Its purpose is to prevent manned and unmanned aircraft from causing conflicts in airspace and airfields. To accomplish these aims, the smallest demonstration UAAV was designed using size predictions. The primary criteria for determining its size are the tasks it must do.

A two-foot wingspan and delta wing arrangement were included into the design to provide simple maneuverability and decrease drag, thereby decreasing the structural weight of the wings. Aluminium, primer-coated plywood, and Fluke boards made up the demonstration airframe. This aircraft can now be easily maneuvered both on land and in the water thanks to the electric propulsion system, which consists of a 2300 kV brushless motor installed on the empennage portion. This section describes the UAAV demonstration development process in depth and emphasizes the problems and solutions that were encountered.

The drone's performance requirements, limitations, and availability were specified throughout the design phase.

1. Things needed for air travel include

P. Joseph D. et al. (Eds.): IEMDST 2024, SPPHY 567, pp. 370–380, 2025.
https://doi.org/10.1007/978-981-96-5863-3_47

- Avg. altitude in the air
- Height restriction
- Rate of speed
- Flight duration
- Manage distance

2. Needs for Submerged Operations

   - Seawater density range, Reynolds number
   - How deep the water is
   - A system for arriving and leaving
   - Dynamic stability as a whole

The mission profile of the amphibious UAV is shown below in Fig. 1 [1].

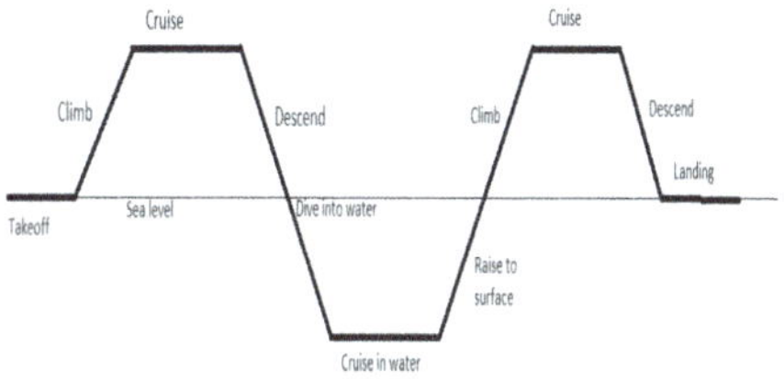

Fig. 1. Mission profile of Amphibious UAV

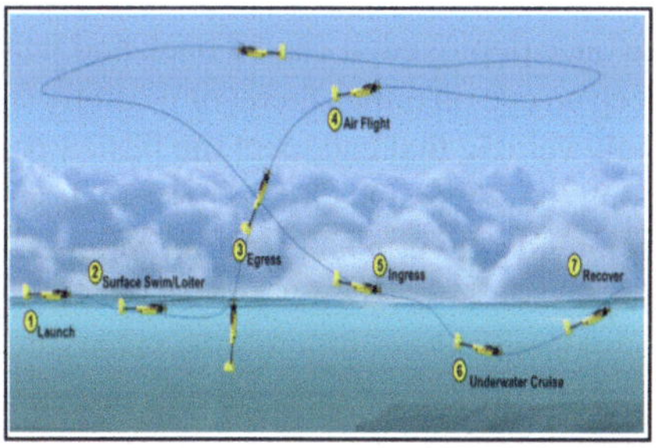

Fig. 2. Concept of Operation

The vehicle's operational concept is illustrated in Fig. 2. Research in the past has shown that UAV-UUV hybrid ideas might reduce drag while moving underwater by folding their wings [2, 3].

1. Setting the vehicle free on the ocean surface initiates a complete mission cycle.
2. After reaching its take-off site, the vehicle halts its forward propulsion and hovers passively with its nose up, thanks to its buoyancy distribution.
3. Egress from the water is initiated when the vehicle is at its nose-up pitch.
4. The next step is to get to a certain height by climbing vertically, and then it's time to start flying forward. Now it's acting like a regular fixed-wing aircraft and heading back to its entry place.

5. After reaching the entrance point, the vehicle will land like any other fixed-wing aircraft, with its wings flaring slightly over the ground and a belly landing. The vehicle then descends by using the engine for forward propulsion and the downward lift from its wings to overcome its limited positive buoyancy.
6. After submerging, the vehicle effortlessly makes its way to its destination. Afterwards, the vehicle has the option to either float back up to the surface or actively navigate to it.
7. The vehicle has two options once it reaches the surface: being retrieved or taking off again to start the cycle afresh.

2 Design Alternatives

In order to decide for the design of the UAV many factors were considered mainly the environment in which it needs to function. As it is an amphibious UAV it needs to function both in water and air unlike common UAV which work only in any one of the mentioned environments. The electronic equipment and circuits need to be encased and secured to avoid contact with the surrounding environment and given the aircraft needs to fly for more than twenty minutes so, a design with good weight to power ratio and that has minimum drag was top in priority.

We take into consideration a vehicle's CONCEPT of OPerations (CONOPS) that combines the features of a UUV with those of a fixed-wing UAV, resulting in a design that can effortlessly move between the water and air domains, transforming swiftly as required. The following steps make up CONOPS: surface swim/loiter, egress, air flight, ingress, undersea cruise, and vehicle retrieval. Persistent maritime operations are a good fit for this vehicle because once deployed, it can sit back and watch. If necessary, it may swiftly switch sites and keep an eye on things there, or it can fly faster than regular UUVs and follow a target. A single vehicle that can meet the basic operational criteria in each domain is necessary to complete this CONOPS. The vehicle must remain undamaged while operating since it is intended to conduct several cycles as required.

The CONOPS that were taken into account necessitate the use of both rotary-wing lift and fixed-wing lift components, depending on the stage of operation [4, 5]. During the escape phase, a rotary-wing lift is used by vertically orienting the vehicle and using a high-thrust propulsion system. Quickly draining as it emerges from the water, the vehicle may thereafter ascend to flight height. To make the most of its airborne range and durability, the vehicle switches to horizontal fixed-wing flying once it reaches flight height. Therefore, unlike seaplane takeoff, this escape technique does not need a lengthy stretch of calm water.

Research in the past has shown that UAV-UUV hybrid ideas might reduce drag while moving underwater by folding their wings. Because the vehicle could switch to the air regime to shift locations quickly, high-speed underwater operations were not necessary for the CONOPS studied. Consequently, a fixed-geometry wing's greater underwater drag and slower underwater movement were seen as a reasonable compromise when contrasted with the complexity and weight gain of a folding wing.

Buoyancy, Lift, and Weight Control: A vehicle's buoyancy and weight are two static forces that must be balanced for efficient underwater operations, hence the vehicle's total density must be extremely near to water's. On the other hand, a less dense airframe improves efficiency by reducing stall speed and wing stress during air travel. Minimizing the vehicle's waterproof compartment capacity to lower its submerged displacement helps reconcile these opposing objectives. While a big enclosed wing would have been efficient in the air, it would have needed ballast to function in the water because of the volume it displaced if a normal wing design had been used. By reducing the vehicle's density while in flight, a variable ballast system that takes into account floodable and drainable quantities may help to alleviate the decline in flying performance. This vehicle serves as a proof of concept and uses a passive approach. To obtain near-neutral buoyancy underwater, the wing volume fills with water during entry. However, upon egressing, the water is swiftly drained.

3 Analysis

Mesh Metrics: You may configure and examine mesh metrics, as well as assess mesh quality, using the mesh settings (Fig. 3 and 4). Mesh quality requirements vary between physics and solvers.

ANSYS offers the following mesh metrics:

a. Skewness
b. Aspect Ratio
c. Orthogonal Quality
d. Warping Factor
e. Mesh Quality

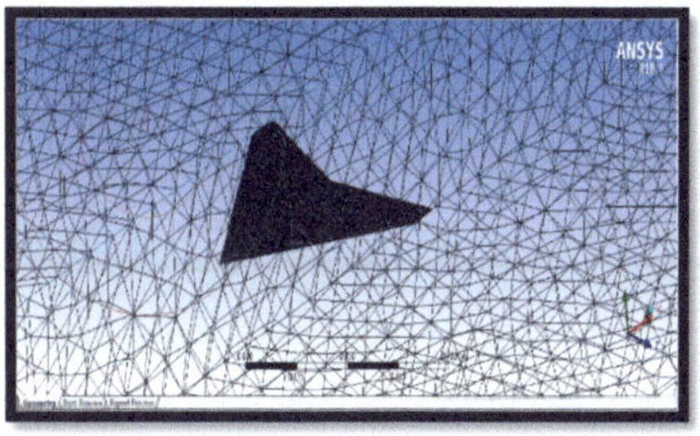

Fig. 3. Unstructured Mesh Domain

Skewness (Fig. 5):

- The skewness of a cell is defined as its form deviation from that of an equilateral cell with the same volume.
- On average, a triangle or tetrahedron mesh should have a skewness value of less than 0.33 and no more than 0.95 in the majority of flows.

Aspect Ratio (Fig. 6):

- Reconstructed quadrangles' aspect ratios are defined by the ratio of their longest to shortest sides. This formula applies to generic triangles and quads as well.

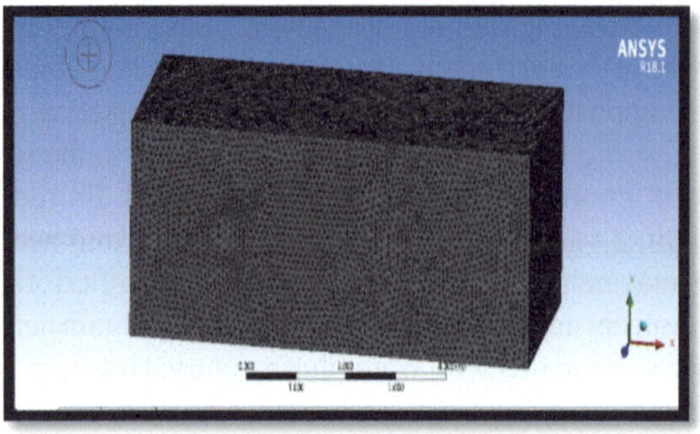

Fig. 4. Wireframe Design

Mesh Metric	Skewness
Min	1.5721e-004
Max	0.97842
Average	0.21407
Standard Deviation	0.119

Fig. 5. Mesh metrics and skewness.

Mesh Metric	Aspect Ratio
Min	1.1668
Max	33.044
Average	1.8166
Standard Deviation	0.46609

Fig. 6. Aspect Ratio.

- For a perfect equilateral triangle or square, it is equal to 1.

Orthogonal Quality (Fig. 7):

- Using vector mechanics, orthogonal quality may be calculated.
- The directions from the cell centroid to each face, the face normal vector, and the vector connecting the centroid of each neighboring cell are used for computations.
- 0 is the worst and 1 is the best.
- Standard Deviations are allowed only up to structure domain size.

Factor for Warping:

Mesh Metric	Orthogonal Quality
Min	2.1578e-002
Max	0.99566
Average	0.78463
Standard Deviation	0.11743

Fig. 7. Orthogonal Quality.

- It is ideal for the nodes of a quadrilateral element to be on a single plane, but this is obviously not achievable when dealing with complex geometric profiles or curvatures. "Measure of out of planeness of a Quadrilateral is Warping Factor or Warping Angle".
- Only quads are subject to this check; the warping computation does not apply to triangular components since three points form a plane.

Mesh Proficiency:

The ability to compute a more accurate solution faster is indicative of a higher-quality mesh. There is a conflict between precision and rate of execution. Although the computational cost goes up, the accuracy goes up every time the mesh size goes down.

4 Results and Discussion

Below depicted figures are results of our UAAV performance under different end boundary conditions and fluid media such as air and water.

Contours under water as fluid:

1. Curve of velocity

   A technique for calculating the discharge of a stream by taking the point velocities, contouring them to get the average cross-sectional flow velocities, and then multiplying the resulting values by the cross-sectional areas.

   Contours under water as fluid are shown in Fig. 8, representing the curve of velocity. Persistent maritime operations are a good fit for this vehicle because once deployed, it can sit back and watch [6].

   Solution:

   UAAV can move with a speed of 6 m per second in water bodies and can also the sustain turbulences in water which are created because of UAAVs movement.

   Minimum Velocity −1.38 m/s

   Maximum Velocity −6.67 m/s

   Velocity of Under turbulence −4.00 m/s
2. Pressure Contour

   Current and proposed pressure contours for the manifold Optimization of volumetric efficiency by flow analysis of a multi-cylinder gasoline engine's exhaust manifold. Furthermore, in order to finish pressure area assessments and pressure contour maps for our water system, the city is considering digital terrain modeling and city contouring. Pressure Contour under water are depicted in Fig. 9.

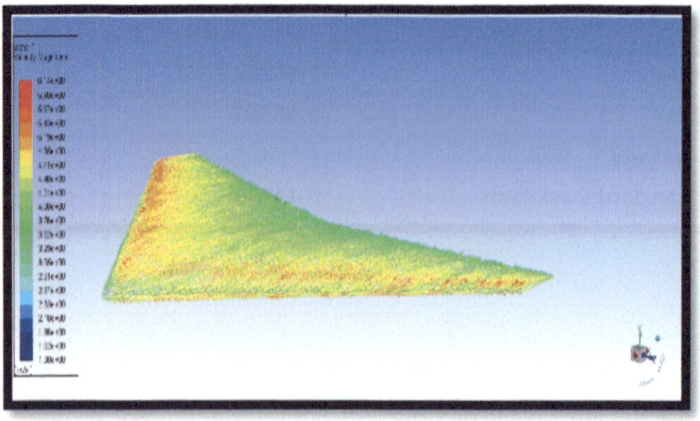

Fig. 8. Velocity Contour

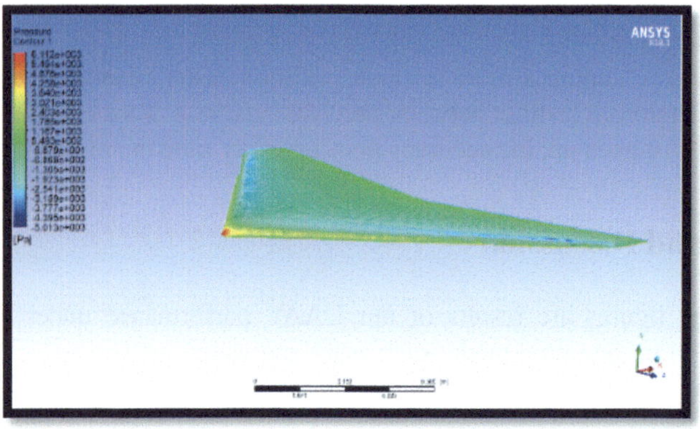

Fig. 9. Pressure Contour

Solution: If the UAAV is traveling at a speed of 6 m/s, it will encounter extra atmospheric pressure of 6, 121 Pa.

Maximum Pressure −6112 pascals +1atm.pr

Minimum Pressure −0549 pascals +1atm.pr

Contours under Air as Fluid:

Alterations to media outcomes have the potential to significantly impact UAAV performance, it leads to even malfunctioning if the design is not efficient [7, 8]. Hence, performance optimization is calculated based on the analysis which is performed in air as well.

1. Velocity contour solution

UAAV can move with a speed of 6 m per second in water bodies and can also the sustain turbulences in water which are created because of UAAVs movement. Velocity contour in air are shown in Fig. 10, where the UAAV's maximum velocity reaches 15.2 m/s [7, 8].

Minimum Velocity −3.11 m/s

Maximum Velocity – 15.2 m/s
Velocity of Under turbulence – 10.4 m/s

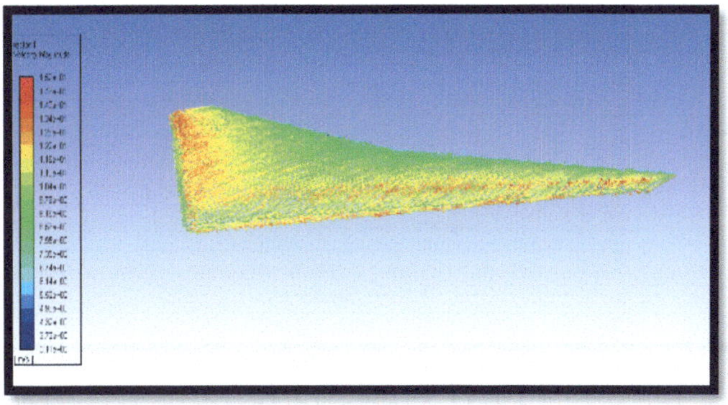

Fig. 10. Velocity Contour

2. Pressure Contour solution
   Pressure contours in air are illustrated in Fig. 11, highlighting the UAAV's performance at varying speeds. In addition to the atmospheric circumstances, UAAV encounters a pressure of 47.82 pascals while in motion at a speed of 12 m/s.
   Maximum Pressure – 47.82 pascals + 1atm.pr
   Minimum Pressure – 27.8 pascals + 1atm.pr

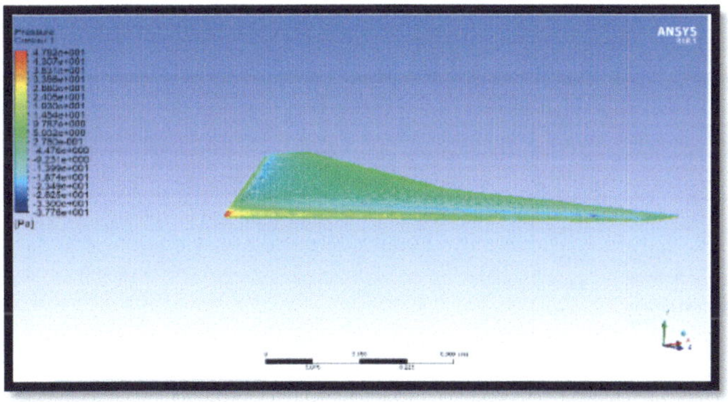

Fig. 11. Pressure Contour.

The lift-to-drag ratio, lift, drag, moment metrics are summarized in Fig. 12 [7, 9], and lift are the primary metrics we extract from the data shown above. Next, we plot these values versus the attack angle and lift. Our understanding of the wings' performance is enhanced by the presented graph. Both the above- and below-water performance outcomes are shown in Figure.

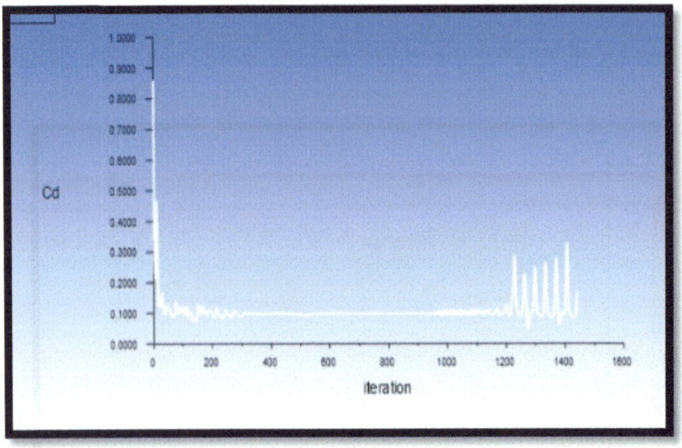

Fig. 12. Coefficient of Drag.

In Fig. 13, 14 and 15, we can see a wing's performance both above and below water. With the exception of the lift-to-drag ratio and the drag polar, most data are in good agreement with one another. Given that the two operational settings differ in density and kinematic viscosity, this is to be anticipated. We calculated the results for air and sea transport using the fixed lift polar type. In this case, the aircraft maintains a constant lift when the angle of attack (AOA) increases; the change in velocity compensates for the fixed lift. At the same angle of attack (AOA), water has a far higher viscous drag than air, even at a slower speed. The reason for this is the very dense water and the enormous pressure found under the surface.

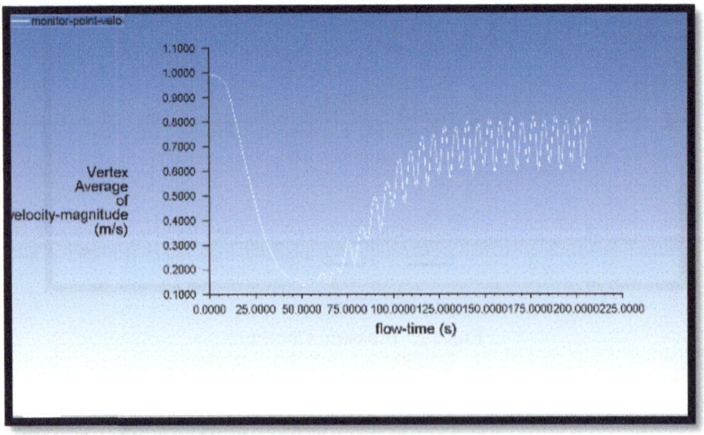

Fig. 13. Velocity Magnitude

The two methods for Entering and Exiting are depicted in Fig. 14, with the final model shown in Fig. 15 [10] are as follows.

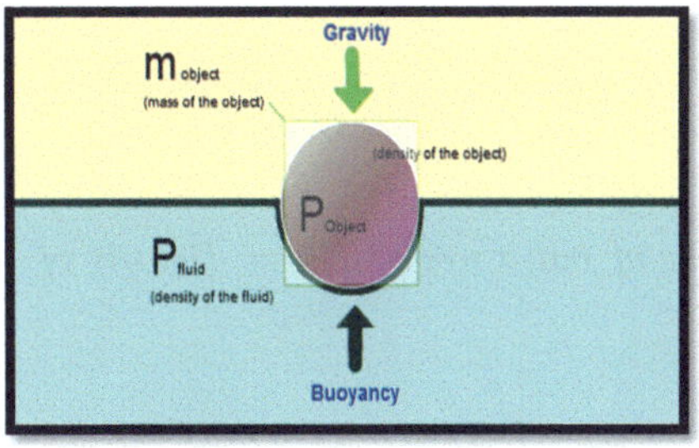

Fig. 14. Buoyancy Force

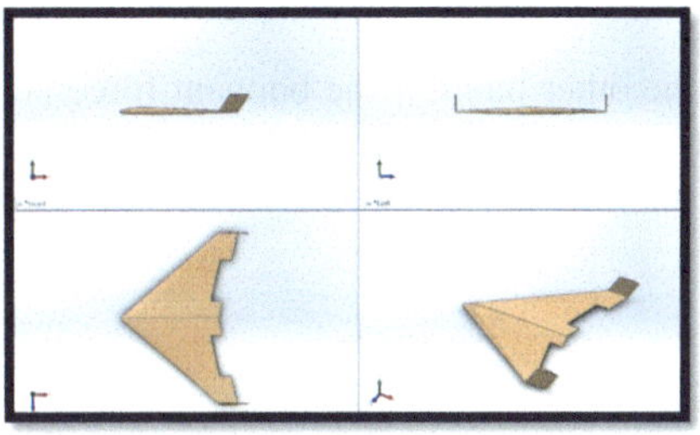

Fig. 15. Final Model

Starting Off: At takeoff, you have the option of either manually launching the wing or employing a bungee catapult launcher.

Mechanism for Entry and Exit: A flying wing has an interior hollow and open wingtips. One use of the wing is to hold ballast. The wing must touch down on the water's surface before it can begin its underwater journey. The wing becomes filled with water as it enters via the open wingtip. Because of this, the wing becomes less buoyant and may enter the water by lowering its pitch and driving itself forward. The wing-mounted engine facilitates inward thrusting, while the differential thrust system facilitates directional changes. The electronic speed controller (ESC) might potentially overheat while submerged in water, thus it's best to avoid using more than 10% of the motor's capacity. When the UAV takes to the air, the spoilerons are shut. Catapult mechanisms are very old enough technique, but here UAAV uses this technique very efficiently.

Mathematical Model: The buoyant force is the upward push that an item experiences while submerged in a fluid. The displacement of a body through a liquid causes a force that is proportional to its mass. Similar to other forces, the buoyant force is measured in Newtons (N).

$$\begin{aligned} F_{buoyant} &= \rho g V_f \\ \rho = density;\ g = gravitational\ force;\ V_f &= volume\ of\ fluid \end{aligned} \tag{1}$$

$$V_s \times D_w \times g = V_b \times D_b \times g \tag{2}$$

$$V_s/V_b = D_b/D_w \tag{3}$$

$$V_s = Volume\ of\ displaced\ liquid;\ D_w = Density\ of\ water;\ g = gravitational\ constant;$$

$$V_b = Volume\ of\ balsa\ wood;\ D_b = Density\ of\ balsa\ wood$$

5 Conclusion

Efficiency: After bringing the weight of the water expelled by the drone and the upward buoyant force to a balance, the model is able to float on any body of water, as shown in the previous calculations.

Ability to scale: On the other hand, if the buoyant force is greater than the weight of the water displaced, meaning the drone weighs more than the volume of the water (density < 1), then the drone will launch. A certain backup needs to be placed.

References

1. Nasir, R.E.M., Kuntjoro, W., Wisnoe, W.: Proc. Technol. (15), 783–791 (2014)
2. Panagiotou, P., Fotiadis-Karris, S., Yakinthos, K.: Aerosp. Sci. Technol. **73**, 32–47 (2018)
3. Sun, C., Song, B., Wang, P., Wang, X.: **9**(6), 693–704 (2017)
4. Pisanich, G., Morris, S.: Proceedings of the 21st Digital Avionics Systems Conference 2002, vol. 2, 8C4 (2002)
5. Eubank, R., Atkins, E., Macy, D.: AIAA Infotech@ Aerospace Conference and AIAA Unmanned... Unlimited Conference, p. 2021 (2009)
6. Fabian, A., Feng, Y., Swartz, E., Thurmer, D., Wang, R.: Hybrid aerial underwater vehicle (MIT Lincoln Lab) (2012)
7. Sobolic, F.M.: Agile flight control techniques for a fixed-wing aircraft (2009)
8. Yang, X., Wang, T., Liang, J., Yao, G., Liu, M.: Prog. Aerosp. Sci. **74**, 131–151 (2015)
9. Gao, A., Techet, A.H.: Oceans **2011**, 1–8 (2011)
10. Reid, B.D.: The Flying Submarine: The Story of the Invention of the Reid Flying Submarine RFS-1. Eagle Editions (2004)

Correction to: Proceedings of the International Conference on Emerging Multifunctional Materials and Devices for Sustainable Technologies

Paul Joseph D., Jayalakshmi V., and K. Thangaraju

Correction to:
P. Joseph D. et al. (Eds.): *Proceedings of the International Conference on Emerging Multifunctional Materials and Devices for Sustainable Technologies*, SPPHY 567, https://doi.org/10.1007/978-981-96-5863-3

The original version of the book was inadvertently published without the series name and volume number, which has now been included. The erratum book has been updated with the change.

The updated version of this book can be found at
https://doi.org/10.1007/978-981-96-5863-3

P. Joseph D. et al. (Eds.): IEMDST 2024, SPPHY 567, p. C1, 2025.
https://doi.org/10.1007/978-981-96-5863-3_48

Author Index

P. Joseph D. et al. (Eds.): IEMDST 2024, SPPHY 567, pp. 381–383, 2025.
https://doi.org/10.1007/978-981-96-5863-3

Y

The manufacturer's authorised representative in the EU is Springer Nature Customer Service Centre GmbH, Europaplatz 3, 69115 Heidelberg, Germany. If you have any concerns regarding our products, please contact ProductSafety@springernature.com

Printed and bound by CPI Group (UK) Ltd, Croydon, CR0 4YY
01/10/2026
02225934-0015